CHEMISTRY CONNECTIONS
Ideas to Interpret Your Changing World

CHEMISTRY CONNECTIONS
Ideas to Interpret Your Changing World

Tom Hughes
Ventura College

KENDALL/HUNT
PUBLISHING COMPANY
Dubuque, Iowa

TO
Ed, Cathy, Charlie, Jean, Nancy, Andy, John, Peter, Michelle, Eddie, Karen, Bill, Sandy, Mary, Bob, and the many others who, often in intangible ways, helped immeasurably to show how far the unknown transcends the things we know, and that we are heirs of the whole world.

Cover art: Dover Pictorial Archive Series

Library of Congress Catalog Card Number: 82–84629

ISBN: 0–8403–2942–3

Printed in the United States of America

B 402942 01

Contents

chapter 4

The Atom: Its Properties and Structure, 56

chapter 5

Chemistry in Action: Electrons, 88

chapter 6

Compounds: Using Names, Formulas, and Equations, 115

chapter 7

Some Chemicals for Life: Water and Its Elements, 131

chapter 8

Gasoline and the Nature of Hydrocarbons, 160

When we try to pick out anything by itself, we find it hitched to everything else in the universe.

—JOHN MUIR

Preface

You may wonder about the title: "CHEMISTRY CONNECTIONS." Many people think chemistry is the study of a lot of facts about chemicals. This is not what it's all about. The major goal of chemistry is to bring out connections. Some of these are the connections between things and between things and your body. But the most important connections are those between your mind and the world. This includes your present experiences and those you will have in the future.

The interesting and surprising fact is that as you begin to see the interrelation of all things you will become more aware of the unity of the universe. And you will intuitively feel yourself as a very important part of nature. When you study nature in detail with the connections emphasized, you will be able, to some degree, to transcend the notion of an isolated individual self, and to feel yourself as a part of a fantastic universe.

The purpose of this preface is to give you help in achieving these goals. I realize that many students starting to study chemistry do so with some fear about their ability to succeed. This fear of failing is common but not at all justified. Within a short time, if you follow the suggestions here and in the early chapters, you should be relatively free of this anxiety. Then you may even enjoy chemistry.

You should realize at the outset, however, that it will take a little time and effort on your part. Especially if your anxiety or fear is unconscious, as it often is.

We live in an instant society—from instant breakfasts, to fast-food lunches, to instant relay of news from all around the globe. But where nature is involved, *things take time.* Stop now and think of the Grand Canyon or remember the rocks you've seen at the sea shore or in running streams. Nature's way is generally slow and easy. When you look at stones on the beach or in the river bed, you find the edges worn smooth by hundreds of years of work by water, wind, and waves. Your mind is a vital part of nature and it is therefore natural that it will also take time to relieve your anxiety. And, when that is relieved, more time will be required to develop your understanding of your world because you will be developing a fresh awareness, and new ways of thinking about the things you experience every day. New habits do not come overnight. So relax. Be patient. And take your time.

Nature is a labyrinth in which the very haste you move with will make you lose your way.

—SIR FRANCIS BACON

There was always more in the world than man could see, walked he ever so slowly. He will see no more by going fast, for his glory is not in going but in being.

—JOHN RUSKIN

1. PSYCHOLOGY COMES BEFORE CHEMISTRY

You will notice that I have been talking above chiefly about your mind and its relation to the world around you. Your mind is the most important part of your chemistry course, so it is logical to emphasize this in the beginning. You already have many ideas in your mind about chemistry and chemicals. But it is not these that I'm referring to now. It's your outlook or mental image of yourself, and your feelings about how well you learn.

Psychologists now recognize that what people *believe about themselves* they *will become.* Consequently, at the start of your chemistry study it would be good to examine your feelings. If you blame bad luck, or fate, your parents, or poverty, or other "forces outside your control" for your state of success, you are programming your brain to continue just the way you are. I'm suggesting that you assume responsibility for your life and your learning. If you take charge of your life, you can no longer put "blame" on yourself or others for any behavior. Instead, you personally take the responsibility for your success (and happiness), and actually *see* yourself as succeeding.

It will be worthwhile to summarize this important section. *Your beliefs about yourself and your world will control your experience. And your images of yourself will become self-fulfilling prophecies. So, begin by seeing yourself succeeding!* Your success in chemistry thus starts in your own mind.

Let me continue on the subject of the mind. This course is based very heavily on psychology. And psychology is essentially the science of consciousness. I am deeply in debt to a particular branch of psychology which covers the mechanisms by which knowledge grows. The function of this branch is to analyze how people pass from lesser states of knowledge to states of more advanced knowledge. In other words, it is concerned with how you "get smarter."

In addition, I'm very intrigued by, and have been heavily influenced by, the findings of recent research regarding brain function. Consequently, your chemistry text is also heavily influenced by these psychological factors. In the text itself, this is not spelled out in words like I'm doing now. However, you may notice these sometimes subtle psychological influences there, most often when you're reviewing or perhaps much later when you think about how you have learned things. To go into detail on these areas would make this a psychology course and your text a book of psychology.

However, it will be worthwhile to review for you a few of the general and practical ideas that will be helpful for you as you learn how to learn chemistry, and as you develop and improve your own personal life through chemistry.

2. YOUR DOUBLE BRAIN: THE RIGHT AND LEFT OF IT

You and I have a brain which resembles the two halves of a walnut—two bumpy, rounded halves connected at the center, just like the walnut. This has long been known by physiologists, and they recognized that there was a complex, connecting cable of nerves between the left and right halves, or hemispheres. Because of this connection, it was long assumed that the brain was a single functioning unit.

It is difficult to summarize the exciting studies on the brain which were conducted mainly at the California Institute of Technology by a team of scientists headed by Roger W. Sperry. In the 1950s they did work on animals, studying the functioning of the two hemispheres of the brain. Then, in the 1960s, they expanded their work to humans, studying especially people who had been completely disabled by severe epileptic seizures. It was found that the epileptic seizures of these patients could be controlled only by surgical separation or cutting of the connections between the two halves of the brain.

Essentially, the findings of Dr. Sperry and his associates indicated that we really do have a double functioning brain, with two ways of knowing. Each half of our brain may handle information in different ways. The *left half* is the *analytical, objective, abstracting* half which *counts, marks time, verbalizes, is logical*. The *right half* is *intuitive, subjective, sees things in a holistic way* rather than step-by-step, is *relational* and *time-free*.

Even though each half of our brain takes in the same information from our senses, they each process the information differently. They may each do their own thing, dividing the job between them. Or, one half, usually the "dominant" left may "take over" and suppress the right half.

> ***The main theme to emerge (from the research) is that there appear to be two modes of thinking, verbal and nonverbal, represented rather separately in left and right hemispheres, respectively, and that our educational system, as well as science in general, tends to neglect the nonverbal form of intellect. What it comes down to is that modern society discriminates against the right hemisphere.***
>
> —*ROGER W. SPERRY*

Two major modes of consciousness exist in Man, the intellectual and its complement, the intuitive. Contemporary science (and, indeed, much of Western culture) has predominantly emphasized the intellectual mode, and has filtered out rich sources of evidence: meditation, "mysticism," non-ordinary reality, the influence of "the body" on "the mind."

—ROBERT E. ORNSTEIN

3. A PRACTICAL SUGGESTION FROM BRAIN RESEARCH

We should involve both halves of the brain, especially the neglected right half, as an aid in getting understanding. This means we cannot use only words in learning. You may remember in the past an occasion when you "memorized" something for a test so you "had it down cold", but later you realized that you never did really understand it even if you "passed" the test.

I suggest that you *not try to memorize* your way in chemistry. You will be remembering words but the major activity which will help you really understand is not just remembering verbal descriptions or definitions but "getting" the idea into your mind. Memory is important but just words without visual or sensory input prior to their use is not effective. Also, this can be very frustrating because you "can't remember *the* definition" etc. So, while you and I will often be using words, I want to emphasize that your experiencing and visualizing are equally, or possibly more, important. You will see many applications in the future of where I suggest activity—both body and mental. Please do it! You will even have "thought experiments" where you should shut your eyes and turn on the "movies" of your mind, experiencing and learning through imagery.

The words of the language as they are written and spoken do not seem to play any role in my mechanism of thought.

—ALBERT EINSTEIN

It is important to realize that I'm not saying that the left half of your brain is not important. Or, that the right is more important. They both are important, and are indeed complementary, just like your left and right hands, even though you may prefer one over the other. We are complete, whole individuals and your course in chemistry is certainly holistic in every way, starting with the brain and you.

4. DRAWING: GIVING ACCESS TO RIGHT BRAIN FUNCTION

You will see many occasions where figures, drawings (like those of atoms which you've heard so much about already), and pictures are given in your text, or used by your instructor in class. You should follow the lead here and make the effort to visualize in your mind drawings, pictures, and models. Also, it is very important that *you draw pictures* of the items which you are trying to understand.

In other words, pictures, models, and drawings are really absolutely essential in learning chemistry. *Please don't neglect pencil and paper for doing these drawings,* even doodles or whatever. Do not worry about the artistry in your drawings. Art perfection doesn't matter since drawings by even the best artists do not picture what is really there. For example, no one has ever seen an atom. It is often frustrating to a teacher to realize how often students waste time by going over and over words when a few drawing experiences would quickly let them "get the idea." You should also consider using colored pencils for the drawings since color seems helpful in engaging the right brain. Also, quiet music in the background seems to help.

Therefore, please recall the suggestion on psychology in the first section of this preface. Your beliefs about yourself will control your experience. So don't believe "I can't visualize, or draw well." You've been visualizing and using your right brain all through your life, although, unfortunately, perhaps not very much in your educational experiences, which may have concentrated on verbal transmission of ideas. Your dreaming, remembering of faces, and events in your past life all depend on your visualization. You are already "good at it." These all show use of your right brain. What we all need is practice in using the right brain. And, like learning any skill, it takes a little time and patience to get the approach more fully operational.

5. SOME EXAMPLES OF YOUR RIGHT BRAIN CAPABILITIES

Let me give you a few more examples of how effective and necessary your right brain is and how you already use it, without perhaps being aware of its ability in helping you deal with everyday experiences.

I'm going to ask you to think of the face of a friend of yours. It may be better if you shut your eyes. Therefore, stop reading and pause while you do this *now before going on to the next paragraph.*

I'm sure you could see your friend's face rather distinctly. And you would recognize the friend at once if that person came into the room. Now I want you to try to describe that friend's face in writing, using words only. *Pause now and take a few minutes* to write the description on a piece of paper.

This was a little harder, was it not? Do you think I could recognize your friend in a crowded room if you gave me your description in words? I think you know if would be difficult. It would be much easier if you gave me a picture of your friend. Haven't you often been asked what a certain friend of yours looks like and you often say: "well, I know what she (or he) looks like but I just can't describe her (or him) very well." This is natural. The right brain is more specialized at recognizing faces, since this a holistic job. The left, verbal, half can't do this particular job nearly as well.

Here's another example for you to try. Pretend you're talking to someone who just landed from some other planet than our earth where they don't wear shoes. Try to tell this person verbally, or write out a description, of how to tie a shoe. *Pause now* and take a little time to try this.

This was hard to do, right? And you know how much easier it is by just demonstrating it. That's the way we all learned to tie shoes. Use of words only (the left brain) is not adequate here. The fact is that you have your shoes tied now and may not be able to remember doing it this morning. The action never went through your conscious verbal hemisphere.

Let me give you one more example. I'm going to ask you to try to describe (to a friend, or an imaginary person from another planet) a spiral staircase, *without using your hands* in a spiral gesture. Stop reading now and take a little time to try this, using words only.

In describing the spiral staircase most people you ask will use some form of spiral gesture, or draw you a picture. This is just one more example of a right brain specialty and should be enough to convince you that words are in some cases inadequate for the jobs, the communicating, and the learning that we have to do.

6. THE KEY TO LEARNING CHEMISTRY

The key to learning chemistry, for non-science majors especially, lies in psychology. It is the twofold approach of providing concrete experiences, and encouraging a shift to using the right side of the brain.

Chemistry, in its essence, has a left-brain orientation. It is predominantly involved with logic, abstraction, verbal and symbolic expression. However, these abstractions are not grasped without first exposure to concrete experiences. Here, encouragingly, recent brain research, some of which is briefly described above, confirms and expands the findings of developmental psychologists, especially the work of Jean Piaget. It was Piaget who spent a lifetime of research and observation on how learning occurs and who came up with the concrete experiential approach as a prior necessity to learning abstract materials, long before the right and left brain studies were made.

> ***Abstraction is only a sort of trickery of the mind if it doesn't constitute the crowning stage of a series of previously uninterrupted concrete actions.***
>
> —*JEAN PIAGET*

In other words, you have assurances from psychology of the extreme importance, indeed, the indispensable value, of experience and activity on your part. You have to experience continually. And you are doing this as a part of living, even though I'm asking you to be much more aware of your everyday experiences.

Also, it is especially important to realize that each chapter of this book starts the exercise section with experiments. You will not need to do all of these, but be sure you realize the importance of *you personally doing* the ones your instructor assigns. You must also *think about* the experiences. Your senses, in the discovery action of the experiment, will get the holistic view to your right brain. Then you will naturally develop connections with your left brain, bringing in whatever intuition your right brain gets from the experience. Use of visual imagery here will be extremely helpful in many cases.

The experiments are often deliberately open-ended and non-specific in demand. This is to discourage and make it difficult for the left brain to attempt to dominate and merely pick out verbal information and abstract "explanations" from the chapter. In other words, the casual description of the experiments leaves a lot up to you and encourages you to use your own brain, both left and right, to develop your understanding and to do the explaining. Many of the exercises at the ends of the chapters have a similar goal in mind. You have as good a brain as any scientist has, and can do science thinking more readily than you now think, as will be demonstrated in the early chapters of the book.

You may recall a quotation of Einstein given previously that words, both written and spoken, seemed to play no role in his mechanism of thought. His creativity came from imagery and intuition (right brain) which he then had to develop in words and symbols (left brain) for communicating his creative insight to others.

7. CHEMISTRY'S HIDDEN BENEFITS FOR YOU

It may be encouraging to you that I have been using these psychological approaches for almost 20 years in helping students learn a little about the world, themselves, and chemistry. This includes students of widely ranging ages, occupational goals, background, and interests—most of them not science majors. I can asure you that the psychological approaches work. This again may aid in relieving you of any anxieties you may have.

The system works best with people who want to learn how to learn. In connection with your learning some chemistry, your whole outlook will be changed. Because you will be using a part of your brain which may have been somewhat neglected in your theoretical studies of the past. Also, you will be sharpening your attention to observing yourself and the world around you. This will have the effect of making your life seem richer because you will be seeing better, experiencing more. You will be using a slightly altered state of awareness which emphasizes your holistic, intuitive, right-brain mode. This will develop your ability to see things freshly and in their totality. You will see the underlying patterns and connections in nature, beneath the surface appearances.

> ***The rare moment is not the moment when there is something worth looking at but the moment when we are capable of seeing.***
>
> —*JOSEPH WOOD KRUTCH*

Hopefully, you may also discover hidden treasures within yourself. And you may feel the interconnectedness of everything in the universe—including yourself and people you know and don't know. You may have thought that you were just going to learn about some chemicals in this course. What I am emphasizing is what many people don't realize. Along with the chemical knowledge may come new ways of looking at things, improved use of your imagination, and the ability to think more creatively in other areas of your life. Creative poets, artists, and scientists each experience nature in a somewhat altered state of consciousness—free from the veil of ordinary, hum-drum phenomena.

> ***Poetry not only awakens and enlarges the mind of the reader, but it strips the veil of familiarity from the world, and lays bare the naked and sleeping beauty which is the spirit of its forms.***
>
> —*SHELLEY*

8. CHEMISTRY CONNECTIONS

You may have noticed in the discussions above that one thing leads to another. Your learning to learn connects with the development of your knowledge of chemistry. This is connected to personal satisfaction and pleasure as you view the world, which in turn encourages you to expand your connections to the whole universe. The fantastic tool for all of this is that amazing and valuable gift which we all possess—consciousness. It is consciousness, with its many facets, which provides your connections.

The scientist does not study nature because it is useful. He studies it because he delights in it, and he delights in it because it is beautiful. If nature were not beautiful, it would not be worth knowing, and if nature were not worth knowing, life would not be worth living.

—*HENRI POINCARÉ*

Every second we live is a new and unique moment of the universe, a moment that never was before and will never be again. And what do we teach our children in school? We teach them that two and two make four and that Paris is the capital of France. We should say to each of them, "Do you know what you are? You are a marvel. You are unique. In the millions of years that have passed, there has never been another child like you."

—*PABLO CASALS*

Since you have that precious possession of consciousness, I am confident of your success as we explore together the chemistry connections. Connections of you to the world, connections of ideas to the cosmos, connections between you and your food (and vice versa), connections to and between everything. Because everything is connected.

Most people are on the world, not in it—they have no conscious sympathy or relationship to anything about them—they are undiffused, separate, and rigidly alone, like marbles of polished stone, touching but still separate and alone.

—*JOHN MUIR*

The student who can begin early in his life to think of things as connected, even if he revises his view with every succeeding year, has begun the life of learning. The experience of learning is the experience of having one part of the mind teach another; of understanding suddenly that this *is* that *under an aspect hitherto unseen; of accumulating, at an ever-accelerated rate, the light that is generated whenever ideas converge. Nothing that can happen to people is more delightful than this, and it is a pity when it does not happen to them as students.*

—*MARK VAN DOREN*

ACKNOWLEDGEMENTS

The Estate of Albert Einstein for permission to use their photograph and quotations (*Out of My Later Years,* Philosophical Library, © 1950); Norma Millay Ellis for permission to use poetry from "The Goose Girl," copyright 1923–1951 by Edna St. Vincent Millay and Norma Millay Ellis, *Collected Poems,* Harper & Row; Charles Scribner's Sons for poetry from *The Poetry of Eugene Field,* Complete edition © 1895–1920; Doubleday & Company, Inc. for work from *Heretics,* Gilbert K. Chesterton, © 1905, and from *The Moon and Sixpence,* W. Somerset Maugham, © 1919, and from *Design with Nature,* Ian L. McHarg, © 1969; Jean Larcher for cover design, from Geometrical Designs & Optical Art, Dover Publications, Inc., New York, © 1974; and to the many others who were so generous in providing material.

I am also deeply indebted to many students and teachers with whom I have been privileged to work over the years. They have often shared ideas with me both verbally and by letters about our common goal of helping people to enjoy the pleasures of learning. I will continue to be grateful for any comments or suggestions from future users of this book.

Tom Hughes
Ventura College
Ventura, CA 93003

chapter 1

Chemistry: Concepts for a Changing World

All religions, arts and sciences are branches of the same tree. All these aspirations are directed toward ennobling man's life, lifting it up from the sphere of mere physical existence and leading the individual toward freedom.

—EINSTEIN

Nature to be commanded must be obeyed.

—FRANCIS BACON

All sciences are moral endeavors. For, as a scientist, one seeks to understand what is true so that he, or others, may better use such understanding in order to obtain more of what is good.

—A. J. BAHM

An eighteen-year-old girl was siphoning gas from her father's car into a container and transferring it to her own car. It was a warm August day. In the process a chemical reaction occurred which turned into a disaster. Here are the headline and lead paragraphs from the newspaper story:

> *GIRL LEAPS TO SAFETY AS FIRE SWEEPS HOME*
>
> A girl leaped from the balcony of a split-level home to safety last night after being cornered by flames.
>
> Karen Smith, 18, was taking gasoline from a car in her garage to use in another vehicle about 7 P.M. when the garage suddenly exploded in flames.
>
> Miss Smith raced upstairs to telephone for

(Courtesy of the Los Angeles County Fire Department.)

help and managed to dial an operator before the flames cornered her. She then ran out to the balcony with the flames right behind her. She leaped to the ground just before the entire back end of the house, including the balcony, was engulfed in flames.

The girl was not hurt when she jumped but suffered minor burns on her arm and face from the exploding gasoline fumes. She told firemen she went into the garage and was using a large bowl to take gasoline from one car to put in the tank of another. The vapors were ignited by a pilot light on the hot water heater, firemen said.

Flames were visible for miles as the home was quickly engulfed. The house was completely gutted. Both cars in the garage were destroyed.

A chemist could call this a planned experiment. If you put all the ingredients together—liquid gasoline, open container, hot garage, source of ignition—you could end up with the flammable vapors doing their thing. Being flammable, they burn. This is the same type of chemical reaction which you get when charcoal is soaked with lighter fluid used as a "fire starter" for quickly preparing burning barbecue coals. The chemistry is basically the same.

Fire is essentially neither good nor bad. When out of control in a house, fire is bad; in the barbecue, good. Applications of chemistry can be either good or bad. Knowing the properties and conditions for the applications of chemistry—and science in general—should help us make good choices and gain control over our lives. This book is therefore for the average person who needs more information on the material environment—the chemical environment—in order to get along better with it, to choose, control and, incidentally, enjoy it more.

Chemistry, very briefly, is the branch of science which studies material things—the stuff we live with, eat, drink, and are ourselves made up of. It is concerned with matter and what happens when matter reacts—as in a gasoline engine, or a fire.

Science itself is not easily defined but a few descriptions may be helpful. Poincaré said: "Science is knowledge not of things but of their relationships." And Einstein: "Science is the attempt to make the chaotic diversity of our sense experience correspond to a logically uniform system of thought." Pasteur noted: "Science is knowledge gained by systematic observation, experiment, and reasoning."

You will see that even these "definitions" by eminent scientists are not really adequate to give you the feel of what science really is. Science is a process, a way of life of certain people who are scientists and whose driving force is the desire to "make sense" out of the world of our experiences. The *end product* of all work of scientists of the past and of the present is a *knowledge of facts and laws arranged in an orderly system.* This is a dictionary definition of science. But this is like explaining poetry as rhymed metrical writing, which is, again, a dictionary definition. And a dictionary definition is only the beginning of thought, never the end of it.

Albert Einstein and his wife Elsa, in Pasadena (1931), when Einstein was Visiting Professor at the California Institute of Technology. Einstein represents one of the greatest scientific minds that the human race has produced and he played a pioneering role in the evolution of modern science. But he was also a warm human being who developed simple insights into the interpretation of nature. In addition, he was a tireless worker for peace and for personal and political freedom. (Courtesy of the Einstein Estate.)

This apple orchard in bloom involves a good deal of chemistry. What makes it especially interesting is that the commercial orchard grows on land reclaimed after strip mining. This is visual confirmation of the potential for maintaining nature's balances by knowledge and determination on the part of people. (Photo from General Dynamics, courtesy National Coal Association.)

WHY STUDY CHEMISTRY?

Before you start your investigation of chemistry, it should be worthwhile to consider a few reasons for studying it. You may come to this subject with the idea that the "usefulness" of science knowledge is the main value of science. The practical, useful value of chemistry is certainly an important aspect of our modern world. Consider the fact that everything you see, feel, taste and touch is chemical. And even things you don't see and often don't realize are there—like the air—are chemicals. You're familiar with water, sugar, salt, drugs, foods, soap, clothes, wood, metals, hair, glass, plastics, cosmetics—all of which are chemicals whose usefulness you will be learning more about.

But you will also be developing your understanding of the underlying unity in all the materials of your environment. Your study of this book will help you feel more at home in the "house" which nature provides. You will be better aware of the chemical basis for ecology—the science or study of the relation of living things to their environment and to each other. (*Ecology* from the Greek: house + study.)

You will appreciate better how you fit into the interrelated and ever-changing systems of nature—especially the general or so-called ecosystems. An ecosystem is the broad group or organization of relationships upon which the life of any particular living organism is based. It includes

There is much chemistry involved in city living in addition to the obvious pollution problems. (Photo by the Los Angeles County Air Pollution Control District.)

such factors as food, air and water supplies, weather, natural enemies. Ecology is essentially a branch of biology and this aspect will not concern us here. But ecology has a fundamental basis in chemistry and without chemical understanding ecology would be lame.

With your knowledge of the underlying structure and composition of nature's material world, you'll be able to deal with other facts—even those not yet known—to be experienced by you at some future time. This may be next week, or next year, or sometime in the distant future when you will make a better choice, or be able to live better, because you had some fundamental experience with the science of chemistry and science thinking.

We humans have a natural bent to investigate and to question: Why? One of the ancient Greeks, Aristotle, said it simply: "Man by nature desires to know." But we are living on the third planet of a star (our sun) which is just one of about 50 billion other stars in our particular galaxy. And the galaxy is but one of many billions of galaxies. And you are reading this in only the last part of the twentieth century. So we can hardly try to know all the facts, or, what is perhaps worse, pretend that we do.

But we are capable of putting things in order. Some of the delight deriving from a fine sports car comes from planned coordination of braking power, transmission, engine organization, bucket seating, lacquer coating, stripes, design, and so on. A part of the thrill from the moon shots, especially to the astronauts and scientists who were involved in the program, was in the beautiful functioning of complex and finely coordinated physical mechanisms. There is also the higher level of coordination involved in forming ideas and thinking through complex interrelationships between them. This is the kind of ordering and coordinating which is done in science and which we hope to experience in this book on the science of chemistry.

Thus in simple terms, a broad purpose of this book is to fulfill the need and love for satisfying experience which every human seeks (even if he's not ready to admit it). This means all humans, you especially. It is not only scientists and poets who desire to know and understand experience. They just do more with it, but all of us have the basic need to know. Otherwise, why go to school? Maybe you think that there are other reasons, and there are—like making a living. But the basic one is the natural human desire to know.

With all the emphasis on these somewhat abstract needs, you may be thinking: "But won't I learn about pollution and drugs?" You may in fact have thought that a modern chemistry book would start right off with these topics. These subjects will certainly get considerable attention later on. But one premise of this book is that you and your intellectual needs are more important even than pollution and drugs.

There is another important reason for starting with you and your mind as an interpreter of experience. The chemicals which you contact today are not necessarily those you will contact tomorrow. Drugs are continually changing, for example. What you need is some way of interpreting the behavior of the chemical world tomorrow as well as today. Therefore you do not need a listing of chemicals to know. What you need is an approach that will fit the lists of tomorrow and beyond.

In the early chapters you will see how scientists make some order out of the material world. You'll get the feeling that scientists are not superior beings to you. They are just specialists in an approach which you yourself already use. You will thus get acquainted with a few basic ways to look at and make sense of all chemicals—on the earth, on the moon, or those we will find on Mars.

In the early chapters you have to spend some effort in getting the fundamental basis for understanding materials which we will consider more specifically later on. And you may be impatient to get into some of the chemicals you've heard about like vitamins, sex hormones, and marijuana. Also you probably will not fully appreciate the need for a little theory until you see the applications in later chapters. But the effort you spend will be more than compensated for by the personal insights which you will have after you establish for yourself an approach and a basis for your thinking.

A BASIC APPROACH: IDEAS TO EXPLAIN THINGS

If there is any theme in all of the chapters to follow, it is this: Thinking, or the use of ideas, can be developed to explain why things behave the way they do.

On the one hand, we will be describing things by their properties or characteristics—how they behave, what they do, how they affect you. On the other hand, we will be getting into reasons why, in whatever way we can get into them. The first aspect is the *operational,* and the second the *conceptual.* In other words, operational descriptions and definitions of materials relate to how they operate, that is, what they do and how they affect your senses. Conceptual descriptions and definitions go deeper than the simple behavior and try to reach a more basic underlying cause or explanation—to whatever degree we can.

For example, you might find that polluted water is not good to drink. It may taste bad, contain sediment, have a foul odor, or a bad color, or in general not behave like water should. These are all operational findings. Conceptual description would begin to get into the kind of materials which may be mixed in the water. This is an attempt to get an idea or concept which might explain what is involved in, say, the water's foul odor. (The term *idea* is synonymous with *concept.*)

You will see later that we can come up with levels of explanation. In other words, we can refine our ideas about a particular set of operational findings. This we can do by investigation and thinking. Thus the conceptual end of the description of material things is open-ended—we do not necessarily get *the* concept which explains an operation completely and perfectly. One idea may lead to another, and that to another without apparent finality. Progress for you personally, and for people collectively, may indeed involve moving from a primitive concept to a more refined one along a whole series of more and more fitting descriptions. These are also more and more mentally satisfying.

Another example may help here. Scurvy is a disease characterized by weakness, swollen and bleeding gums, skin blotches, and prostration. This is an operational description. Invesigation over many years showed first that scurvy was caused by lack of certain foods like citrus fruit. This is a beginning of explanation—a slight move toward the conceptual. Eventually the cause of the disease was shown to be a lack of vitamin C in the diet. (Chapter 10.) This is an even better conceptual description: a disease caused by lack of vitamin C. We could go further and identify the nature of the chemical vitamin C. Then we would have a better conceptual description. There is even more room for conceptualizing or developing ideas, for example, about the mechanism by which vitamin C affects the health.

The generation of concepts to get to the heart of the matter is of course a predominant capability of mankind. It lies behind all expression and transmission of ideas, verbally and in writing. Science by nature seeks out explanations and the underlying causes of things. Since chemistry is the science of matter, we will be especially concerned with ideas on what lies behind the behavior of matter. This is the conceptual side which will be emphasized in the first chapters. You will then find yourself using chemistry concepts, almost unconsciously, to make sense of the way materials in your environment behave. The realm of concept can have profound value for interpreting and changing the operational world. Concepts used by you in relation to materials can make your life much safer, more comfortable, more healthful, and perhaps more meaningful. In fact it is only through concepts about things that we can really understand them, make proper judgments, and take proper action.

We can manage air pollution, for instance, not by just cursing it, or avoiding it, or acting as if it isn't so bad, which was the case for many years. We can find out what is the nature of the atmosphere, sunlight, and unburned fragments of gasoline. We can then learn how these ingredients "cook" in the stagnant air of, say, the Los Angeles basin to form irritating chemicals. With an understanding of the process, we can perhaps control it.

Here is a summary of the two-pronged approach to learning about chemistry. You may want to refer to this as you study the early chapters and develop your own personal insights on how to think about things.

OPERATIONAL	*CONCEPTUAL*
As a description of matter relates to	*As a description of matter relates to*
*How it behaves or operates or what it does	*An idea to explain the characteristics or operational behavior of matter, or why it behaves in a certain way
*The characteristic properties matter has	*Why it has the properties it does
*What are the effects it has on your senses	*Why it acts the way you perceive it
*What you perceive, observe, or experience	*Why it functions the way you find it to
*A practical picturization	*Some theoretical basis for its behavior
*Factual behavior	*Understanding of behavior

Notice that the overall difference is that operational definitions of various forms of matter refer to *what* the matter does. The conceptual definitions or descriptions refer to *why*.

EXPERIMENTS AND EXPERIMENTATION

The stress on how far you can go in the conceptual area should not cause you to lose sight of the fact that chemistry starts with experience. The course of study you are starting is a chemistry course; it concerns materials of your environment and how they react with each other and with you. The nature of these materials can best be appreciated only if you get yourself in conscious contact with chemicals. This is a necessary complement to any coverage you get in books and lectures.

You are, of course, in vital contact with chemicals at all times, but you may not be too aware of them. Here are three suggestions for getting your mind involved with your chemical environment.

1. Be alert, observant, alive to your body, your food, even your breathing—all your contacts with the chemical environment. (Not just the smog, or the unpleasant gas smell when you follow other cars too closely.)

Chemists working in a modern analytical laboratory. Many tests, from raw materials to finished products, are required to assure identity and purity of chemicals involved. (Courtesy of Merck & Co., Inc., Rahway, N.J.)

2. Do some of the simple experiments suggested in this text. Make notes of your observations, trying always to interpret or explain in terms of the concepts and theories which we will be gradually unfolding in the text itself. *And don't be afraid of error in your guesses and explanations.* All guesses are tentative: a great scientist has estimated that science is based on 99 percent error. It is by means of error that science moves closer to truth. In science we continually correct our guesses to bring them more in line with our improved understanding.
3. While you are developing the habit of interpretation, move out into your everyday activities and interpret, or explain to yourself, the happenings of nature that you encounter in terms of simple concepts of chemistry. The emphasis is on "simple" concepts because the soundest explanations are made in terms of the simple.

Your kitchen, your car, your foods, air blowing through your windows, dust settling on window sills, soiled clothing, the evaporation of water from wet clothes—these are all your laboratory where observation and experimentation occur. Chemists use these same materials even though they also build special rooms equipped with more convenient ways of experimenting with environmental chemicals. These special buildings or laboratories have some simple and some rather complex equipment to better study the chemicals of our environment. However, all the world is a laboratory to one who is alerted and observant to the changes continually occurring. The more awareness you, yourself, develop, the more alive you will feel and the more meaning you can apply to your changing self and your changing environment.

You will be surprised as you study chemistry at how improved will be your observation of the natural environment. You will learn what will be a lifelong habit: not to look without seeing, but to consider nature more critically than before; and then to understand in terms of structure, mechanism, and interrelationships. Perhaps you will see how scientific investigation and interpretation of your environment can provide intellectual delight through satisfying your natural need to know. This enrichment of your life will come as you actively develop a greater understanding and appreciation of the order of the world of nature in which you live.

THE ARTS

This book is designed for people interested in the liberal arts—especially for people who may never take any more advanced courses in the physical sciences. The term *arts* does not refer simply to skills, but to branches of learning wherein skills—physical or intellectual—may be attained. Figure 1.1 shows how chemistry fits into the overall scheme of the arts. The distinctions between the various arts are made on basic points of difference in the arts which, of course, are not all-exclusive. The finished carpenter's chair can be beautiful as well as practical. And ideas can have a profound beauty of their own.

The arrows on the right in Figure 1.1 indicate how the liberal arts feed back to both the fine arts and the useful arts. For example, newer acrylic paints—the product of chemical understanding which developed over many years—are adding new dimensions to both the fine arts of

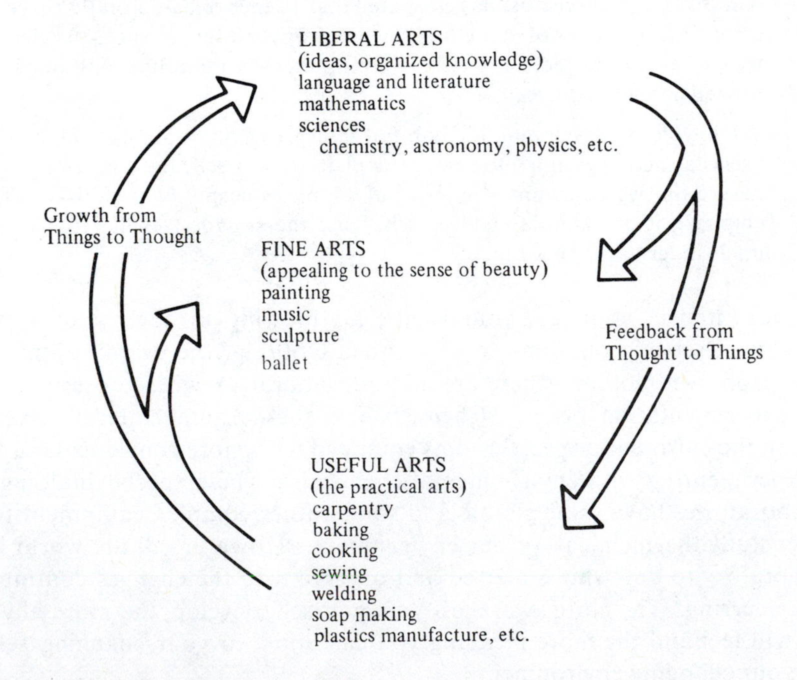

FIGURE 1.1
Relationships among the arts.

landscape and portrait painting and also to the useful art of house painting (chapter 13).

To emphasize only the useful arts as a branch of learning produces a person who earns a living but is not likely to enjoy it. Raising sights to the level of the fine arts adds potential for aesthetic enjoyment. The really educated person is continually being developed on the third level of the liberal arts. Here he can be liberated or freed from the lack of order and the confusion which the world may otherwise present to him. In this connection, no one can call himself educated today who is not acquainted with the scientific approach and the effects of science on our culture. One aim of this book is to help you get the needed understanding.

PROJECTS AND EXERCISES

Experiments

1. This is an experiment in crystal growing. In every household there are available chemicals which can be used to grow crystals. Some of the chemicals in the home will not be as useful as others in demonstrating the growth of large visible crystals. Some good ones are common table salt, epsom salts, and, with care, sugar. If sugar is used, crystals are more easily obtained by hanging a thread with a weight attached (nut or bolt) in the solution. (See Figures 1.2 and 1.3)
   First, obtain some small jars or transparent food dishes. Baby food jars are good. Now select one or two materials which you will use for crystal growth, such as table salt, or epsom salts. Fill a container half full with hot water. Then add the chemical gradually with continuous stirring. Your purpose is to add enough to fully "saturate" the hot water, but not too much so that there is a large excess. In the case of some chemicals you may notice a cloudiness in the solution even though more of the basic material can go into solution. Can you guess a cause? After you have stirred in what you think is all the solution can take, add a little more hot water and stir again. Then let it settle for a day or so until any material suspended in the solution settles out. Pour the *top,*

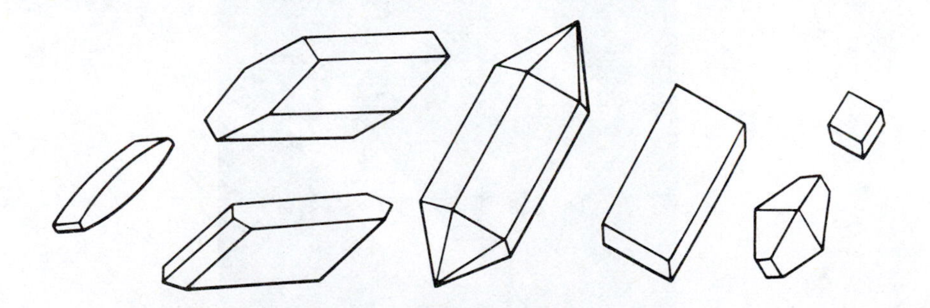

FIGURE 1.2

Examples of some of the varieties of crystal shapes found in nature. Shapes of crystals of the same substance can look different depending on changing conditions of growth, but the angles between their faces will always be the same.

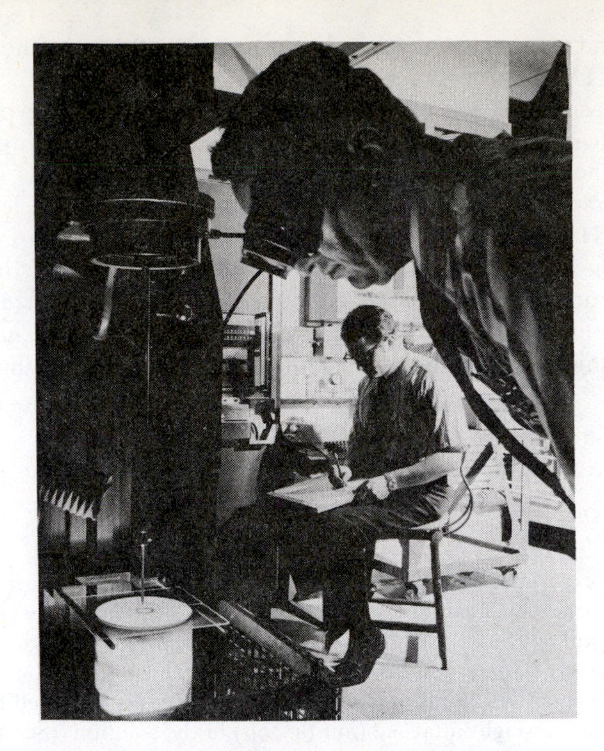

Very specialized equipment is sometimes used to grow crystals. Here scientists are shown working with an automatic crystal growing machine which is computer controlled to grow large, single crystals. Crystals have many uses in communications equipment. (Courtesy of Bell Laboratories.)

Chemicals may be purified and grown into definite geometric shapes. Here is a crystal (lithium tantalate) grown by scientists at Bell Laboratories for use in electronic communications equipment. (Courtesy of Bell Laboratories.)

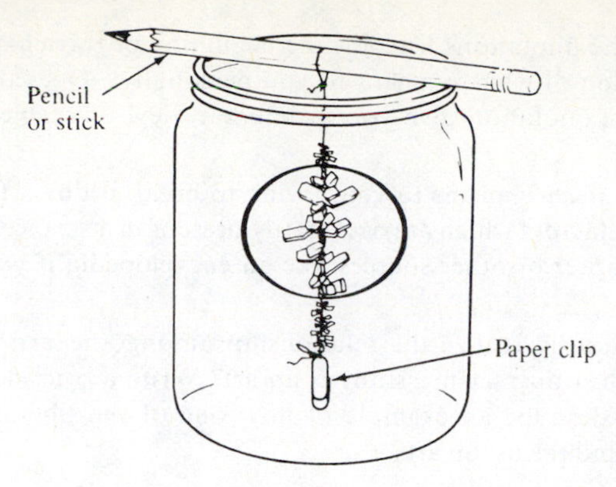

FIGURE 1.3
One way of growing sugar crystals.

clear liquid into another clean, transparent jar or container. (Discard the cloudy layer.) Now label this container and put it in a place where you can observe it daily without moving it. You will need to leave it undisturbed for several weeks. *Remove the cap or cover* when you put the labeled container aside. Make a record on a sheet of loose-leaf binder paper of any details like dates, cloudiness, and so forth. Then leave a column where you may enter subsequent dates along with any information as to the behavior of crystals, if any. You will be getting more information on crystals in a later chapter, but the important procedure now is to observe carefully and to make an attempt to explain any behavior noted.

As the water in your container evaporates, the level will drop. Do not let the water level fall below the solid material that appears in the bottom of the container. Before that happens, stop the experiment. Pour off the liquid that is left and carefully dry the solid material by means of a cleansing tissue. When the solid crystal is dried by wiping off excess liquid you will want to examine it again, very carefully noting any differences between different chemicals. Record your observations. Save the crystals for later reference.

2. Go outside and find some rocks. Examine them carefully, using a small magnifying glass or hand viewer, if available. Look especially for variety. If you live near a building of brick or limestone then observe these also under magnification. In addition, you might look at some unpainted concrete. Pick several hard structural materials like these and observe them carefully. Make a written record of what you find worth noting. Are the rocks uniform? What do you think the inside is like? Look for a split rock and find out. If you try breaking a rock yourself be careful to keep chips from flying into your eyes. How would you describe the composition of rocks from the nontechnical, nonchemical point of view?

Exercises

3. Go into your kitchen and list all the chemicals you can find there. Limit yourself to about ten minutes. Do some exploring and thinking. Then list the chemicals, using only the common name, for example, table salt. This is not a technical assignment but just a chance for you to observe and record, based on your present understanding. Technical details and long-winded chemical names are not wanted.

4. Take one of the quotations listed at the beginning of this chapter and write a short exposition of what it means to you personally. Does the expression of the idea in the quotation give you a different view of science than you had before?
5. Explain what *science* means to you, trying to break it down from the general to the specific factors which are necessarily present in a science. Use the coverage in this chapter or other sources like an encyclopedia if you have no ideas of your own.
6. Did you previously realize the relationship among the arts? What do you think of the idea that science study is an art? Write a brief description of the arts and then describe an example of how one of the liberal arts influences both the fine and the useful arts.

SUGGESTED READING

1. Consult a good encyclopedia for clarification of any ideas or issues raised in this chapter, for example, science, chemistry, fire, chemical industry, rocks. The encylopedia represents a concise, ordered assembly of interesting material on a subject in easy-to-reach fashion. If in the future chapters you need amplification on a subject be sure to check the encyclopedias first. They are usually less involved and less specialized and more readable than the extensive books on a subject. The encyclopedia-checking habit can save you considerable time in getting background on a particular subject.
2. Margenau, H., and Bergamini, D., *The Scientist,* Life Science Library, Time, Inc., New York, 1964.
3. Emmons, H. W., "Fire and Fire Protection," *Scientific American,* 231, No. 1, pp. 21–27 (July 1974).

chapter 2

Scientific Thinking and Molecular Theory

The whole of science is nothing more than a refinement of everyday thinking.

—EINSTEIN

I know that I am mortal and live but for a day; but when I search into the encompassing, ordered circling of the stars, my feet no longer touch the earth, but side by side with Zeus himself, I take my fill of immortality, the very food of the gods.

—PTOLEMY

The eternal mystery of the world is its comprehensibility.

—EINSTEIN

Science thinking is essentially the same as everyday thinking. Briefly it is: *Look at it. Think. Guess. Try it out.*

The girl who decided to put gas into her tank was thinking scientifically. The experience probably went something like this: She got into the car, turned on the ignition key to engage the starter. She heard the usual starting noise, the motor turning over indicating that the battery was working. But the car did not start. She thought about it, then guessed that the gas tank was empty. She checked the gauge which read below empty, thus confirming her guess. This led to her decision to get gas. *How* she did this was disastrous, indicating a need for further knowledge about gasoline.

SCIENCE AND SCIENCE THINKING

The root of the word *science* comes from the Latin for *to know* and it relates to *knowledge*. People by nature develop a way of knowing, and a particular approach to interpreting and understanding the world in which we live, and this is science. It is then less a subject of study than it is an activity of people, an attitude of people, and a method for understanding what they encounter in their active involvement in living. Science is not a body of facts, even though as human activity it leads to a systematic knowledge of facts. And more facts accumulate daily.

The statistics about the facts are themselves astounding. It is estimated that 90 percent of all the scientists who ever lived are alive today (25 percent of all the people who ever lived are also alive today). The amount of technical factual information doubles every ten years and about 50 percent of what we know today about chemistry was learned in the years since 1950. Hundreds of thousands of books and journals are published with a veritable deluge of information. In fact, at the level of present accumulation of information, the huge Library of Congress in a hundred years or so would have to occupy the whole of Washington D.C. That is why hundreds of thousands of books are discarded yearly, and many put on microfilm.

This book then will not try to weigh you down with scientific facts. It will help you develop a feeling for the approach of science. It will give you experience in science thinking, and the methods of scientists, so that you can deal adequately with the facts and activities of your life ten or twenty years from now. These facts of your future are mostly unknown today. Of necessity, however, we will be involved with some present-day facts because thinking is an activity based on sense appraisal of the things of experience. Our thinking must occur in relation to them.

There is no one scientific method, no magic formula for science success, any more than there is one way to write a poem or compose a piece of music. The approach is the thing. In science it is the open-minded observation of things, thinking, guessing to explain or predict, and then trying out the guess. This general approach is refined by scientists but never standardized. They never have the last word or the final, complete explanation. They are always observing, thinking, guessing, checking.

CHEMISTS, CHEMISTRY, AND MATTER

Chemists are scientists with a particular special interest in one aspect of the broad spectrum of science—chemistry.

CHEMISTRY IS THE SCIENCE DEALING WITH MATTER, ITS COMPOSITION, PROPERTIES, AND CHANGES.

Matter is one of the fundamental units of experience and is hard to define in terms of simpler concepts. Matter can be described as anything that occupies space and has weight. It is the stuff we touch and feel and hear and smell in our daily contact with our surroundings. Chemistry makes matter more understandable.

The sciences dealing with nature were originally considered under one heading of "natural philosophy," or "natural science." As knowledge increased and mankind developed more insight into the variety of the natural environment, people became interested in special areas of science. Thus the fields of chemistry, physics, biology, geology and astronomy developed.

LAW, HYPOTHESIS, THEORY

Some special language is used in describing the scientific approach. In general terms, scientific methods involve: Look at it. Think. Guess. Try it out. But the guess in science is given a more formal name—*hypothesis* (Greek: placing under).

A HYPOTHESIS IS A GUESS TO TRY TO EXPLAIN AN OBSERVATION.

If the person then checks (tries it out) and confirms the guess, we have a *theory* (Greek: a sight to see).

A THEORY IS A CONFIRMED HYPOTHESIS, OR AN EXPLANATION WHICH HAS BEEN CHECKED.

The word *fact* is sometimes used to mean an observable event. There is one other general word often used in science in a different way from everyday life. This is a *law* (Old English: something fixed).

A LAW IN SCIENCE IS A SUMMARIZING STATEMENT ABOUT EXPERIENCE.

A law is merely a statement about behavior of nature which has been found to be invariable under the same conditions. It does not mean a rule or set of rules established by a community of people and binding on its members. This is the legal meaning of a law, like "the laws of the United States." It also does not mean a rule set up by people for nature to follow! On the contrary, it is an expression by people about the ways that nature behaves, and the *people follow the rule* or suffer the consequences.

It is important to distinguish carefully between a law and a theory. A law is not an explanation, as a theory is. For example, the law of gravity states that bodies attract each other, and that if you jump from a ten-story building you will fall to the ground. This does not explain *why* you fall. As a matter of fact there is not yet any generally accepted scientific hypothesis or theory to account for the reason why, or to give the cause of gravity. Newton (1642–1727) described gravity but expressly

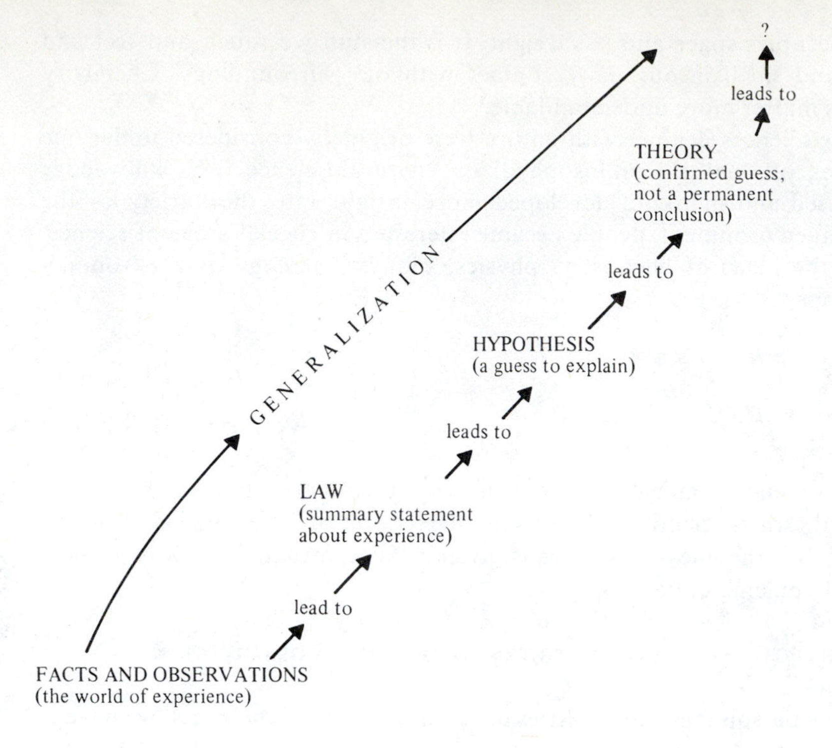

FIGURE 2.1
Levels from the specific world of experience to the more general world of ideas.

refrained from speculating on its cause. We can measure gravity's pull by weighing but so far we have no theory to explain it.

The formulation of law—as a generalization people have distilled from experience—does require considerable thought. But the higher stage of *explanation* requires even more. See Figure 2.1.

THE THREE FORMS OF MATTER: SOLID, LIQUID, GAS

The case of the girl in the garage (chapter 1) is a good example, in the preliminary stage at least, of the general approach of science. Her hypothesis was correct but her working knowledge of the properties of gasoline was not. The fact that gasoline is a liquid which easily becomes a gas, especially in hot places, was a contributing cause of the fire.

We could profitably start our study of matter by trying to make distinctions between solid, liquid, and gas—the three forms or *states* of matter. Water when frozen as ice is still the same basic stuff of matter. It has simply changed its state, not its nature. When it melts we get the same water back again that we put into the ice cube tray of the refrigerator. Under other conditions water becomes a vapor or gas, as when you boil it or when the sun draws it up into clouds. Clouds are merely tiny liquid droplets of water.

At normal room temperature gasoline is liquid. But we use it because of the ease of getting it into the "gas" state since that is what is needed in the cylinder of a car where gasoline, as a gas, mixes with air. Then this gas mixture gives us an explosion when the spark plug is energized. The exploding gas-air mixture pushes the piston which drives the shaft which drives the wheels.

However, nature is consistent. If gasoline vaporizes in a cylinder of a car, it will also vaporize from a bowl, or from the floor of a garage or service station, or from an open gas tank. The way to control it is not by calling the fire department but by keeping it from vaporizing till needed. Just because you cannot see the gasoline vapors (the real "gas" of gasoline) moving down the garage and filling up the garage, is no reason for ignoring this gasoline property—which is to become a gas easily at room temperature. It is the nature of gasoline to do what gasoline does. You cannot change the basic nature of chemicals. All you can do is study their nature. Then you can *control* the gasoline reaction and use it to advantage.

MOLECULES: DISTINCTIVE BITS OF MATTER

You may have wondered:

- Why gasoline burns
- Why some plastics melt while others do not
- How crystals form
- Why rubber stretches
- Why plastics and bottles do not dissolve in water
- Why ice melts
- How soap cleans
- How cake and bread rises
- How iron rusts
- Why glue sticks paper together
- How a match lights
- Why salt dissolves in water

These and all the varied properties and behavior of matter are results of structure—how matter is built up, and what the units of matter are. The building blocks or basic units of matter are atoms and molecules.

Trying to decipher the puzzling nature of the differences between various kinds of matter is not a modern occupation. The ancients recognized distinguishing properties of different materials, such as hardness, color, taste, bendability, and so on. From this first distinguishing between different things begins the chain of explanation, ending in ideas which are the theory side of science.

The ancients were intrigued about whether matter was continuous, like Jello appears to be, or whether it was made up of particles, like a bucket of sand or oranges. Perhaps even a bucket of water is a bucket of particles, on a size level too small to be seen? Most people said "seeing is believing" and assumed that water, at least, is continuous.

The question comes down to this: Is there a line of size running from the bucket of oranges, which we see as definite "particles," to the water, whose particles we obviously cannot see? (Figure 2.2.)

These are the things the Greeks thought about. And they came up with

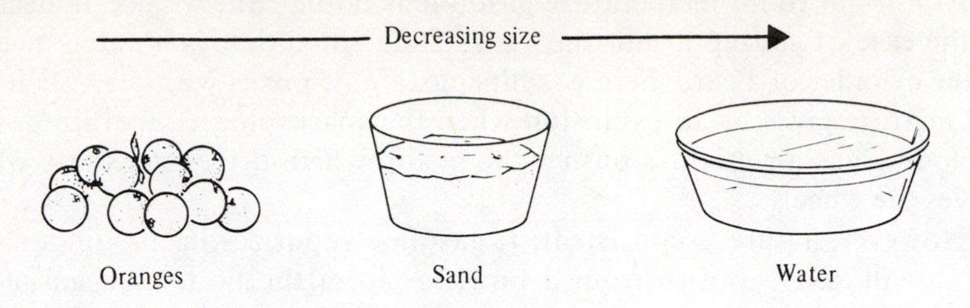

FIGURE 2.2
The Greek question: Are the oranges and sand really continuous, or is the water made up of particles?

some answers. Most thought that all matter is basically continuous. The oranges, if you go down to the basic "stuff" of them, would also, in this view, be continuous like water. Others said matter is discrete, that is, made up of separate particles. According to this view, if you take a piece of gold and keep cutting it in two you would finally get to a piece which you couldn't cut up or divide any more. This is the *atom* (Greek: uncuttable)—the smallest particle of a substance like gold. The divergence of the two views of matter was extreme. Some said that the oranges really were like water obviously was—continuous. Others, the minority, said that water really was like oranges obviously were—discrete, or made of particles.

The direct checking of such hypotheses was not possible in the early days. Yet curious people—scientists among them—did not hesitate to investigate and draw conclusions from indirect observations, as we do all the time. For example, when you examine a frosted light bulb, it is not necessary to break the glass to see if the filament inside is broken or not. You must guess that the filament is intact if the bulb lights when it is placed in a lamp socket. To test your guess or hypothesis you turn on the lamp. If the bulb lights, this confirms your guess that the filament is intact. If not, you probably should guess that the filament is broken, or the bulb is burned out. But be careful! Maybe the lamp is disconnected from the wall and your testing will be misleading—possibly causing you to discard a bulb which is good.

The results of many tests and experiments and checks by many people over hundreds of years have now backed up the guess or hypothesis that matter is made up of discrete particles. All the work of chemists proceeds on that basis. (See experiment 1 at the end of this chapter.)

MOLECULAR THEORY

Indeed, the particle hypothesis is confirmed and is called a theory—the molecular theory of matter. This is an attempt to *explain* behavior, as all theories are, by assuming that matter is ultimately made up of discrete (distinct or separate) particles which cannot be broken into parts without

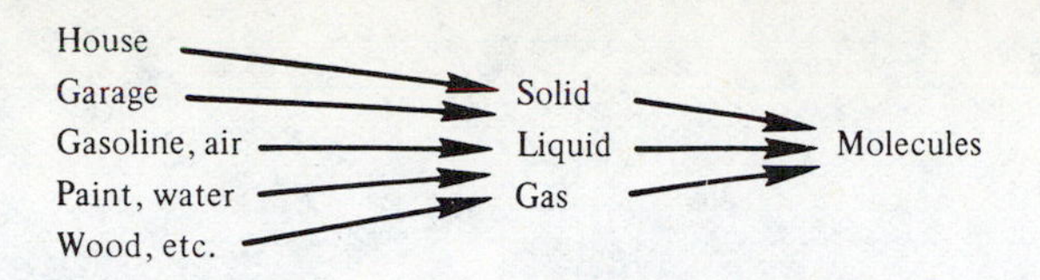

FIGURE 2.3
The direction of thinking, from the operational world of experience to the particles that allow us to explain differences in materials.

changing the properties of the particular material. These tiny particles are such as those of water, sugar, and gasoline. They are called *molecules* (Latin: little mass).

A MOLECULE IS THE SMALLEST PARTICLE INTO WHICH A SUBSTANCE CAN BE DIVIDED WITHOUT FORMING A NEW SUBSTANCE.

The molecule is usually very, very small. For example, a small drop of water contains more than 30 billion billion molecules. Yes, not 30,000,000,000 or just 30 billion but 30 billion times a billion, that is, 30,000,000,000 × 1,000,000,000 which equals 30,000,000,000,000,000,000 molecules. The population of the United States is in the range of 300,000,000. The total population of the whole world is only about 4.5 billion, or 4,500,000,000. No wonder water looks continuous. You cannot see such small particles.

You may have been thinking of atoms, since that term is probably more common than molecule. If so, you are a chapter ahead. We are making our way, via abstraction of a mild degree, from the "real" world of houses, gasoline and water to the theoretical explanation of behavior of these items of your environment. The level of thinking so far is shown in Figure 2.3. You can see already some of the simplification that an organization of thinking can bring. Beyond molecules we will, of course, consider atoms. And then beyond atoms? More abstraction—but all starting from experience, based on reality, and explaining reality. Our purpose is to make sense of your world.

KINETIC-MOLECULAR THEORY: A BROAD GENERALIZATION

The Kinetic-Molecular Theory is one of those generalizations which scientists have developed to help explain the behavior of things. There are not many of these general views, and it would be worthwhile to consider in more detail the meaning and value of the theory which is now thoroughly entrenched in both thought and practice relating to all material things.

From your experience, you realize that materials have different degrees of ability to stay together. (See Figure 2.4.) Solids like ice and wood and iron and gold are made up of molecules held tightly together but still with some space between them. In solids these molecules are generally

Photo of the earth taken by Apollo astronauts from above the moon's surface, 240,000 miles away. A view like this gives a better perspective of the spaceship earth, which is your home. Similarly, a theory in science gives a simpler and more general view of what might otherwise be a complex of varied and confused experiences. (NASA photo.)

arranged in special symmetrical ways and the regularity is what gives rise to crystal shapes of many solid materials like sugar and salt. There must be some electrical force holding these molecules in place to explain the tenacity of solid matter. The forces are varied between various types of materials; for example, we bend iron, flatten gold, and squeeze lead, but glass shatters and wood just gives a little before it snaps.

Liquids do not have the rigidity of solids so we could surmise that there must be some looseness between the molecules to allow them to slip past each other. Here again there are various degrees of looseness: a can of honey can be tossed over and quickly recovered whereas a glass of beer or milk, on tipping, runs all over the table and floor.

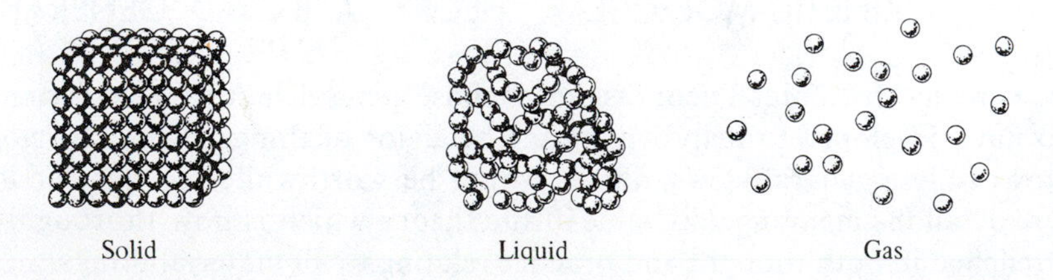

FIGURE 2.4
Casual picturization of major differences between solid, liquid and gas states of matter. The distances cannot be drawn to scale and the rapid motion of molecules also is not depicted.

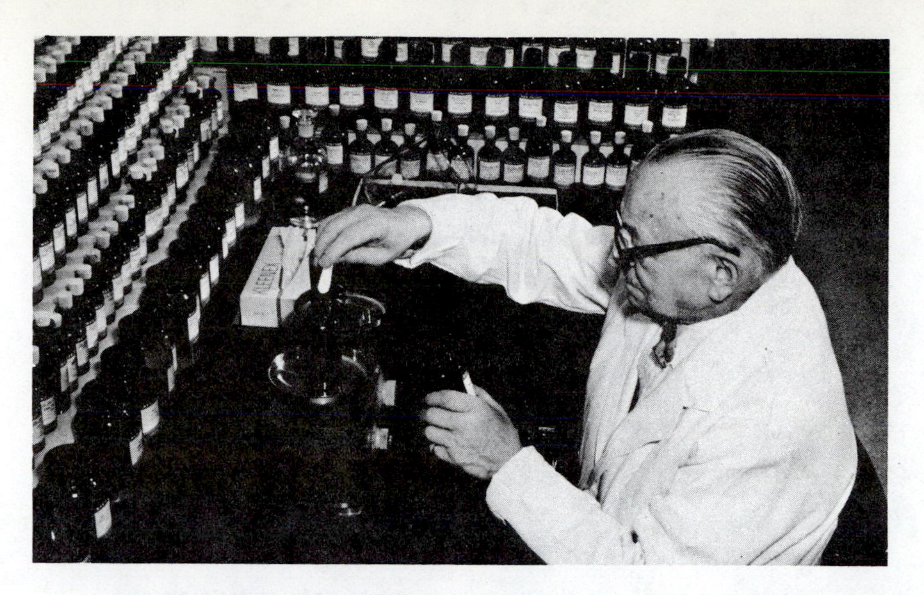

Perfume chemists develop new fragrances in convenient individual consoles. Note the use of caps and corks on all bottles to prevent escape of volatile components in gas form. (Courtesy of Max Factor.)

The molecules of a gas are looser still. If a bottle of perfume or a burning incense stick is left in a room, the aroma, carried by molecules of gas, is soon all over the place. In fact, perfume would be useless if it did not travel around and penetrate the air spaces in the environment of the person using it. Judging from observations like this, we can guess that there must be spaces between the molecules of a gas to permit easy mixing with other gas molecules, in this case air.

We can carry our guessing further. The molecules of a gas which have just previously been a liquid (like perfume or gasoline) must have some attraction for each other since they were recently held together in a liquid form. In fact, we could freeze the molecules and make a solid, recognizing by the hardness of the material that they are even more rigidly held together than before. So what could we do to explain why the particles of the gas perfume no longer collapse into the liquid again? Or why water can easily be frozen to a rigid form, or converted to a gas by putting a fire under it? In other words, how can we explain why one state of matter, or structure of molecules, can be changed into another state of matter, or structure of molecules? The molecular theory of matter, as we have considered it so far, needs improvement.

We need a further idea or concept than just particles. We must additionally guess that the molecules are in motion and by giving them energy they can be moved apart more. And, conversely, by taking energy away we can quiet them down so that their forces of attraction are more effective—producing a liquid from a gas or, by more restriction, producing a solid. The expansion of our guess or hypothesis—confirmed enough to be called a theory—now includes the idea of molecules in motion. This

Recent trends in perfumes involve molecules which mimic natural fragrances like apple, lemon, lime, and strawberry. The operational fragrance effects are obtained by using specific molecules in gas form. Chemists have now identified many varieties of perfume molecules. (Courtesy of Max Factor.)

broader generalization, or improvement on the original molecular theory, is called the Kinetic-Molecular Theory (occasionally referred to as KMT).

Kinetic:	motion
molecules:	particles
theory:	explanation

The Kinetic-Molecular Theory provides a model of moving particles to explain matter's properties. Historically the ideas were developed over many years. The original efforts were related to explaining the behavior of gases, such as our problem of today: why does the gas of gasoline travel down the garage to the water heater? Or, the older problem of which Cleopatra perhaps knew something and which Anthony should have wondered about: why does perfume affect someone at a distance? These are *diffusion* phenomena and the Kinetic-Molecular Theory easily explains them: the gas particles are extremely tiny molecules or bits of the substance which are in very rapid motion and separated from each other. They therefore can move into another gas (like air) because it, too, is mostly space—with tiny molecules banging around in space without any attachments.

HOW THE IDEAS OF KMT DEVELOPED

The thinking of Evangelista Torricelli (1608–1647) gives insight into how the ideas of the KMT developed. He was a student of Galileo, the great Italian astronomer and physicist. Torricelli became interested in the pecu-

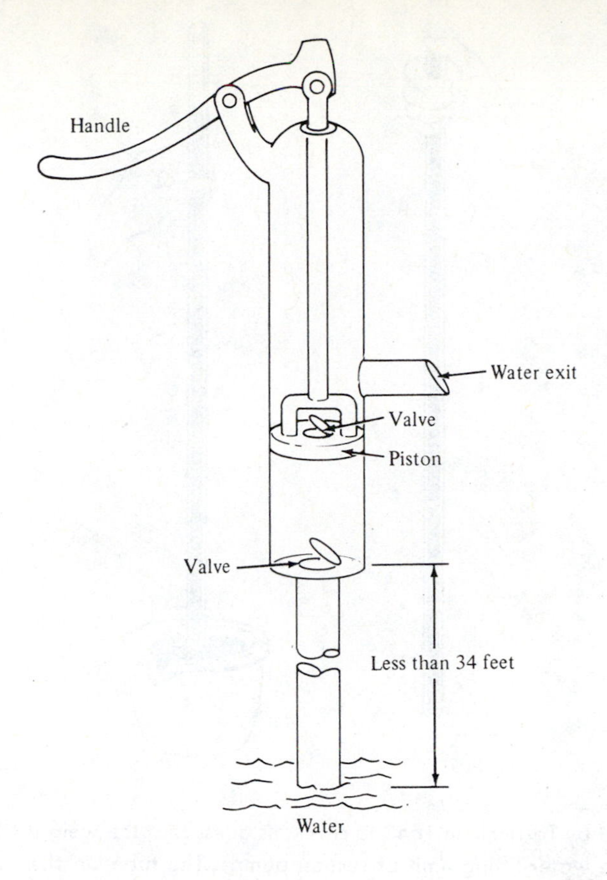

FIGURE 2.5

This is a cross-section of a simple lift pump or suction pump for raising water. The action was described (prior to Torricelli) by saying that the raising of the piston created a vacuum in the cylinder and "nature abhors a vacuum." Consequently, nature forced the water up the cylinder.

liar action of water pumps. Definite practical limits had been found on how high a single lift or suction pump could raise water. (See Figure 2.5.) In other words, water could be raised to a height of only about 33 or 34 feet and no higher. This limit was never adequately explained. People just accepted it as a fact of life and worked around it, as we do in the case of other of nature's limiting ways.

Torricelli wondered about the 34-foot limitation. He guessed that perhaps it was connected with the generally accepted idea that air had weight. Torricelli wanted to try out his guess and he and some of his students came up with the simple idea that liquid mercury, which is about 14 times as heavy as water, could not be pumped up as high as water. In fact, if it was air pressure, from the "sea of air" in which we live, that determined the 34 feet of water in a lift pump, then mercury could probably be lifted or held up only 1/14 as high as water. This would be 1/14 of 34, or somewhere between 2 and 3 feet. Torricelli and his students took a glass tube about 3 feet long, filled it with mercury and inverted it in a pot of

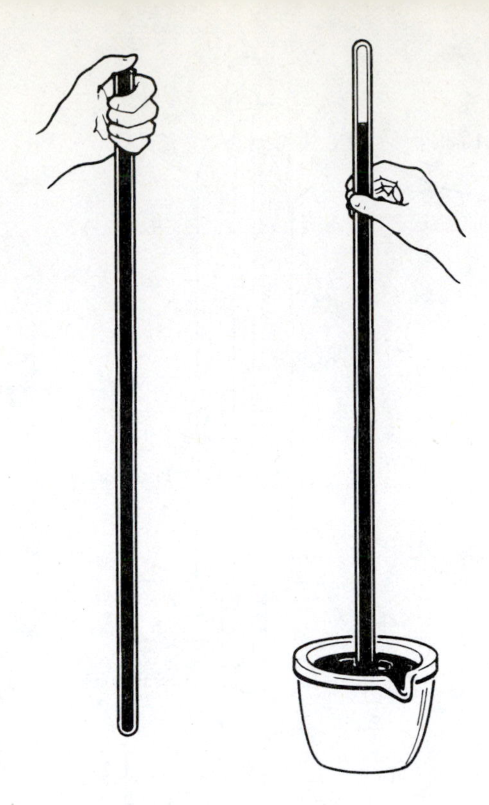

FIGURE 2.6
The experiment tried by Torricelli in 1643 to check his guess that the weight of air might determine the water-lifting limit of suction pumps. The tube on the left is filled with mercury and inverted in the pot of mercury on the right.

mercury with a student's thumb holding it until the open end was under the level of mercury in the pot. (See Figure 2.6.)

The mercury did not run out of the tube completely but fell to a level of between 2 and 3 feet! It was somewhere around 30 inches. The space above the mercury looked empty. The year was 1643, the first time people "looked" at a vacuum in a laboratory. Here was a space from which air particles were surely excluded. In addition, this was proof that the mass of the particles of air above us really has weight and is responsible for the natural limitation of suction pumps.

The Torricelli instrument is of course quoted on every weather broadcast when someone says, for example: "The barometer stands at 29.5 inches and is rising." The barometer (Greek: weight [of air] + measure) tells the pressure of the atmospheric air pressing down on any particular day and is related to weather changes. (See Figure 2.7.)

Robert Boyle (1627–1691) was an Irish chemist who experimented extensively with gases. He developed an effective air pump with which he evacuated containers and tried experiments in the "empty" containers. In 1660 he published a book describing his experiments with his pump in which he especially noted that "there is a spring or elastical power in the

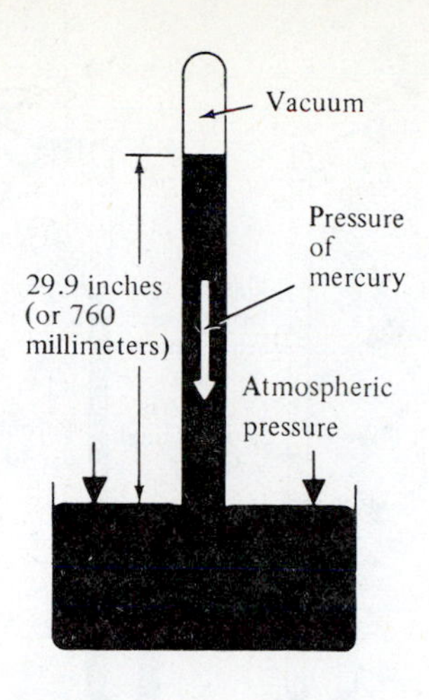

FIGURE 2.7

A barometer. At sea level, the atmosphere pressing down on the mercury holds up a column of mercury about 30 inches. At elevated locations, the weight of air is less and the height of mercury is also less.

air we live in." This is the same "spring" you notice when you push down on a bicycle or basketball pump when the tire or ball is almost fully inflated. The extent of the spring of the air is something special to gases. This is a major, though simple, concept which was evolved from observations of a very simple instrument—essentially a bicycle pump.

Boyle went on to other experiments with gases using simple apparatus like glass tubes in the form of a U and containing different amounts of mercury. He found another regularity in the behavior of all gases. If he doubled the pressure on a gas, its volume was cut in half. When the volume was increased to twice its original value, the final pressure was half that of the original. His generalization was called Boyle's Law and is the foundation of much practical physics and chemistry. (See Figures 2.8 and 2.9.)

Other scientists picked up Boyle's ideas and eventually went on to explain his findings by the idea that gases are mostly empty space with the molecules moving around very rapidly.

Amedeo Avogadro (1776–1856), an Italian physicist, is best known for his principle that equal volumes of all gases under the same conditions of temperature and pressure contain the same number of particles. This was proposed in 1811 and is a far-reaching generalization accepted without question today. However, it was rejected, ignored, or neglected for almost fifty years after Avogadro first proposed it to help explain some of the facts being observed in experiments with gases.

Avogadro pictured the moving molecules as occupying a small portion of the larger space apparently occupied by the gas. However, this was not by any means the dominant view of all scientists in those confusing years of the first half of the nineteenth century. For example, the great Swedish chemist Berzelius (1779–1848) pictured a gas as made up of particles in

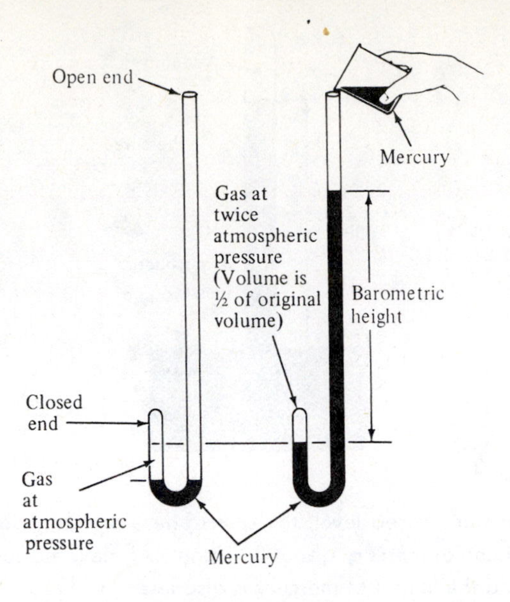

FIGURE 2.8
One of the ways Robert Boyle investigated the "spring of the air." The U-shaped tube is closed at one end. When mercury is poured in the open end, the air in the closed end is visibly compressed. The volume can be measured. In the diagram above, mercury is added to a level equal to barometer height. Then the volume is reduced to half its original value.

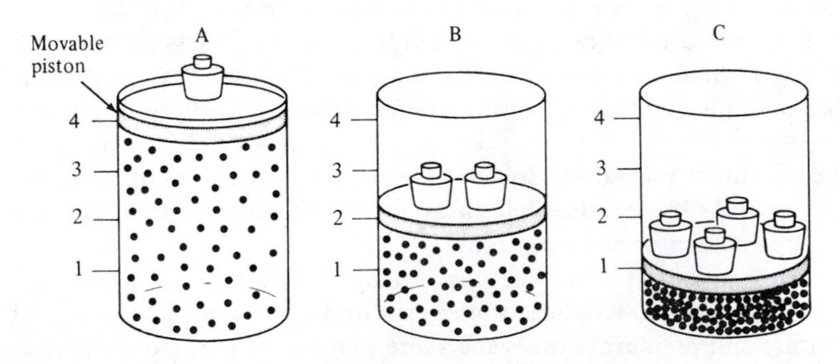

FIGURE 2.9
How the gas model of molecules in mostly empty space explains Boyle's findings. The weights are supported by an essentially weightless piston. When the pressure is doubled in B, the molecules are squeezed closer together, occupying half the original volume and colliding with the piston twice as frequently. In C, the pressure is doubled again, cutting the volume to one half that of B or one quarter that of A. The same original number of molecules is confined to a smaller space, accounting for the higher pressure, which is now four times that in A, or two times that in B.

contact with each other, as in a bucket of balloons. This is in sharp contrast with the Avogadro idea that the tiny particles of a gas were suspended in space like tiny particles of dust you see dancing in a sunbeam.

Scientists at the time generally favored the ideas of Berzelius and this was one of the reasons for the poor acceptance of Avogadro's views. It is interesting that not until the last half of the nineteenth century, a little more than one hundred years ago, did a convergence occur toward one generally accepted model for a gas. It was then Avogadro's ideas which were finally accepted and incorporated into the matured body of the Kinetic-Molecular Theory.

WHAT DOES THE KINETIC-MOLECULAR THEORY MEAN?

The fruitfulness of a theory can be seen if you follow up the ideas it suggests. For example, on the assumption that a gas consists of tiny molecules moving through mostly empty space, the molecules will be colliding with each other and with the sides of the container. They are presumed to have some attraction for each other. This is somewhat like gravitation or magnetic attraction, that is, operating at a distance. However, since the gas does not collapse of its own accord to a liquid, these attractions must not be too effective against the fast movement of the molecules. Thus perfume diffuses easily.

The collisions of the molecules on the walls must be the reason for the pressure in rubber tires. It is the rapid to-and-fro motion of gas particles which keeps the walls distended—just enough to give us a good cushion of a ride. The magnitude of this bombardment pressure depends on how hard the molecules hit, and on how many hit in a particular time. These actions are all invisible but are now the accepted basis for explaining how the gas air in the tire space gives us the overall property (pressure) which we need for this kind of application (the good cushion of a ride). In short, the Kinetic-Molecular Theory *explains* the behavior of gases, like diffusion and pressure.

Many other properties of materials are also explained by this particle-in-motion idea. The theory is thus a model to explain matter in bulk. It was originally applied chiefly to gas behavior because a gas is the easiest of the three states to investigate and explain. However, the same basic theory explains solids and liquids and this is a sort of bonus from the gas studies. The attraction between molecules which is not effective in ordinary gases because of space between the molecules and their speed begins to be more effective in the liquid state.

The attractive force or tendency to hold together of similar particles is referred to as *cohesion.* (Latin: together + to cling.) Like gravitational force between sun and planets or between earth and satellites, cohesion is stronger when the molecules are closer together. In a solid they are very closely packed so that we get rigidity and a definite shape. The rigidity arises from the restriction of the molecules within a tight overall space. This does not mean to imply that they are absolutely still. There is move-

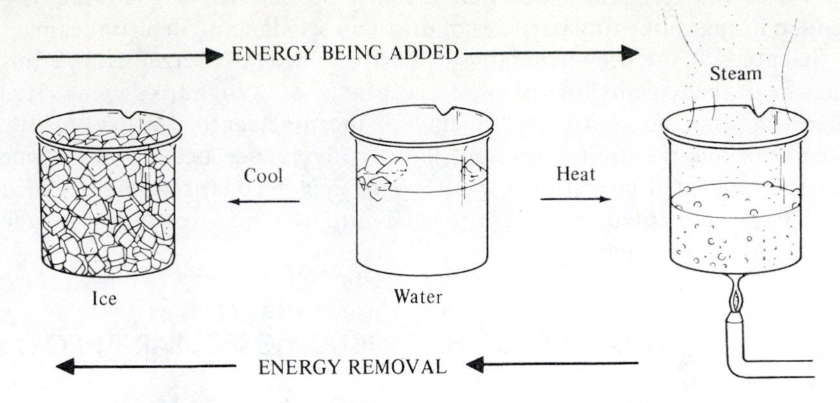

FIGURE 2.10
Physical changes in the appearance and state of water depend on energy.

ment but it is localized and not able to change the average location of a particular molecule.

In a liquid, the molecules are less tightly held and cohesion is less. That is why the liquid has a restricted looseness; water has a definite volume but takes the shape of a glass it's placed in. A gas is looser yet, the molecules having broken away from the cohesive effects. It has neither definite shape nor volume and *fills* the entire container—whether it is a bottle, a garage, or a house.

You have probably seen the large tents used by professional exterminators in the fumigation of a house. The tent is draped over the house and then poisonous gas is released inside. Even the termites hidden away in their tiny cells come to learn that the nature of gas is to fill the container in which it is placed.

When a solid like ice is heated, the heat energy causes the molecules to move faster. Eventually, if enough heat is supplied, the molecules overcome the cohesive forces holding them in a rigid pattern. This is melting. Then they slide past each other—and we have liquid water. Heating water even more, like in a tea kettle, eventually provides enough energy so that the molecules form bubbles and break out into the air. Then we get a gas. This common process is called boiling. Cooling reverses the process by taking away energy. (See Figure 2.10.)

SUMMARY OF KINETIC-MOLECULAR THEORY

The Kinetic-Molecular Theory represents one of the monumental achievements of the human mind. Do not let its simplicity mask its value. Most broad, general, and universal concepts in science are simple. The purpose of theorizing is to explain, and in the process, to bring some simplicity and order to the apparent complexities of our experiences.

It may be worthwhile here to describe the main points of the theory, not for memorization, but as a summary to help you bring together in your mind all of the discussion in this chapter.

The KMT, then, provides a model which explains behavior of the material world through use of the following major ideas:

1. *Matter is composed of very small particles called molecules.* Each kind of matter is made up of its own kind of molecules which are all the same but distinctly and essentially different from those of other kinds of matter.

2. *The molecules are in rapid motion.* That is, they have energy of motion, or kinetic energy.

3. *Molecules have some attraction for each other but the effectiveness of this attraction depends on the molecules and how close they are to each other.* Thus for one kind of matter like water, the molecules have effective attraction in your freezer so that they form a solid wherein movement from place to place is restricted. In the liquid, the movement is more free, but only in the gas state are the molecules far enough apart to practically allow no appreciable effect of the intermolecular attraction. This is because the *gas is mostly space.*

4. *When molecules collide (in gas) there is no overall loss in energy.* This is sometimes said in another way: molecular collisions are perfectly elastic. This is certainly quite different from automobile collisions. It is fortunate for us that our collisions on the large-scale level use up energy and come to a halt. Otherwise cars would be continually and forever bouncing off each other on our freeways.

SOME AMAZING FACTS ABOUT MOLECULES

We will not describe any of the detailed calculations and measurements which scientists have made of the world of molecules. However, it is highly rewarding to see some of the results. They are indeed startling and in some measure almost beyond belief.

The molecules in a gas are not only extremely numerous but extremely active. For instance, the molecules of air, which are chiefly nitrogen and oxygen (chapters 7 and 16), are moving, on the average, at speeds of approximately 1000 miles per hour! That is not a typographical error. The oxygen and nitrogen molecules in air move at speeds *averaging one thousand miles per hour!* If you have some hydrogen molecules in a gas sample, these would be traveling at approximately 4000 miles per hour. Being smaller, they go faster. This is greater than a mile per second!

We previously considered that the number of molecules in a tiny drop of water is approximately 30 billion billion! Now consider heating the drop and boiling the water away so that it becomes a gas, that is, changing the water from the liquid to the gas state. The same molecules are then present in the gas as were originally in the liquid drop. However, the amazing fact is that they would now occupy about 1600 times as much space as the drop did! The expansion is due not to swelling of the water molecules but to a tremendous expansion of the empty space between them.

Next consider taking a small volume of the gas as large, say, as the head of a pin. This volume would still contain about *30 million times* as many molecules as there are people on earth!

You know that gas molecules do not accomplish direct-line, long-distance movement of thousands of miles in an hour since it takes a few seconds after you open a perfume bottle before you smell the aroma. Why? The answer is: collisions. The molecules, being very numerous and also very fast, hit each other and the sides of the container quite often. How often? You would never guess. The number of collisions experienced by any one molecule in your room in one second is about 10 *billion!* A prize fighter punching a bag 3 times per second would have to stand there punching continuously for one hundred years before he hit the bag as many times as one molecule gets hit in one second! Of course the molecules of a gas are not all moving at exactly the same speeds. There is a distribution of velocities and the higher the temperature, the faster the average speed.

No wonder the word *gas* comes from the Greek root word *chaos,* which means confusion and disorder. Out of the Kinetic-Molecular Theory you may come to a little more orderly way of looking at chaos.

You can see clearly from the way KMT developed the broad picture of how science proceeds along different lines at the same time. Many non-scientists assume that there is simply a scientific method which the trained scientist uses and "abracadabra," out comes the correct scientific findings. You can also see that scientists make mistakes, even in their science specialties. In other words, they behave like people do. One of the values of science, however, is that it has a self-correcting device built into its approach: look at it, think, guess and try it out means continually checking and leads to further experimentation. The self-correcting element comes about because scientists consistently ask more questions and check new guesses. In the case of science, too many cooks do not spoil the broth.

Science is a complex human activity. It is a struggle to know and a process of growth in knowledge through frustration, differences of opinion, strong emotion even, but always with the excitement of exploration and a feeling of satisfaction in learning a little more. Science can also be viewed as a part of the collective knowledge of the human race wherein new ideas are continually evolving from experiments and observations. But it is always important to realize that the very heart of science is the process. Science is more a verb than a noun.

PROJECTS AND EXERCISES

Experiments

1. About 400 years before Christ, the Greek thinker Democritus came up with an idea which was apparently contrary to what the average person ordinarily thought about matter. Democritus was one of the earliest people on record

who opposed our first naive impression that matter is continuous. Think of this as you look at water, steel, glass. These things appear to be continuous. If there were holes in the glass, would not the water leak through? Democritus insisted that all material things, even though they look continuous, are really discontinuous, that is, made up of particles like a sandy beach.

You can perform an experiment to help you start thinking about this important problem and its answer. You will need a glass of water and a few crystals of almost any colored dye. You can use concentrated liquid food colors if you have no solid dye available.

Place water in the glass and let the swirling settle down. Then carefully drop a piece of solid dye to the bottom (or, very carefully put a drop of liquid food color or ink into the water). Let the glass stand as you observe it over a period of time. What happens? Does what you observe favor discontinuous or continuous theory of matter? Use your selected theory to explain the behavior you observed. Are there holes in the water? How would you explain the behavior of the dye in water by the Kinetic-Molecular Theory? Does this make sense also of what happens when sugar is stirred in coffee? Would coffee dissolve faster in hot or cold water? How would the Kinetic-Molecular Theory help here? Is the apparent continuity of water not "true" really? Do you know of any other examples where your sense impressions are apparently in conflict with what you *know* to be true?

2. The view of gases in the Kinetic-Molecular Theory would of course indicate that you have to confine a gas if you want to keep it in one place. You know what would happen if you left the top off an expensive bottle of perfume, a felt pen, or a container of gasoline. However, many people do not relate that behavior to what happens when you leave an uncovered dish of food or a piece of a vegetable in the refrigerator.

   Take two similar pieces of lettuce, two equal lengths of celery, and two carrots of the same size. Put one of each into a transparent food bag, then tie the closed and twisted end securely. Put the other pieces of celery, carrot, and lettuce on an open plate and put both the bag and the plate on a shelf in your refrigerator. Do not put them into a drawer.

   Look at these samples occasionally and interpret the results in terms of the KMT. Record your observations. Can a refrigerator serve as a dehydration (water-removing) device? Do you see why manufacturers provide humidity drawers, called "crisper" drawers? How do the drawers serve their purpose?

3. This experiment must be done at a time of year when flies are around. Find a place where several flies are active, preferably near a flat surface where you can swat them (a table, window, or screening). Alternatively, if no flies are around you can use tiny, crumpled pieces of tissue paper. Try landing the fly swatter on these without their blowing away.

   Get a piece of thick paper or thin cardboard. Cut in the shape of the screened or perforated portion of the fly swatter. Then tape it over the screen or perforated plastic so that the holes are completely covered. Now try to swat a few flies. Keep track of your method and your tries so that you can duplicate the effort later. After several trials, and after writing down your experiences, remove the cover of the fly swatter holes and repeat the fly swatting procedure, trying to duplicate as closely as you can your first run with the covered fly swatter. How does your success compare with the first method? Can you provide a reasonable guess as to the cause? How would you apply the KMT to this situation? Describe the mechanism by which the perforations in a fly swatter operate, using the KMT.

Exercises

4. Take one of the quotations listed at the beginning of this chapter and write a short exposition of its personal meaning to you. You may want to indicate what your ideas were formerly and how they compare with your thinking now.
5. Describe briefly scientific thinking. Is there one scientific method? Give from your own experience a personal example of a situation where scientific thinking was involved. Indicate where the thinking was adequate or inadequate for the particular situation.
6. You have seen fogging of a clear food wrap if used over a dish in the refrigerator. Explain this in terms of the Kinetic-Molecular Theory, using any of the distinctions between gas, liquid, or solid described in this chapter.
7. When a bottle of ammonia (or perfume) is opened in a corner of a quiet room, the gas quickly diffuses and you smell it in the opposite corner. (a) Which theory would best explain this, the one that claims matter is continuous, or the one describing matter as particulate (made up of particles)? Explain briefly why or why not. (b) If the molecules of ammonia are moving very rapidly (this is about 1100 miles per hour), why does it take some time for the gas to reach the other side of the room? In other words, why is it not practically instantaneous? (c) If a room could be made into a vacuum, what would the effect be on the movement of the ammonia molecules?
8. Boyle in his discovery of the "spring of the air" was working on the property of gases we call compressibility. The compressibility of gases is extreme compared with that of liquids like water, mercury, or brake fluid. Briefly describe the mechanism, or what really happens, when you compress a gas. Why is a liquid less compressible?
9. In many of the arctic regions the temperature never gets high enough to form liquid water from ice. Do you think that particles of the ice could nevertheless get free as a gas without first becoming a liquid? Use the concepts of the KMT. Think of your environment and try to come up with an example to justify your answer. Does this give any indication as to whether the molecules in a solid have stopped moving entirely?
10. Gases are quite compressible whereas liquids and solids are not. How would you explain the cushion or elastic effect of a "water bed" mattress? Would it be the same mechanism as that in a mattress with springs? Are springs compressible in the same way that we say a gas is? How would the cushion effect of a water bed compare in mechanism with an air mattress used in camping? Use the KMT ideas in your thinking.

SUGGESTED READING

1. Again check your encyclopedia for further broad information on any particular topic covered in this chapter.
2. Alder, B. J., and Wainwright, T. E., "Molecular Motions," *Scientific American,* 201, No. 4, pp. 113–126 (October 1959). Scientific American Offprint No. 265.
3. Young, Louise B., ed., *The Mystery of Matter,* Oxford University Press, New York, 1965, Part 2, "Is Matter Infinitely Divisible?"

If we begin in certainties, we shall end in doubts. But if we begin with doubts, and are patient in them, we shall end in certainties.

—FRANCIS BACON

It seems probable to me, that God in the beginning formed matter in solid, hard, impenetrable, movable particles . . . and that these primitive particles are incomparably harder than any porous bodies made from them. Indeed, they are so hard that they never wear out or break in pieces.

—NEWTON

And all the loveliest things there be
Come simply, so, it seems to me.

—EDNA ST. VINCENT MILLAY

chapter 3

Beyond Molecules: The Atoms

The world is made up of a number of things. Looking just in your kitchen or bathroom, you realize that there's a great variety of materials around. If you follow a natural human trait to make things simpler, or to seek a basic unity behind the great varieties of matter, you would be tempted to go further than just saying the massive numbers of different things in the world are made up of massive numbers of different kinds of molecules. That does not simplify enough. You might be tempted to make another guess or hypothesis.

This is what the early Greek thinkers did. They came up with the idea of elementary matter and elementary particles—atoms—to explain the vast complexities of the material world. In our sophisticated way today, we say that the molecules of different materials are different because they are made up in different ways of basic building blocks, or atoms. This simplifies the idea of great diversity of matter because there are only about one hundred kinds of atoms.

AN ATOM IS THE SMALLEST PARTICLE OF MATTER WHICH HAS DISTINCTIVE CHEMICAL CHARACTERISTICS.

Later we will be able to define it better.

MIXTURES AND SUBSTANCES

Before atoms, we must make an important distinction between types of materials. The strawberry jam you spread on your bread is obviously not a uniform material. There are lumps of strawberries spread throughout the sweet fluid. We call this a mixture, which is material put together in a rather loose way. What about Jello? It looks like a uniform material when it is served. But it, too, is a mixture if you think a little about the preparation from grainy, colored powder and water. What about the ocean, the air, or tap water? These are all mixtures. Then what are they mixtures of? In other words, where do we get to the single substance which is not thrown together from other materials?

A SUBSTANCE IN CHEMISTRY IS A UNIFORM KIND OF MATERIAL, OR A DEFINITE VARIETY OF MATTER, ALL SPECIMENS OF WHICH HAVE THE SAME PROPERTIES.

A MIXTURE IS A NONUNIFORM MATERIAL WHICH RESULTS FROM PUTTING SUBSTANCES TOGETHER. IN OTHER WORDS, IT IS A VARIETY OF MATTER, ALL SPECIMENS OF WHICH DO NOT HAVE THE SAME PROPERTIES.

ANY MATERIAL IS EITHER A SUBSTANCE OR A MIXTURE.

In referring to a substance as uniform, sometimes the term *homogeneous* is used (Greek: of the same kind, uniform throughout). Most materials which you meet in your environment are mixtures. It is possible to separate a mixture into the homogeneous components or individual substances which were thrown together to make it up. Such a separation depends on the fact that different substances can be separated from each other by using their differences in properties, such as color or crystal shape.

You could mix sand and salt so that the overall appearance is somewhat uniform. But under a magnifying glass you could pick out particles of sand and particles of salt. Or you could put the sand-salt mixture into water, shake it to dissolve the salt, and then pour off the water-salt mixture from the sand.

You can see that it is not always easy to separate a mixture. More importantly, it is often difficult to *recognize* that you are dealing with a mix-

A variety of rocks showing various degrees of nonuniformity. (Photo by Gary R. Smoot.)

ture. For example, if someone mixes table salt and sugar, it appears to be a uniform material. But if you used the mixture, without knowing, for either sugar or for salt, you would recognize that it is not what you thought it was.

A few years ago a failure to recognize a salt-sugar mixture resulted in the deaths of many newborn infants in a hospital maternity ward. In that case someone inadvertently dumped a bag of salt into the sugar bin used in preparing infant formulas. The fact that the material—a mixture mostly of salt with a little sugar—*looked* like sugar led to its use in the formulas, and newborn infants were fed excess salt, a poison in this case. Many died. What is worse, no one was able to recognize the nature of the problem immediately. Guesses or hypotheses were formed by doctors, nurses, and health-investigating teams. The obvious things were guessed at first, like "staph" infection. Investigation showed no support for the hypothesis based on known types of germ infection. A strange new disease was contemplated. The deaths continued and the maternity section was closed down for further investigation. Then one of the nurses took a coffee break and she fortunately did not drink her coffee black. She put a teaspoon of "sugar" from the bin into her cup. In one gulp, a new guess or hypothesis was formed and, of course, the case was solved.

Other examples of mixtures which have less tragic potential are available in your everyday experience. Cast iron appears to be a uniform sub-

Even without a magnifying glass, you can see that granite is a mixture. (Photo by Gary R. Smoot.)

stance but again a magnifying glass shows the presence of two different kinds of particles. Upon careful observation of rocks like granite, you will see the presence of various particles.

The decision on uniformity or nonuniformity of a material depends on the ability to see different parts. Success in differentiating the parts is closely tied in with the method used to examine the specimen of material. Thus the salt-sugar mixture in the nursery was thought to be homogeneous because the scale used to observe it was too large.

In your mind you may see that the gradual refinement of separating power that could apply to a material may not yet have reached a limit. In other words, if we have "finally separated" a material into individually uniform parts, might there be a newer instrument which still might show these parts to be nonuniform? The final answer to this baffling and interesting question is still a very vital part of the frontiers of physical science. However, in chemistry we can get a clear picture of matter without necessarily reaching the ultimate in resolution of differences. The molecular approach helps here to provide more conceptual distinctions.

A SUBSTANCE IS A MATERIAL WHICH HAS ALL THE SAME KIND OF MOLECULES.

A MIXTURE IS A MATERIAL WHICH DOES NOT HAVE ALL THE SAME KIND OF MOLECULES.

We need not say "pure" substance because the very definition of substance assumes no mixture or contamination with other substances—which is the general technical meaning of pure. In practice, however, pure is relative and we may accept a material as "practically 100 percent," and therefore one substance, even though some slight amount of a different

substance may be present. For example, if you demanded table salt to be absolutely 100 percent, you would probably have to pay $30 for a pound. If you accept a few minor impurities (different substances mixed in), then you can get it for about 30 cents a pound. The acceptable relativity on purity here is dependent on particular circumstances. In the infant formula a little salt would have been all right—even healthful.

Because a mixture is a mechanical arrangement of at least two or more different substances, the proportion of either of the substances may vary. The salt-sand mixture may contain mostly salt, mostly sand, or any composition in between. A substance by definition contains only one kind of molecule, and its composition therefore cannot be varied.

BASIC BUILDING BLOCKS: ATOMS

Let us return to the problem of finding a more simple building block for the many varieties of substances. By using certain chemical processes, scientists have been able in some cases to break a substance into two or more other substances with new and different properties. If it can be so divided, they call the original substance a *compound.* If not, it is called an *element.*

A COMPOUND IS A SUBSTANCE WHICH CAN BE BROKEN INTO TWO OR MORE OTHER SUBSTANCES BY CHEMICAL MEANS.

AN ELEMENT IS A SUBSTANCE WHICH CANNOT BE BROKEN INTO OTHER SUBSTANCES BY CHEMICAL MEANS.

$$\text{Materials} = \begin{matrix}\text{Substance} = \begin{matrix}\text{Element}\\ \text{Compound}\end{matrix}\\ \text{Mixture}\end{matrix}$$

You will recall from our discussion of molecular theory that the molecule is the smallest particle into which a substance can be broken down and still retain the original operating nature and properties of the substance. The molecule is the operating unit of the substance. Added to this is the fact that there are *degrees of simplicity* within substances. These degrees of simplicity are expressed by atoms—the units which some Greek thinkers said you could not cut or divide.

AN ATOM IS THE SMALLEST PARTICLE OF AN ELEMENT.

Atoms can enter into definite groupings or combinations with atoms of other elements to form the molecules of a compound. A molecule of a compound then must consist of two or more different kinds of atoms. A molecule of an element usually consists of one atom. In some cases, however, the operating unit of an element—the molecule—may be made up of two or more atoms of the same kind of substance. (See Figure 3.1.)

Hydrochloric acid, which you have in your stomach, is a compound

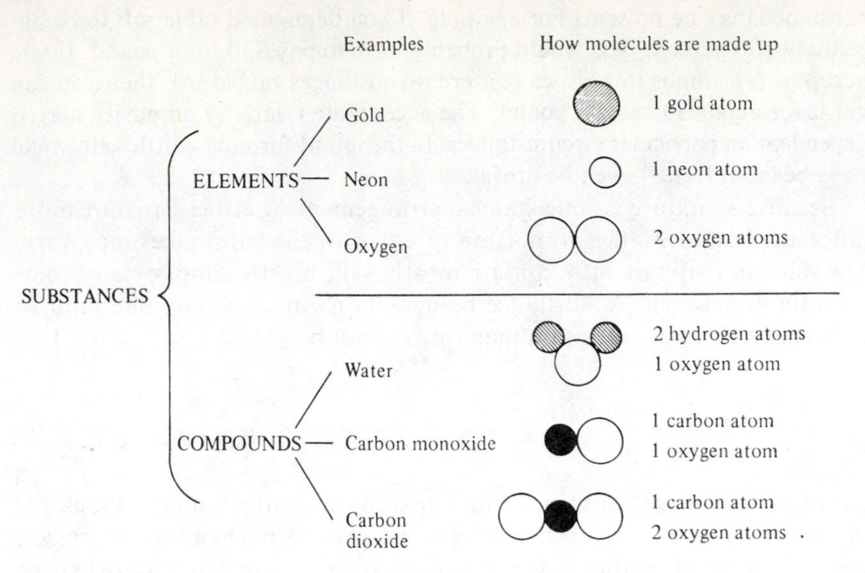

FIGURE 3.1
How the idea of atoms helps in describing different substances.

made up of molecules which contain one hydrogen atom and one chlorine atom. In other words,

Hydrogen + Chlorine ⟶ Hydrochloric acid

This is one of the ways a chemist indicates what the elements are that yield or produce (→) the compound. We will amplify this symbolism in chapter 6.

RECYCLING: THE LAW OF CONSERVATION OF MASS

In early times, when a candle burned, people wondered at the "disappearance" of its matter. Today we live in the age of recycling and everyone is aware that matter just doesn't disappear. Chemists have learned the fundamental basis for all recycling endeavors. They call it the Law of Conservation of Matter, or more generally, the Law of Conservation of Mass. This means that matter does not change in amount when it undergoes changes of various kinds. The law gets its name from the fact that mass or weight is the measure of the amount of matter present in any sample. Another way of stating the law is: Matter can neither be created nor destroyed.

This law is a statement about experience and does not attempt to explain why matter really doesn't disappear. The Law of Conservation of Mass is the result of years of observation and the work of many people.

The case of a burning candle. Matter "appears" to be lost. (Photo by Gary R. Smoot.)

Measurements were made in the early investigations which showed that the total weight of gases from a burning candle were equal to the original weight of the candle plus the weight of air consumed. (See Figure 3.2.)

Some of you may wonder whether the Law of Conservation of Mass is still valid because of the atomic bomb, where matter was said to be "destroyed." But what happened with the advent of atomic energy was an expansion of the law to include a more generalized grouping of "mass-

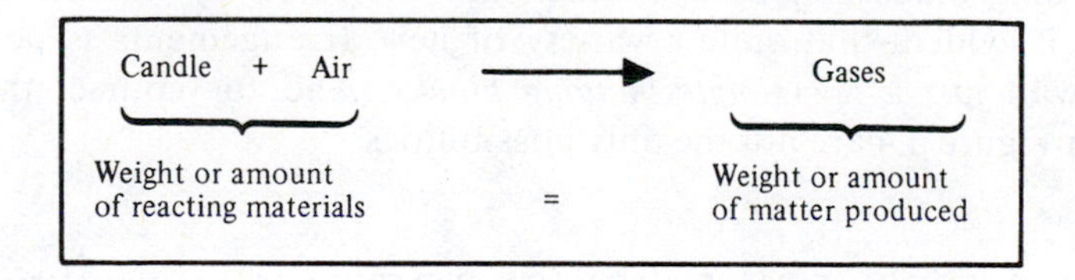

FIGURE 3.2
Matter is not lost when a candle burns.

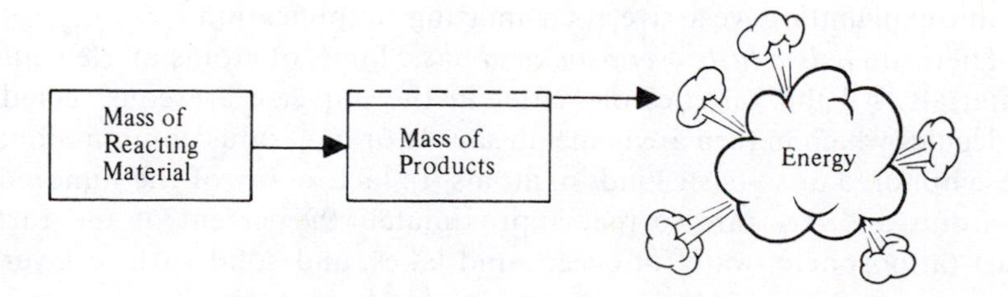

FIGURE 3.3
The modern version of the conservation law includes both mass and energy.

energy." Now the law we formulate may be called the Law of Conservation of Mass-Energy which means that the (mass + energy) of reactants equals the (mass + energy) of the products. (See Figure 3.3.)

In other words, mass and energy are interconvertible (under very special conditions), as pointed out by Einstein. Or, mass and energy are merely different aspects of the same entity—energy being a more refined form of matter. In any case, nothing "disappears" into nothing.

THE BASICS OF CHEMICAL CHANGE

The Kinetic-Molecular Theory cannot explain what happens with a burning candle. KMT explains molecular properties, like perfume diffusing in a room, or pressure in a balloon, but it is not basic enough to explain what happens when molecules get together and "disappear" to form molecules of different substances. We need to explain how the wax molecules in a candle provide us with molecules of other substances—carbon dioxide and water. This is referred to as *chemical change,* where different chemicals are formed from starting chemicals (candle + air). The fact that a solid material like wax gives loose, "light," gaseous products indicates that molecules of the wax must have some way of changing into those of the gas products and therefore must have internal structure.

We are inevitably led to the atom—those particles which we now could guess are used to make up molecules. The Law of Conservation of Mass, confirmed over and over in measurements and tests, does not allow us to assume that matter of molecules is destroyed in burning. The only possibility is that the properties of the molecule are altered. Figure 3.4 gives examples of how this could happen, assuming now that molecules must have building blocks, or atoms, which are indicated by different sized circles. It is evident that quite a variety of new arrangements is possible starting with just a *few simple building blocks.* And, or course, the examples in Figure 3.4 are not the only possibilities.

MANY MATERIALS FROM FEW BUILDING BLOCKS

The known kinds of chemical compounds now run into several millions. Many thousands of new compounds are being made each year. Yet in our atomic explanation we arrive at an amazing simplification.

There are *only a little over a hundred* basic kinds of atoms or elementary materials. All the varieties of matter in the universe are constructed of molecules which in turn are made up simply or in various combinations of these hundred or so basic kinds of atoms. In fact, many of the hundred or so atoms are very rare so that approximately 99 percent of the earth's crust (atmosphere, water of oceans and lakes, and solid surface layer) is made up of only ten of the most common kinds of atoms. (See Figure 3.5.)

Only a limited number of elements are found on the earth in the free or elemental form. Among these are the gases oxygen, nitrogen, helium

(1) The original molecules break down into smaller molecules.

(2) The molecules build up into larger, more complex molecules.

(3) The molecules rearrange their parts into new types of combinations.

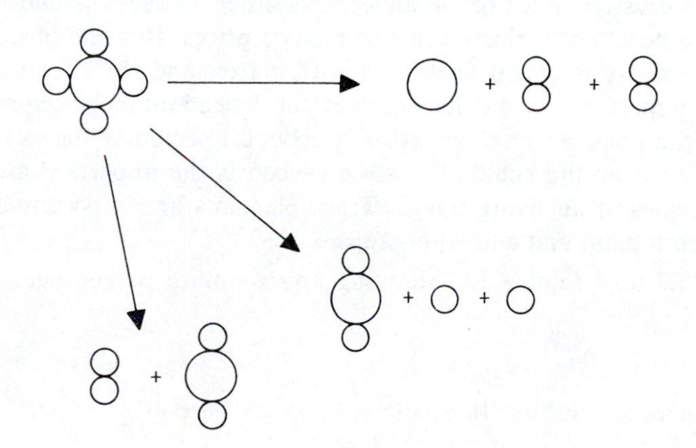

FIGURE 3.4
Examples of the way molecules can undergo changes. The changes observed with materials strongly suggest that molecules are made up of parts—the atoms.

and argon, the metals gold, silver, copper and platinum, and the nonmetal elements sulfur and carbon. Most of the elements are found combined with other elements in compounds. The more important metals are often found combined with oxygen or sulfur. For example, iron in the ore *hematite* is present as a compound with oxygen; and in the material called *iron pyrites,* or *fool's gold,* it is present in combination with sulfur. Alu-

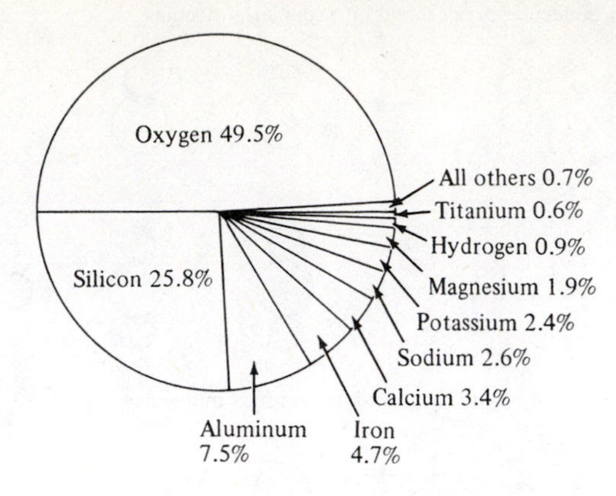

FIGURE 3.5
The relative abundance of elements in the earth's crust, including atmosphere, waters and surface layer. The percentages are calculated by weight and are, of course, approximate figures.

minum, a very common metal, is found combined with oxygen and silicon chiefly, along with other elements in clay deposits.

It is interesting that the approximate value of all the chemicals in the average adult human can be calculated, based on current prices, at about $6.75. (See Table 3.1.) Einstein and Lincoln and your mother, as seven-pound babies, would have been worth thirty cents at today's prices. But one obvious limitation of chemistry is that it deals only with matter and the changes it undergoes. The importance of an element does not depend on its price or its abundance. For example, carbon comprises only about 0.1 percent of the earth's crust, but without it no life could exist since carbon is the important atom present in molecules of all living things. Trace elements are important for functioning of both plant and animal organisms.

Both Table 3.2 and Figure 3.6, showing approximate percentages of

TABLE 3.1
Average Elementary Composition of the Human Body in Weight Percent

Element	*Percentage*
Oxygen	65
Carbon	18
Hydrogen	10
Nitrogen	3
Calcium	1.5
Phosphorus	1
Other elements including traces of gold and silver	1.5

TABLE 3.2

Some of the Many Elements in Ocean Water

Element	*Percentage*
Oxygen	85.7
Hydrogen	10.8
Chlorine	1.9
Sodium	1.0
Magnesium	0.1
Sulfur	0.09
Calcium	0.04
Bromine	0.007
Iodine	0.000006
Iron	0.000001
Silver	0.000000004
Gold	0.0000000004

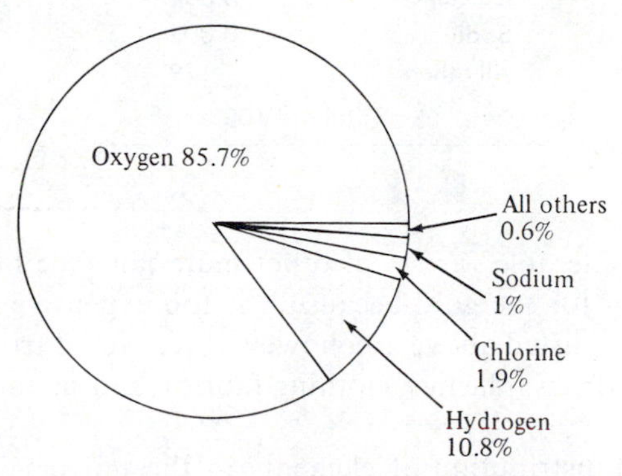

FIGURE 3.6

The relative abundance of elements in sea water.

some elements in the ocean, necessarily emphasize the presence of the compound of hydrogen and oxygen, which is water. But there are vast and valuable resources mixed with water in the ocean. If a cubic mile of ocean water were evaporated we would get about 140 million tons of solid residue. This would consist mostly of table salt, which is a compound made from sodium and chlorine. However, the small percentages of even trace materials would provide amounts not easily fathomed by the data given in the table above.

In that cubic mile of sea water we would have

Chlorine	81 million tons
Sodium	43 million tons
Magnesium	5.5 million tons
Bromine	280,000 tons
Iodine	200 tons
Gold	Over 50 pounds

TABLE 3.3
Approximate Cosmic Abundance of Elements

Element	Percentage
Hydrogen	75.2
Helium	23.3
Oxygen	0.6
Neon	0.3
Nitrogen	0.2
Carbon	0.08
Iron	0.07
Silicon	0.04
Magnesium	0.04
Sulfur	0.02
Argon	0.01
Aluminum	0.005
Calcium	0.004
Sodium	0.002
All others	.129
Total	100%

along with a considerable variety of other materials. We literally do mine the ocean but not for the gold because it is too expensive to recover. All the other materials listed above are, however, recovered from the ocean to make medicines, drugs, plastics, clothing fabrics, and metal alloys, among other things.

The estimated distribution of elements in the universe is indicated in Table 3.3. You will note the emphasis on hydrogen here since this is a primary constitutent of the stars. Also you can see that 99 percent of all the elements in the universe consist of the three elements hydrogen, helium and oxygen.

ATOMIC THEORY

The culmination of the idea of atoms, in an attempt to *explain* observations like the Law of Conservation of Mass, was reached when John Dalton (1766–1844), an English schoolteacher, spelled out in detail an atomic hypothesis—later given the name: Dalton's Atomic Theory. The atomic theory is a basic conceptual generalization—one of the guiding overviews which the human mind is always seeking as a goal to make sense out of the confusing mass of countless material things.

Dalton proposed the following essential ideas around the year 1803.

1. All elements are composed of small, indivisible particles called *atoms*.

2. The atoms of each element are *alike* but they differ from atoms of every other element.
3. When two elements *combine,* a small fixed number of atoms of one element join with a small fixed number of atoms of the other element to form every particle, or molecule, of the compounds.
4. Atoms of two elements may combine in different ratios to form more than one compound.

The fact that Dalton is given major credit for the atomic theory does not mean that it was the sudden discovery of one man. Dalton just effectively tied up many laws and quantitative findings which had been building up over many years from the time of the ancient Greeks. In fact, Isaac Newton (1642–1727), one hundred years before Dalton, had given strong support to the same idea which Dalton, a chemist, finally summarized more completely. Newton was predominantly a mathematician, astronomer, and physicist. He surely discovered enough for one man in his work on gravity, motion, and light. Yet it is interesting that his broad mind also involved itself in the "atomic" hypothesis. The basic unity of nature is no more obvious than here in the convergence of ideas from many disciplines to a single view of matter.

According to the Dalton summary, all molecules of any one substance are identical since they are each made up of the same kinds of atoms. The properties of the molecules—how they behave—then is dependent on the atomic constitution which is the same for every molecule of a particular substance. It is different from that of molecules of other substances. In other words, sugar is sugar because of its molecular architecture. Salt is different because it has a different atomic makeup.

We can now refine our definitions of elements and compounds.

ELEMENTS ARE SUBSTANCES WHOSE MOLECULES CONTAIN ONLY ONE KIND OF ATOM.

COMPOUNDS ARE SUBSTANCES WHOSE MOLECULES CONTAIN MORE THAN ONE KIND OF ATOM.

SUMMARY OF IDEAS ABOUT MATTER

Most materials in your environment are mixtures, and most of these are mixtures of compounds. Molecules of elements are usually composed of one kind of atom, but may be made up of more than one atom of the same kind. For example, oxygen gas molecules, which make up about 21 percent of air, are composed of two atoms of oxygen. Oxygen molecules are an important part of the atmosphere (chapter 16). The molecules of compounds have to be composed of at least two different kinds of atoms. A summary of the ideas about material things that we have discussed so far is given on the next page.

All Material Things May Be Broken Down Into

	Substances which are either		*Mixtures*
	Elements	*Compounds*	
Examples:	Sulfur Gold Silver Oxygen	Table salt Sugar Water Gasoline	Sand-salt Sugar-salt Air (oxygen, nitrogen and other gases)
How Made Up:	Simple primitive matter; not made up of anything but itself. Can be monatomic, diatomic, etc.	Combination or definite linkage of two or more elements in definite fixed proportion by weight.	All mixed up—can contain elements thrown together (not "combined"), or compounds thrown together, or both. In all cases, no definite proportions.
Identity of the Components:	No different components; therefore no identity problem.	Components lose identity. Properties of resultant combination do not resemble those of elements from which formed, e.g., water a compound from gases hydrogen and oxygen, table salt from metal sodium and poison gas chlorine.	Components still retain some identity and properties when they are intermingled, giving a composite with particles of each component spread around but still with properties unchanged.
Separation of the Components:	No separation problem.	Separation involves not simple physical or mechanical means but chemical changes, i.e., breaking molecules apart and rearranging.	Separation is possible by simple physical or mechanical methods, using the different properties of the different components of the mixture.

The number of names we need to know in a study of the composition of the environment is considerably reduced by using atoms as units. Consider the size of a telephone book—yet all the names of all those interesting and different people are made up of only twenty-six letters! Or consider the words and stories in countless books all composed of only the same twenty-six letters. The atomic approach simplifies understanding of the millions of chemicals by showing they are all made up from only small numbers of different basic units—the elements.

The names of the elements are now as casual as your own name even though originally they may have been a description of the particular element. For example, iodine is a name taken from a Greek word meaning *violet-like* because iodine has a violet tinge in solid form and gives a violet vapor when heated. Chlorine is a name derived from the Greek word for green since the gas is yellowish-green in color. Some elements are named after places, like polonium after Poland and californium after California. Others were given names to honor people, like einsteinium after Einstein, or curium after Madame Curie who discovered polonium and radium. She named the latter in recognition of the strong radiations emitted.

Chemists have adopted a convenient system of abbreviations to represent the names of the various elements. This symbol shorthand consists of using a single capital letter or a capital letter followed by a small letter for each element. The symbols for most elements consist of the initial capital letter of the English name of the element or the initial capital letter followed by another small letter from the name. Some common elements and their symbols are given below.

Element	*Symbol*	*Element*	*Symbol*
Oxygen	O	Nitrogen	N
Hydrogen	H	Neon	Ne
Helium	He	Carbon	C
Sulfur	S		

Since the element silicon could not have the symbol S which means sulfur, we use the symbol Si to show that we mean silicon. Similarly, since C is the symbol for carbon, the symbols for calcium, Ca, and chlorine, Cl, must have a second, small letter to distinguish them.

The symbols can represent more than one level of abstraction; they can give more information than the name of the element. For example, hydrogen is symbolized by the letter H and oxygen by the letter O. The letters thus identify or stand for these particular substances. But if a chemist wants to spell out more detail about the composition of the compound water, it is not enough to say that water is composed of hydrogen and oxygen. A more complete picture is given if we say there are 2 atoms of hydrogen and 1 atom of oxygen combined in water. Hence the chemist says the water molecule is H_2O. This combination of symbols which rep-

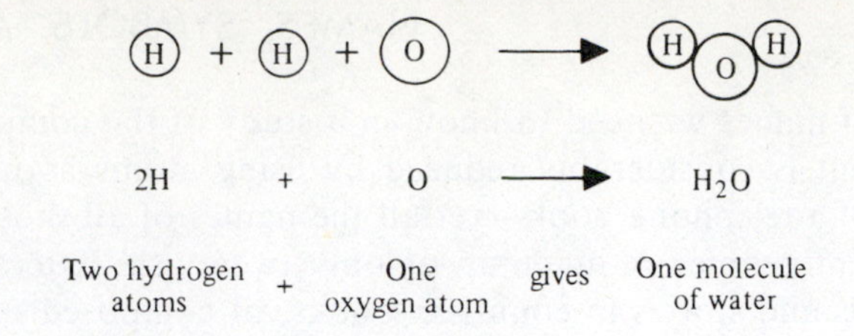

FIGURE 3.7
Use of symbols to show the nature of water.

resents a compound is called a *formula.* The formula for water is then H_2O, where the symbol has the more important additional value of standing for one atom of the particular element. If there is more than one atom in a molecule this is indicated by the subscript *following* the symbol. If only one atom of a particular element is present, like oxygen in water, the "1" is assumed by the symbol alone.

The formula for the compound water is shown in Figure 3.7. Other examples of compounds and their formulas are:

NaCl	Table salt, made from 1 atom of sodium (symbol Na) and 1 atom of chlorine (symbol Cl).
SO_3	Sulfur trioxide, made from 1 atom of sulfur and 3 atoms of oxygen.

The fact that symbols represent atoms of elements and thus express the composition of particular compounds can be expanded further. The chemist is concerned often with the amounts or quantity of matter, and since each atom is different from other atoms we might presume they are different in weight or size. Thus a symbol can also represent the amount of matter represented by the particular weight of that kind of atom.

Table 3.4 lists the most common elements and the symbols most commonly seen in designations of chemical compounds. In order to be able to interpret the meaning of the shorthand symbol combinations for various compounds, you should memorize the symbols of the most common elements. You will then be able to quickly and easily understand the composition of many more complicated molecules. Memorizing chemical symbols is a little like learning the alphabet so that you can read a letter, a book, or a poem.

The upper limits of a person's vocabulary is estimated at approximately 5000 words, although exceptional people do have usable vocabularies quite a bit higher. Shakespeare was an exception. He used 10,000 words. But we are talking about countless compounds—more being made daily—running up in the many millions. And all of these can be described in terms of the small variety of atoms which make up their molecules. You have already seen some examples, like H_2O (water) and NaCl (table salt).

Most of the symbols are related directly to the common English name

TABLE 3.4

Symbols of the Most Common Elements

Element	*Symbol*	*Element*	*Symbol*	*Element*	*Symbol*
Aluminum	Al	Fluorine	F	Phosphorus	P
Antimony	Sb	Gold	Au	Platinum	Pt
Argon	Ar	Helium	He	Potassium	K
Arsenic	As	Hydrogen	H	Radium	Ra
Barium	Ba	Iodine	I	Silicon	Si
Beryllium	Be	Iron	Fe	Silver	Ag
Bismuth	Bi	Lead	Pb	Sodium	Na
Boron	B	Lithium	Li	Strontium	Sr
Bromine	Br	Magnesium	Mg	Sulfur	S
Cadmium	Cd	Manganese	Mn	Tin	Sn
Calcium	Ca	Mercury	Hg	Titanium	Ti
Carbon	C	Neon	Ne	Tungsten	W
Chlorine	Cl	Nickel	Ni	Uranium	U
Chromium	Cr	Nitrogen	N	Xenon	Xe
Cobalt	Co	Oxygen	O	Zinc	Zn
Copper	Cu				

Chemist using a scanning electron microscope to study hair structure at Clairol's Research Laboratory. Even the high magnification of electron microscopes is not sufficient to show the tiny atoms that make up hair. A photomicrograph of hair is shown in chapter 10. (Courtesy Clairol.)

TABLE 3.5
Symbols that are Unrelated to Common English Names of Elements

Common Name	Symbol	Symbol Source*
Antimony	Sb	Stibnum
Copper	Cu	Cuprum
Gold	Au	Aurum
Iron	Fe	Ferrum
Lead	Pb	Plumbum
Mercury	Hg	Hydragyrum
Potassium	K	Kalium
Silver	Ag	Argentum
Sodium	Na	Natrium
Tin	Sn	Stannum
Tungsten	W	Wolfram

*These are old Latin names from which our English names are derived and are for reference only.

but some are derived from an early name of the element, for example, Ag for silver, from the Latin Argentum. Another example is Pb for lead, from the Latin Plumbum (related directly to the name *plumber,* who works with lead pipes). Table 3.5 lists separately the symbols derived from non-English names.

ATOMIC STRUCTURE AND CHEMICAL BEHAVIOR

There is a compound of carbon and oxygen which you know as *carbon dioxide.* You breathe it out with every breath and it is also the product from the combustion of materials containing carbon in their molecules, such as wood, cellulose, gasoline, and trash. You also read occasionally about someone, inadvertently or otherwise, dying from asphyxiation caused by *carbon monoxide* poisoning when a car is left running in a closed garage. Carbon monoxide is also a compound of carbon and oxygen. The fact that a carbon-oxygen compound is fatal in one case and a necessary part of animal respiration in another is easily explained by showing that two different substances are involved, one CO and the other CO_2. They have different properties because they are different molecules. The structure of the molecules determines their character. Structure is a major concern in our study of matter because it explains behavior and thus allows us to make choices—on many of which life itself depends.

Breathing carbon monoxide at concentrations as low as 0.05 percent produces headaches, dizziness, and unconsciousness in a few hours. Auto exhaust fumes contain about 7 percent of carbon monoxide and are fatal in a few minutes.

TABLE 3.6

Atomic Description of Some Common Chemicals

Substance	*Formula*	*Each Molecule Contains*
Carbon monoxide	CO	1 C atom; 1 O atom
Carbon dioxide	CO_2	1 C atom; 2 O atoms
Sulfur oxides		
Sulfur dioxide	SO_2	1 S atom; 2 O atoms
Sulfur trioxide	SO_3	1 S atom; 3 O atoms
Nitrogen oxides (designated NO_x in auto exhaust studies as a general term to include all nitrogen oxides)		
Nitric oxide	NO	1 N atom; 1 O atom
Nitrogen dioxide	NO_2	1 N atom; 2 O atoms
Dinitrogen trioxide	N_2O_3	2 N atoms; 3 O atoms
Dinitrogen tetroxide	N_2O_4	2 N atoms; 4 O atoms
Table salt (sodium chloride)	$NaCl$	1 Na atom; 1 Cl atom
Water	H_2O	2 H atoms; 1 O atom
Table sugar (sucrose)	$C_{12}H_{22}O_{11}$	12 C atoms; 22 H atoms; 11 O atoms
Oxygen gas	O_2	2 O atoms
Nitrogen gas	N_2	2 N atoms
Neon gas	Ne	1 Ne atom
Methane gas	CH_4	1 C atom; 4 H atoms
Propane gas	C_3H_8	3 C atoms; 8 H atoms
Octane (one compound in gasoline, a mixture)	C_8H_{18}	8 C atoms; 18 H atoms

Table 3.6 gives many other examples of how materials you may have heard about are described more specifically by their atomic makeup. The maximum exhaust emission standards now being established to control air pollution from autos involve some of these chemicals. The major standard emissions are usually listed in a sticker on new cars, with letter designations as follows:

HC for hydrocarbons, compounds of hydrogen and carbon only, such as methane, and octane.

CO for carbon monoxide.

NO_x for the general designation of all the nitrogen oxides (chapter 16).

The value of a chemical formula in showing the constitution of a substance cannot be overemphasized. It provides the beginning of an understanding of the properties of materials. Compare the revelation and precision you get from the formulas NaCl, H_2O, and $C_{12}H_{22}O_{11}$ with the unrevealing names, salt, water, and sugar. Further refinements are possible and you will see some of these later.

OPERATIONAL AND CONCEPTUAL LEVELS

This chapter shows how we deal with our environmental materials on two levels—operational and conceptual. First we see how materials behave, what their properties are—how they operate. We thus distinguish between mixtures and substances by observing the properties which they exhibit. For example, we say that salt-sand is a mixture because it can be physically separated into salty and sandy materials.

Then we are led to the concept of atoms and molecules to *explain* and differentiate the behavior of materials. For example, sand-sugar is a mixture because it contains two materials with distinctly different molecules. You can find descriptions of atoms, elements, compounds, and so on in this chapter first on an operational and then on a conceptual level. The level of concepts is where the heart of learning lies.

The emphasis in this book is on the relation between the operating world and the conceptual world. Ideas are born from the experiences you have—the heat of a fire, the smell of food or perfume, the refreshment of an ocean breeze, the suffocation of smoggy city air. And the ideas can be related back to things, or developed further on their own. Herein lies the solution to many of our physical problems. You can hardly solve a problem until you know what it is. And this means concepts.

PROJECTS AND EXERCISES

Experiments

1. Get a magnifying glass or small hand magnifier (5X up to about 20X magnification). Then go outside and examine some natural environmental materials, for example, rocks, granite, concrete, asphalt, limestone, and so forth. Determine what you can about the apparent constitution of several samples. Do you note signs of mixtures? Can you see how some of these materials might be considered substances when examined casually and without magnification? Can you therefore relate the resolving power of the instrument used in examination to the determination of the homogeneity of a sample? That is, can you see how you might call a well-mixed assemblage of two different kinds of matter a homogeneous material if the instrument you use to examine it is insufficiently refined to detect the individual particles of different kinds? If you were asked to conceptually visualize a mixture that would look homogeneous by all but the ultimate in resolving instruments, how far down in subdivision of the individual pieces of each substance would you have to go? Can you think of any examples of this kind of a mixture?
2. Your ordinary tap water looks like a substance, in that it appears uniform. However, you know now that uniformity of appearance is sometimes misleading. Did you ever see a pan from which tap water had evaporated? Was there any residue? What does this suggest? How could you devise a simple experiment to determine if your tap water is a substance or a mixture? Try out your method. What conclusions can you draw?

3. Why is the term *pure substance* redundant? You often hear it used, however. Explain what it usually means and then describe briefly how you would make sense to a nonscience student why *pure* is not necessary as a qualifying adjective to the word *substance*.
4. When a car first starts up, especially on a cool morning, you note liquid condensing from the exhaust. This is actually water. Would you say that the material which exits from the exhaust is a mixture or a substance? What other indications do you have from reading about auto exhaust that it is either a mixture or a substance? How many materials in auto exhaust can you name? Where do you think the water comes from?
5. This exercise requires you to think and then give operational and conceptual definitions of several terms. Remember that definitions are not just memorized statements like quotations of poetry. They are ways of defining or making clear in words the meaning of a thing. Your words are as good as anyone's. List three columns on your paper and in the first put down the words as shown below, followed by a column for the operational definition and then by a third column for the conceptual definition.

	Operational Definition	*Conceptual Definition*
Compound		
Element		
Substance		
Mixture		
Chemical change		

6. Go to a local new car salesroom and look at the descriptive stickers on the autos that refer to the emission standards. If there are none, ask the salesperson for information on the performance of the car relating to emission standards. Describe what these tell you about the nature of the chemicals in auto exhaust.

SUGGESTED READING

1. Young, Louise B., ed., *The Mystery of Matter,* Oxford University Press, New York, 1965, Part 2, "Is Matter Infinitely Divisible?," "The Atomic Idea," "Early Atomic Theory."

chapter 4

The Atom: Its Properties and Structure

The beginnings of all things are small.

—CICERO

Father, Mother and Me,
Sister and Auntie say
All the people like us are We,
And every one else is They.

—RUDYARD KIPLING

If I have seen farther than other men, it is because I stand on the shoulders of giants.

—NEWTON

You may already have questions on the adequacy of the atomic theory as we have described it so far to explain the behavior of materials. For example, why does water have the composition H_2O, with two hydrogen atoms to one atom of oxygen, when table salt, NaCl, has a one-to-one ratio? Or, what makes the atoms of each element different from those of other elements? Historically, many people were involved in the more refined guesses that led to our present model of the internal structure of atoms. This chapter will be limited to a brief summary of our present views with emphasis on a few examples of how scientists have developed the ideas from experience with materials.

RELATIVE WEIGHTS OF ATOMS

If the atoms of each element are identical to each other and different from those of other atoms, then it might be possible to prepare a

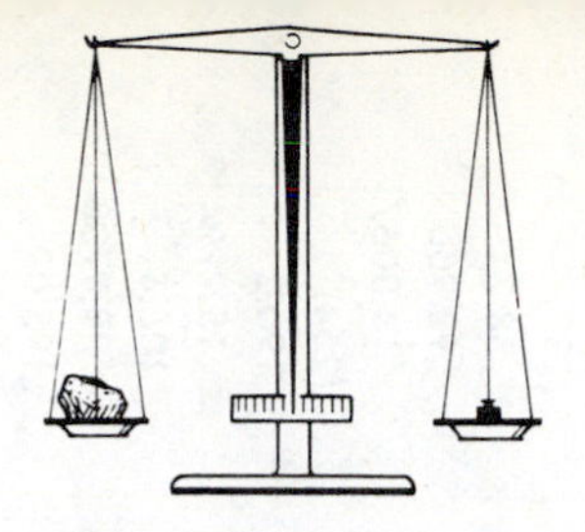

FIGURE 4.1
A simple beam-type of balance of the kind originally used to determine weight or mass of chemicals. The chemical is placed in the left pan and standard weights are added to the right pan until a balance is obtained.

list of atoms and their properties. If you think about this, it appears that there are very few properties which we can list, and these are basic, simple properties.

One important property is weight or mass—meaning the quantity of matter in the atom. An atom, being a definite quantity of a particular element, must have a definite weight. This is the one basic physical property of an atom which chemists have emphasized. Size in terms of volume, or space occupied, can also be estimated, but is not as precise a measurement as mass or quantity usually determined by weight. (See Figures 4.1 and 4.2.)

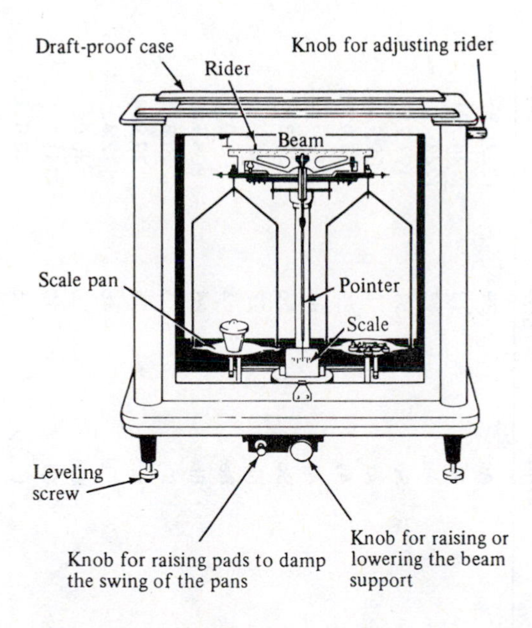

FIGURE 4.2
A more advanced analytical balance of the type used during the nineteenth century when people made much progress in chemical theory by precise weighing of chemicals.

TABLE 4.1
Table of Atomic Weights (Based on Carbon-12)

Name	Symbol	Atomic No.	Atomic Weight	Name	Symbol	Atomic No.	Atomic Weight
Actinium	Ac	89	(227)	Mercury	Hg	80	200.59
Aluminum	Al	13	26.9815	Molybdenum	Mo	42	95.94
Americium	Am	95	(243)	Neodymium	Nd	60	144.24
Antimony	Sb	51	121.75	Neon	Ne	10	20.183
Argon	Ar	18	39.948	Neptunium	Np	93	(237)
Arsenic	As	33	74.9216	Nickel	Ni	28	58.71
Astatine	At	85	(210)	Niobium	Nb	41	92.906
Barium	Ba	56	137.34	Nitrogen	N	7	14.0067
Berkelium	Bk	97	(247)	Nobelium	No	102	(254)
Beryllium	Be	4	9.0122	Osmium	Os	76	190.2
Bismuth	Bi	83	208.980	Oxygen	O	8	15.9994
Boron	B	5	10.811	Palladium	Pd	46	106.4
Bromine	Br	35	79.904	Phosphorus	P	15	30.9738
Cadmium	Cd	48	112.40	Platinum	Pt	78	195.09
Calcium	Ca	20	40.08	Plutonium	Pu	94	(242)
Californium	Cf	98	(251)	Polonium	Po	84	(210)
Carbon	C	6	12.01115	Potassium	K	19	39.102
Cerium	Ce	58	140.12	Praseodymium	Pr	59	140.907
Cesium	Cs	55	132.905	Promethium	Pm	61	(147)
Chlorine	Cl	17	35.453	Protactinium	Pa	91	(231)
Chromium	Cr	24	51.996	Radium	Ra	88	(226)
Cobalt	Co	27	58.933	Radon	Rn	86	(222)
Copper	Cu	29	63.546	Rhenium	Re	75	186.2
Curium	Cm	96	(247)	Rhodium	Rh	45	102.905

Dysprosium	Dy	66	162.50
Einsteinium	Es	99	(254)
Erbium	Er	68	167.26
Europium	Eu	63	151.96
Fermium	Fm	100	(253)
Fluorine	F	9	18.9984
Francium	Fr	87	(223)
Gadolinium	Gd	64	157.25
Gallium	Ga	31	69.72
Germanium	Ge	32	72.59
Gold	Au	79	196.967
Hafnium	Hf	72	178.49
Helium	He	2	4.0026
Holmium	Ho	67	164.930
Hydrogen	H	1	1.00797
Indium	In	49	114.82
Iodine	I	53	126.9044
Iridium	Ir	77	192.2
Iron	Fe	26	55.847
Krypton	Kr	36	83.80
Lanthanum	La	57	138.91
Lawrencium	Lr	103	(257)
Lead	Pb	82	207.19
Lithium	Li	3	6.939
Lutetium	Lu	71	174.97
Magnesium	Mg	12	24.312
Manganese	Mn	25	54.938
Mendelevium	Md	101	(256)
Rubidium	Rb	37	85.47
Ruthenium	Ru	44	101.07
Samarium	Sm	62	150.35
Scandium	Sc	21	44.956
Selenium	Se	34	78.96
Silicon	Si	14	28.086
Silver	Ag	47	107.868
Sodium	Na	11	22.9898
Strontium	Sr	38	87.62
Sulfur	S	16	32.064
Tantalum	Ta	73	180.948
Technetium	Tc	43	(99)
Tellurium	Te	52	127.60
Terbium	Tb	65	158.924
Thallium	Tl	81	204.37
Thorium	Th	90	232.038
Thulium	Tm	69	168.934
Tin	Sn	50	118.69
Titanium	Ti	22	47.90
Tungsten	W	74	183.85
Uranium	U	92	238.03
Vanadium	V	23	50.942
Xenon	Xe	54	131.30
Ytterbium	Yb	70	173.04
Yttrium	Y	39	88.905
Zinc	Zn	30	65.37
Zirconium	Zr	40	91.22

Since the chemists could not measure the weight of a tiny atom directly, they determined the *relative* weight compared to some other atom. They did this by comparing how much of an element was needed to combine with a definite weight of another element in forming a compound. There were complications at first, but by dogged persistence and the cooperation of many people in many countries, the early chemists laid a fairly good foundation. Over a period of several generations, the relative weights were refined and values were fairly well accepted.

Today we find the relative weights as the fundamental property in every listing of the elements. Table 4.1 shows the values accepted today, based on the relative weight of the particular atoms compared to carbon as a standard at 12. You can see that most are known to a high degree of accuracy. The table lists all the elements including some of the rather rare and unstable ones for which an exact and meaningful relative weight is not obtainable. This is the reason for the parentheses around the approximate values given in some cases.

Using this system, no element has an atomic weight (or atomic mass) of less than 1. Hydrogen is the lightest element with a relative weight of approximately 1 compared to carbon at 12. Oxygen has a relative weight of 16 compared to carbon at 12.

GROUPING THE ELEMENTS

When chemists had assembled considerable information on the relative weights of three or four dozen elements, by the middle of the nineteenth

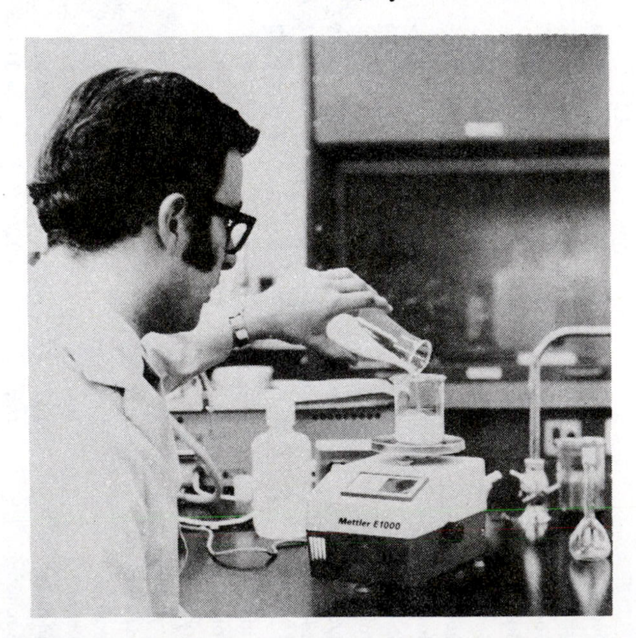

A compact type of modern balance designed to give fast weight readings. (Courtesy of Mettler Instrument Corporation.)

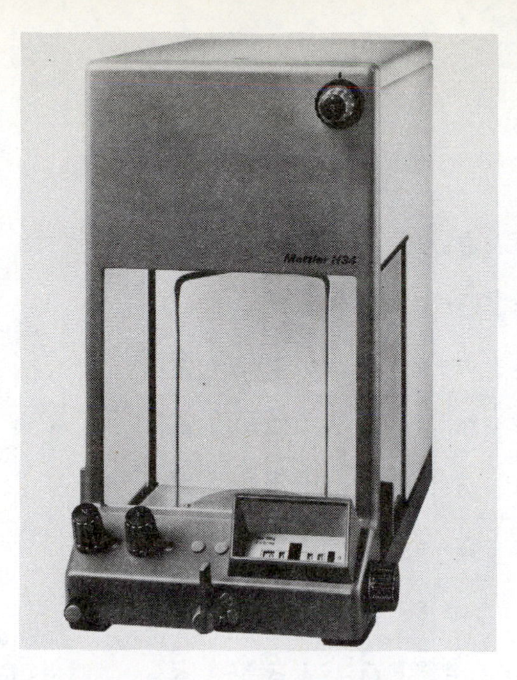

A modern analytical balance capable of high precision weighing. (Courtesy of Mettler Instrument Corporation.)

century, various schemes were tried to arrange the elements into families or natural groupings of similar elements. The fundamental drive again was the human one of looking for more simplicity, or more order, in the growing complexity of the data on the various elements.

The Russian chemist Mendeleyev (1834–1907) is recognized as having established the modern idea of a special arrangement of the elements called the Periodic Table. Mendeleyev ingeniously allowed spaces in his table for elements not yet discovered. In addition, because of similar properties in various family groupings, he was able successfully to predict approximate properties like atomic weights for some unknown elements. The table has now been completed as far as all the blank spaces are concerned and it has been extended to include over one hundred elements.

A modern form of the Periodic Table of the Elements is shown in Table 4.2. In the beginning you will naturally be somewhat confused by this listing. In fact, no chemist knows everything the table contains. It is strictly a reference tool, like a telephone book but much less complex. The Periodic Table lists not just all the elements that make up all the people that are listed in all the telephone books in the world, but lists all the kinds of atoms which go to make up the whole material universe! There are no others, on Mars or on Pluto.

The reason for the name *Periodic* needs further clarification. You will note that the elements are numbered in consecutive order in the table. The consecutive number of the element is also used as the identifying number of the element.

TABLE 4.2

Periodic Table of the Elements

Key: Atomic Number — Name — Symbol — (APPROXIMATE) ATOMIC WEIGHT (example: 11 / Sodium / Na / 23)

Period	Group IA	IIA	IIIB	IVB	VB	VIB	VIIB	VIII	VIII	VIII	IB	IIB	IIIA	IVA	VA	VIA	VIIA	Noble Gases
1	1 Hydrogen H 1																	2 Helium He 4
2	3 Lithium Li 7	4 Beryllium Be 9											5 Boron B 11	6 Carbon C 12	7 Nitrogen N 14	8 Oxygen O 16	9 Fluorine F 19	10 Neon Ne 20
3	11 Sodium Na 23	12 Magnesium Mg 24											13 Aluminum Al 27	14 Silicon Si 28	15 Phosphorus P 31	16 Sulfur S 32	17 Chlorine Cl 35.5	18 Argon Ar 39.9
4	19 Potassium K 39.1	20 Calcium Ca 40	21 Scandium Sc 45	22 Titanium Ti 48	23 Vanadium V 51	24 Chromium Cr 52	25 Manganese Mn 55	26 Iron Fe 56	27 Cobalt Co 59	28 Nickel Ni 58.7	29 Copper Cu 63.5	30 Zinc Zn 65	31 Gallium Ga 70	32 Germanium Ge 73	33 Arsenic As 75	34 Selenium Se 79	35 Bromine Br 80	36 Krypton Kr 84
5	37 Rubidium Rb 85	38 Strontium Sr 88	39 Yttrium Y 89	40 Zirconium Zr 91	41 Niobium Nb 93	42 Molybdenum Mo 96	43 Technetium Tc 99	44 Ruthenium Ru 102	45 Rhodium Rh 103	46 Palladium Pd 106	47 Silver Ag 108	48 Cadmium Cd 112	49 Indium In 115	50 Tin Sn 119	51 Antimony Sb 122	52 Tellurium Te 128	53 Iodine I 127	54 Xenon Xe 131
6	55 Cesium Cs 133	56 Barium Ba 137	* 57 Lanthanum La 139	72 Hafnium Hf 178	73 Tantalum Ta 181	74 Wolfram W [Tungsten] 184	75 Rhenium Re 186	76 Osmium Os 190	77 Iridium Ir 192	78 Platinum Pt 195	79 Gold Au 197	80 Mercury Hg 201	81 Thallium Tl 204	82 Lead Pb 207	83 Bismuth Bi 209	84 Polonium Po 210	85 Astatine At 210	86 Radon Rn 222
7	87 Francium Fr 223	88 Radium Ra 226	** 89 Actinium Ac 227															

Series	Period														
* Lanthanide Series	6	58 Cerium Ce 140	59 Prase-odymium Pr 141	60 Neodymium Nd 144	61 Promethium Pm 145	62 Samarium Sm 150	63 Europium Eu 152	64 Gadolinium Gd 157	65 Terbium Tb 159	66 Dysprosium Dy 162.5	67 Holmium Ho 165	68 Erbium Er 167	69 Thulium Tm 169	70 Ytterbium Yb 173	71 Lutetium Lu 175
** Actinide Series	7	90 Thorium Th 232	91 Protactinium Pa 231	92 Uranium U 238	93 Neptunium Np 237	94 Plutonium Pu 242	95 Americium Am 243	96 Curium Cm 247	97 Berkelium Bk 247	98 Californium Cf 249	99 Einsteinium Es 254	100 Fermium Fm 253	101 Mendelevium Md 256	102 Nobelium No 256	103 Lawrencium Lr 257

THE ATOMIC NUMBER IS THE CONSECUTIVE ORDER NUMBER OF THE ELEMENT IN THE PERIODIC TABLE STARTING WITH THE LIGHTEST ELEMENT, HYDROGEN, WHICH IS ELEMENT NUMBER 1.

This is an operational description of the atomic number. When the elements were thus arranged in consecutive order, generally in the order of increasing atomic weight, it was noticed that periodically, or every so often at some interval, similar properties recurred. This is like the periodic change in the tides due to the moon. In other words, elements with similar properties appear at fairly regular intervals. Consequently, instead of stringing the list of elements out into a long line or a long column, we can group them in columns and let elements with similar properties fall under each other. These groupings in vertical columns are the natural "families" of elements and this is the basis of the Periodic Table.

FAMILIES OF ELEMENTS

You will note that on the far right of Table 4.2 there is a column containing elements called *noble gases.* Starting from the top, this column contains elements designated by symbol and number (using the number as a subscript before the symbol) as follows: $_{2}He$, $_{10}Ne$, $_{18}Ar$, $_{36}Kr$, $_{54}Xe$, and $_{86}Rn$. These are, respectively, the elements helium, neon, argon, krypton, xenon, and radon. The main point about all these gases is that they are very unreactive, that is, they do not get together with other elements to form compounds readily. In fact, they form a "family" of unreactive atoms compared to any other atoms, and they are therefore called the "noble" gases—presumably because they are like the old-time nobility which did not form links with the common people.

Another example of a family is the elements in the first column beginning with lithium: $_{3}Li$, $_{11}Na$, $_{19}K$, $_{37}Rb$, $_{55}Cs$, and $_{87}Fr$. These are lithium, sodium, potassium, rubidium, cesium, and francium, respectively. You probably now know the symbols and names for Li, Na, and K, since these are the common ones in this family which is called the alkali metal family because the main source originally was plant ashes (Arabic: alquali = ashes). They also form alkalis with water which is a special form of chemical compound that neutralizes acids (chapter 7). Like the other families, the alkali metal family consists of similar elements: they are all soft and malleable (easily flattened with a hammer); also they are the most active of all the metals chemically, that is, they tend to form compounds very readily.

Another of the families which is often singled out is the halogen family, consisting of those elements in the column next to the noble gases or second from the right side of the table. These are $_{9}F$, $_{17}Cl$, $_{35}Br$, $_{53}I$, and $_{85}At$; fluorine, chlorine, bromine, iodine, and astatine, respectively. Again, all these are not common, the main ones being chlorine, bromine, and iodine. They are called *halogens* which is a name derived from the Greek word meaning *salt-former*. The name derives from the fact that they all form compounds called salts, when reacted with metals. You are very familiar with at least one example of a salt, NaCl, or table salt. Other similar salts are: KCl, potassium chloride; NaBr, sodium bromide; and KI, potassium iodide.

The latter salt, KI, is added in small amounts by the manufacturer to your iodized table salt. This addition of the different salt, KI, to table salt, NaCl, in very small percentage of only 0.01 percent (about 1 pound of KI to 10,000 pounds of salt) has essentially eliminated in the United States a disease of the thyroid gland called goiter (chapter 10).

A further convention of the Periodic Table is to refer to the horizontal rows or elements as *periods* while the vertical columns are called *families.* The families are sometimes given Roman numeral designations as shown in the Periodic Table. For example, the first period contains only 2 elements, H and He. The second period contains 8 elements, Li, Be, B, C, N, O, F, and Ne. The third period again contains 8 elements running from number 11, Na, through number 18, Ar. After this period we get into more complexities in arrangement, for example, the fourth period contains 18 elements from number 19, K, to number 36, Kr, krypton.

THE STRUCTURE OF THE ATOM: THE ELECTRON

Dalton's original formulation of the atomic theory does not explain why some elements form groups with similar properties. The Dalton atomic model also does not explain why water has the formula H_2O while table salt is NaCl. We need some kind of structure for the atom or some kind of "parts list" for the atom's makeup. A parts list might help clarify the problems by postulating that there is something about the arrangement of the parts in the atom which could account for the special behavior of atoms.

Much of our understanding of the structure of atoms came from ex-

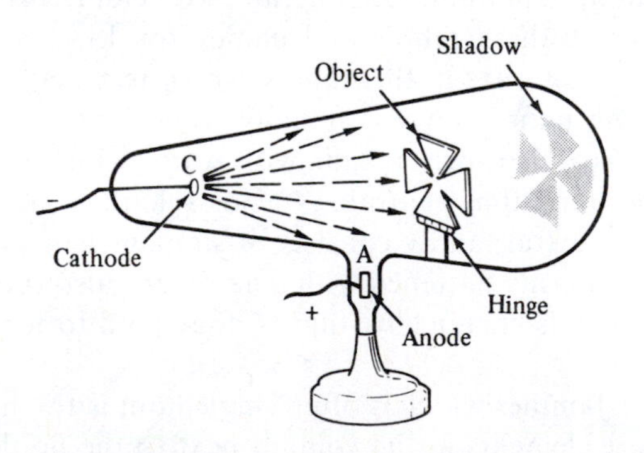

FIGURE 4.3
A standard Crookes tube. The streams of particles coming from the cathode, called cathode rays, hit a luminiscent coating on the inside of the glass and cause it to light up. Shadows are cast by anything in between which stops the particles.

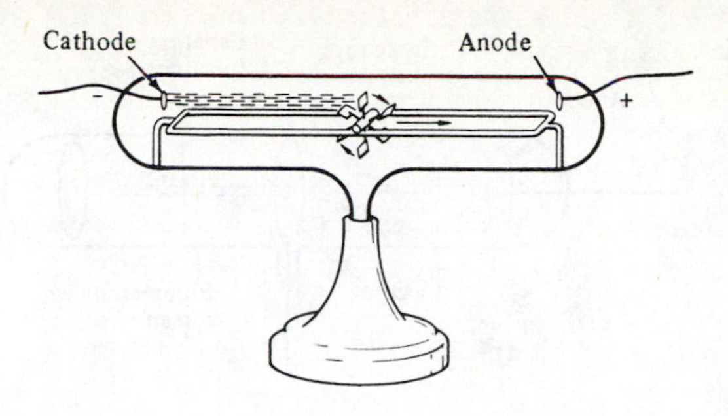

FIGURE 4.4

A special tube made by Crookes to show that the particles coming from the cathode can make pinwheels whirl as if blown by a strong wind. Reversal of the current causes the pinwheel to whirl in the opposite direction.

periments on conduction of electricity through tubes containing only small numbers of gas molecules. Around 1878, the English scientist William Crookes, using special tubes which were highly evacuated, showed that there was a stream of some kind of matter coming from one of the electrical contacts, or electrodes, in the tube. This was the negative electrode, called the cathode. (See Figures 4.3 and 4.4.)

Then in 1897 the English scientist J. J. Thomson showed that the cathode ray beams were made of particles of negative charge called *electrons*. (See Figure 4.5.) By using some equations of classical physics, Thomson determined the ratio of charge to mass for the electron. The charge was later measured by Millikan at the University of Chicago. The mass was then calculated. It turns out that the electron is about 1/2000

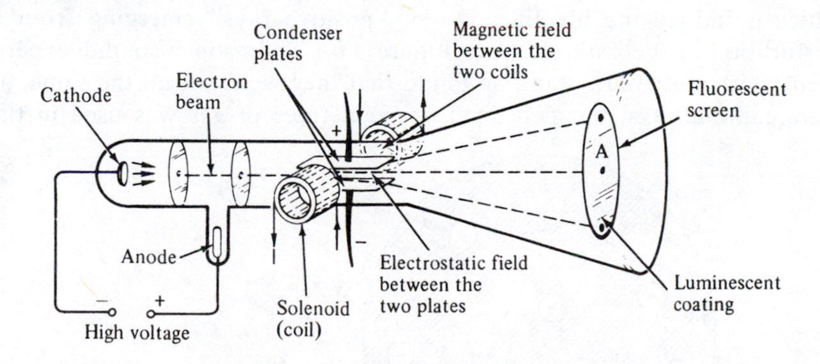

FIGURE 4.5

Diagram of the cathode ray tube used by Thomson. This is essentially a TV tube. The screen on the right side is coated with a material which emits visible light when an electron beam strikes it. This beam of electrons normally strikes the screen at the point labeled A, giving a bright dot (like the spot on your TV screen when you turn it off and before the electron gun cools off). Thomson applied electrostatic and magnetic fields around the cathode ray (as your TV tube does to move the beam). One of his findings was that the beam was attracted toward the positively charged plate.

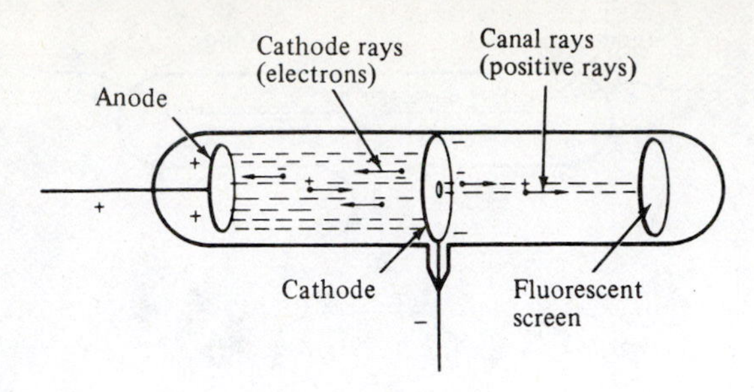

FIGURE 4.6
A diagram of the type of tube developed by Goldstein.

of the mass of one hydrogen atom. In other words, it would take about 2000 electrons to total the weight of one tiny hydrogen atom, the tiniest atom that exists!

The nature of the electrons were found to be the same irrespective of (1) the material of which the cathode is made; (2) the type of trace gas present in the evacuated tube; (3) the kind of metal wires used in bringing the current to the cathode; and (4) the materials used to produce the current. These findings certainly pointed to the fact that the *electron was a fundamental particle* of all material things, and not just like another unknown element.

No smaller charge than that on the electron has ever been found. Consequently, for convenience this charge is assigned a value of -1, and is the basic unit of electrical charge.

In 1886 the German physicist Eugen Goldstein developed a different type of discharge tube containing a cathode perforated with a tiny canal which permitted the identification of "positive rays" emerging from a perforation in the cathode. (See Figure 4.6.) Thomson also did experiments with positive rays and he found that they were not all the same, as were cathode rays, but depended on what trace of gas was used in the

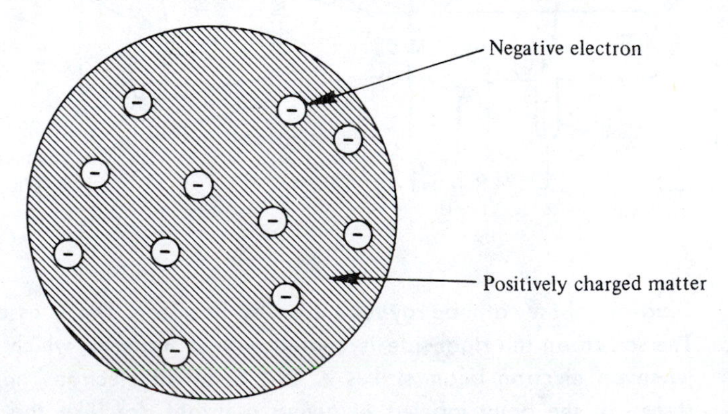

FIGURE 4.7
Thomson's model of the atom. The positive mass is spread throughout the atom's volume, with negative electrons stuck in it at various spots. This model was proved to be incorrect by Rutherford's experiments.

evacuated tube. Thomson identified them as made up of positive particles, varying in weight according to the gas used in the tube. In no case has it been possible to isolate a positively charged particle of smaller weight than that of the hydrogen atom. The positively charged hydrogen atom was later given the name *proton* (Greek: first) since it is the first and simplest of the atoms.

Thomson proposed a model for the atom's structure. It is sometimes called the "plum pudding model" after a special English dessert with bits of raisins and fruit spread throughout the pudding. It could just as readily be called a "watermelon model." (See Figure 4.7.)

THE STRUCTURE OF THE ATOM: THE NUCLEUS

Further evidence against the "indivisible atom" of Dalton was the discovery of radioactivity by the French scientist Becquerel in 1896. *Radioactivity* is the spontaneous emission of particles and radiation by atoms. Later Marie Curie discovered radium, a strongly radioactive element.

Radium samples were made available to Ernest Rutherford (1871–1937), a New Zealander working in England. He and many other scientists were interested in determining the nature of the radiations. Becquerel

Ernest Rutherford, a physicist, who did important experimental and theoretical work on the nature of the atom. Work in Rutherford's laboratory laid the foundation for the nuclear atom. (Courtesy of the Rutherford Museum, McGill University.)

had identified some of the radiation as electrons. Rutherford found a less penetrating beam which he called *alpha rays,* and he gave the name *beta rays* to the electron stream. Meanwhile a third component of the "Becquerel rays" from radioactive decay was identified as electromagnetic radiation, similar to light but of much higher energy (described below). This was given the name *gamma rays.* (The rays were named after the *a, b,* and *c* letters of the Greek alphabet, alpha, beta, and gamma.) The nature of the rays initially was quite mysterious. By a procedure typical in science (give the mysterious thing a name or a number, experiment and play with it for a while), Rutherford laid important foundations for our present understanding of the atom. He found that the alpha rays were attracted toward the negative plate in an electrostatic field and hence were positively charged. (See Figure 4.8.) In addition, in 1909 he isolated the component from the alpha rays and identified the units as atoms of the number 2 element, helium, with a double positive charge. This was an amazing confirmation of the idea that each atom must be built up from the same parts list since the alpha particles were born in the radioactive breakdown of the larger atoms of radium!

Rutherford pushed on in trying to probe the interior structure of heavy atoms by shooting alpha particles at thin foils of metal such as gold. He had measured the speeds of the alpha particles coming off the radioactive sample and found they were very high. The gold foil he had chosen for the experiment was very thin. Calculations based on Thomson's model showed that the alpha particles should go right through the foil. Most of them did. However, he and his coworkers found that some alphas were bounced back toward the direction from which they came.

Here is how Rutherford described the experiment:

> Then I remember Geiger [the inventor of the Geiger counter who was working with Rutherford] coming to me in great excitement and saying, "We have been able to get some of the alpha particles coming backwards!" It was quite the

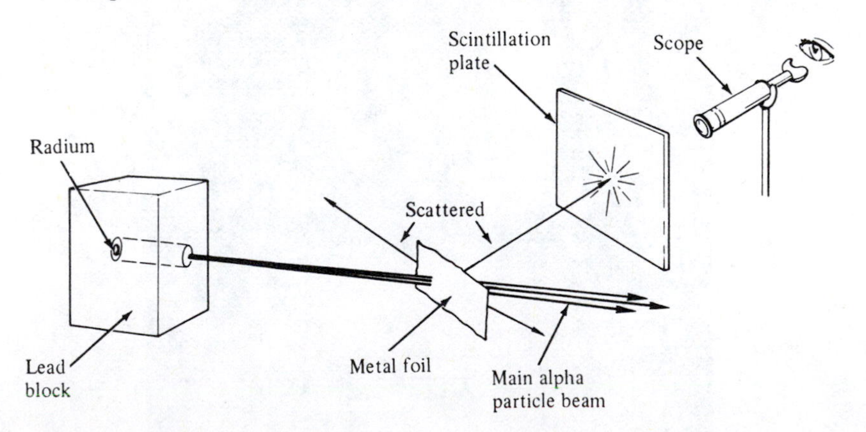

FIGURE 4.8
A simplified diagram showing the experiment which Rutherford's group performed to probe the structure of atoms. The lead block permitted a beam of alpha particles to be shot at the foil target.

most incredible event that has ever happened to me in my life. It was almost as incredible as if you fired a 15-inch shell at a piece of tissue paper and it came back and hit you. On consideration I realized that this scattering backwards must be the result of a single collision and when I made calculations it was impossible to get anything of that order of magnitude unless you took a system in which *the greater part of the mass of the atom was concentrated in a minute nucleus.* [Italics added.]

Here was born the theory of the nuclear atom. It is not easy for us to fully appreciate the impact of the rebounding alpha particles. Rutherford had, of course, detailed calculations showing that the alphas were very fast and massive projectiles compared to the spread-out positive mass of the Thomson model of the atom. The alphas were over 7000 times as heavy as the electrons and were traveling at very high speed. This meant that strong forces had to be applied to deflect them in their paths or occasionally to bounce them backwards. Rutherford was thus forced to imagine a new picture of the atom composed of a tiny nucleus (Latin: little kernel or nut) in which the atom's positive charge and most of its mass are concentrated. (See Figure 4.9.)

Since the atom *is mostly empty space,* it is easy to see why most of the alpha particles went right through the thin foil. When an alpha occasionally runs into a nucleus, the positive charges repel each other and the alpha is then scattered. If a "head-on" collision occurs with a heavy nucleus which has a very high positive charge, like one of gold, then the alpha rebounds. The electrons in the atom, being so light in weight, have no appreciable effect on the path of the heavy alpha particles.

The full impact of Rutherford's theory, initially proposed in 1911, is so great that it is hard to get an immediate grasp of its implications. All of your material world is mostly empty space. The "solid" wood of chairs and tables, the steel beams used in skyscrapers and the "solid" steel in your car, the concrete roadways—everything is mostly empty space, even

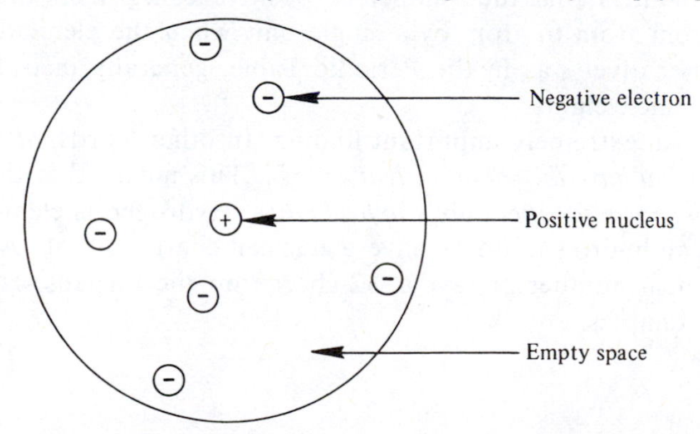

FIGURE 4.9

Rutherford's nuclear model of the atom showing that the atom is mostly empty space. The particles are not drawn to scale. If drawn to scale, you would not be able to see the nucleus at all. Like all pictures of the atom, this one also requires that the actual picture or concept be formed in your mind.

you yourself. If all the electrons and nuclei in your body could be squeezed closely together, eliminating the space in the atoms, then you would become a speck visible only with a magnifying glass.

Rutherford's visualization of the atom as a nucleus surrounded by tiny moving electrons farther out from the nucleus comparatively than the planets are from the sun helps to explain some common experiences. The static electricity you sometimes feel if you take off a polyester or nylon garment is accounted for by the easy removal of some electrons in the vast and spacious cloud surrounding the nucleus. An ordinary atom is electrically neutral, the total positive charge on the nucleus being equal to that of the surrounding electrons. This means rubbing objects together like your body and the polyester shirt, or a comb through your hair, can cause loose electrons to separate from the clouds in one material and collect in another. The place where electrons are scarce is more positive and the place where they are in excess is more negative. It is commonly known that opposite charges attract each other and that like charges repel. If you've had experiences with static electricity like these, then you have been involved in casually checking out the structure of atoms—even if you didn't realize it.

ELEMENTS AND THE NUCLEAR CHARGE

Rutherford's concept of the nuclear atom laid the theoretical groundwork for explaining both the individuality and the variety of the elements. Scientists found ways of estimating the charges on the nucleus by the amount of deflection of the alpha particles, assuming that larger positive charges would have larger deflecting power on the positively charged alphas. Also, they investigated electromagnetic radiation called *X-rays* obtained by bombarding various target atoms with electrons. The X-rays suggested the idea that the number of positive charges on the nucleus increases from atom to atom by a single unit when the elements are arranged consecutively, as in the Periodic Table, generally in order of increasing atomic weights.

This was an extremely important finding. In other words, *all atoms of any one element have the same nuclear charge.* This means that the *nuclear charge is the basis for the atom's individuality.* Hydrogen is element number 1, and all hydrogen atoms have a nuclear charge of +1. All helium atoms (element number 2) have a +2 charge on the nucleus. Here are a few other examples:

Element Number	1	2	3	....	79
Charge on Nucleus	(+1)	(+2)	(+3)		(+79)
Name	Hydrogen	Helium	Lithium		Gold

The positively charged nucleus is surrounded in each case by electrons equal in number to the plus charge on the nucleus.

> THE ATOMIC NUMBER OF AN ELEMENT IS THE NUMBER OF POSITIVE CHARGES ON THE NUCLEUS.

This corresponds directly with the number order of the elements in the Periodic Table. We can now get a more basic conceptual definition of an element.

> AN ELEMENT IS A SUBSTANCE ALL OF WHOSE ATOMS HAVE THE SAME NUMBER OF POSITIVE CHARGES ON THE NUCLEUS.

THE STRUCTURE OF THE ATOM: THE PROTON

One of the simple instruments used in the fundamental studies that led to our picture of the atom was the cloud chamber invented by Charles Wilson in 1907. (See Figure 4.10.) The Wilson cloud chamber makes the path of a moving charged particle "visible" by a mechanism similar to fog formation by condensation of water molecules on small dust or smoke particles in the air.

It was by means of the cloud chamber that Rutherford was able in 1919 to definitely identify a second fundamental particle of atoms—the proton, with a +1 charge. Rutherford guessed that protons are located in the

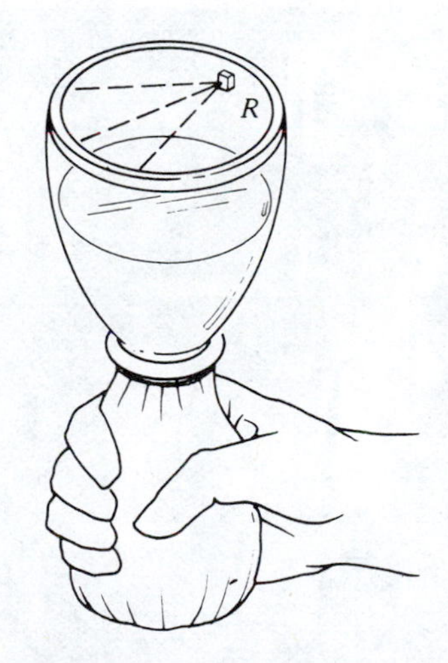

FIGURE 4.10
A simple Wilson cloud chamber. The bulb and chamber contain water. If the bulb is squeezed the temperature of the air rises and more water molecules evaporate. If the bulb is then suddenly released, the temperature drops and the air is momentarily supersaturated. A radioactive substance at *R* gives off charged particles which knock electrons off air molecules. This leaves a path of charged particles on which water molecules condense and form a fog track.

nucleus. The positive charge on the nucleus is now known to be made up of proton units.

THE ATOMIC NUMBER OF ANY ELEMENT IS THE NUMBER OF PROTONS IN THE NUCLEUS OF ITS ATOM.

This is an even more conceptual definition than that given earlier. The number 1 atom is hydrogen, with a nucleus containing a proton, and way out there in the space around it, the electron is revolving in circular orbit, according to Rutherford's model. The number 2 element, helium, as a neutral atom, has two protons in its nucleus, with two electrons in the large volume outside. Silver, number 47, has 47 protons in the nucleus, giving a total charge of +47, with 47 electrons occupying that vast empty space around the nucleus.

Following Rutherford's suggestions, it was evident that the ordinary physical and chemical characteristics of the elements manifest properties of the surrounding electron system, since this is so large compared with the tiny nucleus. But, according to the nuclear model, the total electron system depends on the total electric charge on the nucleus. This is quite an amazing simplification. All the diversity of properties between the elements in the Periodic Table is now expressible as a function of the proton structure of the nucleus. In other words, *the proton structure is now considered the basis for an element's individuality.*

A cloud chamber used in laboratory studies. (Courtesy of the Lawrence Berkeley Laboratory, University of California, Berkeley.)

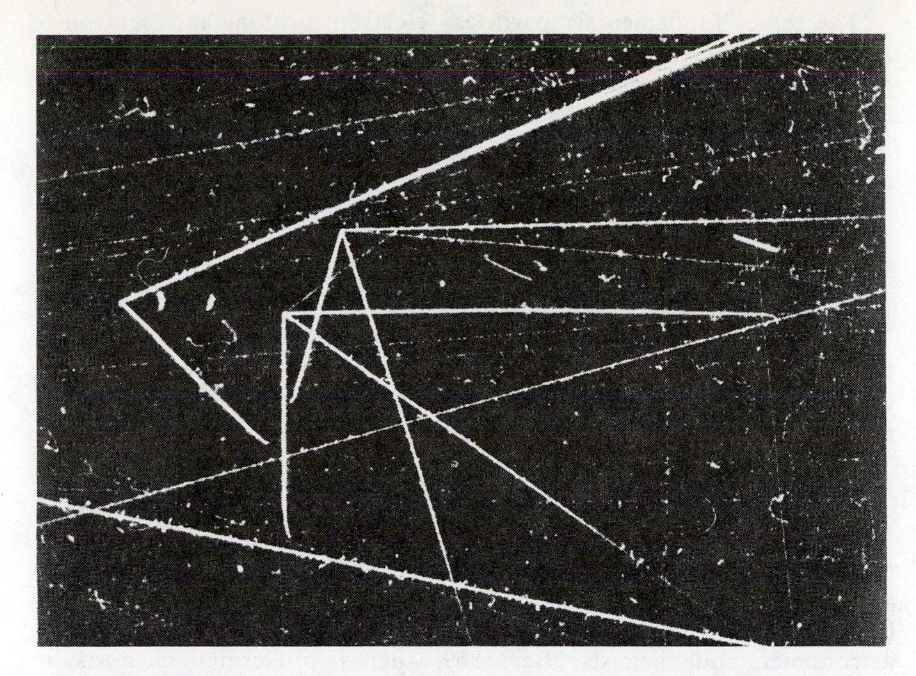

The bold geometric pattern of this cloud chamber photograph was made by the disintegration of carbon and oxygen nuclei by high energy neutrons from the 4000-ton, 184-inch cyclotron at the Lawrence Berkeley Laboratory. The heavy tracks are made by heavy positively charged nuclear particles propelled from the exploding nuclei. (Courtesy of the Lawrence Berkeley Laboratory, University of California, Berkeley.)

THE STRUCTURE OF THE ATOM: THE NEUTRON

The third "fundamental" particle was discovered in 1932 by James Chadwick, who worked in Rutherford's laboratory. This was a difficult particle to find because it is neutral and therefore makes no direct tracks in the cloud chamber. Some peculiar tracks had been found in cloud chamber photographs, starting in the middle from no apparent collision or charged particle source. These were shown by Chadwick to be caused by *neutrons* hitting a nucleus of an atom to give charged pieces.

TABLE 4.3
The Chemist's Three "Fundamental" Particles that Go to Make Up All the Matter in the Universe

Particle	*Usual Symbols*	*Electrical Charge*	*Relative Weight*
Electron	e^-, beta ray	−1	1/1837
Proton	p, H^+	+1	1
Neutron	n	0	1

The three "fundamental" particles—electron, proton, and neutron—are all we need to explain properties and behavior of atoms. (See Table 4.3.) Neutrons exist inside the nucleus along with protons in some kind of special stability which we do not understand. There have been indications of quite a few other kinds of particles in the nucleus which are described in books on atomic physics.

LIGHT FROM ATOMS THROWS LIGHT ON ATOMS

You have seen how people develop concepts about matter by observing natural behavior and making guesses. To understand this development up to our modern concepts about matter we will need to investigate another type of signal by which nature gives us as a hint of her inner mechanisms. The signal is light of various colors.

Isaac Newton in 1666 found that a beam of white light when passed through a glass prism was broken into a rainbow of colors called a *spectrum.* (See Figure 4.11.) This effect was later used by astronomers and incorporated with a telescope into an instrument called a *spectroscope* (seeing the spectrum). The spectroscope was used with great success by astronomers and chemists after 1859 when two German chemists at Heidelberg, Kirchhoff and Bunsen, discovered that, in the gaseous state, each chemical substance emits its own characteristic spectrum. (See Figure 4.12.)

This was a startling and very practical discovery. It showed a sort of fingerprint of each substance. Kirchhoff and Bunsen were so successful at identifying an element by means of its fingerprint or spectrum that they discovered two elements at that time unknown—cesium and rubidium. Subsequently, Marie Curie also used spectrum studies to identify the new elements she discovered—polonium and radium.

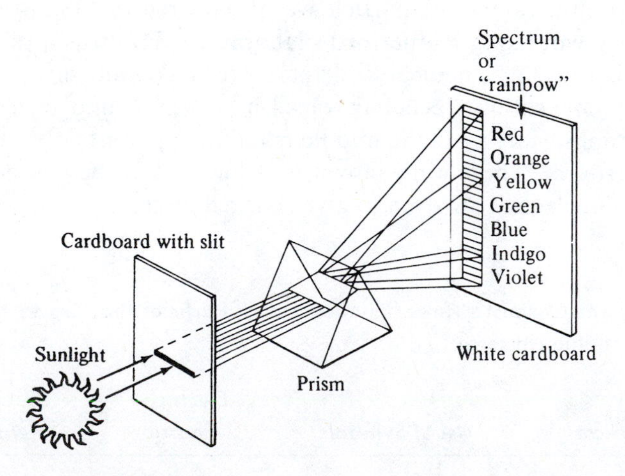

FIGURE 4.11
Diagram showing how a prism breaks ordinary white light into a continuous spectrum of colors.

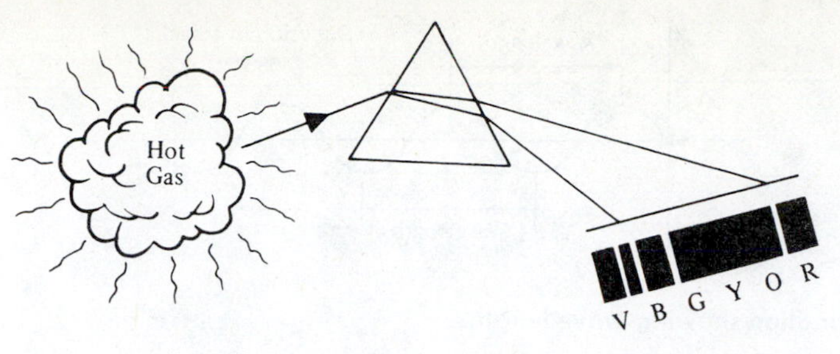

FIGURE 4.12
Outline diagram showing how a highly energized substance in gas form emits a special series of colored lines. Every element has its own spectrum.

A majority of the sun's elements have been identified, the method of analysis being the spectroscope. In 1868 the spectrum of the sun's atmosphere was found to contain lines which had never been identified with an element then known on earth. This element was called *helium* from the Greek word for sun. In 1895, twenty-seven years later, William Ramsay in England discovered helium in very small amounts in air.

It is an interesting fact that the method used by Rutherford to identify alpha particles as helium nuclei was examination of the spectrum. Without this simple tool much of his work would not have as easily advanced to its subsequent conclusions. The amazing thing is that neither the discoverers of the practical use of spectra nor Rutherford at the time of his identification knew how the distinctive lines were originated in the atom. Rutherford only knew that every sample of helium always gave the special distinctive lines which no other element provided. He did not know why, nor had he time to find out. He just used this practical knowledge and put off finding out more about it till later. This is again a common human trait and a truly scientific procedure. Put off till tomorrow what you don't need today. But, keep working on something and, above all, keep thinking. In other words, be lazy in a practical and energetic way.

The Rutherford model of the atom, with a tiny nucleus surrounded by electrons in orbits like planets around the sun, naturally raised further questions, as any good model or theory will. In particular, how are the electrons arranged around the nucleus in empty space? The basic breakthrough to this problem was provided by the Danish physicist Niels Bohr who in 1912 asked to join Rutherford's group. It might be worthwhile to digress to consider the nature of light which Bohr used successfully to give an abstract idea of electron behavior in the atom.

THE NATURE OF LIGHT

Before 1900 there had been much discussion on the nature of light, with varying acceptance of the "corpuscular" or "particle" theory and the "wave" theory. The wave model implied that light was a radiation con-

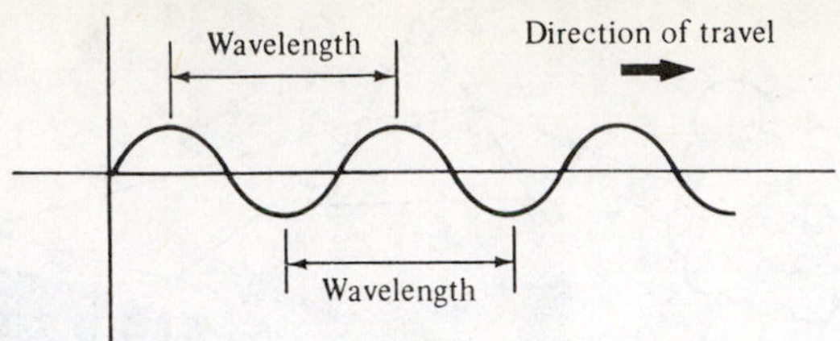

FIGURE 4.13
Diagram of wave motion showing wave length.

sisting of electromagnetic waves (both electrical and magnetic in their properties) which were emitted continuously by matter at high temperatures. In 1900, Max Planck, a German physicist, reported theoretical studies he made of hot objects. He discovered that he could explain the behavior of light given off by hot bodies only if he assumed that atoms emit radiation in energy bundles, that is, not continuously as had been previously supposed. Planck himself had difficulty believing this startling conclusion from his calculations. Subsequently, Einstein in 1905 showed that light energy is both emitted and absorbed in discrete bundles, or pulses, which he called *quanta* (Latin: how much). The quantum is then the smallest amount of energy capable of existing independently. It is the "particle" of energy. Thus Planck's idea became the foundation stone of modern physics which views light as a sort of duality of both particle character and wave character. The amount of energy in each quantum depends on the length of the light wave which is dependent on the number

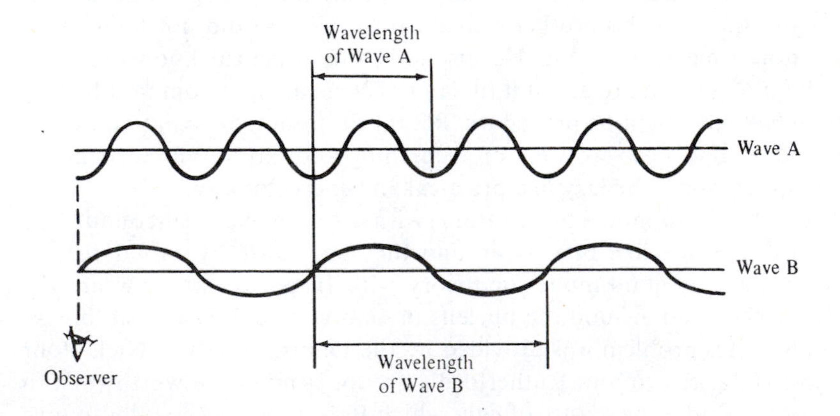

FIGURE 4.14
Pictorial representation of two waves of radiation. Wave *A* is vibrating twice as frequently as Wave *B*, the frequency being the number of waves that pass a given point per second. In any space interval there are twice as many waves in the case of *A* as there are of *B*. Since all electromagnetic waves travel forward at the same speed, then the radiation with the highest frequency will have shorter waves, or technically, the wavelength—the distance from one point to a similar point—will be shorter.

of vibrations made in a second. The greater the number of vibrations per second the higher the energy. Thus if we have two different electromagnetic radiations (like visible light, X-rays, gamma rays, or radio waves) they will differ in the frequency of vibration, or the number of vibrations per second. The general nature of a wave is indicated in Figure 4.13. A good way to visualize waves is to tie a piece of rope to a tree or fence or door knob and then to move the end up and down to form waves in the rope. You will find that if more energy is put into the up-and-down movement, then the waves will have shorter wavelengths. (See Figures 4.14 and 4.15.)

The wide range of waves is indicated by Figure 4.16 which represents the most important regions of the electromagnetic spectrum. The waves that stimulate the retina of your eye are only a very small part of the complete spectrum. The diagram is arranged in the order of increasing wavelength from the top to bottom. For example, the gamma rays which come off in radioactive decay of certain elements represent very high energy radiation with very short wavelength. The X-rays have somewhat longer wavelength, and then we move to the ultraviolet rays which represent the wavelengths just shorter than we can see. These are the rays which cause suntan and also produce Vitamin D in the skin.

The visible range represents wavelengths from the violet, which are

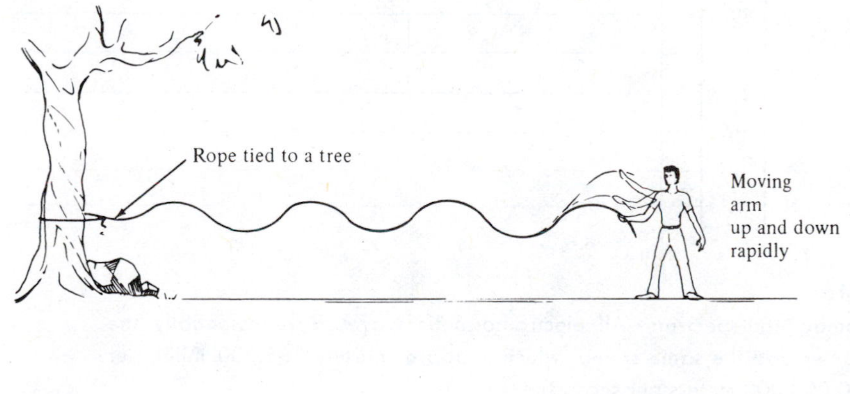

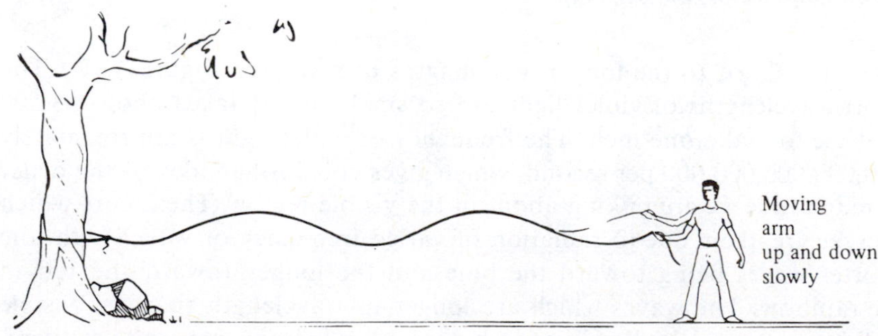

FIGURE 4.15

A practical demonstration of how higher energy input produces shorter wavelengths.

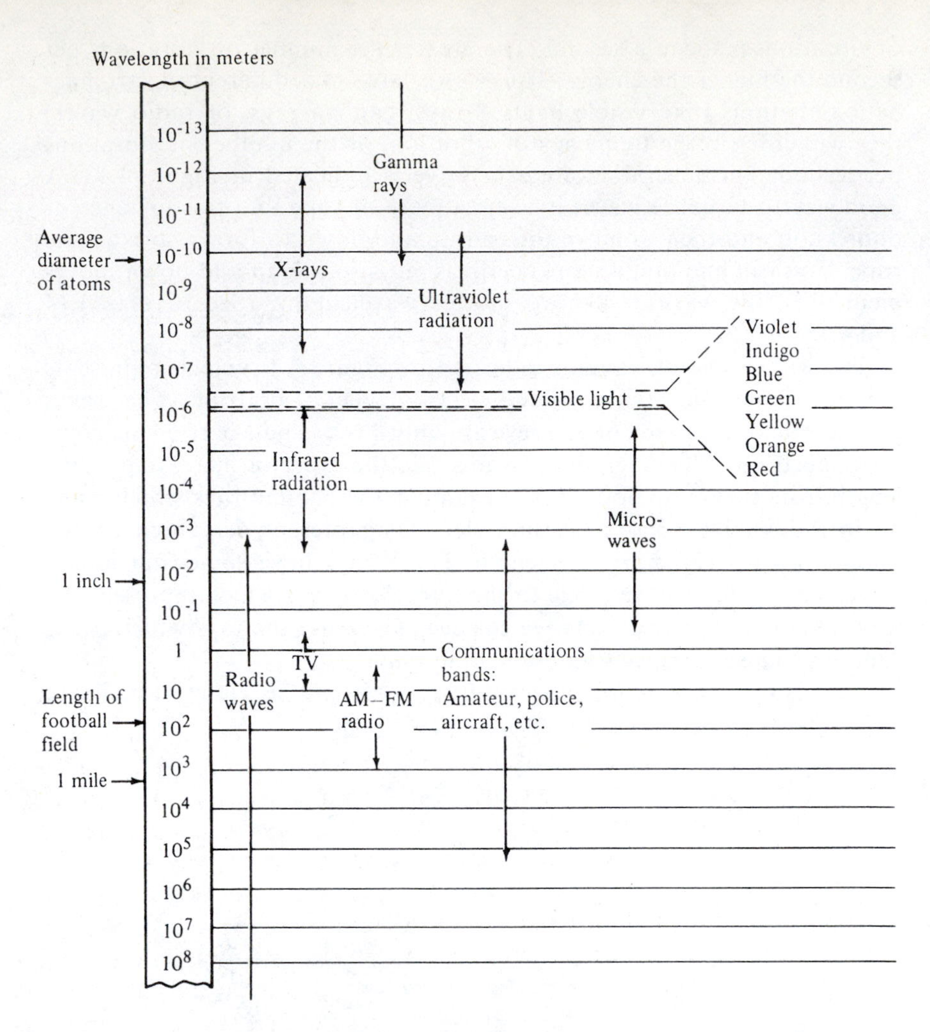

FIGURE 4.16
The electromagnetic spectrum. All electromagnetic waves have essentially the same character and the same speed, which is approximately 186,000 miles per second (300,000,000 meters per second).

relatively short, to the longer wavelengths of red. (See Figure 4.17.) The short wavelengths of violet light are so small that it takes about 63,500 of these to make one inch. The frequency of violet light is approximately 750,000,000,000,000 per second, which gives you a rough idea of the order of magnitude we are talking about in the visible region. The colors which you see are, then, due to radiation of varied frequency or wavelength, the shorter waves being toward the blue and the longer toward the red in the rainbow. The waves which are longer in wavelength than the visible radiant energy are "below" red in having a lower frequency and are therefore called *infrared* rays. These rays are given off by all warm objects like stoves and fireplaces. They usually accompany other waves in the visual light range in these cases. Microwaves are used in radar and in ovens.

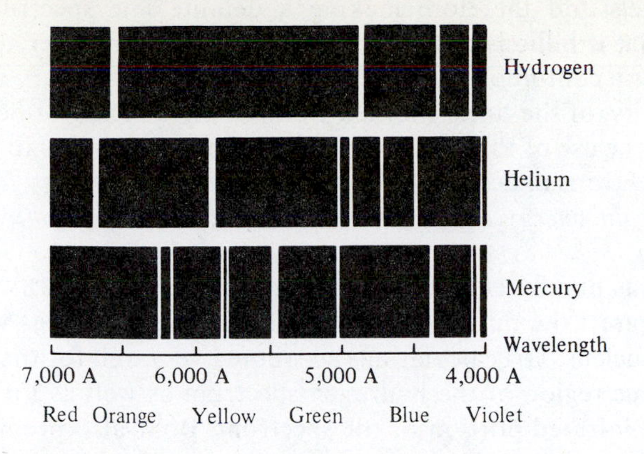

FIGURE 4.17

Parts of the spectra of a few elements. The original spectrum of each element contains sharp lines of varied colors, as indicated by the lower scale designations shown above. The line pattern is definite for each element and always exactly reproducible. Each element has its own distinctive spectrum which is different from that of all other elements. This is why it is often referred to as a "fingerprint" of the element.

Radio waves on which radios and TV depend represent a wide range of wavelengths. When you switch stations you merely select a certain wavelength for the station you want, each station operating on an assigned wavelength.

The broad range makes it necessary to use various measuring units for the different waves. The longest have wavelengths measurable in miles while the shortest run to billionths of an inch. The length of waves in the visible region is often designated in a unit called an Angstrom, which is approximately 4 billionths of an inch. A = Angstrom in Figure 4.17.

ELECTRON BEHAVIOR IN THE ATOM

Based on the examination and measurement of the wavelengths of lines from photographs of spectra such as those shown in Figure 4.17, Bohr in 1913 suggested that atoms normally have their electrons in places of relatively low energy which are in the orbits or levels nearest the nucleus. When disturbed by heating or bombarding with other high energy electrons, the electrons in the atom can be forced in a direction outward from the nucleus, but the levels they can enter are definite and depend on the particular atom. Bohr guessed that the electron is not like a planet or a satellite which can presumably orbit anywhere. There are definite limits in the atom. When an electron falls back into a lower level again, a definite package or quantum of energy is given off in the form of radiation. If it is in the visible region of the electromagnetic spectrum we see a bright line.

Of course, one single electron would not make a very bright line, so what we see is due to billions of electrons all jumping between definite energy levels and therefore making a definite line spectrum. You can imagine that if billions of electrons were jumping all over the place you would have a continuous spectrum—not lines. It is the very nature of the discontinuity, or the lines, that led Bohr to the concept of definite energy levels and the use of Planck's quantum idea of light to explain the arrangement of electrons in definite levels. *The fact that each energy level is definite means that the energy difference between levels is also definite.* (See Figure 4.18.)

Bohr thus numbered the levels as shown in Figure 4.18, beginning at the first nearest the nucleus as 1, then 2, 3, 4, and so on as we move out from the nucleus. His calculations accounted very well for the known lines in the visible region of the hydrogen spectrum as well as for the series of lines in the infrared portion of the spectrum. Bohr also predicted another series of lines in the ultraviolet portion and these lines were soon discovered, giving considerable notoriety to Bohr's work.

Bohr's work laid the foundation for a picture of the atom more refined than Rutherford's. But, like every model, this one was modified by other workers who found needs for even more sophisticated views, especially relating to atoms more complex than the hydrogen atom with its single electron.

The workers following Bohr were oriented very heavily toward mathe-

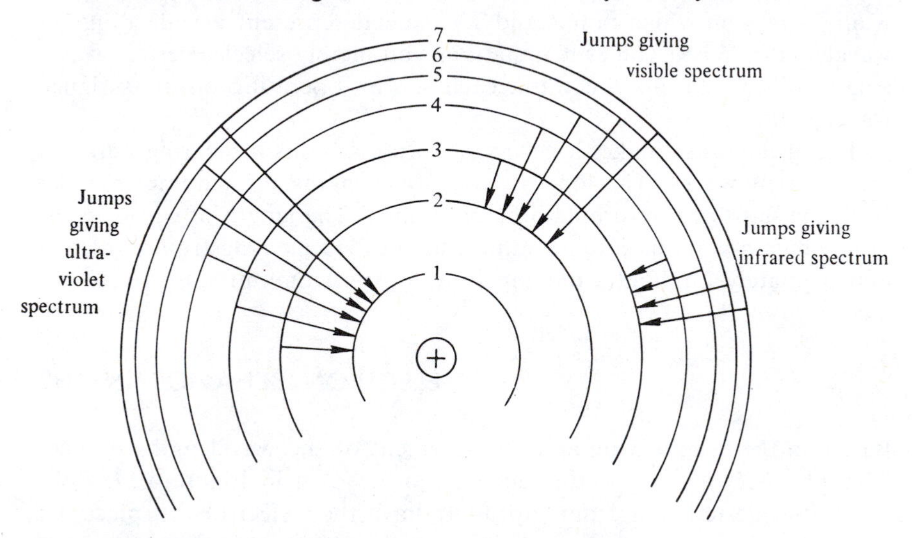

FIGURE 4.18
Schematic diagram of the electron orbits or levels in the hydrogen atom according to the Bohr model. The level occupied by the single electron under normal conditions is represented by the circle nearest the nucleus. The other circles are possible levels of higher energy, that is, the electron in these would have greater energy than in the "ground state," level 1. This is because the electron in the higher levels is farther from the nucleus and energy is required to push it out there. The arrows show some of the electron jumps that are possible to the first three levels. Each of these jumps represents a definite quantum of energy and corresponds to a definite line in the hydrogen spectrum.

matical models and their findings are based on complicated equations of probability and wave motion. The study of quantum theory as expanded and applied to the theoretical explanations of atomic structures is even given the special name of *quantum mechanics.*

SUMMARY OF THE MODERN VIEW OF THE ATOM

The major results of quantum mechanics involve us in a visualization based on Bohr's original ideas, but with the energy levels expanded into a more "fuzzy" or "smeared" version of electron operation. In this book we will use chiefly the basic approach of Bohr to explain behavior of atoms. However, it may be interesting here to list a summary of the modern view of the atom, keeping in mind that this relates to the most complex, theoretical area of chemistry and may be difficult to understand. The limits in the structure of the atoms which the modern view describes give us a direct insight into why the atoms behave the way they do. We will see this in a simpler way in the next chapter.

1. Electrons can operate only within definite energy levels or "stationary states." These are numbered beginning nearest the nucleus with 1, 2, 3, and so forth. (The numbers are called principal quantum numbers.)
2. Electrons can jump or move only between these levels or else leave the atom completely. The lines in an element's spectrum are caused by the electrons falling from a higher energy level to a lower one.
3. The electrons in any quantum level are limited. In other words only so many electrons can be in a particular level. There can be *less* than this limit but *not more.* For example, the first level can have a maximum of 2 electrons; the second level can have a maximum of 8 electrons; the third level can have a maximum of 18 electrons; the fourth level can have a maximum of 32 electrons; the fifth level can have a maximum of 50 electrons, and so on. (The limiting number of electrons, which is the maximum capacity of an energy level, is equal to $2n^2$ where n is the quantum number for the particular level.)
4. The operations of electrons cannot be exactly restricted to circular line-like orbits or even elliptical orbits but rather must be considered as occupying a "smeared out" volume where we cannot pinpoint an electron but use the shading of the smear to show where the electron most probably spends its time. The "picture" (and verbal descriptions) are limited as all pictures are. They can also be thought of as showing "population density" as on a city map even though the people are certainly never tied down to one spot. (See Figure 4.19.)
5. Each quantum level may be broken into sublevels, the number of sublevels being equal to the principal quantum number. For example, the first level can have only one level since $n = 1$. The second level has 2 sublevels, the third 3 sublevels, and so on. This means that the electrons within the second level actually do not all have exactly the same energy, some having a slightly different energy from others in that level.

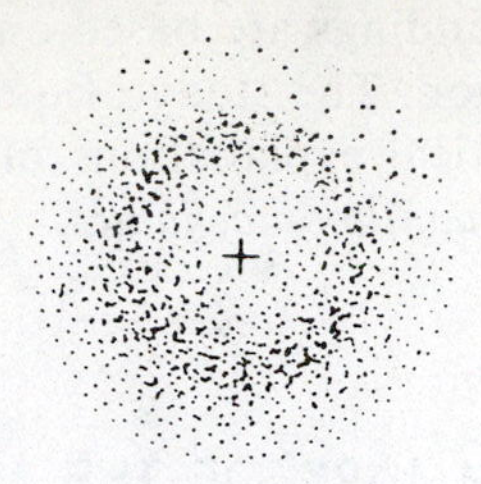

FIGURE 4.19
Time-average view of the hydrogen atom showing the electron operations around the positive nucleus. The darker section shows where the electron is most probably found. The whole cloudy ball is referred to as an electron cloud.

6. The electrons which occupy various levels and sublevels operate in what are called orbitals ("like an orbit" or "orbitlike," but not the same thing as orbit). The word change here emphasizes that the electron is not just zipping around in a linear orbit like the earth around the sun or like an artificial satellite around the earth. Its energy gives it lateral as well as straight-line motion which means that it occupies a certain volume which is spread out and given the term *orbital.* An orbital then is an operations volume for an electron, like a football field is the operations volume for the quarterback.
7. The shapes of the orbitals, or operating volumes occupied by electrons, assume certain symmetries. The simplest orbital is spherical. The orbital volumes are usually shown by a smeared or shaded rendering of the electron cloud shape in an artistic attempt to picture a localized cloud. (See Figure 4.20.)

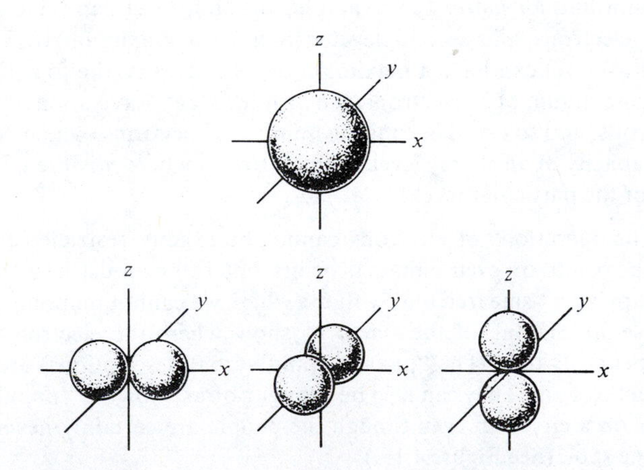

FIGURE 4.20
Examples of some orbital shapes as derived from quantum mechanics. The top has spherical symmetry and is designated as an *s* orbital. The lower orbitals are three similar orbitals of dumbbell symmetry but oriented in different directions, as shown by the *x*, *y*, and *z* axes. These are designated as *p* orbitals.

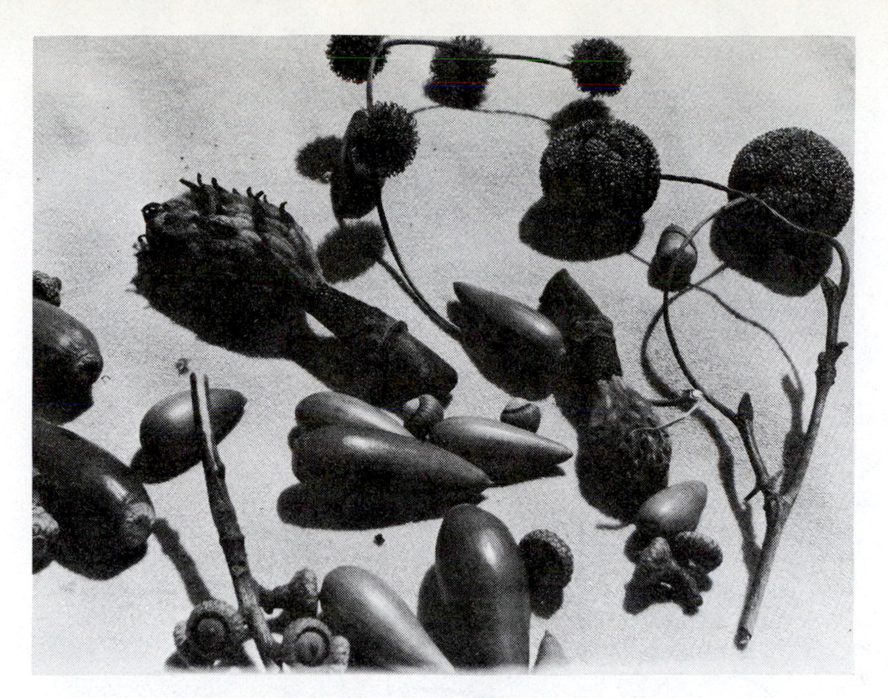

Some varieties of shapes found in natural objects. Spherical shapes are very common in nature, and it is not too surprising that this shape also occurs down deep in the atom. Some dumbbell-type symmetry is also found in large natural objects. (Photo by Gary R. Smoot.)

8. The electrons in each energy level will be found to occupy definite orbitals—only 2 electrons to any orbital—and these start from the simple spherical shape and lead up to more complicated ones. The first level has no sublevels and only one type or orbital, the spherical. In the second main energy level there is a repeat of the spherical type of orbital, which again holds 2 electrons. But the second main level can hold a total of 8 electrons, and the 6 additional electrons can go into 3 of the dumbbell-shaped orbitals, 2 electrons in each of the possible orbital distributions shown above. In the third level we have a total of 18 electrons. The first 2 go in a spherical orbital, then 6 go into the third-level dumbbell type, leaving 10 electrons which go into 5 orbitals of even more complex shape which we need not consider here.

THINKING ABOUT THE ATOM

Figure 4.21 shows several of the ways the atom has been pictured since Dalton's original formulation of atomic theory. Also you can see how theories are altered as each one improves upon the one before.

In addition to having a mental picture of the atom's structure, it is useful to think of the size of the atoms which we discuss so casually in chemistry so that we do not lose sight of the vast difference in dimensions from the world of everyday experience.

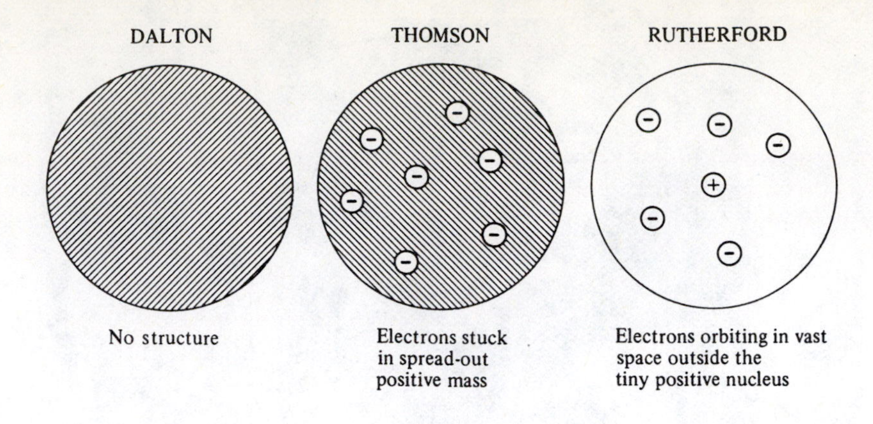

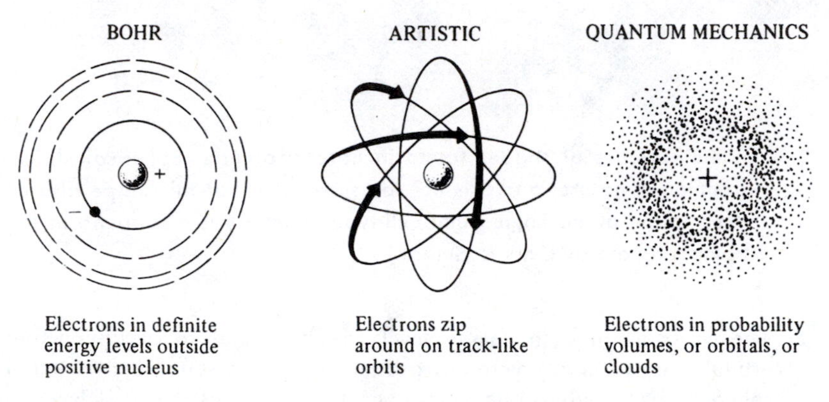

FIGURE 4.21
The progress of atomic structure models.

An atom is a very tiny thing. There are about 100 billion billion atoms in a tiny drop of water. If you took a single grain of sand and enlarged the atoms so that they were as big as the head of a pin, then all the atoms in that grain of sand would make an enormous pile one mile long, one mile high, and one mile wide.

An atom is only (on the average, since they differ slightly in size) about one hundred millionth of an inch in diameter. In other words a millionth of an inch is a fairly long distance on the atomic scale. It is about one hundred times the average atom's size. If an atom were enlarged as big as a golf ball, the atoms in an inch would stretch from New York to San Francisco. It would take more than one million atoms, edge to edge, to match the thickness of this page of paper. Another analogy: there are more atoms in your hand than grains of sand in all the beaches of the world.

Inside this tiny thing, moreover, there are mostly endless expanses of space (on an atomic scale). A typical atom has a diameter about 100,000 times larger than its nucleus. In other words, if a hydrogen atom's nucleus were enlarged to the size of a golf ball, its single electron would be operating generally about a mile away. Yet this electron moves around so fast that it makes over one hundred million billion circuits every second. It is this everywhere-at-once character of the electron which accounts for the solidity of material things and the cloudlike representation of orbitals, which you have seen above.

The volume of a typical atom is 1,000,000,000,000,000 times greater than the volume of the nucleus. The density of matter in the nucleus is therefore enormous—in the range of a million million million times more dense than the atom as a whole. In other words an atomic nucleus the size of a grain of sand would weigh 200,000,000 pounds.

It is into this strange world that we have been able to probe with simple devices like cloud chambers and gas discharge tubes. Mankind's success in measuring the minuteness of the atom and fashioning "pictures" of what goes on in there stands as one of the greatest experimental and conceptual achievements of all time.

Photomicrograph of paper like the page this is printed on (100X). Now that you know the tiny size of atoms, you can appreciate that these do not show here. More details on how atoms are involved in paper molecules will be given in chapter 10. (From the Electron Microscopy Department of The Institute of Paper Chemistry.)

PROJECTS AND EXERCISES

Experiments

1. This will be an experiment to demonstrate simply the easy generation of electrical charges and the fact that forces are exercised between various charges. Assemble from around the house a few long plastic objects of various sorts like a plastic straw, a plastic stirrer like those used in restaurants for stirring highballs, a long piece of plastic from some toy, or some combs, a toothbrush, or similar long plastic articles. Then get some dry paper towels, or garments made of synthetic fibers like polyester, a silk scarf, an old nylon stocking, or other pieces of fibrous material like blankets. Also get several lightweight objects like salt, sugar, thread, hairs, small bits of dry paper, powdered coffee. Assemble the small lightweight objects in various areas on the table. Then hold one of the long plastic objects like a straw or stirring rod of plastic over the materials on the table and slowly move in close. Next rub the straw or rod or other object briskly with the paper towel or other fabric material. (Try several combinations during the course of the experiment.) Then bring the plastic rod down just above the small specimens on the table. Do this one at a time, rerubbing as necessary between tries. Do you observe a different behavior than before you rubbed the rod? Try this with other combinations. Some combinations will be found to be much better than others, and you can best determine this by trial and error.
   When you get a combination of a long plastic object and a fabric for rubbing which seems to develop a good charge, try other tests like holding it just above the fine hairs on your arm or on a friend's head. What happens now? Hold the freshly rubbed plastic piece closer and closer to your ear. What do you feel? Do you hear anything? This may or may not work, depending on the dryness of the day. All of the above tests work best on a dry day. If the weather is damp or rainy, it would be better to wait till a better day.
   Describe briefly what happens in your varied rubbing experiments. Then try to give a conceptual explanation, using what knowledge you now have about the makeup of matter. Does this experiment make any more real to you the basic nature of matter, as now visualized by science? Do you have to see the electrons to believe that they may be involved?
2. This is a chance for you to "see" radioactivity and the breakdown of atoms into parts. When you go to bed some night and are not too tired, lie there for a while with the light out until your eyes get accustomed to the dark. Beforehand have a watch or clock with a luminous dial by the side of your bed, and also a magnifying glass or hand lens if possible. (The lens is not absolutely essential.) Then after you are adjusted to the dark, hold the watch or clock close to your eyes so that you see tiny flashes. Here is a startling thing, the apparently continuous nature of your luminous dial is actually discontinuous or pulsing. It consists of tiny flashes of light, which when combined with many others gives the effect of continuous light. The luminous dial consists of a radioactive material which gives out alpha particles and a phosphor coating which gives a flash of light when a positively charged alpha particle knocks electrons out of atoms of the phosphorescent coating. This is strong evidence for the atomic theory since each of the flashes represents an atom breaking up. Look at the flashes with the lens. What are your impressions? Do you feel a little closer to the atom here? (Most dials work in this experiment. If yours doesn't, try to borrow another.)

3. Take one of the quotations at the beginning of this chapter and describe briefly what it means in terms of the concepts covered in this chapter.
4. The modification and "updating" of Dalton's atomic theory points up how progress occurs in science. No theory is perfect but a good theory is gradually refined by making newer distinctions without actually abandoning the central concept. Briefly describe how the basic assumption of Democritus and Dalton of discontinuous matter, or discrete particles, is still a part of our thinking. The unit of the discontinuity is shifted downward. What could you say about our present views, that is, do we favor the discontinuous or continuous nature of matter? Do we have a final, unchangeable theory now?
5. Alpha particles form fairly straight paths in a cloud chamber. How could you describe the action of the alpha particles in causing the visible paths? Do you "see" the alpha particles? Explain what you mean. Does the alpha hit a nucleus every time it meets an atom? What is the implication for structure here?
6. Using the Bohr theory of the atom, explain how you could account for the appearance of bright lines of various colors in the spectrum of a gas.
7. Rutherford's nuclear model of the atom is now accepted by scientists. This says that the atom is mostly empty space. If this is a true theory, then why, when you place your hand on a table, or squeeze an orange, or otherwise touch a material object, doesn't your hand go right through the empty space?
8. Consider the comparison often made between the Bohr model of the atom and the solar system. List as many similarities and differences as you can think of between the Bohr atom and the solar system.
9. The original arrangement of the elements into families with similar characteristics was based by Mendeleyev on the sequence order of increasing atomic weights. The basis for the Periodic Table has been refined so that we now recognize the basic ordering system of atomic number as more fundamental. Would this drastically change the order of the elements? Can you find any place in the table where some confusion would result if we rigorously followed the increasing order of atomic weights?

SUGGESTED READING

1. Andrade, E. N. D., *Rutherford and the Nature of the Atom,* Anchor Books, Doubleday & Co., Inc., Garden City, New York, 1964. Paperback.
2. Schrodinger, E., "What is Matter?," *Scientific American,* 189, No. 3, pp. 52–57 (September 1953). Scientific American Offprint No. 241.
3. Young, Louise B., ed., *The Mystery of Matter,* Oxford University Press, New York, 1965, Part 3, "Is Matter Substance or Form?," "Radioactivity," "The Structure of the Atom."
4. Romer, A., *The Restless Atom,* Anchor Books, Doubleday & Co., Inc., Garden City, New York, 1960. Paperback.

chapter 5

Chemistry in Action: Electrons

Science does not mean contemplation of knowledge that has been achieved, but rather it means continuing labor and constant forward-moving change.

—MAX PLANCK

So in each action 'tis success
That gives it all its comeliness

—WILLIAM SOMERVILLE

The best part of beauty is that which a picture cannot express.

—FRANCIS BACON

Until the beginning of the present century there was a large accumulation of information on chemical reactions but no overall theory or model to explain why the reactions occur. Then the foundation for an explanation was established by atomic structure concepts such as those discussed in the last chapter. You have therefore a rather good background on atomic structure—better than any scientist had in the 1870s.

The next step is to consider how atoms link up in forming compounds, and this requires a little more detail on how each atom is built up. It is advisable for you to read through this chapter rather quickly first to obtain an over-view which will make the details more under-standable. Then come back over the details again (and again, if necessary) so that you can see the way the individual atom is built up within the definitive limits we find in nature. This procedure will make it easier to understand how compounds are formed from the elements. Without a basis in atomic structure, it is impossible to understand the formation of compounds.

Keep in mind that the shortcut visualizations which chemists use are not actual pictures of the atoms. In the step from probabilistic orbital operations of electrons to visualizing how electrons get involved in chemical compounds, we will mentally pin the electron down even though this is a physical impossibility. The shortcut visual approaches "pinning down" electrons are logically correct, but not at all physically possible. For example, we will be making "pictures" of electrons statically set in certain orbits, even though you know this does not truly represent the buzzing clouds of electrons.

BUILDUP OF ATOMS: THE ONION-SKIN STRUCTURE

We know that the electrons around an atom's nucleus are limited to certain energy levels, and the number of electrons in each level is also limited. The easiest way to see how the atoms differ from each other, using the basic limits found experimentally through spectra and now formalized in quantum theory, is to "build up" the Periodic Table starting from the simplest atom, hydrogen. This is a good procedure for understanding the electron arrangement in atoms. We start, of course, with the basic idea that the proton structure of the nucleus is what makes an element what it is. (The Periodic Table was shown in Table 4.2.)

There are many methods of showing the way electrons go into "shells" or levels as we build up atoms successively by adding more and more protons in the nucleus. A simple method of depicting atoms uses the older Bohr version of electrons in levels, like skins of an onion enclosing the nucleus. The capacity of the skins increases as we move out from the center, just as the larger onion skins have more cell spaces. The method is applied to the first twenty elements of the Periodic Table in Figure 5.1.

For hydrogen, the nucleus must contain 1 proton, giving a charge of +1. The neutral hydrogen atom then has 1 electron outside in its cloud to balance the nuclear charge. We place or locate this single electron in the first level—or the first "onion skin."

The second element is helium, which has 2 protons in the nucleus and therefore 2 electrons in the cloud. These go into helium's first level as shown. When we get to the third element, lithium, we have 3 protons in the nucleus, which requires 3 electrons around it to balance the charge and provide a neutral atom. We now run into the limiting restriction which does not allow more than 2 electrons in the first level. You will recall from the last chapter that the maximum number of electrons in the levels is as follows:

Level Number	*Maximum Number of Electrons*
1	2
2	8
3	18
4	32, etc.

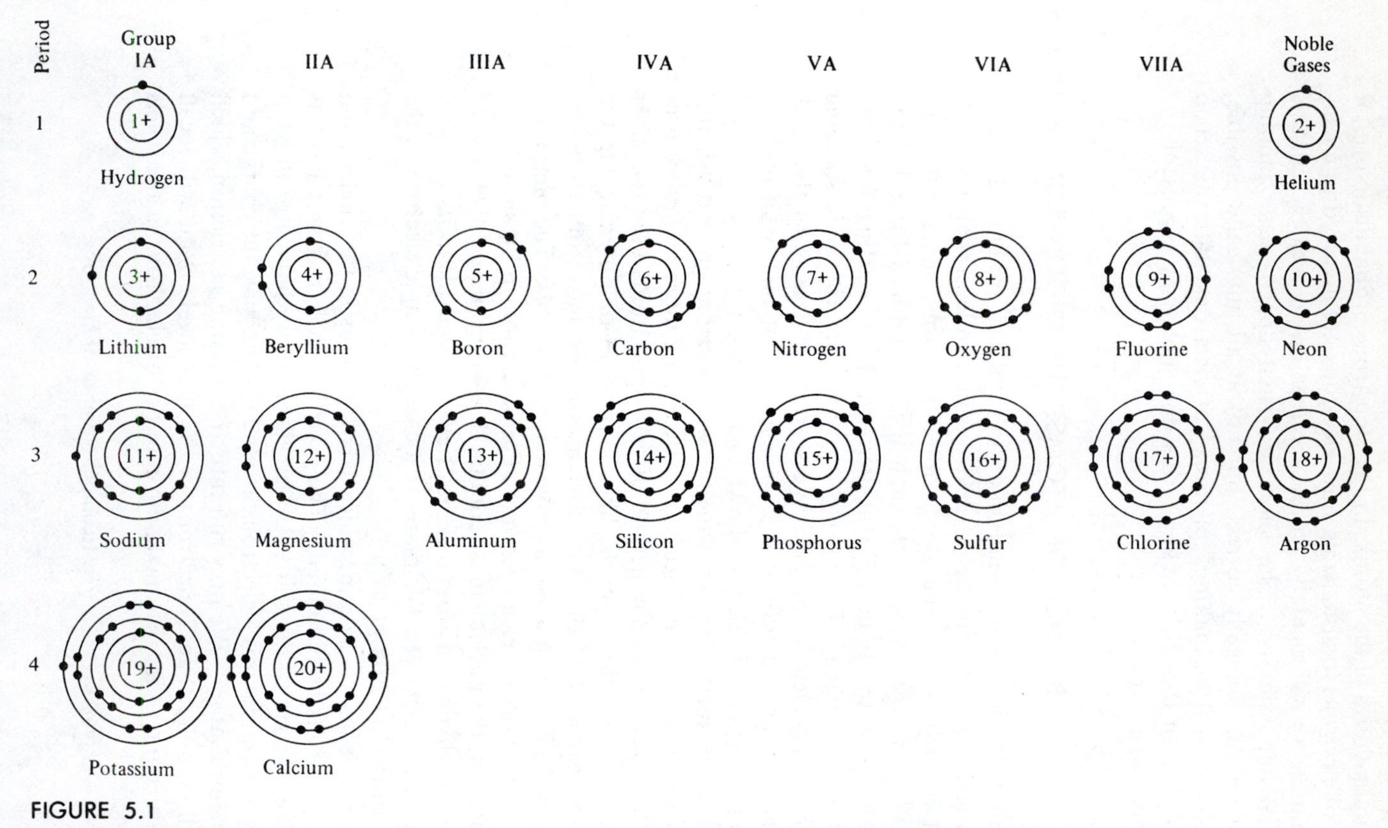

FIGURE 5.1

An "onion-skin" view of the first twenty representative elements showing total electrons in each energy level. Note the buildup of one electron at a time with each element following the addition of one more proton in its nucleus over the previous element. Limits for each level develop similar outer electron structures in the vertical groups or families.

For lithium therefore we place 2 electrons in the first level and then we must place the third electron in the next level, level number 2. The electron is shown in the second onion skin.

Element number 4, beryllium, has 4 protons in its nucleus so it requires 4 electrons outside and the fourth electron is added in the second shell. We continue in this way, adding one proton and one electron to form elements number 5, boron; 6, carbon; 7, nitrogen; 8, oxygen; 9, fluorine; and 10, neon. When we arrive at neon, which is another of the noble gases like helium, we get to a completely filled second level with 8 electrons. Therefore, when we build up element number 11, sodium, with 11 protons, we add the eleventh electron not in the second level, which is full, but in the third level.

Look along each row or period and note how a single electron is being added, gradually filling the outer shells with electrons, one at a time. This balances the one-at-a-time change in character of the elements by proton buildup in the nucleus.

You can see more clearly here how the first row or period contains only 2 elements because the first shell or level can hold only 2 electrons. The "periodicity" of the elements in column arrangement is obviously due to similar electron arrangement in the outer electron levels. You will note that ${}_{1}H$, ${}_{3}Li$, ${}_{11}Na$, and ${}_{19}K$ are in the same vertical column because the single outer electron gives them similarity in properties. Hydrogen, while having some similarity, is a little more special and is not considered an alkali metal.

The halogen family, second from the right, also has a similar outer electron structure, each element—${}_{9}F$ and ${}_{17}Cl$—having 7 electrons in the outer level. If we added the other halogens, ${}_{35}Br$, ${}_{53}I$, and ${}_{85}At$, we would also find 7 electrons in their outer levels. This points out the reason for inclusion of these elements in the same family group, or vertical column.

A CHEMICAL FAMILY IN THE PERIODIC TABLE HAS ELEMENTS WITH SIMILAR EXTERNAL ELECTRON STRUCTURE.

You can also note that neon and argon both have 8 electrons in their outer levels. This structure represents an extremely stable electron arrangement and these elements are often said to be "inert" because they are unreactive. Indeed, no compounds of neon or argon have ever been prepared. This extreme inertness carries over to helium also with its 2 electrons in the outer level. It happens that helium's outer level is complete or closed with only 2 electrons because that is the natural limit of the first level.

The localization of the last electron in potassium requires further discussion. You might expect that this electron in potassium would go into the third energy level which can hold a total of 18 electrons. However, by spectral studies, this has been found not to be the case. Instead, the electron goes into the fourth level because the energy state of this level is slightly less than that of the third sublevel which is still unfilled. The available space in the third level is thus temporarily skipped over. The next electron in the buildup with calcium also goes into the fourth level. Then, beginning with element number 21, scandium, the buildup returns to filling up the empty part of the third level which was temporarily skipped. We will not need to go into further detail in this book other than to say that most of the other elements following the first twenty representative elements are built up similarly to those we have considered above, with some interesting variations.

BUILDUP OF ATOMS: PARTIAL CIRCLE DESIGNATION

Sometimes the onion-skin diagrams are shortened for convenience. The designation of total electron makeup of each level is then indicated by a number in partial concentric rings drawn out from the nucleus, with a small "e" standing for electrons. Here are a few examples:

H (+1) 1e

He (+2) 2e

Li (+3) 2e 1e

C (+6) 2e 4e

Na (+11) 2e 8e 1e

Or the "e" designation for electrons may be left out entirely:

K (+19) 2 8 8 1

A more abbreviated form shows the total electrons in each level, with a comma between levels:

Element	*Electron Structure*
Na	2,8,1
K	2,8,8,1
Mg	2,8,2
Ca	2,8,8,2
Cl	2,8,7
Br	2,8,18,7

BUILDUP OF ATOMS: ELECTRON DOT PICTURIZATION

Another more sophisticated method of shorthand to show electron arrangement is called the electron dot notation. In order to emphasize the electron population in the outermost level, first the symbol for the element is used to represent both nucleus and all electrons except those in the outer level, that is, nucleus + inner shell electrons. This is called the kernel of the atom and is represented by the standard atomic symbol. This symbol is then surrounded by dots representing the electrons in the *outer level* only, with no special significance as to location around the symbol. Here is a comparison of the electron dot method with others:

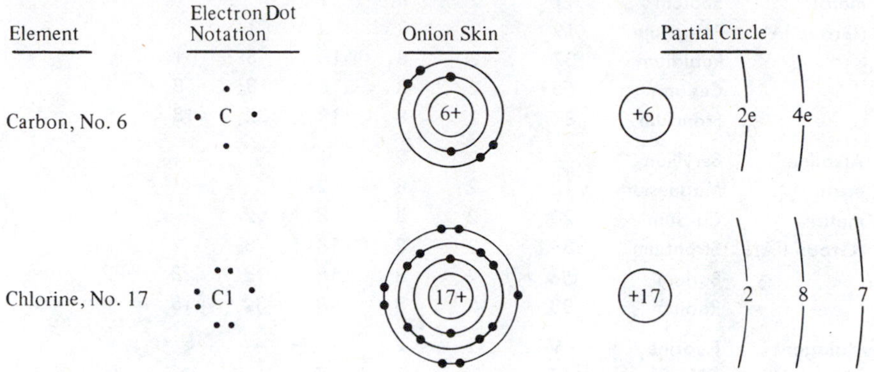

Element	Electron Dot Notation	Onion Skin	Partial Circle
Carbon, No. 6	· C ·	6+	+6 2e 4e
Chlorine, No. 17	:Cl·	17+	+17 2 8 7

The electron dot notations for the first twenty representative elements are shown below:

H·							He:
Li·	Be:	·B·	·C·	·N·	·O:	:F:	:Ne:
Na·	Mg:	·Al·	·Si·	·P·	·S:	:Cl:	:Ar:
K·	Ca:						

The number of electrons in the outer level of the atom is the crucial determining factor in the behavior of the atom. Elements have similar properties when they have similar outermost electron structures. The outer electron structure is of course not independent. The total character of the element is determined basically by the number of protons in the nucleus—the atomic number. Once that is determined, then the consequent buildup of electrons—subject to the limits set by nature—takes on a pattern which is the mechanism by which specific behavior is transmitted to the particular atom. However, because of the vast expanse of the atom, the "contact areas" between atoms of necessity has to be the outer electrons. Indeed, these outer electrons are called *valence* electrons because, for all practical purposes, they set the *value* of the atom, or its combining capacity, in a way which we will consider shortly.

Table 5.1 shows a few of the most common families along with the electron configuration of each of their elements. Note especially the similarity of outer electron structure. Within each family the numbers of electrons

TABLE 5.1
Electron Structure for Atoms in a Few Common Chemical Families

Family	Element	Atomic Number	Electron Configurations: Number of Electrons in Principal Energy Levels 1	2	3	4	5	6	7
Alkali metals (Group IA)	Lithium	3	2	1					
	Sodium	11	2	8	1				
	Potassium	19	2	8	8	1			
	Rubidium	37	2	8	18	8	1		
	Cesium	55	2	8	18	18	8	1	
	Francium	87	2	8	18	32	18	8	1
Alkaline earth metals (Group IIA)	Beryllium	4	2	2					
	Magnesium	12	2	8	2				
	Calcium	20	2	8	8	2			
	Strontium	38	2	8	18	8	2		
	Barium	56	2	8	18	18	8	2	
	Radium	88	2	8	18	32	18	8	2
Halogens (Group VIIA)	Fluorine	9	2	7					
	Chlorine	17	2	8	7				
	Bromine	35	2	8	18	7			
	Iodine	53	2	8	18	18	7		
	Astatine	85	2	8	18	32	18	7	
Noble gases	Helium	2	2						
	Neon	10	2	8					
	Argon	18	2	8	8				
	Krypton	36	2	8	18	8			
	Xenon	54	2	8	18	18	8		
	Radon	86	2	8	18	32	18	8	

in the outermost shell are identical except for helium which has a complete shell at 2 electrons. You may want to check this table with the full Periodic Table to see the family correlation of similar electron structure in the highest occupied energy levels.

ISOTOPES

In our consideration of electronic structure we have said nothing about the neutron structure of the nucleus. This is because the proton structure of the nucleus determines the character of a particular element. However, if you checked the listing of atomic weights you would see that the total weight of an atom usually requires the presence of some neutrons in the nucleus along with the protons. For example, the relative mass of a proton is approximately 1, and helium has 2 protons in its nucleus, which would not explain its weight as approximately 4. Since 2 electrons have a negligible weight compared to a proton or neutron, we have to assume the helium nucleus normally contains 2 neutrons along with the 2 protons to give a total relative weight of approximately 4. This has indeed been confirmed.

Scientists have been able to measure the mass or weight of atoms fairly accurately using various methods related to the measurements on positive particles in gas discharge tubes. (See chapter 4, pp. 66–67.) They have found that atoms of the same element may contain different numbers of neutrons and thus have different weights. We know that an atom of a particular element must have only one value for the positive charge on its nucleus, that is, it has only one possibility for the number of protons in its nucleus. However, there is nothing in our theory which requires every atom of a particular element to have the same number of neutrons.

> ISOTOPES ARE ATOMS OF THE SAME ELEMENT WHICH HAVE DIFFERENT MASSES OR WEIGHTS BECAUSE THEY CONTAIN DIFFERENT NUMBERS OF NEUTRONS.

The word *isotope* is derived from the Greek meaning "in the same place," because the isotopes naturally have the same place in the Periodic Table, since this is determined by the atomic number or the basic character of the element. Most elements found in nature have 2 or more isotopes. Extensive investigations of naturally occurring elements have shown that the percentages of the isotopes for a given element are generally constant.

Isotopes of neon were the first ones identified. This was in 1913 in J. J. Thomson's laboratory. Now three naturally occurring isotopes of neon are known. (See Figure 5.2.)

There is a further unrealistic convention in our "drawings" of the atom. Chemists usually draw a solid circle around the proton and neutron symbols (or just the proton number if that is the emphasis). The circular enclosure has of course no basis in reality. The protons and neutrons are not *in* the nucleus. They *are* the nucleus.

It is interesting that nature throws a little variety at us in the form of isotopes. Nature, like us, is not perfect. However, before 1913 scientists

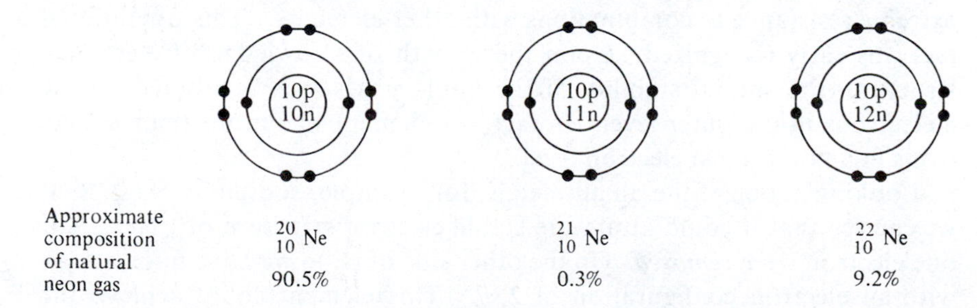

FIGURE 5.2

Diagrams of the three isotopes of neon. The standard shorthand symbolism uses a subscript for atomic number and a prefix superscript for the total number of protons and neutrons in the nucleus. Neon now has an accepted atomic weight of 20.183 which is an average weight of all neon atoms in a sample.

were considerably puzzled by, in some cases, the very great deviations of atomic weights from whole numbers. There had been before this a general acceptance of the theory that all atoms were made up of the same basic parts, essentially protons and neutrons of approximate relative weight of "one" and electrons, with a weight approximately 1/2000 of this. However, there remained a nagging suspicion that the theory was not adequate because the atomic weights did not come as close to whole numbers as they should if atoms were made up of just these basic parts. After isotopes were discovered the major reason for confusion was uncovered.

Three isotopes of hydrogen are known. The usual hydrogen atom has 1 proton in the nucleus. The second atom type has of course the 1 proton also or else it would not be hydrogen. However, it also has a neutron associated with the proton in the nucleus. It is called deuterium or heavy hydrogen (Greek: second). A third very rare isotope contains 2 neutrons along with the proton. It is called tritium (Latin: three or third).

"Heavy water" is water containing the heavy hydrogen isotope, deuterium, combined with oxygen. Some interesting experiments have been carried out using heavy water for living things. Certain kinds of bacteria, algae, and molds have been grown in heavy water. No higher plants or animals so far have been found to survive if fed heavy water. Numerous experiments have been performed using heavy water for animals, mainly in connection with possible use of the deuterium isotope for control of tumor growth. However, when more than about one-third of the hydrogen in the body is replaced by deuterium, the animal, whether it is a dog, mouse, or rat, dies.

CHEMICAL REACTIONS: THE OCTET RULE

You have already seen the recurrence of an outer electron structure of 8 electrons in all the noble gases from neon onwards. As mentioned before, the noble gases are sometimes called inert gases because of their extreme resistance to combinations with other elements. This operational fact was early recognized. Upon the growth of knowledge of electronic structure, the special stability of the noble gases was attributed to the 8 electrons in the outer level. In fact, no element has more than 8 electrons in an outermost electron level.

Looking at one of the alkali metals, for example, sodium, $_{11}Na$ 2, 8, 1, we can see that it could attain the stable electron structure of $_{10}Ne$ 2, 8 if one electron were *removed.* On the other side of neon we have fluorine, $_{9}F$, with an electron configuration of 2, 7. This element could achieve the stable 8 electron outer level by *gaining* one electron. This is indeed exactly what happens if fluorine is allowed to contact sodium. A violent and explosive reaction occurs which indicates the great driving force of both sodium and fluorine to achieve the stable 8-electron configuration in the outer level. A compound, the salt NaF, is formed. (See Figure 5.3.)

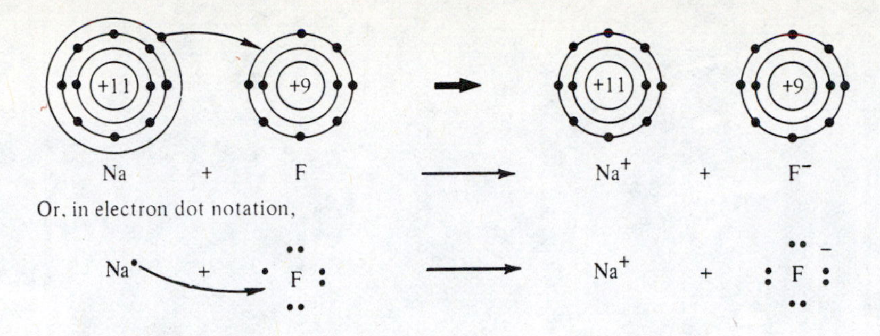

FIGURE 5.3

How the achievement of a stable outer level of 8 electrons helps explain the reaction of sodium and fluorine.

This is then one way that elements achieve the octet or 8-electron outer ring. It illustrates what seems to be a basic driving force of chemical reactions—the tendency to attain the stable electron configuration of 8 electrons in the outermost level.

THE OCTET RULE, OR "RULE OF EIGHT," SAYS THAT ELEMENTS REACT GENERALLY TO ACHIEVE THE STABLE OUTER SHELL STRUCTURE OF EIGHT ELECTRONS.

Elements near helium achieve stability by getting 2 electrons in the outer level, but the idea is essentially the same. The complete or closed outer shell with the very small atoms contains only 2 electrons since that represents the maximum allowable in the first shell.

CHEMICAL REACTIONS: IONIC BONDING

The arrow in Figure 5.3 is a shorthand indication that the two atoms "give" or "yield" the two new species on the right, each of which now has 8 electrons in the outer shell. However, since the sodium atom has *lost* one electron, the unit that results has a charge of $+1$, while the fluorine atom which has *gained* one electron now has a charge of -1. Justify this yourself by counting up the total negative charge against the nuclear positive charge which of course remains constant.

AN ION IS A CHARGED ATOM OR GROUP OF ATOMS.

The word *ion* is from the Greek for "goer," because the ions "go" under the influence of applied electric charge. The nature of the charged particle or ion is indicated by a superscript, for example,

$$Na^{+1}, Cl^{-1}, K^{+1}, I^{-1}, \text{ or just } Na^{+}, Cl^{-}, K^{+}, I^{-}$$

with the "one" assumed.

Initial emphasis on theory and the tiny size of atoms and molecules should not obscure the fact that chemical reactions can be carried out on a rather large scale. Here are shown large reaction vessels and complex pipeline systems in a modern drug manufacturing facility. (Courtesy of Parke, Davis & Company.)

The compound formed in the example above is the salt sodium fluoride, symbolized as NaF. The sodium ion is now bound or "bonded" to the fluoride ion by electrostatic attraction between + and − particles. In solid form, which is the normal condition for NaF, there are many ions of each kind intermingled in a regular ordered array but still with the overall NaF formula.

AN IONIC BOND IS THE UNION OR BOND BETWEEN OPPOSITELY CHARGED IONS. IT RESULTS FROM A TRANSFER OF ELECTRONS BETWEEN ATOMS.

Let us consider the reaction of sodium with chlorine. This should be similar to the reaction with fluorine because chlorine is in the same family. The electrons are arranged a little differently just to emphasize that there is no special significance to where you place the electrons in the atoms.

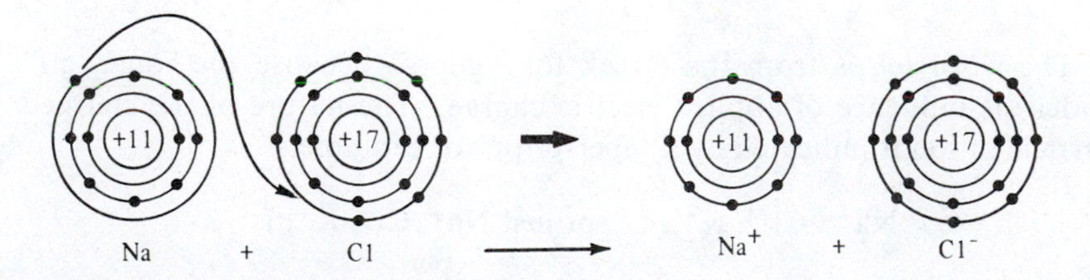

In electron dot notation the reaction is:

$$\mathrm{Na\cdot} + \mathrm{:\ddot{\underset{\cdot\cdot}{Cl}}\cdot} \rightarrow \mathrm{Na^+} + \mathrm{:\ddot{\underset{\cdot\cdot}{Cl}}:^-}$$

AN EXAMPLE OF IONIC BONDING: CRYSTALS OF TABLE SALT

The transfer of electrons in ionic bonding accomplishes a profound change in the materials present. This is unfortunately often missed by the beginning student who may see little difference between the free metal sodium and the positively charged sodium ion present in salt. The difference between these is extreme—it can amount to a life and death situation if you happen to mix them up in practice.

For example, chlorine is a greenish gas at room temperature and was the first poison gas used in World War I. Sodium is a soft silvery metal which if placed in contact with water reacts violently, releasing hydrogen gas and melting the sodium metal. Yet the reaction product of these two is the salt you sprinkle on eggs and meat to give them added flavor. You require Na^+ ions in your blood but if you had Na metal inserted into your veins you would literally explode! Yet the difference is that sodium metal loses one tiny electron. But it acquires quite drastically different properties. The Na^+ ion is, of course, closely related to the metal Na atom. (In writing the formula NaCl, do not miss the fact that the product is an entirely new substance which is neither Na nor Cl.)

When a sufficient number of sodium atoms have given over their electrons to chlorine atoms, a beautiful structure is formed. Then we have the positive nuclei hidden deeply in each ion (+11 for Na and +17 for Cl) but shielded from each other by clouds of negative electrons which are held in place by the positive pulls of the nuclei. The overall effect is to leave

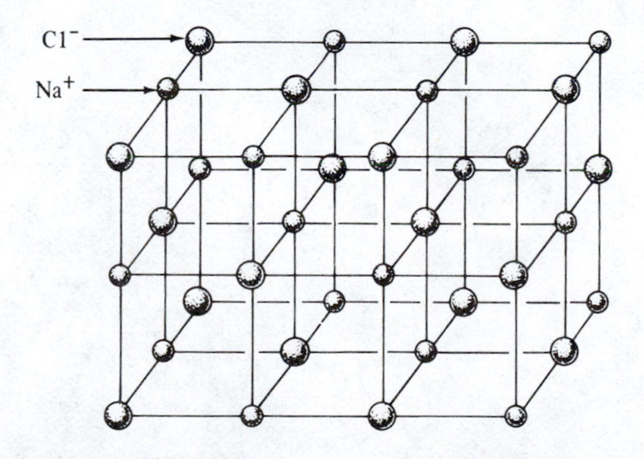

FIGURE 5.4

A schematic ball-and-stick model of crystalline NaCl. This shows the nicely coordinated network of ions which makes the material a hard, rigid, solid of fairly high melting point. You can see why we do not locate molecules of NaCl in this interconnected solid. This is true of all ionic compounds.

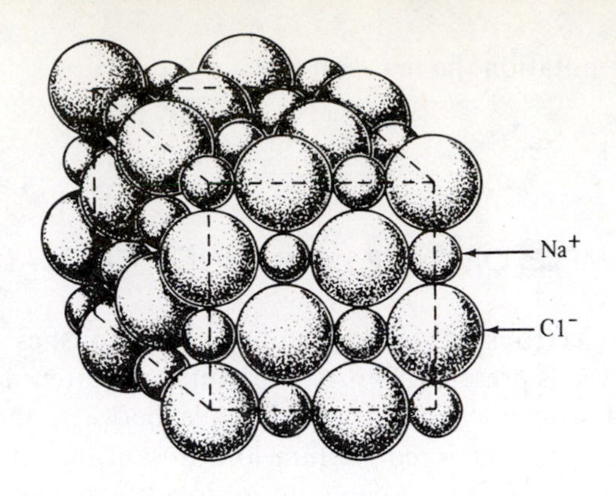

FIGURE 5.5
A more realistic model of NaCl crystal showing relative spaces occupied by the ions and the general arrangement into cubic crystals. Each Na^+ ion is surrounded by 6 Cl^- ions and each Cl^- ion is surrounded by 6 Na^+ ions, making a strong interlocking network. The overall formula is thus NaCl even though it is not possible to isolate in the solid a particular NaCl molecule.

the Na^+ ion with a deficiency of the buzzing electrons and the Cl^- ion with an excess. This arrangement is so uniform and coordinated that the thousands upon thousands of units then mutually attract each other and keep the whole buzzing confusion together in a beautiful and orderly

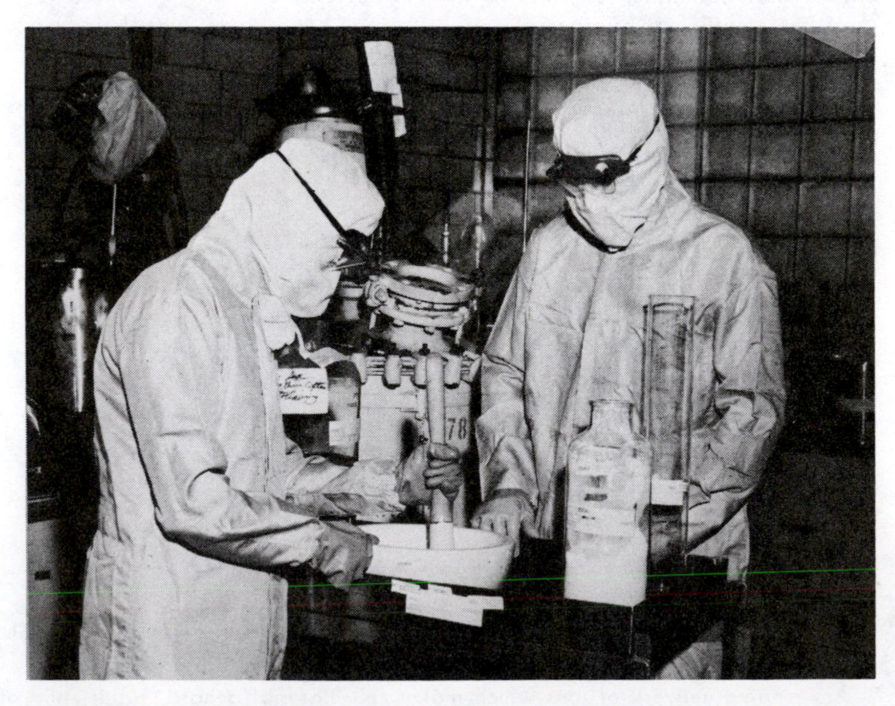

Some chemicals have to be handled in special ways. Here a drug is being compounded under sterile conditions. (Courtesy of Parke, Davis & Company.)

crystalline array of ions. This is table salt and is simply the result of chemical bonding. Figures 5.4 and 5.5 show two models of the result.

MORE EXAMPLES OF CHEMICAL REACTIONS

The electron similarity of all the elements in the halogen family (F, Cl, Br, I, At)—all with 7 electrons in the outer shell—would indicate similar reactions with sodium, and you have just seen two examples above. In experiments we do find that the halogens all form salts like NaCl—NaF, NaBr, NaI, and NaAt. Also, the other members of the alkali metal family react similarly to sodium since they all have the lone electron which they tend to lose easily in their drive toward stable external electron structures.

We say that atoms like fluorine and chlorine are "electron hungry" or have high affinity for electrons, while atoms like sodium are more likely to lose an electron easily or have little affinity for more electrons. It is much easier for sodium to achieve an outside octet by losing one electron. Think of the difficulty that would be involved by bringing in 7 electrons to complete the third level in sodium, when like charges repel each other.

Here are a few more examples showing the similarities of behavior:

(+19) 2e 8e 8e 1e (+9) 2e 7e → (+19) 2e 8e 8e^{+1} (+9) 2e 8e^{-1}

K atom + F atom → K$^+$ ion + F$^-$ ion
(or the salt KF)

(+3) 2e 1e + (+9) 2e 7e → (+3) 2e^{+1} + (+9) 2e 8e^{-1}

Li atom + F atom → Li^{+1} ion + F^{-1} ion
(or the salt LiF)

(+19) 2 8 8 1 + (+53) 2 8 18 18 7 → (+19) 2 8 8^{+1} + (+53) 2 8 18 18 8^{-1}

K atom + I atom → K$^+$ ion + I$^-$ ion
(or the salt KI which is used as an additive to table salt to prevent goiter)

You will notice in all the examples above that the elements form compounds in the ratio of one atom to one atom. In order to get an idea of how compounds form with different ratios than 1 : 1, we will consider some of the alkaline earths, the family next to the alkali metals, and designated Group IIA in the Periodic Table.

One of the most common members of this family is magnesium. When magnesium metal is brought in contact with fluorine a rather energetic reaction occurs, indicating some "eagerness" on the part of these elements to get together. Analysis shows that the white product formed is the salt

MgF_2. Why are there 2 atoms of fluorine here when all along above we have been writing NaF, KF, and so on, with only 1 atom of F to 1 of the metal? The electron structure gives an easy explanation:

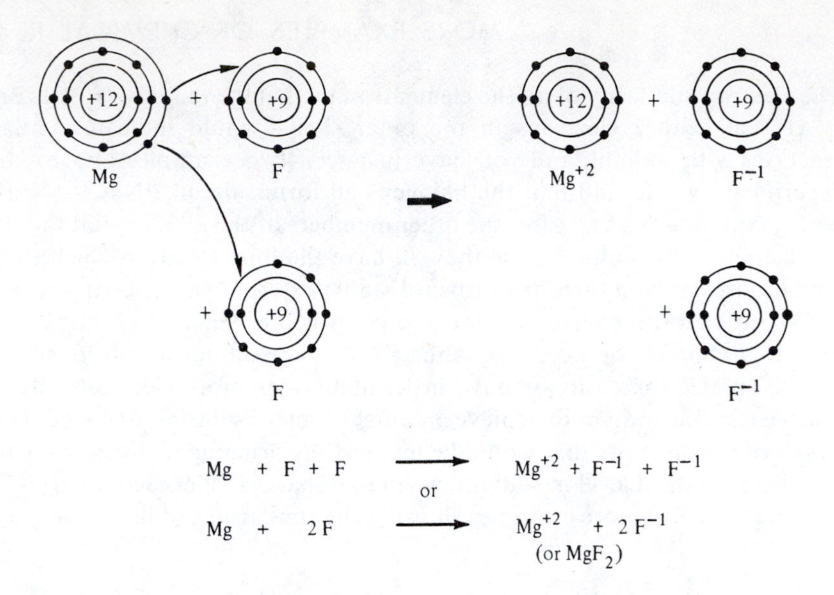

Follow this reaction carefully. Notice that a fluorine atom can take only 1 electron to attain the stable outer structure of 8. But magnesium has to lose 2 electrons to attain the stable 8. Therefore, Mg gives 1 electron to *each of 2* F atoms making a stable octet in Mg and also in each of the two F atoms. Nature is still following the same rules. It would be contrary to the rules to try to force 2 electrons into 1 atom of fluorine.

Let us look at another example showing that other members of the same families behave similarly:

$$(+20)\ 2\ 8\ 8\ 2 \ + \ (+17)\ 2\ 8\ 7 \ \rightarrow \ (+20)\ 2\ 8\ 8^{+2} \ + \ (+17)\ 2\ 8\ 8^{-1}$$

$$+ \ (+17)\ 2\ 8\ 7 \qquad\qquad + \ (+17)\ 2\ 8\ 8^{-1}$$

$$Ca \ + \ Cl + Cl \rightarrow Ca^{+2} \text{ ion} \ + \ Cl^{-1} \text{ ion} + Cl^{-1} \text{ ion}$$

or

$$Ca \ + \ 2Cl \rightarrow Ca^{+2} \ + \ 2Cl^{-1}$$

or

$$Ca \ + \ 2Cl \rightarrow CaCl_2$$

Again notice that all atoms have achieved the stable external shell of 8 electrons. Here is the electron dot notation for this reaction:

$$\text{Ca}: + \cdot\ddot{\underset{\cdot\cdot}{\text{Cl}}}: + \cdot\ddot{\underset{\cdot\cdot}{\text{Cl}}}: \rightarrow \text{Ca}^{+2} + :\ddot{\underset{\cdot\cdot}{\text{Cl}}}:^{-1} + :\ddot{\underset{\cdot\cdot}{\text{Cl}}}:^{-1}$$

(or $CaCl_2$)

or

$$\text{Ca}: + 2\ \cdot\ddot{\underset{\cdot\cdot}{\text{Cl}}}: \rightarrow \text{Ca}^{+2} + 2\ :\ddot{\underset{\cdot\cdot}{\text{Cl}}}:^{-1}$$

Note that we indicate 2 atoms or ions of a particular element by a 2 *in front of* the symbol. When incorporated in a compound we indicate the number combined by a subscript *following* the symbol, for example, $CaCl_2$. This is like saying that for a motorcycle assembly we need to take one chassis and two wheels but when assembled we have only one motorcycle, with the two wheels tied in to the unit. Again in the case of ionic materials there is a large array of regularly arranged units, the shape and arrangement depending on the particular ions involved.

Some other examples of chemical reaction through formation of ionic bonds are given below:

Formation of LiBr, lithium bromide:

$$\text{Li}\cdot + \cdot\ddot{\underset{\cdot\cdot}{\text{Br}}}: \rightarrow \text{Li}^{+1} + :\ddot{\underset{\cdot\cdot}{\text{Br}}}:^{-1}$$

Li atom + Bromine atom → Li ion + Bromide ion

(or salt LiBr)

Formation of MgO, magnesium oxide:

$$\text{Mg}: + \cdot\ddot{\underset{\cdot}{\text{O}}}: \rightarrow \text{Mg}^{+2} + :\ddot{\underset{\cdot\cdot}{\text{O}}}:^{-2}$$

Mg atom + O atom → Mg ion + Oxide ion

(or MgO, the oxide of Mg)

Note the switch to a 1 : 1 ratio in the compound MgO, explained by the fact that a Mg atom needs to lose 2 electrons to attain the stable octet of neon and an oxygen atom has to gain 2 electrons to attain, in this case, the exact same stable electron structure. Check the table of electron shells (Figure 5.1) to assure yourself that this is true.

Here is one more tricky example. What happens when aluminum reacts with oxygen? Here the aluminum atom has 3 electrons in excess of the octet, and oxygen atoms each have 6 external electrons so that they need only 2 electrons to complete the outer shell. Can you guess what happens here? Think about it for a little while before looking at the reaction. Perhaps you have figured out that some kind of arrangement has to be made to satisfy all the atoms involved. That is correct and here is the way the arrangement is made:

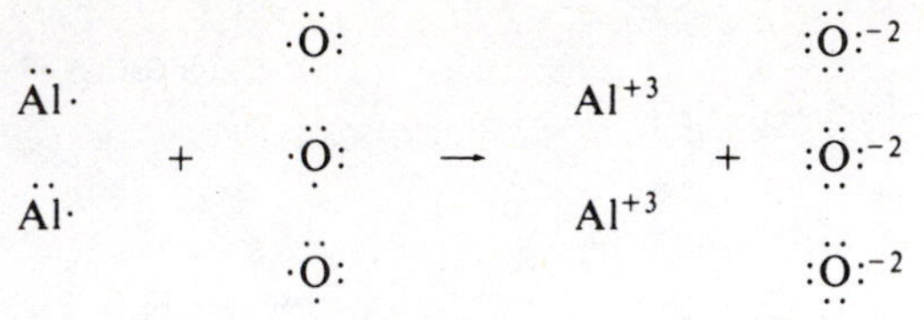

$$\begin{array}{ccccccc} & & \cdot\ddot{\underset{\cdot}{O}}: & & & & :\ddot{\underset{\cdot\cdot}{O}}:^{-2} \\ \ddot{Al}\cdot & & & & Al^{+3} & & \\ & + & \cdot\ddot{\underset{\cdot}{O}}: & \longrightarrow & & + & :\ddot{\underset{\cdot\cdot}{O}}:^{-2} \\ \ddot{Al}\cdot & & & & Al^{+3} & & \\ & & \cdot\ddot{\underset{\cdot}{O}}: & & & & :\ddot{\underset{\cdot\cdot}{O}}:^{-2} \end{array}$$

2 Al atoms + 3 O atoms $\longrightarrow$ 2 Al ions + 3 O ions

(or the oxide, Al_2O_3)

The stable octet has been emphasized in the reactions above because it conveniently and simply explains so much in the way of chemical behavior. Some exceptional compounds do form without the stable octet in the outer electron level but these are unusual cases that we will not consider in this book.

VALENCE AND DISTINCTIONS BETWEEN METALS AND NONMETALS

We noted earlier that the electrons in the external shell of an atom are called the valence electrons because they set a value on a particular element's combining capacity. From the examples we have now considered you will see that a particular element generally shows a consistency in its combining capacity.

VALENCE IS THE COMBINING CAPACITY OF AN ELEMENT.

One atom of sodium always combines with 1 atom of chlorine to form NaCl. You have now seen why—the electron arrangements require a 1 : 1 ratio. Magnesium always combines with 2 atoms of chlorine to form $MgCl_2$, magnesium chloride. In other words, Mg forms two bonds, one with each chlorine atom. It has a value of 2, usually expressed as an ionic valence of +2, because it forms an ion with a charge of +2. The valence can also be thought of as the number of atoms of an element with a valence of 1, like Na or Cl, which will combine with 1 atom of the element under question. Thus aluminum forms $AlCl_3$ which means the ionic valence of Al is +3, or, more casually, it has a valence of 3. Chlorine here has an ionic valence of −1. (See the Periodic Table on page 62.)

You will notice something quite regular about how the various elements assume their valences. Those on the left side of the Periodic Table are of the type which have 1 or 2 or 3 electrons in the outer electron level and they therefore *lose* these to expose a completed outer shell which lies just inside the valence shell electrons. On the right side of the table, because of the natural buildup of electrons in the outer levels, the elements are closer in the outer shell to attaining a full level, or octet. Here they have 5 or 6 or 7 electrons in their outer level and therefore tend to *add* electrons to achieve the completed octet.

Look again at the Periodic Table and note this general arrangement. You will notice a stairstep line toward the right side of the table. This diagonal, zigzag line separates the metals on the left of it from the nonmetals on the right. The real distinction between metals and nonmetals is in the electron structure. Since this structure changes gradually as we move in the buildup from left to right across the table, there is no abrupt division between metal and nonmetal. Indeed the elements on the border line are called *metalloids*, because they show both metallic and nonmetallic character. Examples of metalloids, or metal-like elements, are germanium, Ge; arsenic, As; and silicon, Si. These are used in solid state electronic systems because of their special electronic character of being somewhat like both metals and nonmetals.

Notice a large group of metals in the central block of the Periodic Table. These ten groups numbered from IIIB to IIB are called *transition elements*. In the buildup of electrons they are involved in filling up inner electron levels. These tend to form positive ions also but some have more than one valence state due to loss of electrons in stages from outer levels and then from inner levels. We will not need to consider electron structures of these in any detail in this book.

CHARACTERISTICS OF METALS

The general characteristics considered to be metallic properties, like shininess and good conduction of heat and electricity, are caused by the easy availability of electrons in the thinly populated outer levels of metals. These loosely held valence electrons are thus responsible for the attainment of + valence states and also for high conduction of both heat and electricity.

Applying voltage to a metallic wire causes electrons to flow easily from one end of the wire to the other. The battery or other source of electricity simply pumps electrons in at one end of the wire and out at the other. The electrons in the metal of the wire are then easily displaced and push other electrons from neighboring areas. There is a chainlike movement of electrons very similar to the "bucket brigade" sometimes used to fight fires in remote areas away from water supply pipes. Heat conduction involves a similar mechanism with the kinetic energy of motion transferred through the mobile electrons.

The model of a metal our theory provides us thus involves the concept of loose valence electrons under the influence of an array of positive kernels in such a way that the electrons are said to be somewhat delocalized and are free to move about in the metal crystal. This is in contrast with nonmetals like sulfur which has tightly held electrons and prefers to get more electrons to achieve the octet. The metal atoms, when

Aluminum-finned tube for air conditioning heat exchanger. The easy availability of electrons in aluminum metal provides the mechanism for good conduction of heat. (Courtesy of Reynolds Metals Company.)

Steel-supported aluminum conductor wire being installed on a transmission line. Use of aluminum, which is a very abundant element in the earth's crust, allows conservation of the more valuable copper for other purposes (chapter 12). (Courtesy of Reynolds Metals Company.)

The easy bendability of metals is emphasized in this photo. Aluminum auto bumpers are formed from the straight aluminum extruded bar shown on the left. (Courtesy of Reynolds Metals Company.)

they get together, seem to be able to form easily a sort of commune where every atom achieves an octet by letting the outer electrons loose to occupy the overall volume of the metal. This also accounts for the metallic luster or shiny surfaces of metals. The valence electrons are again responsible. They are loose in the outer high energy orbitals with close proximity to other slightly higher energy orbitals. When light quanta strike them, the electrons are forced to jump up to the nearby levels and then they fall back again, giving out light energy as quanta of visible light which gives the metallic luster. Here is an excellent example of how conceptual understanding gives us a good explanation of the operational behavior and characteristics of metals.

Other operational properties of metals, such as bending easily rather than breaking, malleability (can be hammered into sheets), and ductility (can be drawn into wires), are also explainable by the conceptual model just described. The positively charged ions, which have donated their electrons to the community pool, are easily forced to slide over each other in the fluid, delocalized electron stream. Indeed, the alkali metals are relatively soft and can be cut with a knife. Most of the metals you come into contact with, like iron and chromium, have more complex arrangements or impurities to hold the atoms together a little more tightly. But they all are relatively bendable compared to, say, sulfur. You can see that structure and shape are important determinants of how materials act.

You may have wondered whether all atoms are held together by ionic bonds, and if you think about a few examples you will realize that there must be some other type of bonding. In fact, most of the material world, especially the living organic materials of animals and plants, is formed with a different type of bonding.

Consider the hydrogen atom, which has a single electron in its outer shell. Hydrogen's first shell can hold 2 electrons. In other words, 2 electrons would make for a complete outer shell. And we find that hydrogen as it exists in nature is H_2, that is, 2 atoms in a molecule—indicating that hydrogen has completed the outer shell by a neat trick of having two hydrogen atoms come together. Then each of the atoms contributes its electron to make a pair, which they then both *share* between them. You could not say one hydrogen atom transfers its electron to the other because there is no reason why one should have more pulling power than the other. Consequently, we say a bond is formed in the hydrogen molecule by a sharing of electrons.

A COVALENT BOND IS A UNION OR BOND FORMED BY THE SHARING OF A PAIR OF ELECTRONS BETWEEN TWO ATOMS.

What we say happens is that the two hydrogen atoms come close enough together for the positive nuclei to pull on both electrons and provide a molecule with a pair of electrons occupying the volume mostly between the two atoms, but also having some lesser probability of being on other sides of the two nuclei. We can represent it by the electron dot notation:

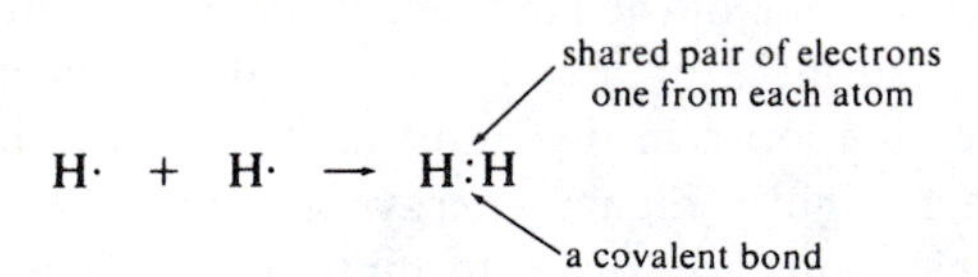

Sometimes a dashed line is used to represent the pair of shared electrons; thus H—H is the same as H : H. A more realistic but still far from "true" picture would show electron clouds "overlapping," as in the following:

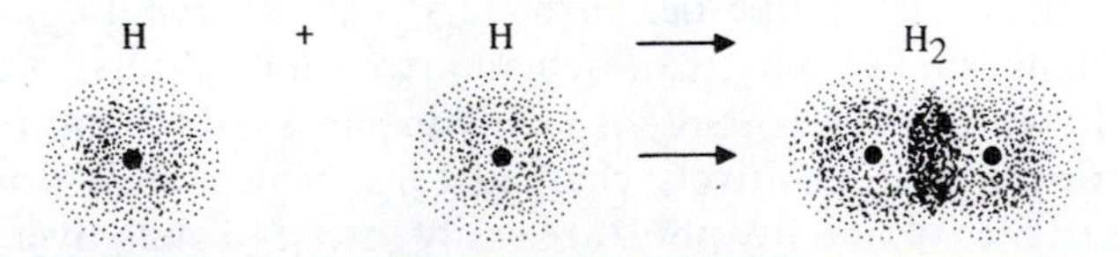

What about chlorine? It too is found in nature as a diatomic or two-atom molecule, Cl_2. A similar situation occurs:

:C̤̈l· + :C̤̈l· → :C̤̈l:C̤̈l:

or Cl—Cl or Cl_2

Thus two chlorine atoms join by sharing a pair of electrons. Each atom then can be considered to have a completed outer shell of 8 electrons. You can count the electrons in each shell above and find the 8, with the shared pair counting in each atom's outer level.

Other examples of covalent or shared-pair bonding are the following:

Bromine:

:Br· + ·Br: → :Br:Br:

or Br—Br or Br_2

Nitrogen:

·N· ·N· → :N⋮⋮N:

or N≡N or N_2

Oxygen, O_2, the gas we breathe and which is present in air to the extent of approximately 21 percent, can be shown as follows, although there are indications that the bonding is not quite as simple as this.

Oxygen in the air:

:O::O:

or O=O or O_2

POLAR MOLECULES: UNEQUAL SHARING

All the examples of covalent bonds given so far involve sharing between two like atoms. There can be no question as to whether each atom shares equally because the atoms being the same, there has to be equal sharing.

When two *different* atoms share electrons there can be an *uneven* pull by the atoms on the pair, giving one atom a larger share than the other and making it a little more negative because of the uneven distribution of charge. For example, when hydrogen and chlorine react to share a pair of electrons, we get the acid called hydrogen chloride or hydrochloric acid, HCl. In the covalent bond between the atoms, the chlorine atom pulls the pair closer and gives the molecule an overall unbalance.

H· ·Cl: H· → ·Cl: H :Cl:

Shift of pair toward chlorine atom, giving a "polar" bond

A POLAR BOND RESULTS WHERE THERE IS UNEQUAL SHARING OF ELECTRONS BETWEEN ATOMS SO THAT THE NEGATIVE CHARGE CLOUD IS SHIFTED TOWARD THE ATOM WITH GREATER ATTRACTION FOR ELECTRONS. THIS ATOM HAS AN EXCESS OF NEGATIVE CHARGE, AND THE OTHER ATOM AN EXCESS OF POSITIVE CHARGE.

This can result in a polar molecule.

A POLAR MOLECULE HAS NEGATIVE CHARGE AND POSITIVE CHARGE NOT SYMMETRICALLY DISTRIBUTED SO THAT ONE END IS SLIGHTLY POSITIVE AND THE OTHER END SLIGHTLY NEGATIVE.

Polarity thus refers to a bond or molecule where there is charge separation. This situation is very common. In fact, a good part of your body chemistry, as well as the special value of water to living things, depends on such polarity in molecules.

You might logically ask: How can you tell which atom pulls the electron more? The answer has been worked out in a listing of what is called *electronegativity* values. (See Table 5.2.) These are arbitrary, relative numbers running from the most electronegative element, fluorine, which is given the number 4.0, to the least electronegative, the alkali metals francium and cesium, which have the value 0.7.

ELECTRONEGATIVITY IS THE RELATIVE PULLING POWER OF AN ATOM ON A SHARED ELECTRON PAIR.

The measurements to provide these numbers are rather sophisticated and still being refined. The results however are very helpful in deciding how the electron pairs will be distributed.

You can note now that the nonmetals have high values of electronegativity and the metals low values, which is consistent with our previous findings that nonmetals tend to pull in electrons while metals tend to let them go. This means also that a bond between a metallic element and a nonmetallic element would result in the electron pair tending very much toward the nonmetal. In the ionic bond the pull is so great that we say the pair goes to the nonmetal and we get positive and negative ions (e.g., Na^+, Cl^-). This is the extreme case of polarity. However, most bonds are formed between atoms not quite so far apart in electronegativity. Examples are the bonds in HCl and also the bonds between hydrogen and oxygen in H_2O. In these cases we do not get a complete transfer of electrons but just a sort of "hogging" of the pair by the most electronegative element. This is what gives a slight negative end on the molecule (with a corresponding equal slight positive end opposite).

The two major kinds of bonds (ionic, by transfer of electrons, and covalent, by sharing of electrons) are the extreme or "pure" types. Most chemical bonds are neither completely ionic nor completely covalent, but something in between. That is, most covalent bonds are similar to those in HCl where the shared electrons are really spending more time closer to one atom than the other. There are no separate ions formed, but just an unequal distribution of charge within the molecule.

Thus hydrogen chloride, more commonly called hydrochloric acid, which is the acid in your stomach, is not

$$H:\ddot{\underset{..}{Cl}}:$$

TABLE 5.2
Electronegativity Values

H 2.1																	He —
Li 1.0	Be 1.5											B 2.0	C 2.5	N 3.0	O 3.5	F 4.0	Ne —
Na 0.9	Mg 1.2											Al 1.5	Si 1.8	P 2.1	S 2.5	Cl 3.0	Ar —
K 0.8	Ca 1.0	Sc 1.3	Ti 1.4	V 1.6	Cr 1.6	Mn 1.5	Fe 1.8	Co 1.8	Ni 1.8	Cu 1.9	Zn 1.6	Ga 1.6	Ge 1.8	As 2.0	Se 2.4	Br 2.8	Kr —
Rb 0.8	Sr 1.0	Y 1.2	Zr 1.4	Nb 1.6	Mo 1.8	Tc 1.9	Ru 2.2	Rh 2.2	Pd 2.2	Ag 1.9	Cd 1.7	In 1.1	Sn 1.8	Sb 1.9	Te 2.1	I 2.5	Xe —
Cs 0.7	Ba 0.9	La 1.1–1.2	Hf 1.3	Ta 1.5	W 1.7	Re 1.9	Os 2.2	Ir 2.2	Pt 2.2	Au 2.4	Hg 1.9	Tl 1.8	Pb 1.8	Bi 1.9	Po 2.0	At 2.2	Rn —
Fr 0.7	Ra 0.9																

and is not

$$H^{+1} \quad :\ddot{\underset{..}{Cl}}:^{-1}$$

but something in between,

$$H \quad :\ddot{\underset{..}{Cl}}:$$

At this point we have established basically why elements combine and why bonds are formed between atoms, holding them together. In the next chapter we will extend this background on the formation of compounds and consider some of the short cuts to writing and interpreting formulas of compounds, with special reference to inorganic compounds. The term *inorganic* means "not living." Materials like rocks, salts, minerals, and water are inorganic substances. In later chapters we will see more of the organic compounds, which are the special types made up chiefly of covalent bonds and found in all living organisms.

An example of completely inorganic materials. This is a view of a rocky crater near Stone Mountain on the moon, taken by U.S. Apollo astronauts. As predicted by scientists, the same elements we know on earth were found to make up the compounds on the moon. (NASA photo.)

Experiments

1. The early workers in atomic structure, like Bohr and Einstein, often used what are called "thought experiments" in order to consider the possibilities of various ways of looking at things. These thought experiments could actually not be carried out at all in the physical realm but the very thinking about them and considering the possibilities often clarified concepts and forced the acceptance of one postulate and the discarding of another.
   Try a thought experiment on the circling electron cloud. Think of the possibility of either of two situations. Either the electrons, in circling the nucleus and when in contact with electrons of neighboring atoms, are likely to lose energy (like a ping-pong ball does when you drop it on the floor) or else they are contacting their neighbors without losing any energy at all. Take each of these possibilities and think of the consequences. How could the electrons contact each other? Do you think they could rub off even a slight fraction in the contact? If no physical effect is caused, then how does one substance "contact" another? Think of other possibilities like these. Which of the possibilities do you favor? What would be the consequences for matter if the alternate were the case?
2. This is another chance to do a thought experiment. Think about the sodium atom and its change when it becomes an ion. Now ask yourself, does the formation of the ion produce a larger or a smaller particle? Visualize the situation in terms of electrons, proton structure in the nucleus, and so on. Think of each possibility: either the ion is smaller or it is larger or it is the same size as the atom. What would your thinking indicate according to the mechanism of formation of the ion? Consider each case and come to a conclusion. Only one of them is a "true" possibility. Justify your thinking by explaining the operation in terms of cause and effect.

Exercises

3. The word *curious* means "eager to know" and is derived from the Latin for "full of care, inquisitive." It has some undesirable implications, as when you have a curious neighbor. Discounting this implication, do you think adults should be curious? In what way? Is adult curiosity different from that of a child? If so, how? If not, why not? The origin of the word *wonder* is not known, but to wonder relates to feeling surprise or astonishment or to feeling curious. Do you think that science will eventually eliminate wonder? Think and give a few examples and reasons for your answer.
4. Now that you have more knowledge of atomic structure you can provide a more refined conceptual definition of several items we have considered so far. For the following list, make up two columns and in one put an operational description or definition and in the second column give a conceptual description as you now visualize it.
   (a) Atomic number
   (b) Element
   (c) Families of chemical elements
   (d) Isotopes
   (e) Chemical reaction

(f) Metals and nonmetals
(g) Table salt

5. Use the Periodic Table to determine
   (a) How many protons there are in an oxygen atom;
   (b) How many electrons are in a neutral oxygen atom;
   (c) How many electrons are in the oxygen ion;
   (d) How many protons there are in a potassium atom.
6. Draw an "onion skin" or shell model of any atom in what is called the normal or ground state. Then show how you could picture it if it was bombarded by particles or heated so that the electrons were excited. What would happen if it now were allowed to return to the original condition (which atoms do automatically, since they want to be at the low-energy ground state level)?
7. Work out partial circle designations for the following elements:

   (a) O (d) Ne (g) Na (j) Ar
   (b) N (e) B (h) K
   (c) Cl (f) C (i) F

8. Give the electron dot notation for each of the following elements by consulting only the Periodic Table:

   (a) H (d) Ca (g) Ar (j) I
   (b) O (e) Na (h) Ne
   (c) N (f) K (i) Cl

9. Indicate the number of protons, neutrons, and electrons in neutral atoms of each of the following isotopes of chlorine:

$$^{35}_{17}Cl \qquad ^{37}_{17}Cl$$

SUGGESTED READING

1. Asimov, Isaac, *A Short History of Chemistry,* Anchor Books, Doubleday & Co., Inc., Garden City, New York, 1965. Paperback.
2. Young, Louise B., ed., *The Mystery of Matter,* Oxford University Press, New York, 1965, Part 5, "Is the Universe Asymmetric?," "Symmetry in Nature."
3. Weisskopf, V. F., *Knowledge and Wonder,* Anchor Books, Doubleday & Co., Inc., Garden City, New York, 1966. Paperback.

Father calls me William, sister calls me Will,
Mother calls me Willie, but the fellers call me Bill!

—EUGENE FIELD

What's in a name? that which we call a rose
By any other name would smell as sweet.

—SHAKESPEARE

He must often change who would be constant in happiness or wisdom.

—CONFUCIUS

chapter 6

Compounds: Using Names, Formulas, and Equations

Although formulas and equations are not a major concern in this book, they have a definite utility. For example, you hear about "phosphates in detergents" and "iron in your blood." You certainly don't think that there is steel wire running up and down your veins. To really understand these commonly used terms, you will have to become acquainted with how the element phosphorus links up with oxygen to form a material containing phosphorus and oxygen atoms combined in a special way. This is called a phosphate, which we will consider shortly. And you will have to find out how iron loses electrons to form Fe^{+2} and Fe^{+3} ions, which in turn form neutral compounds with other negatively charged ions. The iron compounds are the carriers of the ions and are present in foods like raisins and lettuce. (They cause lettuce to turn brown when exposed to the oxygen in the air.)

Now you know that there are only a little over one hundred different elements in the whole universe and many of these are scarce. Most of your environment is restricted to a rather limited number of common elements. But these form some combinations consistently. A formula is a shorthand way of showing the combination of elements. So when you learn some of these common forms of the elements, you will look at new formulas with an understanding of what the shorthand is all about. After all, phosphates have been around much longer than man, and the particular combination of phosphorus and oxygen in these compounds will always be the same. Even if we get all the phosphate out of our detergents, it will still be in your teeth and bones in the basic form that has been the same since the world began.

Which of the elements in the Periodic Table could we expect to combine with each other? And, if they combine, what kind of bond would form between them? And how many atoms are involved? Questions of this kind can now be at least partially answered. The combinations must depend on the ability of the atoms to arrange external electrons into stable structures, essentially the stable octet. You can determine, by a careful consideration of the Periodic Table electronegativity values, that (1) an ionic compound will likely be formed between a metal and a nonmetal; (2) a covalent compound will likely form between two nonmetals; and (3) two metals will likely form a mixture, without much of a compound possibility.

FORMULAS: IONIC VALENCE

We can now begin to see how formulas are set up. You realize from the last chapter that NaCl is the only logical formula for table salt and why it is not $NaCl_2$ or Na_2Cl. The electronic outer structure of a single electron for sodium gives it a valence of +1 while a 7-electron outer level for the chlorine atom causes it to accept only 1 electron in assuming the stable octet, giving it a valence of −1. Thus we have a balance between the two valences, or combining capacities. In Na^+Cl^- the total positive charge is equal to the total negative charge, giving a neutral unit, NaCl.

In the case of $MgCl_2$ we already have seen how the 2 electrons in the outside level of a magnesium atom require 2 chlorine atoms to provide the happy combination all around of stable octets. Here again Mg has a valence of +2 and the balance is obtained in $Mg^{+2}(Cl^{-1})_2$, giving a +2 and −2 for neutrality in the compound $MgCl_2$.

Considerations of valence leads to an automatic and easy way to write formulas. If we have a knowledge of ionic valence we can write the formula for an ionic compound by merely providing the balance through a "crossing over" of valences. The system is outlined below:

1. Write the elements with valence as superscripts, for example,

$$Mg^{+2} \quad Cl^{-1}$$

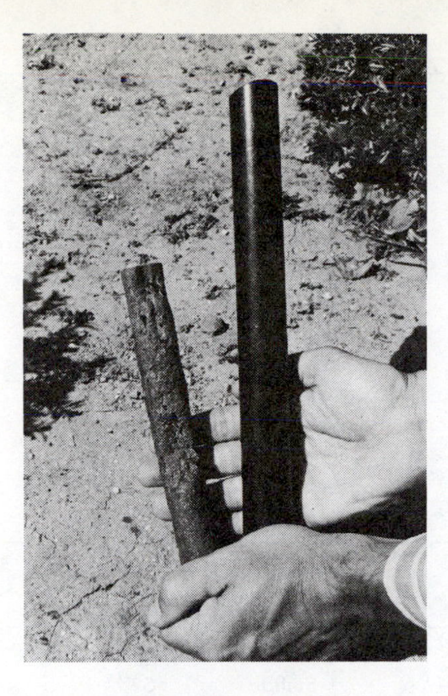

Rusted iron water pipe compared with plastic pipe. The iron metal forms the compound Fe_2O_3 by reaction with oxygen in the air. Plastic piping does not rust. (Photo by Gary R. Smoot.)

2. Cross over the number to the opposite atom as a subscript:

$$Mg_1^{+2} \times Cl_2^{-1}$$

3. This indicates the formula, which you rewrite using subscripts only:

$$MgCl_2$$

Note that the "1" is taken for granted by simply writing the symbol for Mg. In this compound the Mg has a combining capacity of 2 and it has therefore joined to it 2 atoms of Cl, each of which has a combining capacity of only 1.

Take another example:

$$Al_1^{+3} \times Cl_3^{-1} \qquad \text{which is } AlCl_3$$

The reasoning is the same as before. Aluminum, having 3 electrons, transfers these to 3 chlorine atoms.

In order to simplify the system of formula writing, chemists have developed a table of ionic valences, giving the values for common elements and groups of elements which act as units in chemical compounds. This is sometimes called an *ionic valence chart*. However, when an element like

TABLE 6.1
Common Ionic Valences or Oxidation Numbers*

+1		+2		+3	
Hydrogen	H^{+1}	Magnesium	Mg^{+2}	Aluminum	Al^{+3}
Lithium	Li^{+1}	Calcium	Ca^{+2}	Boron	B^{+3}
Sodium	Na^{+1}	Barium	Ba^{+2}	Iron(III)	Fe^{+3}
Potassium	K^{+1}	Iron(II)	Fe^{+2}	(ferric)	
Silver	Ag^{+1}	(ferrous)			
Copper(I)	Cu^{+1}	Zinc	Zn^{+2}		
(cuprous)		Copper(II)	Cu^{+2}		
Mercury(I)	Hg^{+1}	(cupric)			
(mercurous)		Mercury(II)	Hg^{+2}		
Ammonium	NH_4^{+1}	(mercuric)		+4	
		Tin(II)	Sn^{+2}	Tin(IV)	Sn^{+4}
		(stannous)		(stannic)	

−1		−2		−3	
Fluoride	F^{-1}	Oxide	O^{-2}	Phosphate	PO_4^{-3}
Chloride	Cl^{-1}	Sulfide	S^{-2}		
Bromide	Br^{-1}	Sulfate	SO_4^{-2}		
Iodide	I^{-1}	Sulfite	SO_3^{-2}		
Hydroxide	OH^{-1}	Carbonate	CO_3^{-2}		
Cyanide	CN^{-1}	Silicate	SiO_3^{-2}		
Nitrate	NO_3^{-1}	Hydrogen phosphate	HPO_4^{-2}		
Nitrite	NO_2^{-1}				
Bicarbonate (hydrogen carbonate)	HCO_3^{-1}				
Bisulfate (hydrogen sulfate)	HSO_4^{-1}				
Dihydrogen phosphate	$H_2PO_4^{-1}$				

*These represent some of the most important ions found in the environment.

iron rusts in the air, it is really reacting with O_2 and is thus "oxidized" to form the compound designated Fe_2O_3, with iron now at a +3 valence state. Consequently the table is sometimes also called an *oxidation state table* or *table of oxidation numbers*. For our purposes, it is essentially a convenient listing of ionic valences to enable easy and quick writing of a chemical formula. Table 6.1 is a list of commonly encountered and useful ionic valences. It is the basis for easily writing formulas for ionic compounds, using the method of "crossing over" outlined above.

Let us consider a few examples of how the table is used.

1. What could be the predicted formula for barium chloride, a compound of barium and chlorine? From the table barium has a valence of +2 and chlorine −1. Then, setting up the ions, with the positive one first,

$$Ba^{+2} \quad Cl^{-1}$$

we next cross over the valence numbers, giving

$$Ba_1^{+2} \times Cl_2^{-1} \qquad \text{or } BaCl_2$$

2. What is the formula for silver chloride, a compound of silver and chlorine?

$$Ag^{+1} \times Cl^{-1} \qquad \text{which gives AgCl}$$

3. What is the formula for a compound of silver and oxygen, the oxide of Ag?

$$Ag_2^{+1} \times O_1^{-2} \qquad \text{which gives } Ag_2O$$

4. What is the formula for aluminum compound with bromine?

$$Al_1^{+3} \times Br_3^{-1} \qquad \text{which equals } AlBr_3$$

5. What is the formula for the oxide of aluminum, Al combined with O?

$$Al_2^{+3} \times O_3^{-2} \qquad \text{which gives } Al_2O_3$$

6. What is the formula for sodium bicarbonate?

$$Na_1^{+1} \times HCO_3{}^{-1}{}_1 \qquad \text{which is } NaHCO_3$$

7. What is the formula for calcium bicarbonate?

$$Ca_1^{+2} \times HCO_3{}^{-1}{}_2 \qquad \text{which is written } Ca(HCO_3)_2$$

These last two require a little further explanation. You will note in the table a few combination units made up of two or more elements. The group called *bicarbonate* is one of these, and an important ion which you may have heard of before. These combination ions represent commonly occurring stable ionic combinations of various elements in a group which acts like a unit in many compounds with other ions. You have surely heard of someone "taking a bicarbonate" for upset stomach. This refers to the compound $NaHCO_3$, sodium bicarbonate, or "baking soda," which is probably in your kitchen cabinet. It may be called by the older name, bicarbonate of soda. This common compound consists of Na^+ ion in combination with the negatively charged combination ion HCO_3^-. You recognize that this ion contains 1 atom of hydrogen, 1 atom of carbon, and 3 atoms of oxygen. Such an ion is sometimes called a complex ion in contrast to a simple ion like Na^+ of only 1 atom.

A COMPLEX ION IS A GROUP OF ATOMS HAVING EITHER A POSITIVE OR NEGATIVE CHARGE, AND BEHAVING LIKE A SINGLE UNIT IN MANY CHEMICAL COMPOUNDS.

Complex ions can, under some circumstances, be broken into their individual atoms, but in many reactions they just stay together and move from one compound to another intact. That is why they are given special status and listed in tables with the other common ions found in compounds.

Here are a few more examples from Table 6.1.

8. What is the formula for the oxide of magnesium, magnesium oxide?

$$Mg^{+2}_{2} \quad O^{-2}_{2}$$

which by our system of crossing over gives Mg_2O_2. This is always reduced to lowest terms as the formula MgO. Here you could have written the formula MgO immediately after noting that a +2 charge for magnesium would be balanced by a −2 charge on 1 atom of oxygen. Intuitively, this is what a chemist would do, being familiar with the relative charges on each ion. However, while you are learning to handle these ionic valences, it may be necessary to first use the crossing-over technique, giving the Mg_2O_2 formula which then you would note means 2 magnesium atoms to 2 oxygen atoms, more simply written MgO.

9. What is the formula for magnesium sulfate?

$$Mg^{+2}_{2} \quad SO^{-2}_{4\,2}$$

which first gives $Mg_2(SO_4)_2$, and is then reduced to $MgSO_4$. Here again we could note that the +2 charge is balanced by 1 unit of −2 and we could write directly $MgSO_4$. The "1" is assumed whenever a symbol appears, this compound containing 1 atom of magnesium, 1 atom of sulfur, and 4 atoms of oxygen.

10. What is the formula for iron(III) oxide, or ferric oxide?

$$Fe^{+3}_{2} \quad O^{-2}_{3} \qquad \text{which gives } Fe_2O_3$$

This is our first encounter with the possibility of an element having more than one different valence state, a possibility which exists with only a few of the elements in the table. Several metal elements can exist as ions with different charges because of a peculiar loss of electrons in stages. The conditions of the chemical reaction would determine which one forms in a particular case. We need not bother with those details here. We just want to be able to recognize the possibility of different ionic valence and pick out the one designated.

The modern way of naming the compounds with these elements with variable valence is to follow the name or symbol with a Roman numeral for the valence state, for example, iron(II) is Fe^{+2} and iron(III) is Fe^{+3}. Older, less revealing names are still used. These are ferrous for Fe(II) and ferric for Fe(III). Usually we are dealing with only two valences and the low valence form has the *ous* name, the high valence, the *ic* name.

Another example is copper(I) and copper(II) which are, by the older naming system, called cuprous and cupric, for Cu^{+1} and Cu^{+2}, respectively. These are listed in Table 6.1.

11. Give the formula for copper(II) chloride or cupric chloride.

$$Cu^{+2}_{1} \quad Cl^{-1}_{2} \qquad \text{which gives } CuCl_2$$

12. What is the formula for copper(I) chloride, or cuprous chloride?

$$Cu^{+1}_{1} \quad Cl^{-1}_{1} \qquad \text{which is } CuCl$$

13. Give the formula for sodium phosphate.

$$Na^{+1}_{3} \quad PO_4{}^{-3}_{1} \qquad \text{which is } Na_3PO_4$$

This has also been called trisodium phosphate to emphasize that one of the other phosphate complex ions containing hydrogen is not involved (see Table 6.1). This is the "TSP" (for *tri*sodium *p*hosphate) which is found in many washing powder formulations and is one of the phosphates responsible for the recent furor over detergents. (Chapters 13 and 15.)

14. What is the formula for ammonium chloride?

$$NH_4{}^{+1}_{1} \quad Cl^{-1}_{1} \qquad \text{which gives } NH_4Cl$$

Here we have the common complex ion *ammonium* which is related to the household product ammonia—a gas NH_3, ammonia, dissolved in water and usually written as NH_4OH, ammonium hydroxide. The ammonium complex ion is very stable and has the 4 hydrogen atoms tied up very tightly with 1 atom of nitrogen, the overall combination having a positive charge of 1.

15. What is the formula for ammonium sulfate?

$$NH_4{}^{+1}_{2} \quad SO_4{}^{-2}_{1} \qquad \text{which gives } (NH_4)_2SO_4$$

Note this compound very carefully for the meaning of the subscripts and the use of parentheses to indicate that two of the whole ammonium units ($NH_4{}^+$) are in the compound with a unit of the sulfate ($SO_4{}^{-2}$). It can be analyzed as shown on the next page.

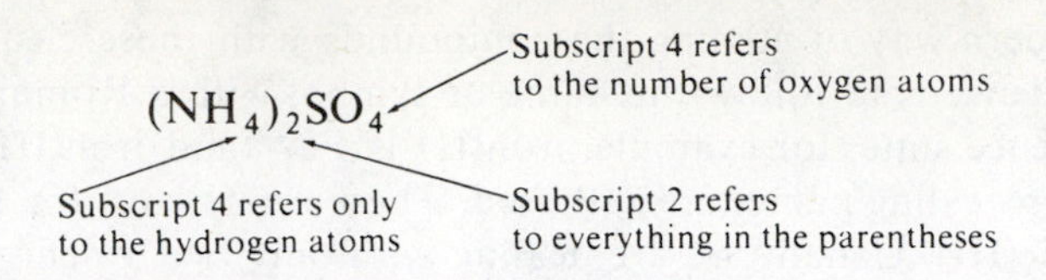
Subscript 4 refers
to the number of oxygen atoms
$(NH_4)_2SO_4$
Subscript 4 refers only
to the hydrogen atoms
Subscript 2 refers
to everything in the parentheses

You can see that if we left out the parentheses there would be confusion. $NH_{42}SO_4$ would indicate 1 atom of N, 42 atoms of H, 1 atom of S, and 4 atoms of oxygen, which would obviously be wrong since there are only 8 atoms of hydrogen in this compound (along with *2 atoms* of nitrogen, 1 of sulfur, and 4 of oxygen).

From these examples you can see how useful the table of ions is in putting down the proper formula for common ionic compounds. It would be a slow and tedious chore to go back to the Periodic Table and the electronic structure each time to determine the basis before writing a formula for a compound. The time invested in memorizing the major important ions with their charges from the table of ionic valences is repaid by the easy handling of chemical formulas.

NAMING COMPOUNDS

Many ways are available for naming a particular combination of atoms in a compound. There are all the old names which came into use long before people even knew what atoms were involved, or even that atoms were involved at all. Then there are also the more systematic, modern ways of naming complex compounds. But even these can be rather complicated, so that chemists working with compounds often use shorthand or abbreviated descriptive terms like DDT, TSP, penicillin, or 2,4-D.

Water is certainly never going to be displaced by any name like dihydrogen oxide. However, it may be worthwhile noting some of the standard conventions used for simple compounds. We will summarize here for reference some of the naming methods you may occasionally run across.

1. Common or "trivial" names are names that have been given to products in common usage without regard to composition. Sometimes they are just like family names with no bearing on what the person looks like. Other times they may be like nicknames people often pick up based on some particular quality or idiosyncrasy, like "shorty." For example, quicksilver is used for the element mercury, found in some thermometers. The name comes from the fact that mercury is silvery in appearance and runs quickly all over the place if the thermometer breaks (and it is a poison, incidentally, which should be swept up and removed at once from habitable areas). Some other trivial names are listed in Table 6.2 along with their chemical formulas.

2. Systematic naming of compounds containing two elements usually follows a general approach similar to some examples we have already run across, like sodium chloride for NaCl. We will use this method wherever necessary. It is formally standardized in naming compounds in scientific work and is preferred over the use of trivial names. For the inorganic

TABLE 6.2
Examples of Trivial Names and Their Formulas

Trivial Name	*Formula*	*Trivial Name*	*Formula*
Water	H_2O	Alumina	Al_2O_3
Baking soda (bicarbonate of soda)	$NaHCO_3$	Ammonia	NH_3
Salt (table salt)	$NaCl$	Lime	CaO
Sugar (cane sugar, beet sugar, sucrose)	$C_{12}H_{22}O_{11}$	Slacked lime	$Ca(OH)_2$
		Marsh gas (firedamp)	CH_4
Alcohol (grain alcohol)	C_2H_5OH	Galena	PbS
Lye	$NaOH$	Marble (calcite, limestone)	$CaCO_3$
Milk of Magnesia	$Mg(OH)_2$	Epsom salts	$MgSO_4 \cdot 7\,H_2O$
Washing soda	$Na_2CO_3 \cdot 10\,H_2O$	Gypsum	$CaSO_4 \cdot 2\,H_2O$
Plaster of paris	$(CaSO_4)_2 \cdot H_2O$	Fool's gold (pyrites)	FeS_2
Blue stone (blue vitriol)	$CuSO_4 \cdot 5\,H_2O$	Wood alcohol	CH_3OH
Hypo	$Na_2S_2O_3$	Quicksilver	Hg
Muriatic acid	HCl	Oil of vitriol	H_2SO_4
Laughing gas	N_2O	Cream of tartar	$KHC_4H_4O_6$
Benzene	C_6H_6	Borax	$Na_2B_4O_7 \cdot 10\,H_2O$
		Soda ash	Na_2CO_3

Chemists run continual checks on chemical reactions to determine that the proper compounds are being formed. Here a syringe is used to withdraw a sample of experimental rubber from the bottom of a large reaction vessel. (Rubber is covered in chapter 9.) (Courtesy of The Goodyear Tire & Rubber Company.)

compounds we usually consider the compound to consist of two parts, the positive and negative. The positive part is named first, followed by the negative part name. Examples follow.

A. Compounds of only two elements where the metal has only one valence are named according to the pattern

Metal Name + *Root of Nonmetal Name* + ide

CaF_2	Calcium fluoride
MgO	Magnesium oxide
KI	Potassium iodide

B. Compounds of only two elements where the metal has variable valence are named according to the pattern

Metal Name + *Roman Numeral Valence* + *Root of Nonmetal Name* + ide

$FeBr_2$	Iron (II) bromide (also, ferrous bromide)
$FeBr_3$	Iron (III) bromide (also, ferric bromide) (The ferrous and ferric come from the Latin *ferrum* for iron from which the symbol Fe is also derived.)
SnF_2	Tin (II) fluoride (also, stannous fluoride, one of the toothpaste additives)

C. Compounds of only two nonmetals are named using prefixes wherever needed for clarity. The prefixes used are these:

Mono	1	Tetra	4	Hepta	7	Deca	10
Di	2	Penta	5	Octa	8		
Tri	3	Hexa	6	Nona	9		

The pattern is as follows:

Number Prefix + Least Electronegative Element + Prefix +
Root of Most Electronegative Element + ide

CO	Carbon monoxide
CO_2	Carbon dioxide
PCl_3	Phosphorus trichloride
CCl_4	Carbon tetrachloride
N_2O	Dinitrogen oxide
N_2O_4	Dinitrogen tetroxide
NO	Nitrogen oxide
N_2O_3	Dinitrogen trioxide
NO_2	Nitrogen dioxide
N_2O_5	Dinitrogen pentoxide

3. Compounds of more than two elements make use of the special and unique names for the complex ions involved according to the general pattern of metal name + complex ion name, but they sometimes use prefixes for emphasis.

K_3PO_4	Potassium phosphate
Na_3PO_4	Sodium phosphate (or trisodium phosphate for emphasis)
Na_2HPO_4	Disodium hydrogen phosphate
NaH_2PO_4	Sodium dihydrogen phosphate
NH_4Cl	Ammonium chloride

EQUATIONS

In discussing how atoms combine (see chapter 5), we have occasionally used what could be called primitive equations, for example:

$$Na + F \rightarrow Na^+ + F^- \quad (\text{or } NaF)$$

A CHEMICAL EQUATION IS A SHORTHAND STATEMENT OF CHEMICAL CHANGE, USING SYMBOLS OF ELEMENTS AND FORMULAS OF COMPOUNDS.

This is a natural development of the basic symbolism of chemistry. Just as it is convenient and timesaving to represent an element by a symbol, and a chemical compound by a formula, so it is useful to represent a chemical reaction by an equation. Its value is as a condensed shorthand way of providing information. We could write the same thing an equation tells us but it might take several paragraphs. Also, equations tell us how much material we are likely to get from certain starting chemicals. So equations save considerable time and help us see what is really happening in a reaction, which extended description might hide.

Let us consider another equation which we have seen in primitive form, showing the reaction which occurs when sodium and chlorine react:

$$Na + Cl \rightarrow NaCl$$

Now we would like our equations to represent, as far as possible, the actual reaction that occurs. You know that chlorine exists in nature as the Cl_2 molecule and not as free chlorine atoms. So to start with, then, we know what the starting materials are, Na and Cl_2, and the ending material after reaction, NaCl. The common elements that exist as two-atom molecules are H_2, N_2, O_2, F_2, Cl_2, Br_2, and I_2. For all other elements, like Na, we use the symbol alone in equations.

The starting materials shown in equations are often called the *reactants* because they are the substances which react. The ending materials are called the *products* because they are produced in the reaction. Therefore the equation should read:

Na	+	Cl_2	$\rightarrow$	NaCl	(Symbol equation)
sodium	+	chlorine	$\rightarrow$ gives	sodium chloride	(Word equation)

The arrow means *gives, yields,* or *produces.* The symbol equation tells us much more than a word equation.

If you look at the above equation you will notice that it is not completely true as written because it says you start with 2 atoms of chlorine and end up with only 1. You know that this is contradictory to the Law of Conservation of Mass which we considered in chapter 3. You cannot have an atom disappear into nothing. Chemists would say that the equation as now written is "unbalanced." To balance it we need to add a 2 in front of the NaCl, indicating that we have 2 particles of NaCl, and this will account for the 2 chlorine atoms:

$$Na + Cl_2 \rightarrow 2\,NaCl$$

But 2 particles of NaCl also means 2 atoms of sodium are needed and therefore we have to add a 2 in front of the Na on the reactants side. Then we have a "balanced" equation:

$$2\,Na + Cl_2 \rightarrow 2\,NaCl$$

The numbers we add before the symbols and formulas are called *coefficients.* We cannot balance the equation by simply adding a chlorine atom to NaCl as $NaCl_2$ because of the electron requirements of sodium and chlorine atoms, even though it would provide a mathematical "balance." Also, notice the subscript following the chlorine symbol, Cl_2. We cannot alter this either since chlorine atoms achieve the octet by sharing a pair of electrons. There is again a basic electronic reason for it. Thus neither

$$Na + Cl_2 \rightarrow NaCl_2 \quad \text{nor} \quad Na + Cl \rightarrow NaCl$$

are correct balanced equations following the way nature behaves in her definite planned approach to building aggregates or molecules from atoms.

Notice again the difference between subscripts and coefficients in equations. A subscript indicates the makeup of a molecule or group of atoms. A coefficient merely says you take so many of the basic units that follow it. If a single unit of a substance is involved, the "1" is omitted because it is assumed you mean one of the unit when you put it down. See the example above where "1" is assumed before the Cl_2.

Here is another example. Remember that the identity of products and reactants must first be determined before any equation can be balanced. This involves laboratory experimentation and investigation, but here we will assume this has been determined. Presume we are given the following experimental data. Magnesium is the metal in certain photo flash bulbs. The bulb also contains oxygen gas, and when energized by an electric current the magnesium metal reacts with evolution of energy as light and heat, giving the white powder which is found to be magnesium oxide.

The steps in balancing this equation then can be outlined as follows:

Step 1: Write the formulas of reactants and products. This is sometimes called the *skeleton equation.*

$$Mg + O_2 \rightarrow MgO$$

You note that there are 2 atoms of oxygen on the left and only 1 on the right. We cannot shape nature to our ends and write either MgO_2 for MgO, or O for O_2 in order to achieve a balance. Instead we place a 2 in front of MgO, indicating we have two MgO units. This takes care of 2 atoms of oxygen on the right which equals the 2 on the left.

Step 2:

$$Mg + O_2 \rightarrow 2\,MgO$$

However, now we need 2 atoms of Mg on the left or reactant side and so we add this in.

Step 3:

$$2\,Mg + O_2 \rightarrow 2\,MgO$$

Looking over the equation we can now count up the atoms on each side of the "yields" arrow and find:

	Left: Reactants	Right: Products
Totals:	2 Mg atoms + 2 O atoms	2 Mg atoms + 2 O atoms

Therefore, the equation is balanced as far as atoms are concerned.

In some cases, we may want to add to the equation the fact that energy is also involved. If you have ever seen or felt a flash bulb after it goes off, you know we get energy. This is usually indicated by adding the word *energy* in the equation, but chemists can also measure the amount and report it as calories (the calorie being the heat needed to raise the temperature of a measured amount of water a definite amount). We do not need this fine detail, so we can finally show the equation as

Step 4:

$$2\,Mg \; + \; O_2 \; \rightarrow \; 2\,MgO \; + \; \text{energy}$$

Whether or not we want to show the energy change depends upon our interest in the energy requirements for a particular chemical change. All chemical reactions involve energy changes but most of the time we may not need to be concerned with this aspect.

A reaction where heat energy *comes out* is said to be *exothermic* (Latin: heat going out). If in the overall reaction heat energy is *put into* chemicals to give the products, the reaction is called *endothermic* (Latin: heat going in). For example:

$$\text{energy} \; + \; 2\,HgO \; \rightarrow \; 2\,Hg \; + \; O_2$$

When all is said and done, equation balancing is not the "be all and end all" of chemistry. It is a minor concern for you unless you go on to study chemistry in more depth. Even the professional chemist does not spend much of his time balancing equations. For simple equations like the above, we "balance by inspection," using trial and error in a somewhat intuitive, puzzle-solving exercise. For more complex equations there are special procedures to make balancing fast and easy. However, you can appreciate that the fundamental rationale for balancing is the Law of Conservation of Matter. And this may be especially important in some

This is an obvious exothermic reaction, as the Apollo/Saturn V space vehicle lifts off from the Kennedy Space Center in Florida to start astronauts on their way to the moon. (NASA photo.)

cases such as pollution, where the quantity of gases produced is determined by the quantities of gasoline or trash burned. Hence quantity considerations become important in understanding gross environmental changes brought about by chemical change.

You have now essentially covered the basic concepts involved in understanding chemicals and their reactions. In subsequent chapters we will apply this knowledge to your environment, which is made up of a vast variety of chemicals. It would be worthwhile at this point for you to go back over the materials of these first six chapters to review and deepen your understanding of the conceptual framework which you will apply to the many practical matters in the chapters to follow.

PROJECTS AND EXERCISES

Experiments

1. This is a "thought experiment" like those we suggested in chapter 5. In other words, think about the situation and try to come to a conclusion by developing supporting or contradictory arguments. Consider the usual way of picturing the nucleus of an atom by drawing a circle and putting the protons and neutrons inside. We have used this picturization method many times. The question you should consider is whether there is a sort of bag or skin holding the nuclear particles, somewhat like a bag of marbles or like the circles we have been drawing. Do you support or not support the possibility of a bag or sphere holding the nuclear particles? Think about this, using some conceptual model, and develop your mental arguments pro or con. Then state your conclusion, giving reasons for it and other reasons against the alternate view.
2. This is another thought experiment. (Remember the "thought-experiment" approach is not kid's stuff. It was used often by the theoretical physicists and chemists who laid the groundwork for our present understanding of the atom.) In the seventeenth century Robert Boyle (1627–1691) proposed an alternate operational definition of an element which briefly says: "An element is a substance which will always increase in weight when it enters into chemical change." Try thinking about this definition and develop some conceptual framework to deal with it. Think about the possibilities, writing down on paper some of your "thoughts" or "doodles" since thought experiments do not preclude the use of paper. In fact, it is on paper that we generally express many of our thoughts, however inadequately. The paper approach may help to organize your thinking. Come up with a general conclusion, supporting, turning down, or possibly giving conditional support to Boyle's definition.

Exercises

3. Give the formulas and names of two compounds involving copper and oxygen.
4. Give the formulas for the following compounds:
   (a) Silver bromide
   (b) Potassium sulfide
   (c) Calcium iodide
   (d) Barium chloride
   (e) Barium sulfate
   (f) Calcium phosphate
   (g) Sodium sulfate
   (h) Calcium carbonate
   (i) Calcium bicarbonate

5. Write the formulas for the following compounds:
   - (a) Ammonium chloride
   - (b) Ammonium sulfate
   - (c) Ammonium phosphate
   - (d) Ammonium dihydrogen phosphate
   - (e) Dinitrogen pentoxide
   - (f) Nitrogen dioxide
   - (g) Calcium hydroxide
   - (h) Sodium hydroxide
   - (i) Sodium cyanide
6. Give the formulas for the compounds which are possible in the following table, assuming the positive ions shown are combined with the negative ions along the top row. The salt KI is shown as an example in one box.

	I^{-1}	CO_3^{-2}	OH^-	SO_4^{-2}	NO_3^{-1}
K^+	KI				
Ca^{+2}					
Fe^{+2}					
Fe^{+3}					
Al^{+3}					

7. Methane, or marsh gas, has the formula CH_4. This is one of the components of natural gas obtained from underground pockets in the earth. When it burns in your gas stove or gas water heater, it combines with oxygen to give water and carbon dioxide. Write out the equation and balance it. Would this be an exothermic or an endothermic reaction? Why? Indicate this by inserting the energy term in your balanced equation.
8. When hydrogen and oxygen gases are mixed and a spark applied, there is an explosion and water is formed. Write out the equation and balance it.
9. The rusting of iron may be conveniently represented as the combination of iron with oxygen in the air in the presence of moisture to give a red product iron (III) oxide or Fe_2O_3. Write out and balance the equation for the reaction. Do not show the water in this equation.

SUGGESTED READING

1. Hughes, D. J., *The Neutron Story,* Anchor Books, Doubleday & Co., Inc., Garden City, New York, 1959. Paperback.
2. Curie, E., *Madam Curie,* Doubleday & Co., Inc., Garden City, New York, 1939.

chapter 7

Some Chemicals for Life: Water and Its Elements

I chatter, chatter, as I flow
To join the brimming river,
For men may come and men may go;
But I go on forever.

—TENNYSON

The sun's a thief, and with his great attraction
Robs the vast sea; the moon's an arrant thief,
And her pale fire she snatches from the sun;
The sea's a thief, whose liquid surge resolves
The moon into salt tears; the earth's a thief
That feeds and breeds by a composture stolen
From general excrement: each thing's a thief.

—SHAKESPEARE

The tear down childhood's cheek that flows,
Is like the dewdrop on the rose;
When next the summer breeze comes by,
And waves the bush, the flower is dry.

—SIR WALTER SCOTT

Water is an extremely important chemical. It is the most common substance on earth. These are trite statements but true just the same. You probably know that 70 percent of the earth's surface is covered with water.

The rivers, large and small, the springs that feed them, the seas and the oceans, together with the vast amount of underground waters, form a nearly continuous water shell of the earth. This is called the *hydrosphere.* We live in it and through it, just as individually each of us lives for a time with his own changing water sphere which, surprisingly, is close in percentage composition to the water coverage of the

A fantastic view of the sphere of the earth, photographed by U.S. astronauts on their way to the moon. The photo extends from the Mediterranean Sea area to the Antarctica south polar ice cap. The predominance of water is evident here in the ocean surface, ice, and the cloud cover in the Southern Hemisphere. Almost the entire coastline of the continent of Africa is shown. The Arabian peninsula can be seen at the northeastern edge of Africa, with the Red Sea in between. (NASA photo.)

earth itself. Your blood is greater than 90 percent water. Your muscles average 75 percent water and even your bones are 22 percent water. On the average you are about 65 percent water by weight. (Someone once said that the crowded surf at a New York City beach in summer was two-thirds people and one-third water.)

This water, evaporating from the surface of your body and breathed out as gas in every breath, must be continually replaced. In a way each of us is a custodian of water which is certainly not our own. Some of the water molecules in your body may have flowed in the veins of Caesar or run down in the tears of Cleopatra. Some of the same water molecules you now possess may have evaporated from the skin of Alexander the Great or Mahatma Gandhi. They may at one time have cascaded down Victoria Falls on the Zambezi River in Africa. And now they are "yours" for a time. The cycling of water is considered in more detail in chapter 15. It might be worthwhile to refer to Figure 15.1 (page 431), where the water cycle is described.

WATER REQUIREMENTS

The quantity of water you need is quite large. This water requirement is not limited to the water you take in as food and drink which amounts to only about 2 quarts a day. You benefit from other much larger water uses which are not easy to recognize. The major ones are industrial and agricultural, with minor uses in transportation, recreation, and power. For example, 150 gallons of water are needed to make the paper for one Sunday newspaper. About 32 gallons are needed to make a pound of steel, and 160 gallons for a pound of aluminum for your car, radio, tools, record player, cans, spoons, and the like. Approximately 1000 pounds of water are cycled through the processes to make one pound of sugar or corn, while about 115 gallons go into growing the wheat for one loaf of bread.

On the average, then, the water requirements for each person in the United States is *1600 gallons per day.* This includes only 70 gallons per person for water used in the home. (It takes 3 gallons to flush a toilet, 30 to 40 gallons to take a bath, and 5 gallons for each minute in the shower.)

In an average lifetime a person may drink around 8000 gallons of water. But his total usage of water would be about 41 million gallons. (And each drop of water contains about 30 billion billion molecules. It is not as surprising then that you have probably mingled in your system a molecule from Cleopatra's tears.)

WHAT IS IN WATER?

Water is made up from H_2 (hydrogen), a very light, explosive gas, and O_2 (oxygen), another slightly heavier gas, but a gas nonetheless. Then how do we get "watery" properties from these gases? You can see that water is not a mixture of oxygen and hydrogen. The basic answer is "chemical reaction." This is the origin of an entirely "new" product even though made from parts which have different properties. No one claims to completely understand this change in properties—but it may now be a little more acceptable since you have already seen how the tiny charged particles make up the varied atoms, and how atoms combine to make up molecules of compounds with quite different properties. We will see how many of the properties of water are explainable in terms of the concepts of electronegativity and atomic bonding.

PROPERTIES OF HYDROGEN

Hydrogen, the lightest and simplest of all the elements, is a tasteless, colorless, odorless gas. It is the primary element in the universe and constitutes the bulk of the outer atmosphere of the sun. The name *hydrogen* is from the Greek for "maker of water." This refers to one of the major properties of hydrogen—it burns in oxygen to give water:

$$2H_2 + O_2 \rightarrow 2H_2O + \text{Heat}$$

This reaction occurs only very slowly at ordinary temperatures. However, at high temperatures, such as that provided by a spark, it may give a dangerous explosion.

Hydrogen was once used in balloons and lighter-than-air craft, sometimes called airships, or dirigibles, or "blimps." This use of hydrogen ended rather abruptly when the large German airship Hindenburg exploded in flames when coming in for a landing at Lakehurst, New Jersey, in 1937. The Hindenburg had been built in 1936 and contained 7,063,000 cubic feet of hydrogen gas. It had made fifty-four flights, thirty-six of them across the Atlantic Ocean. When the hydrogen-filled gas bag exploded, thirty-six passengers and twenty-three crewmen were killed.

The Hindenburg explosion. (Courtesy Wide World Photos.)

Yet when the same reaction between the two gases is carried out under controlled conditions the large heat output is an advantage. The "oxy-hydrogen torch" uses the reaction of H_2 and O_2 in combining to form H_2O. (See Figure 7.1.) The difference is that it is carefully controlled.

You are aware of another dramatic example of the controlled use of the energetic reaction between hydrogen and oxygen. This is in large rocket engines. (See Figure 7.2.) The Saturn V rocket which lifted the U.S. astronauts on the way to the moon landings had three stages.

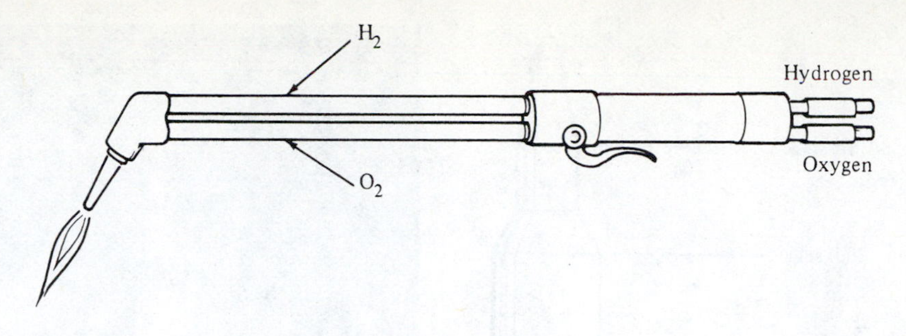

FIGURE 7.1
A diagram of the oxyhydrogen torch used for cutting and welding heavy steel sheets. The gases are not allowed to contact each other until they are fed into the flame tip, where they are ignited.

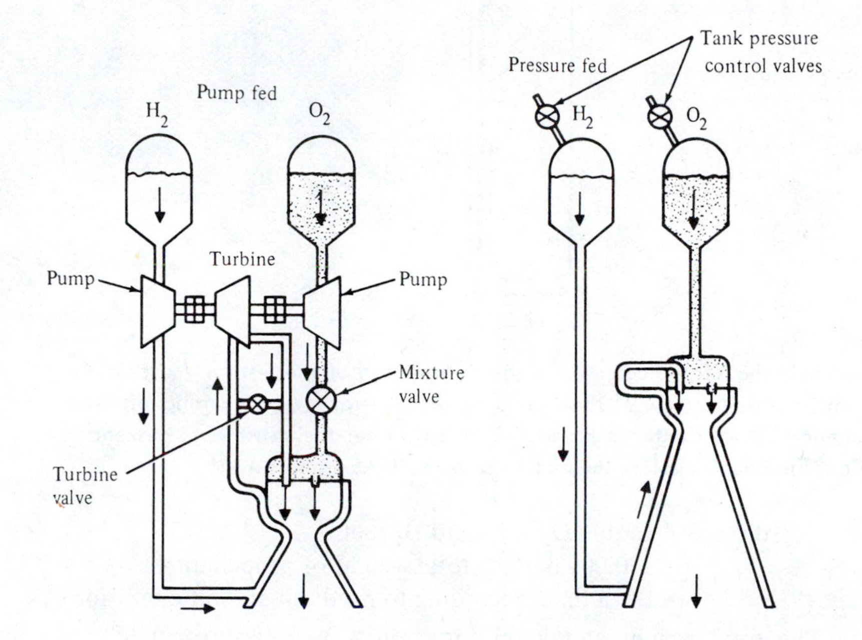

FIGURE 7.2
Two different flow systems for a chemical rocket based on the use of liquid hydrogen as "fuel" and liquid oxygen as "oxidizer." Note how the potential of explosion is prevented by not allowing the gases to mix until they are in the rocket chamber where the exothermic reaction is desired.

1st stage: Liquid O_2 – Kerosene fuel.
4,400,000 pounds total weight of propellants.
Propellants consumed at rate of 15 tons/second. (See Figure 7.3.)
Burn time: $2\frac{1}{2}$ minutes.

2nd stage: Liquid O_2 – Liquid H_2 fuel.
945,000 pounds total weight of propellants.
Burn time: 6 minutes, 40 seconds.

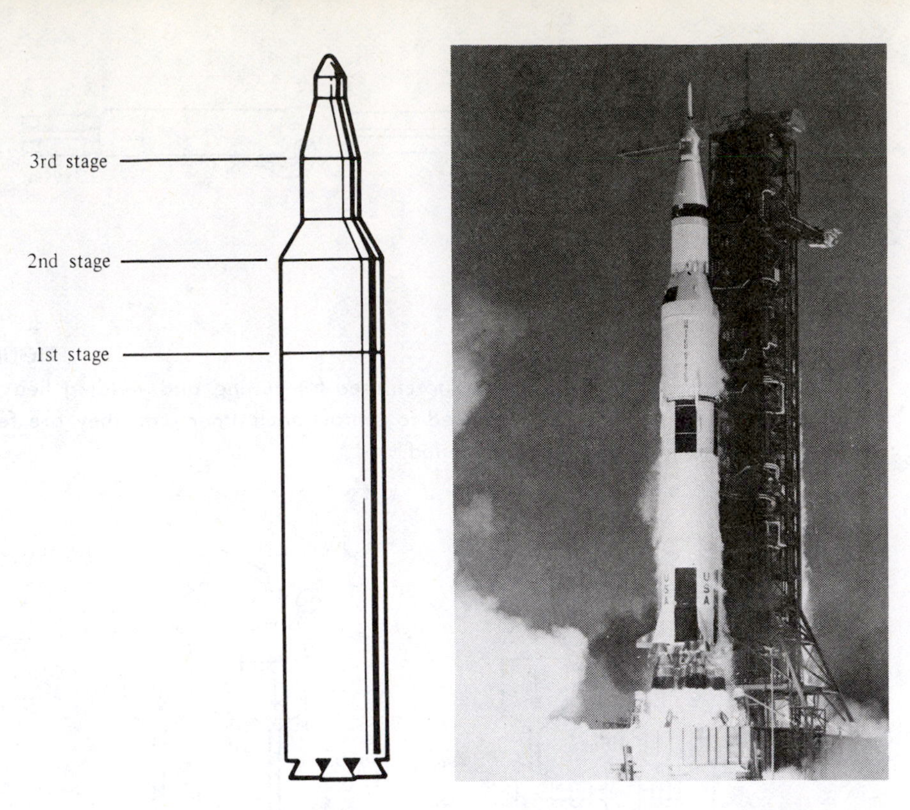

FIGURE 7.3
The sketch on the left indicates the stage designation of the Saturn V rocket. On the right, a close-up view of the Apollo/Saturn V space vehicle lifting off from the Kennedy Space Center in Florida. Four days later the astronauts descended in a lunar module to explore the moon's surface. (NASA photo.)

3rd stage: Liquid O_2 – Liquid H_2 fuel.
230,000 pounds total weight of propellants.
Burn time: according to need; on and off operation.

The total weight at takeoff for Saturn V is approximately 6 million pounds.

Hydrogen molecules achieve a full shell of electrons by equal sharing of an electron pair (chapter 5):

$$H:H$$

Therefore the molecules are not polar and have little attraction for each other. There is not much to hold the light molecules down. In fact, you have to slow the molecules considerably by cooling before you can get hydrogen gas to form a liquid. The temperature required is −253°C (−423°F). (The C stands for the centigrade or Celsius scale for measuring temperature. The F stands for the Fahrenheit scale. The temperature at which water freezes is given the designation 0° on the Celsius scale [32°F]. The boiling point of water is 100°C [212°F]. Your normal body temperature is 37°C [98.6°F]. See Figure 7.4.)

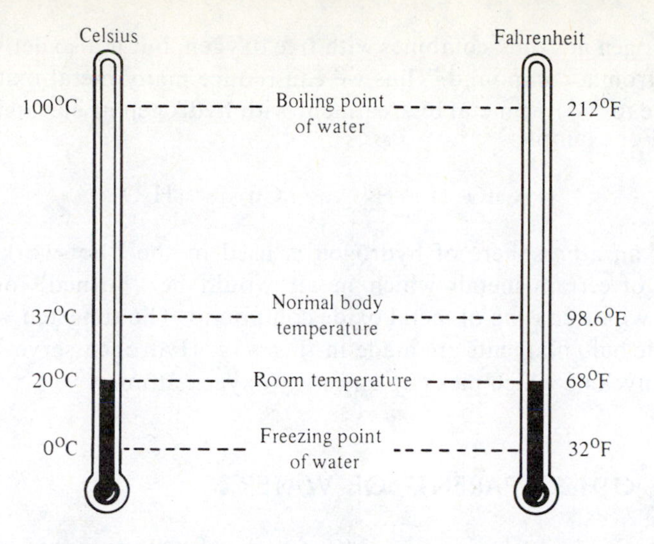

FIGURE 7.4
A comparison of the Celsius (or centigrade) and Fahrenheit temperature scales.

Liquid hydrogen will freeze at about −259°C (−434°F). These very low temperatures have been reached in the laboratory using special apparatus. Normally hydrogen, with so little attraction between molecules, is a gas—with molecules free to roam the great empty spaces.

WHERE DO WE GET HYDROGEN?

Hydrogen does not exist in any significant amount free on the earth's surface. The place to get it then is in compounds. One of the best sources is water.

You have seen in chapter 3 that the simplest hydrocarbon or compound of hydrogen and carbon is CH_4, methane. A common industrial source of hydrogen uses this compound along with water:

$$CH_4 + 2H_2O \rightarrow CO_2 + 4H_2$$

The reaction is carried out in special vessels containing a metal like nickel which speeds up the reaction. Such a material is called a *catalyst* (Greek: to loosen completely). Here the nickel is not consumed in the reaction but serves to provide a special surface for intermediate combinations of molecules which then rearrange to give the product which we want.

Along with hydrogen we get a "by-product" of carbon dioxide gas. The by-product is obtained "beside or along with" the main product. Finding markets or uses for by-products often presents challenges to the chemist and chemical engineer. In the present case, these gases can be separated and both used to advantage. Hydrogen is a common gas of great use in other chemical reactions, for example, in making oleomargarine from vegetable oils (chapter 10). Carbon dioxide has many uses, such as in fire extinguishers and carbonated beverages.

Hydrogen not only combines with free oxygen, but it also actively takes oxygen from a compound. Thus we can reduce many metal oxides found in nature to the free metal by treatment with hydrogen at elevated temperatures. For example:

$$CuO + H_2 \rightarrow Cu + H_2O$$

Also, an atmosphere of hydrogen is used in the "hot-working" and shaping of certain metals which in air would be "burned" or become covered with a coating of metal oxide compound. The tungsten wires used in electric bulb filaments are made in this way. Hydrogen serves as a protective envelope which picks up any stray oxygen atoms.

OXYGEN: THE OTHER "PARENT" OF WATER

Oxygen gas, like hydrogen, consists of molecules made up of two atoms of the same kind (nonpolar molecules). Oxygen atoms are also relatively light. Consequently we have to go down to a temperature of about −183°C (−297°F) at normal atmospheric pressure to form oxygen into a liquid. This is called the boiling point. A lower temperature of −218°C (−360°F) is required to form solid oxygen. Only by taking the kinetic energy out of the gas and slowing the molecules down considerably can we get some degree of "holding together" in liquid or solid form.

Oxygen was discovered in the 1770s by heating an oxide of mercury:

$$\underset{\text{Mercuric oxide}}{2HgO} \xrightarrow[\Delta]{} \underset{\text{Mercury}}{2Hg} + \underset{\text{Oxygen}}{O_2}$$

The small triangle (a Greek *delta*) under the arrow indicates one way the chemist symbolizes heat added. Or we could write:

$$\text{Heat} + 2HgO \rightarrow 2Hg + O_2$$

Theoretically all compounds can be broken apart, that is, chemically decomposed. Practically this is possible only with those which do not require very high temperatures. The decomposition of mercuric oxide is an endothermic reaction (chapter 6): heat enters or goes into the formation of the new products. Most reactions which you encounter are exothermic where heat is given off to the environment.

The chief use of oxygen is in its property of combining with other substances to give various products and more heat energy. Your breathing involves an important oxidation reaction. You breathe about twenty times per minute, taking in approximately eight-tenths of a quart of air each time. In one day you take in over 50 pounds of air. This is about ten times your total intake of food and water combined. Your air requirement comes to more than 20,000 quarts per day—to provide oxidation of foods for energy and heat.

Oxygen combines readily with all the elements except noble gases and certain metals like gold and silver.* Here are some examples:

$$2Cu + O_2 \rightarrow 2CuO + \text{Heat}$$

$$4Al + 3O_2 \rightarrow 2Al_2O_3 + \text{Heat}$$

$$S + O_2 \rightarrow SO_2 + \text{Heat}$$

$$C + O_2 \rightarrow CO_2 + \text{Heat}$$

The last equation represents the "barbecue reaction" where you burn charcoal briquets.

Some reactions with O_2 take place slowly but relentlessly. These cause tremendous economic losses each year. The rusting of iron occurs in a complex series of reactions wherein water is a catalyst. The overall result is:

$$4Fe + 3O_2 \rightarrow 2Fe_2O_3 + \text{Heat}$$

It is estimated that over a billion dollars a year are required in painting to frustrate this reaction.

Coal and coke are mostly carbon. Consequently, the reaction is essentially the same as the one given above for the barbecue reaction. These reactions are considered useful because they provide power, heat, or light. Another is the burning of methane, which is the major component of natural gas:

$$CH_4 + 2O_2 \rightarrow CO_2 + 2H_2O + \text{Heat}$$

Oxygen is obtained commercially by liquifying air and separating the components by distillation (chapter 8). The oxygen is stored in steel tanks under pressures greater than one-hundred times the pressure of the atmosphere.

OXYGEN: IS MORE BETTER?

There is a very unusual history of oxygen usage which demonstrates how careful one must be in judging the application of chemicals. During the early 1940s a strange disease surfaced. It was first reported when a Boston eye specialist was called in to examine babies going blind from some unknown cause. The disease defied diagnosis and no treatment or cure was evident. All through the 1940s, from various U.S. cities and also in some places in Europe, reports grew of the strange cases of blind

*These are often called "noble metals" in reference to this quality of lack of rusting and aloofness from the common behavior of other metals which eventually mingle in the universal oxygen-containing compounds of the earth's crust.

babies. Many of these babies who survived are now around thirty years old and still blind.

A large number of scientists, pediatricians, and physiologists eventually got involved in trying to solve the mystery of the new disease. Unfortunately, it took about fourteen years and much ingenious scientific detective work before the picture was clear and the cause pinpointed.

The disease was characterized by the growth of a curtain of fibrous scar tissue behind the lens of the baby's eyes. Hence it was called retrolental fibroplasia—formation of fiber behind the lens. You have probably never heard of it and fortunately it's no longer a common problem.

It was noticed early that the disease most often struck premature babies. As the number of cases grew, a turning point was reached in the early 1950s when studies in the U.S., France, England, and Australia seemed to indicate a connection between the blindness and oxygen supply. It appeared that the babies likely to become blind were those who had needed extra oxygen shortly after birth. An oxygen deficiency problem was somehow involved. No one yet suggested that oxygen itself might be the culprit. Then in 1951 a guess was made by a French pediatrician that excess oxygen might be responsible. It is not clear how he arrived at this idea but reports had been made previously among aviators and mountain climbers that breathing pure oxygen might cause inflammation and burns in the respiratory tract.

To check the guess, controls were instituted and checks against oxygen usage were made. Dramatic drops in the disease were noted by simply using oxygen sparingly, that is, cutting the percentage used in incubators from over 80 percent down to about 40 percent (compared with air at 21 percent). The solution came in the mid 1950s and the cause of the disease was shown to be the high percentage of oxygen in incubators.

Oxygen—is this chemical good or bad? Obviously, it depends. This is a good example of the general idea that because something is good for us in small amounts does not mean it is good in larger amounts. The reverse should also be considered. Just because something causes disease or cancer in massive doses does not mean it is toxic in smaller quantities; it may in fact be essential.

OZONE: GOOD OR BAD?

Occasionally we find an element which exists in more than one form. Ozone is a second form of oxygen. It has 3 atoms in the molecule and is rather unstable, breaking down to give O_2. The overall reaction may be indicated as:

$$2O_3 \rightarrow 3O_2$$

Ozone has a sharp odor which you can detect around electric switches and motors where we get electric sparks. You can also smell it in the air after a thunderstorm, and around electronic devices which depend on passage of electrons through the air, such as an electronic stencil cutter.

Ozone is a good example of the "Jekyll-Hyde personality" of chemicals. Although it can be useful in certain conditions and certain concentrations, it can be harmful in other circumstances. It is considered a toxic or poisonous chemical.

The high energy radiation from our sun converts some O_2 in the upper atmosphere to O_3. The mixing and reacting of O_3 is such that air at ground level averages only about 1 part of ozone per million parts of air (ppm).* About fifteen miles up, closer to where the O_3 is produced, the concentration runs about 6 ppm. This shields the earth from much of the sun's ultraviolet light which would do damage to living tissues of plants and animals. Without this protective ozone layer we would not be able to live on the earth. However, at ground level one of the major problems in pollution is a growing production of ozone, which is harmful (chapter 16).

WATER: AN UNUSUAL COMPOUND

Water is so common that most of us would not regard it as unusual. But actually its properties and behavior are very special and quite unlike those of any other liquid. Let us consider a few of these special qualities.

Water exists on earth as a solid, liquid, and gas. No other substance appears in all three forms within the earth's normal operating range of temperatures.

Water, in forming ice, displays an unusual characteristic. Most substances contract when they are cooled. The molecules of most liquids when the temperature is lowered to freezing become more densely packed and occupy less space. Not so for water. When liquid water is cooled to 0°C (32°F), it expands as it freezes into ice. This is the reason ice floats on water. If water followed the normal mode here and contracted on freezing, the ice would be greater in density (weight per unit volume) than the liquid. In other words, any volume of ice would be heavier than an equal volume of liquid water. It would therefore sink. In winter new liquid layers would be continually exposed to the cold air temperature. Eventually more and more ice would form in the bottom of lakes, rivers, and oceans. In summer the sun would not reach the depths to melt all the collected ice. Then in succeeding winters more ice would build up. Water life would die and in time all the waters of the earth would be turned to ice except for a slight top layering of water in summer. Life as we know it would vanish from the earth's surface.

Water is the only common substance which is liquid at ordinary temperatures. (Mercury and some oils, which are much less common, are also liquid at ordinary temperatures.) However, if water followed the usual

*The expression *ppm* is used when we have one substance mixed in with another in relatively small amounts. It refers to the concentration of the contaminant in parts-per-million parts of the major ingredient.

trend with substances that have a similar structure it would be a gas at ordinary temperatures.

The boiling point of compounds of hydrogen and other elements is related to how heavy the individual atoms are. In other words, the molecules with low weight atoms are more likely to get out of the liquid state at a given temperature than those molecules containing the heavier atoms. For example, NH_3 is a gas (ammonia) containing in a molecule 1 nitrogen atom with the relative weight of 14, and 3 hydrogen atoms, each with a relative weight of 1. Also the compound CH_4, methane, a gas commonly encountered in marsh gas and natural gas, has only 1 carbon atom with a low relative weight of 12, along with the 4 hydrogens, at 1 apiece.

We find similarly in the oxygen family that the hydrogen compounds of sulfur, selenium, and tellurium (H_2S, H_2Se, H_2Te) are all gases at room temperature. Hence the hydrogen compound with oxygen, H_2O, should even more likely be a gas. For example, water molecules should be easier to separate from a liquid and push into the gas form than H_2S molecules because the oxygen atom's weight is only 16 compared to 32 for sulfur. However, H_2S is a gas at ordinary temperatures, whereas H_2O is a liquid, which does not boil to become fully a gas until it is heated to 100°C (212°F). This is very odd.

Large Energy Input Needed to Heat Ice and Water

Heat capacity is a term used to describe the amount of heat required to raise the temperature of a quantity of a substance by one degree (Celsius). It is really a measure of the ability of a substance to absorb heat without becoming much warmer itself. Water is unusual in having an extremely high heat capacity.

A comparison will illustrate this better than the operational description above. Imagine having a pound of gold and a pound of solid ice both at −272°C (−458°F). If we heat both the gold and the ice *so that each gets exactly the same amount of energy,* we would find the gold beginning to melt at 1063°C (1945°F) while the ice would still be quite frozen, having reached the temperature of approximately −90°C (−130°F). (See Figure 7.5.)

What are we saying when we state that H_2O has a very high heat capacity? We are saying that water can store more energy with less atomic and molecular motion than almost any other substance. The motion of particles is what temperature really measures. And temperature does not rise as rapidly when water is heated as when other materials are given similar treatment.

If you have ever walked on hot sand at the beach you have experienced the effect of differences in heat capacity. The same amount of sun energy falling upon sand and water will increase the temperature of sand about five times as much as water! When you walk on the wet sand near the water's edge you know that you are not bothered by a hot surface. This

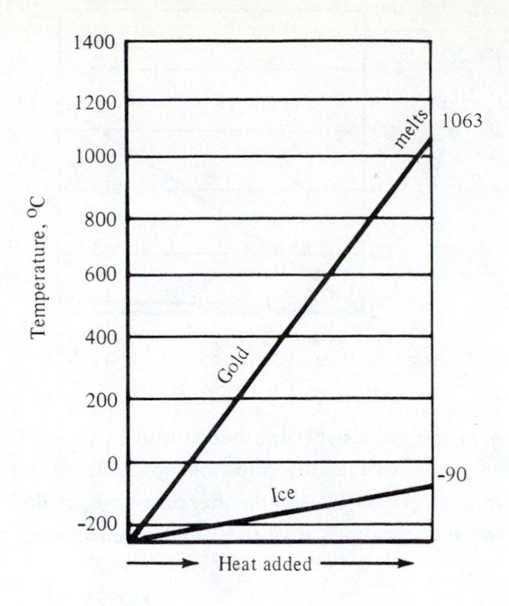

FIGURE 7.5
The unusually high heat capacity of ice or frozen water is shown here. One pound of ice and one pound of gold, both at −272°C, are each provided with the same amount of heat. When the gold melts at 1063°C, the ice is still a solid and the temperature has gone up only to −90°C.

is because water has a heat capacity about five times that of sand—which means it stays cooler *even when receiving the same amount of energy.*

Imagine again one pound of liquid mercury and one pound of water at 0°C (32°F). If you heat these liquids so that each absorbs the same amount of energy, the mercury will boil at 359°C (678°F). (See Figure 7.6.) But the liquid water, *which absorbed the exact same amount of heat energy,* would have gone up in temperature only to approximately 12°C (54°F). Water is an unusual liquid!

Turning Liquid Water Into Gas Form

All liquids can be converted into gas by providing enough energy to separate the molecules. You know that the molecules of a liquid are close to each other compared with those in a gas. When water is heated to 100°C (212°F) it will boil or go to steam at normal sea level atmospheric conditions. However, the water must be given an enormous energy input, compared with other liquids, to make it into a gas, that is, break the cohesive forces which hold the molecules close to each other.

The heat required to change a liquid to a gas (or a solid to a liquid) is referred to as *latent heat* (Latin: hidden). Water requires a large amount of heat energy compared with other liquids.

Even ancient peoples recognized the principle of the evaporative cooler.

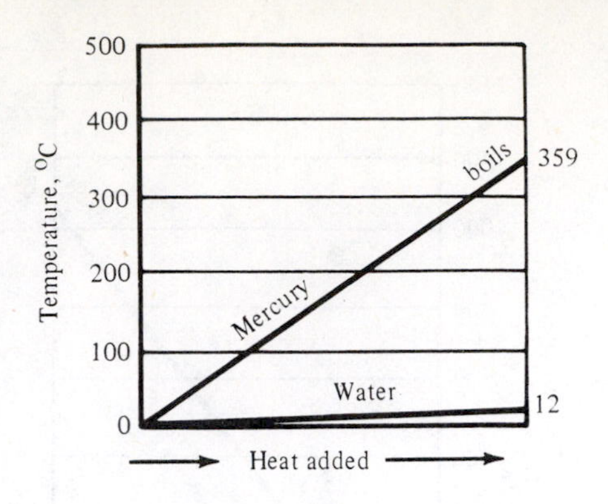

FIGURE 7.6
Here is dramatic evidence of the extremely high heat input required to raise the temperature of water. One pound of mercury and one pound of water are each provided with the same amount of heat. When the mercury boils at 359°C, water has gone to only 12°C even though it has absorbed the same amount of energy as the mercury.

In hot climates they cooled drinking water by storing it in a porous crock. The heat energy evaporated some of the water molecules into gas form which in turn cooled the container of water. Today in some desert areas a porous vessel is still used for this purpose. The whole outer surface gets damp as the water makes its way into the air, evaporating continuously from the porous surface. Your body uses the same principle in controlling its temperature.

Another advantage provided by water's high heat energy requirement comes from the fact that moisture (H_2O gas) in the air has a supply of latent heat which was absorbed in becoming a gas. When the sun goes down the moisture gives up its latent heat energy and returns to earth in the form of dew, thus preventing a wider variation in temperature. If you have ever camped in a dry desert area you know how much colder it gets at night simply because there is not enough water vapor in the air. The climate near lakes and oceans shows more moderate temperature changes.

There is another side to the coin, however. If water requires so much extra energy to separate its molecules to gas form, then it will return the energy any time it "condenses" or comes back to liquid form. This is the reason steam burns are so severe. If you get your hand or arm across the open end of a kettle of boiling water, the burn is worse than if you contact just the boiling hot water because the steam on condensing always gives off the energy it obtained in becoming a gas. Is this good or bad? Obviously, it depends.

Water's Wettability

Water is far ahead of any other common liquid material in its wetting ability and ability to dissolve other substances. In fact, water is the basis

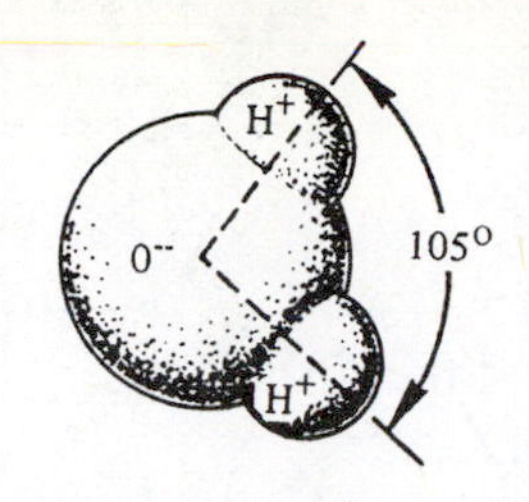

FIGURE 7.7
Representation of uneven charge distribution in water molecule making the oxygen end slightly negative and the hydrogen ends slightly positive.

for most common fluids, like blood, saliva, fruit juices, soft drinks, and so on. Water is a great leveler. It even dissolves rocks. It dissolves nutrients and gases for plants and fish. Also it dissolves the foods you eat and brings needed substances to your cells.

WHY IS WATER SO UNUSUAL?

The basic reason for water's unusual properties is polarity of bonds (chapter 5). The molecule of H_2O involves two polar covalent bonds between hydrogen and oxygen atoms and an angle of approximately 105° between the two hydrogen atoms. (See Figure 7.7.) From the relative electronegativity values O = 3.5 and H = 2.1, you can see that the oxygen atom pulls the pairs of electrons closer to it. The hydrogen atoms acquire a partial positive charge because of the unbalanced distribution. The oxygen "end" of the molecule thus has a partial negative character.

In the photo on the left are three different models for water molecules floating in a dish of water. The model at left of photo is considered the most "realistic." In the photo on the right are models of two water molecules, showing where the hydrogen bond joins the positive end (hydrogen) of one molecule to the negative end (oxygen) of another molecule. The bonding is extended in three dimensions.

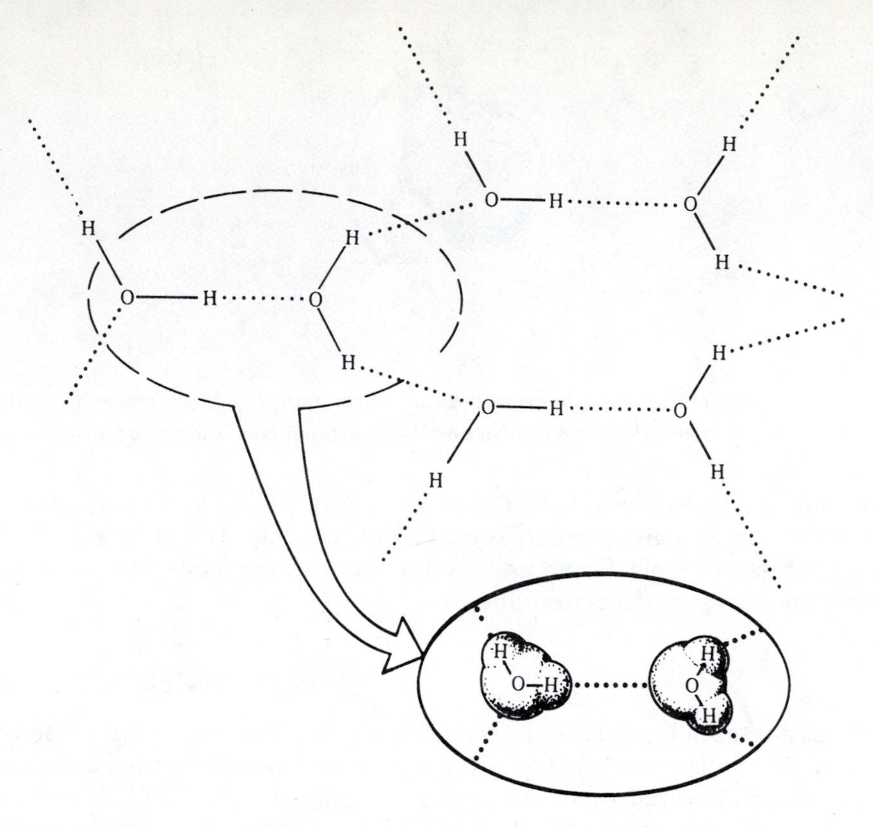

FIGURE 7.8
The "holding together" of neighboring molecules by the pull of positive to negative ends. The dotted lines represent the bonds that keep the H_2O molecules interlocked to each other. These are called hydrogen bonds. The actual bonding is in three dimensions and more irregular.

We have then a basic explanation for water's peculiar properties. As in all scientific explanations, we utilize our theoretical principles to explain why water sticks to itself so strongly. The very high heat capacity and very high heat requirements to convert the liquid water to a gas are merely reflections on the operational level of a basic underlying cause. The conceptual explanation is simple. The water molecules stick together because of the negative and positive charges attracting molecule to molecule. Indeed we end up with a strongly interlocked network. (See Figure 7.8.)

THE *HYDROGEN BOND* IS THE BINDING BETWEEN THE POSITIVE, HYDROGEN *END OF ONE MOLECULE* AND THE NEGATIVE *END OF AN ADJACENT MOLECULE*. IT IS AN ATTRACTION *BETWEEN MOLECULES, NOT WITHIN THE MOLECULE*.

The hydrogen bond is weaker than ordinary ionic or covalent bonds, which are of course involved in the basic chemical combination in compounds. However, it is certainly a very strong influence. This we could surmise from the "unusual" properties which water has. Hydrogen bonds exist between molecules which have hydrogen combined with strongly electronegative elements like oxygen, nitrogen, and fluorine. This type of bonding is important in nylon plastics and in your body protein structures (chapters 9, 10).

Wetting and Dissolving

What about water's wetting ability? And dissolving ability? These are also related to the polar nature of the molecules. Water is the "universal solvent." This means it dissolves a little bit of everything. But water dissolves polar materials the best. Salts, which are an extreme case of polarization, are good examples. As you know, they are made up of positive and negative charged ions. Figure 7.9 illustrates water's effect in dissolving NaCl.

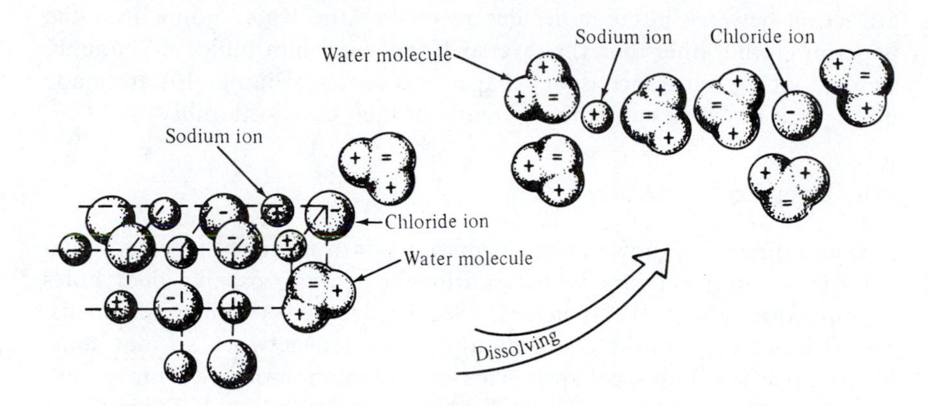

FIGURE 7.9
How NaCl dissolves in water. The negatively charged chloride ions are approached by H_2O molecules which attach their positive ends. This weakens the bond holding the Cl^- ions on the crystal edge. Then a breaking away occurs as the H_2O molecules surround the ions, keeping them separated in the solution. The action with positively charged Na^+ ions is similar. Here the negative ends of H_2O molecules are involved.

Water is not as good a solvent for materials that do not have polarity in their molecules. We say operationally that "like dissolves like." Oil and kerosene (chapter 8) are not polar materials. Consequently water does not dissolve much gasoline or kerosene.

A solution can be defined operationally as a uniform mixture formed by dissolving one material in another.

A *SOLUTION* CONCEPTUALLY IS A HOMOGENEOUS MIXTURE OF MOLECULES, ATOMS, OR IONS OF TWO OR MORE DIFFERENT SUBSTANCES.

Since a solution is a mixture, the proportions of each substance are variable. Solutions play an important role in many environmental operations.

Plants and animals use water-based solutions for carrying nutrients and waste products. Medicines operate through various solution processes.

Most familiar solutions are liquids. However, there are important gaseous solutions like air and solid solutions like alloys. Yellow gold is a solid solution of copper and gold. It contains 90 percent gold, 10 percent copper. Sterling silver is a solid solution of about 92 percent silver and 8 percent copper. You can see that the basic *concept* of a solution leaves open the possibility of the mixture being either solid, liquid, or gas, even though common solutions are liquids.

The phenomenon of wetting is close to that of dissolving. A substance must be wet by water before it dissolves. Some materials like paper and clothing are easily wet by water but do not dissolve. Here again we have polar molecules of water attaching themselves to polar molecules in the cloth or paper which are too large to be separated into solution. Consequently, the water builds up a film on the surface. In other words, the attraction between polar molecules results in the water going into the paper or cloth, rather than the reverse. If the water film builds up enough, we say that the substance is wet. Paper and cotton (chapter 10) are made up of polar molecules and this accounts for their easy wettability.

Why Does Ice Float?

X-ray studies of ice show that the hydrogen bonds arrange the water molecules into a stiff network with repeating hexagonal, or six-sided, holes running throughout the structure. (See Figure 7.10.) When ice melts, energy input shakes up the stiff hydrogen-bonded network so that some of the open structures collapse. This means more molecules move into these formerly open areas and we have a more dense liquid. There is less empty space in liquid water than in ice. This is quite unusual because most materials expand on melting. Again, it is a consequence of the polarity of molecules and the strong, rigid interconnections built up into the hex-

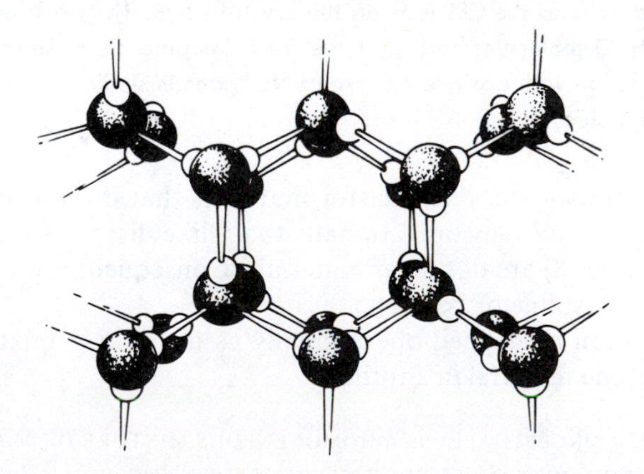

FIGURE 7.10
The open hexagonal structure of an ice crystal. The large shaded spheres of oxygen are each combined with 2 atoms of hydrogen. The spikes represent hydrogen bonds.

agonal network. This three-dimensional open network builds up as we remove energy from the moving molecules of liquid water and freeze it into ice, which is less dense and floats.

In the unusual water property of expansion on freezing we have the basic protective device for preserving life in our rivers, lakes, and oceans. However, if you leave your car out overnight in freezing weather you will likely end up with a "cracked block." Here the water expands as it freezes in the car's cooling system. We avoid this not by changing the polarity of the H_2O molecules, which cannot be done, but by putting other molecules in between the water molecules. We make a solution of an "antifreeze" like an alcohol with water and thus prevent the easy association of H_2O to H_2O by an intermolecular mixing with the alcohol molecules in between.

ACIDS, BASES, NEUTRALIZATION

Among the important water solutions are those of acids and bases. Some chemicals when dissolved in water have a "sour" taste and change the color of certain natural dyes a definite way. For example, litmus, a blue coloring matter obtained from certain plants, turns to a red color. This type of material which changes litmus to red is called an *acid* (Latin: sour). Vinegar, one of the earliest recognized acids, was known from antiquity as the result of fermentation of wines and apple cider. Other materials have a "bitter" taste and turn litmus from red back to blue. These reverse the effect of the acids and are called *bases.* Acids and bases are two important classes of chemical compounds.

AN *ACID* IS A COMPOUND WHICH YIELDS HYDROGEN IONS WHEN DISSOLVED IN WATER. A *BASE* YIELDS OH^- OR HYDROXIDE IONS IN WATER.

The interesting fact here is that if an acid and a base are brought together in equal amounts, the H^+ from the one will react with the OH^- from the other and give us water:

$$H^+ + OH^- \rightarrow H_2O$$

This is called a *neutralization,* that is, the H^+ of an acid is neutralized or made unavailable by the OH^- of the base. (Or we can say that the reverse is true, since this is relative.)

An example of acid is the hydrochloric acid in your stomach, HCl. In water we indicate the following as a simple description of what happens:

$$HCl \rightarrow H^+ + Cl^-$$

The actual mechanism involves a water molecule picking up a proton or H^+ ion from HCl.

$$HCl + H_2O \rightarrow H_3O^+ + Cl^-$$

The H_3O^+ is really a proton with a water molecule attached. All acids in water provide protons which actually exist in "hydrated" form. However, for simplicity, and because all chemists assume it is always hydrated, the extra water is generally left out.

The most common base is lye or sodium hydroxide, NaOH. This is a common article of commerce and is used to make soap, among other things.

Here is how sodium hydroxide neutralizes hydrochloric acid:

$$HCl + NaOH \rightarrow H_2O + NaCl$$

This is a typical neutralization reaction. Water and a salt are formed. Other salt examples were given in chapter 5.

The accepted usage in connection with acids is to write the compound with the H (or H's, if more than one) first to emphasize that they give H^+ ions in solution. For example:

H_2SO_4 for sulfuric acid

$HC_2H_3O_2$ for acetic acid (This is the vinegar acid. Vinegar is about 5 percent acetic acid and 95 percent water. Only one hydrogen atom is active to give H^+.)

H_2CO_3 for carbonic acid (This is the acid formed from $CO_2 + H_2O$, which is important in your breathing mechanism.)

The bases are written with the OH^- shown following the particular metallic or positive ion. Examples are:

NaOH for sodium hydroxide (the "lye" of houshold use)

$Ba(OH)_2$ for barium hydroxide

$Ca(OH)_2$ for calcium hydroxide (in water solution, called "lime water")

Your stomach is naturally acid, containing HCl. Some parts of your system are basic, the blood for example, which is slightly basic. However, the amount of acid or base available varies with the body's needs and with the changes involved in metabolism. Another word for base is *alkali* (Arabic for "plant ashes" where bases were originally found). Thus NaOH is an alkali. It happens to be a strong alkali which means it provides a large quantity of OH^- in water. There are other weaker alkalis which are used in medication. For example, $Mg(OH)_2$ is the product "milk of magnesia" (chapter 14). It is not very soluble in water and therefore has a milky, pasty appearance.

The acid or basic (alkaline) nature of water solutions is very important biologically. Consequently, an arbitrary method has been developed to express the relative acidity or alkalinity of solutions. This is the so-called pH scale. In the commonly encountered ranges the pH scale covers numbers from 0 to 14. We need not be concerned here with the mathematical derivation or conversions. However, a little background will be helpful since the method used to refer to acidity or alkalinity of solutions is usually pH. It is applied to solutions from water in swimming pools to blood plasma.

The hydrogen ion concentration in pure water is very small and comes about from the breakdown of some water molecules:

$$H_2O \rightarrow H^+ + OH^-$$

In pure water only about 2 molecules in a billion will break up into H^+ and OH^-, as shown. There is of course a dynamic interchange going on. From the equation, the amount of OH^- in a certain volume of water would be equal to that of H^+.

The pH scale is based on a standard method of expressing the hydrogen ion concentration in a solution. In pure water, the hydrogen ion concentration, or concentration of H^+ ions, is usually expressed as 0.0000001 (or 10^{-7} in a method using an exponential notation). A Danish brewing chemist got tired of writing all the zeros in describing the concentrations of H^+ ions so he adopted a naming system for a pure water sample with the concentration of 0.0000001 of H^+ ions as having a pH of 7. A pH of 7 would be neutral since the same quantity of OH^- must be present in the volume used, both the H^+ and OH^- coming equally from ionization of H_2O molecules.

Under this system 6 to 0 would be increasingly acid pH; and 8 to 14 increasingly basic pH. Thus we have a scale as follows:

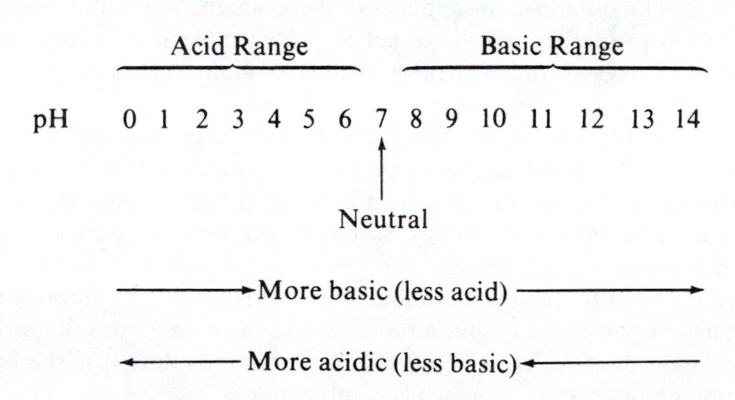

It is important to note that a solution of pH 6 is ten times as acidic as one of pH 7; and a solution with pH 5 is ten times as acidic as one with pH 6, or one hundred times as acidic as one of pH 7, and so forth. Also, a solution with pH 8 is ten times as alkaline or basic as one with pH 7; a solution of pH 9 is ten times as alkaline as one of pH 8, and one hundred times as alkaline as one with pH 7, and so on. Table 7.1 gives the pH for some common solutions.

TABLE 7.1
The pH for Some Common Solutions

Solution	pH
Hydrochloric acid (HCl), 4%	0
Gastric juice (stomach)	1
Lime and lemon juice	2–2.3
Vinegar	2.4
Orange and grapefruit juice	3.2–3.5
Bottled cola drinks	3.5
Peaches; apricots	4–4.5
Urine	6
Saliva	6.3–6.9
Pure H_2O	7
Blood	7.3–7.5
Natural waters	6–9
Ocean water	8
Pancreatic juice	8
Milk of Magnesia ($Mg(OH)_2$)	9
Borax solution, 4%	9.2
Household ammonia, NH_3	11.6
Washing soda (Na_2CO_3), 1%	11.6
Lye (NaOH), 4%	14

COMBUSTION AND FIRE

It is interesting that the oxygen in the air is involved in supporting fires and that oxygen is also present in water, which is important in putting out many fires. The difference in application here again points out the great difference in properties which we get in water molecules, compared to molecules of oxygen and hydrogen (which combine chemically to give water).

When a chemical reaction occurs rapidly with large heat and light output we call it a combustion. Most combustions you are familiar with involve oxygen. However, reactive chemicals other than oxygen (for example, fluorine) can give high energy output in the form of heat and light. Then they are also combustions.

Oxygen is said to "support combustion." This means it combines with a combustible substance to give a rapid release of energy—usually as both heat and light. It is incorrect to say oxygen is combustible. It is the fuel—wood, oil, gasoline, paint—which is combustible.

You may recall the disastrous fire of 1967 in which three astronauts died during a test of a space capsule. A flash fire occurred inside the Apollo I command module which was then being prepared for the first moon landing. The module contained a 100 percent oxygen atmosphere which contributed to the speed of the fire. Nitrogen gas does not at normal temperatures support combustion and this is a tremendous slowing influence on ordinary fires. As you know, air is a mixture of about four-fifths nitrogen gas and only about one-fifth oxygen gas. It is indeed fortunate that we do not live in an atmosphere of only oxygen.

Spontaneous Combustion

You have heard of "spontaneous combustion" which may occur, for example, when old rags soaked with oil or paint are piled in a corner. Oxygen from the air slowly oxidizes the oil in the rags and in the process gives off the usual heat output. This does not immediately start a fire. However, if the heat builds up in the rags (see feedback arrow in Figure 7.11), and more oxygen is available from the air, then the mass eventually

FIGURE 7.11
Spontaneous combustion involves buildup of heat from an exothermic chemical reaction.

may get hot enough to burst into flame. Indeed the buildup of heat makes the whole reaction proceed faster. This is a phenomenon which can easily be avoided by either discarding oily rags at once or storing them in a covered metal can where air is excluded.

What Is Fire?

Fire is the flame, heat, and light resulting from a combustion or a rapid chemical reaction. The classic "triangle concept" of fire (Figure 7.12) involves oxygen, fuel, and heat as the basis of the chemical reaction. However, research in recent years has shown that the chemical reactions involved are not quite as simple as the fire triangle indicates. There really is an underlying "fourth dimension." This is the *reaction chain* which is the basic mechanism of the fire. The reaction chain occurs through the breakup of molecules of the fuel and the recombination of the parts (atoms or groups of atoms) with oxygen.

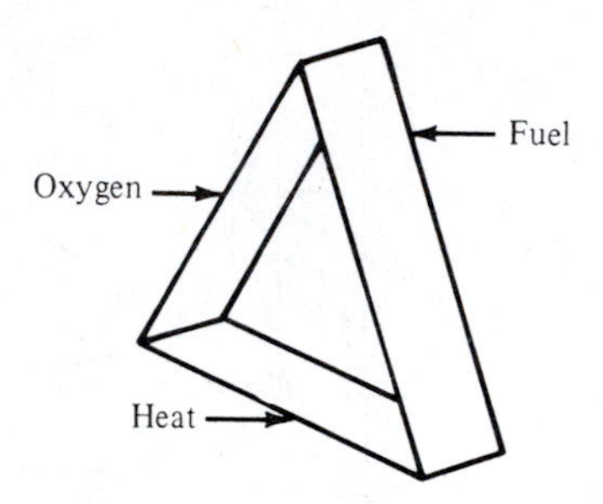

FIGURE 7.12
Visualization of the triangle concept of fire. Fire takes place when the three basic ingredients are present. Opening the triangle by removal of one leg will "put out the fire." Keeping any one factor from joining the other two will prevent the fire from starting.

Consider for instance the gasoline component octane, which has the formula C_8H_{18} (chapter 8). Here is how the reaction chain works. The molecules of octane that are close to the heat source vibrate and move about at enormously increased speed. Then they break apart into fragments that combine with oxygen. The combination gives off energy. Some of this released energy feeds back to other octane molecules which break into fragments, atomic and otherwise. The reaction chain thus continues until (1) all the octane is gone or (2) the oxygen is excluded in some way or (3) the heat is dissipated and the chemicals cooled. The same kind of mechanism is involved with more complex solid material like paper or wood.

This modern concept suggests that we modify the fire triangle into a fire tetrahedron. A tetrahedron is by definition a figure having four sides. (See Figure 7.13.)

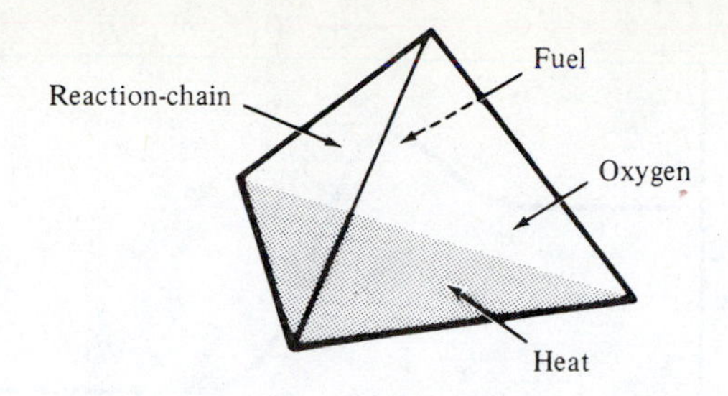

FIGURE 7.13
A more complete visualization of the factors involved in a fire, using the tetrahedron concept.

How do we explain what a fire is? You have the basic ideas in the fire tetrahedron and the fact that light is emitted when electrons fall from higher to lower energy levels in the atom (chapter 4). The flame requirement for fire is then explained by the idea of a reaction chain which shatters molecules and excites atoms by high input of energy. These excited atoms give off light by the electrons falling back to lower levels after being forced out. The incandescent carbon particles in a flame apparently favor a slight yellow glow, and sodium has a specially predominant yellow line in its spectrum. Other colors may at times be emphasized depending on the presence of additional elements. For example, potassium gives a predominant purple light, strontium a red, and calcium a sort of orange. In fireworks the special colors are obtained by adding to gunpowder compounds of elements that give desirable visible lines in their spectrum. Thus sodium compounds are added to give yellow, strontium compounds for red, copper compounds for blue, and barium compounds for green. These compounds are mixed in with the gunpowder base. Modern gunpowder, usually called simply black powder, is a mixture of approximately 75 percent KNO_3, 15 percent carbon as charcoal, and 10 percent powdered sulfur.

What Starts a Reaction Chain?

We can think of a pile of firewood as a stored energy system which waits to be activated. How does the reaction chain start in order to release this energy? Figure 7.14 is a diagram relating the reaction chain in the barbecue reaction. The curve indicates that the reactants ($C + O_2$) have a certain amount of inherent energy. Also, the product carbon dioxide has a basic energy value, but it is definitely *less* than the energy of the carbon + oxygen. Therefore, when carbon dioxide forms from these, we have to account for energy just as we do for mass or quantity of matter. The extra energy comes off as heat. You know this from your barbecue experience. However, you also know that you have to start the fire with a match and kindling, or a match and "fire starter" liquid which burns and then sets

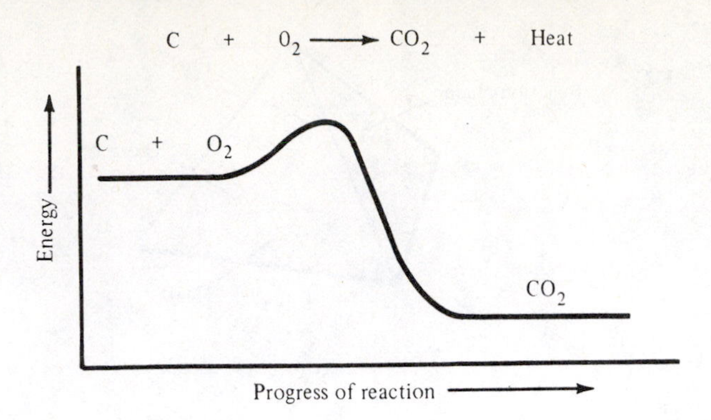

FIGURE 7.14
The reason why charcoal or firewood waits until we need it. Energy must be added to activate the reaction.

the charcoal on fire. What you really do is provide the energy to go up to the top of the curve in Figure 7.14. Then the reaction proceeds and the energy you put in is returned along with extra energy represented by the difference in levels for ($C + O_2$) and (CO_2).

We say that *activation energy* must be supplied to get to the top of the energy hill. This sets up the preliminary reaction-chain agents or intermediates—whatever they are. We know in some cases what these intermediates are; in many others we do not. In any case, the wood or charcoal has to receive an investment of energy first which enables us to then get a larger energy output.

The principle of activation energy applies to all chemical reactions and has been found very fruitful in developing controls of various reactions and in understanding mechanisms of reactions. The mechanism is the way the reaction goes, or the series of intermediate steps through which the molecules reach the final products.

Using the concept of activation energy, you can more readily understand the idea of "kindling" temperature. The *kindling temperature,* or *kindling point,* is the temperature to which a substance must be heated to burst into flame. It is also called the *ignition temperature.*

Kindling temperature is a rough measure of the activation energy required to get a material + air up to the top of the curve so that burning can occur. The activation energy curve is a natural limitation built into all materials to protect us from having everything combustible continually bursting into flame and providing us with unwanted heat.

What Is an Explosion?

Very rapid burning may cause an explosion, as in gunpowder and dynamite. Here oxidation is made to take place very rapidly by providing in the mixture an internal source of oxygen. Then great volumes of gas and much heat are generated in a very short time. The hot gases require many hundreds of times as much space as was formerly occupied by the gunpowder. An *explosion,* then, is a sudden increase in volume caused by gas formation, usually from rapid burning.

A combustible liquid or solid burns relatively slowly and quietly because only a small quantity of molecules at the surface are combining with the oxygen in the air. A candle gradually vaporizes the wax through the wick and the gas provided burns nicely in air simply because the "liquid wax" really doesn't burn. It is the gas that does. (See Figure 7.15.) If we

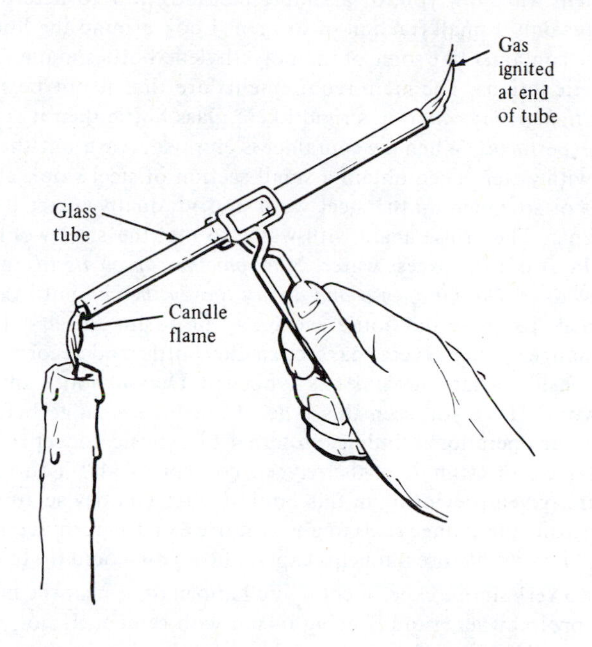

FIGURE 7.15

A simple experiment to show that gases from the wax are involved in a burning candle.

have a combustible gas like hydrogen, however, the surface cannot be limited. You know about the empty spaces in a gas and can understand why gas mixtures react so readily, for example, gasoline or paint solvent vapors and air. If coal, or wood, or flour in dust form is mixed with air we can also get an explosion. This is because of the ease of combining between the now "spread out" fuel and the oxygen in the air. However, mixtures of gases explode most readily.

In this chapter you have studied water, its components, and fire. You can see how certain concepts are extremely important in understanding and properly handling the operations of your environment. Also, you have no doubt noted that astounding changes in properties occur when elements combine to form compounds. You may have also noted an important limitation of science. We can come up with explanations and models for behavior for the materials of our environment; but we do not presume to understand all the causes or "whys" of this behavior. Now perhaps you can understand Einstein's statement, "The eternal mystery of the world is its comprehensibility."

PROJECTS AND EXERCISES

Experiments

1. This experiment will show you, by a simple method, how to determine that the air contains only a small fraction of oxygen. Look around the house for a flexible plastic container like some of the polyethylene bottles made for shampoo or cosmetic lotions. The main requirements are that it not be rigid and that it have a *tight fitting cap.* If it is rigid like a glass bottle then it is not suitable for this experiment. When the container is emptied, wash out the remaining material with water. Then obtain a small section of steel wool, about two square inches or so. Open up the steel wool and gradually insert it into the plastic container. Then rinse again with water so that the steel wool is thoroughly wet. Pour out the excess water. Now *put the cap on tightly* and set it aside for a few days. *Do not open it and do not loosen the cap* until the experiment is finished. Examine the bottle every day or so for the first few days, noting any changes. After several days open the bottle and record your observations. Consider what you think has happened. Do you notice any changes in the steel wool? Have you seen this type of chemical change before? Can you explain your operational findings in terms of a basic concept? Can you give a rough guess or estimate of the oxygen content of air? You can check this with figures given previously in this book. Could you devise a method of actually measuring the change so as to give a more exact percentage figure for oxygen in air? Draw a picture if it helps explain how you would try to do this.
2. This could be a very simple experiment if you happen to be near the beach or a construction project where sand is being mixed with cement. If not, you have probably observed in the past that wet sand has more cohesion than dry sand. You can notice this if you shovel sand. Very dry sand runs off the shovel easily whereas the wet sand can be shoveled in heaps and mounded high on the shovel blade. How would you account for this behavior? Is it a property of the sand alone? Cohesion involves the sticking together of parts of the same substance, adhesion refers to one substance clinging to a different one. Explain the wet sand phenomenon on a conceptual basis—using these ideas along with reference to what really goes on. Go beyond the visual observation that sand clumps more when wet. Could you guess something about the possible internal structure of the sand?

Exercises

3. It has been said that water contributes considerably to the breakup of rocks to provide eventually the soil materials needed by plants. Think about this and give two reasons why water can break up rocks.
4. Describe the mechanism by which water forms a solid on freezing which is less dense than the surrounding water. Then list and describe briefly some of the consequences of this phenomenon.

5. Put ice into a dry glass and pour yourself a cold drink. Watch the formation of that refreshing wet surface on the outside of the glass. How can you account for this in terms of the Kinetic-Molecular Theory and what you now know about the atmosphere?
6. Take one of the quotations at the beginning of this chapter and write a brief paragraph on what it now means to you, especially showing how it can have different but related meanings.
7. When someone has a fever the nurse often places a wet towel or face cloth on the forehead. Operationally this "cools the person off" considerably. How could you give a more basic conceptual explanation?
8. What would you say is the cause of a match going out when you blow it? If your first guess is that the oxygen is blown away, consider that there is such a thing as the mouth-to-mouth method of artificial respiration. If you still cannot think of a good lead here, then look carefully at the fire triangle again.
9. *Surface tension* is a descriptive term for an operational behavior of liquids. The surface acts like an elastic film. Water has a rather high surface tension. A small dry sewing needle laid carefully on water will not sink. (If you have a steady hand and patience, you might try this experiment.) Also certain insects can walk on the surface of water, supported by the "skin." How would you give a brief conceptual explanation of surface tension? Make it simple enough to be explainable to someone who hasn't studied chemistry. If you have to use basic theoretical concepts, explain what they are.
10. Consider the situation where someone says to you that hydrogen and oxygen are dangerous because the Hindenburg blew up. Explain to him why this is not always true. Give some conceptual explanation as to how explosions of these two gases may occur and how they can be prevented. Also give some operational requirements for the proper control of the reaction between hydrogen and oxygen.
11. Write three equations showing the reaction of oxygen with various substances. Are these exothermic or endothermic reactions?
12. If a bright eight-year-old asked you to explain how a fire started in a garage where a painter had thrown old paint rags in a corner, how would you answer?
13. Explain what a fire is. Do you have a more satisfactory understanding of fire since studying chemistry?
14. Pick out two unusual properties of water and give a clear explanation of the theoretical background which makes sense of these properties.
15. Consider where you could have heard the term pH used. Explain briefly what the application involves, including reference to how the pH of a solution may be changed.

SUGGESTED READING

1. Davis, K. S. and Day, J. A., *Water, The Mirror of Science,* Anchor Books, Doubleday & Company, Garden City, New York, 1961. Paperback.
2. Buswell, A. M. and Rodebush, W. H., "Water," *Scientific American,* 194, No. 4, pp. 76–89 (April 1956). Scientific American Offprint No. 262. (Also in *Scientific American Resource Library, Physical Sciences & Technology,* Vol. 2, p. 507.)
3. Revelle, R., "Water," *Scientific American,* 209, No. 3, pp. 92–108 (September 1963).

chapter 8

Gasoline and the Nature of Hydrocarbons

How many ages hence
Shall this our lofty scene be acted over
In states unborn and accents yet unknown?
—SHAKESPEARE

The causes of events are ever more interesting than the events themselves.
—CICERO

We call it black diamonds. Every basket is power and civilization. For coal is a portable climate. It carries the heat of the tropics to Labrador and the polar circle.
—R. W. EMERSON

Carbon is so distinctive in its chemical properties that the study of carbon compounds is given a special status. It is called *organic chemistry* because it began with substances of living organisms like plants and animals. It was originally thought that these special carbon compounds could be made only in living, "organized" cells of plants or animals. Hence the word *organic*.

In the early part of the nineteenth century scientists found that there was no "vital force" of organic life necessary to form carbon into the special combinations called organic molecules. Organic chemistry is now simply the chemistry of carbon compounds. It is still well named since most organic compounds do originate with living organisms. As an example, consider the varied types of plastics which are used in so many common products from polyethylene containers to nylon stockings. We make these from chemical molecules which are derived from petroleum or coal. The petroleum or coal were themselves derived from the structure of plant and animal organisms millions of years ago.

The distinction between organic molecules (like gasoline, butter) and inorganic materials (like rock, minerals, water) is still quite useful. There is something special about carbon compounds.

THE VARIETY OF CARBON COMPOUNDS

If you examine the Periodic Table you will note that carbon is in the center of the first period of eight elements. Leaving out the inert gas neon, which doesn't react, we find carbon exactly in the center of the seven elements:

$$\mathrm{Li\cdot} \quad \dot{\mathrm{Be}} \quad \cdot\dot{\mathrm{B}}\cdot \quad \cdot\dot{\underset{\cdot}{\mathrm{C}}}\cdot \quad \cdot\ddot{\underset{\cdot}{\mathrm{N}}}\cdot \quad :\ddot{\underset{\cdot}{\mathrm{O}}}\cdot \quad :\ddot{\underset{\cdot}{\mathrm{F}}}:$$

The elements on its left tend to lose outer electrons. Those on carbon's right tend to gain electrons to satisfy the octet rule. Carbon is exactly in the middle and it would be hard to either remove or add 4 electrons. Therefore it does neither Instead it forms covalent bonds, that is, it shares electrons. The carbon atom normally forms 4 covalent bonds, thereby achieving the octet and avoiding the high energy requirement for either removing or adding 4 electrons.

There are millions of organic compounds—more than all the compounds of all the other elements combined! Gasoline represents one type in this vast number. Some of the organic compounds which affect you daily are those of your body: proteins, carbohydrates, fats, enzymes, vitamins, hormones. The chemical industry provides many more, including familiar items of your environment such as fibers for clothes and upholstery, paint, varnish, photographic film, drugs, glues, detergents, dyes, rubber, and so on. Let us look at the way carbon gives such variety.

CHEMISTRY AND GEOMETRY OF CARBON ATOMS

The enormous number of organic compounds starts with a carbon atom with 4 valence electrons:

$$\cdot\dot{\underset{\cdot}{\mathrm{C}}}\cdot$$

When carbon forms 4 covalent bonds, each bond consists of a shared pair of electrons. One electron in each bond comes from carbon and 1 from the atom which is forming the bond. CH_4, which is methane gas, is an example of this type of compound. Methane occurs by decomposition of vegetable matter in the earth. It is sometimes observed in flash burning at night over marshes in warm weather (called marsh gas). It is the major component in natural gas used for house heating and cooking. We could write the electron dot formula for methane as

H
H:C:H
H

However, like all formulas, this gives a very inexact picture of the actual molecule. Methane is not a flat, planar molecule. The 4 pairs of electrons in covalent bonds represent 4 regions of high electron density around the central carbon atom. Since the pairs represent concentrations of negative charge they would be expected to repel each other and take up probability "locations" as far apart from each other as possible. Imagine 4 points lying on a sphere, placed as far apart from each other as possible. You might intuitively conclude that the 4 points are located at the corners of a regular tetrahedron (described in chapter 7). You can get a better perception of this if you work with models, like a styrofoam ball and 4 round toothpicks (Experiment 2).

Geometry study and X-ray studies of organic molecules confirm our concept of a tetrahedral structure for carbon chemistry. Figure 8.1 illustrates this model.

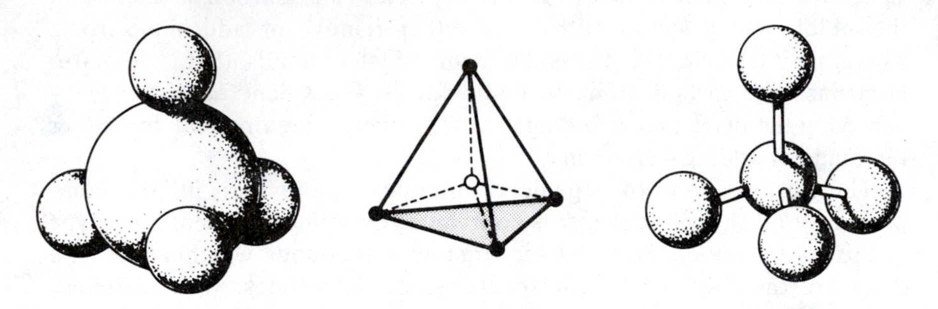

FIGURE 8.1
The basic structure of carbon chemistry. On the left, 4 small balls loosely attached to the surface of the large sphere are considered to be repelling each other. The central picture shows more clearly the tetrahedral or four-sided figure assumed by 4 points equally spaced from each other. The central carbon atom is connected to 4 other atoms by dotted lines representing the 4 covalent bonds. On the right is a ball-and-stick model for methane, CH_4.

Here are some of the ways the chemist pictures the structure of methane.

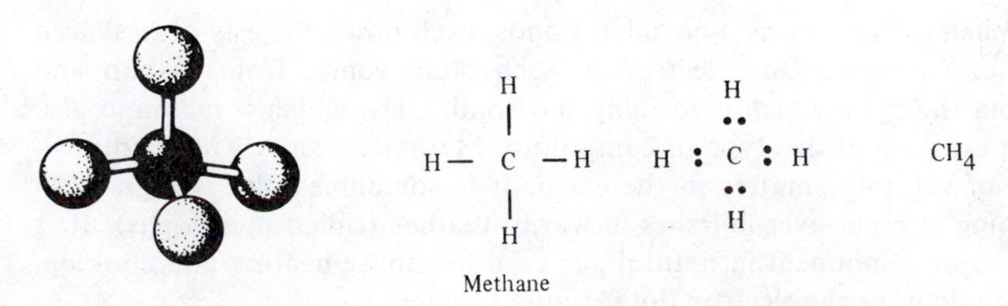

All four representations mean the same thing. Yet none are really what is there in the methane running through the gas lines feeding the fire in your stove or hot water heater.

GROWING CHAINS OF CARBON ATOMS

There is another reason carbon forms so many compounds. Each carbon atom can form bonds with other carbon atoms. Consider one of the covalent bonds of methane. Join it to another carbon atom instead of to a hydrogen atom. Then we have the beginning of the next member of the methane family. This compound containing 2 carbon atoms has the trivial name ethane:

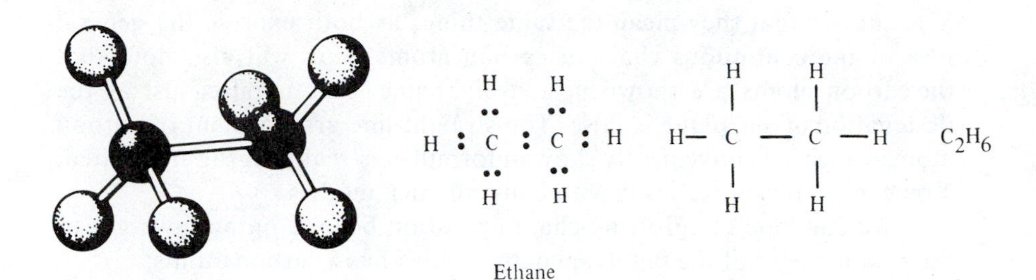

Ethane

You will note that the other covalent bonds of each carbon atom are satisfied by shared electrons with hydrogen atoms. Both methane and ethane are called *hydrocarbons* because they are made from only hydrogen and carbon. There are thousands of additional possibilities of hydrocarbons by adding more carbon atoms to the growing chain.

For example, we may add another carbon atom in place of one of the hydrogen atoms in ethane. Then we get propane. You have heard about this chemical in various "bottled gas" applications. It is of course necessary to also add hydrogen atoms to satisfy carbon's need for four covalent bonds.

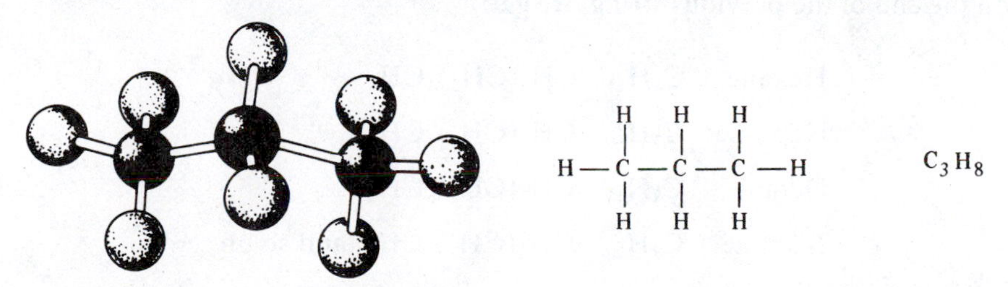

Propane

If we add another carbon in the continuing growing chain we can get a compound containing 4 carbon atoms in a row. This is butane, which is also used in bottled gas heating applications.

```
     H  H  H  H
     |  |  |  |
  H—C—C—C—C—H        C4H10
     |  |  |  |
     H  H  H  H
```

Butane

The chain formula shown for butane is called a structural formula because it shows in simple fashion the structure of the molecule. For even more convenience, a condensed structural formula is sometimes used. For butane this could be:

$$CH_3CH_2CH_2CH_3 \quad \text{or} \quad CH_3(CH_2)_2CH_3$$

Compare these with the full structural formula of butane given above. You can see that they mean the same thing, as both express the general idea of the continuous chain of carbon atoms. You will also note that the carbon atoms are shown in a straight line. This is false, just as the designation in one plane is false. The straight-line arrangement of carbon atoms which we conveniently show in formulas is really zigzag or twisted. You can see this more easily with ball-and-stick models.

If we continue the growing-chain operation beyond butane, we get to the next member of the family, pentane, which has 5 carbon atoms:

```
     H  H  H  H  H
     |  |  |  |  |
  H—C—C—C—C—C—H
     |  |  |  |  |
     H  H  H  H  H
```

$CH_3CH_2CH_2CH_2CH_3$ $\quad$ $CH_3(CH_2)_3CH_3$

Structural formula — Condensed structural formulas

Pentane (Molecular formula C_5H_{12})

You can see here that the so-called molecular formula, which indicates the number of atoms of each kind of element in the molecule, does not describe their arrangement.

Continuing the growing chain operation, adding one carbon at a time on the end of the previous string, we get:

Hexane	C_6H_{14}	$CH_3(CH_2)_4CH_3$
Heptane	C_7H_{16}	$CH_3(CH_2)_5CH_3$
Octane	C_8H_{18}	$CH_3(CH_2)_6CH_3$
Nonane	C_9H_{20}	$CH_3(CH_2)_7CH_3$, and so on

WHAT IS GASOLINE?

Gasoline is a mixture of hydrocarbons chiefly in the range having 5 to 10 carbon atoms per molecule, sometimes designated as C_5 to C_{10} range. Such a description leaves open the possibility of a great variety in the mixture. The range of carbon atoms is one reason why there are so many grades of gasoline. There are two other major reasons for the varieties of gasoline: (1) there are different kinds of molecules in gasoline having the *same molecular formula;* and (2) manufacturers use various additives to get desired characteristics.

Different Chemicals—Same Formula

The idea of different chemicals having the same formula sounds a little weird when you first hear it. The inorganic compounds you have considered so far have formulas like NaCl, or $MgCl_2$, or H_2O, and each of these represents only one chemical substance. However, when you say butane is C_4H_{10}, the situation is not the same. It has been discovered in the laboratory that there are actually two different butane compounds having different properties but with the same molecular formula, C_4H_{10}! For example, the two chemicals have quite different boiling points and melting points, as shown below.

	Butane Number 1 (normal butane)	*Butane Number 2 (isobutane)*
Melting Point	−138°C	−160°C
Boiling Point	−0.5°C	−11.7°C

The existence in the operational world of these two different chemicals having the same formula can be explained easily by using the basic conceptual approach. There are two different ways of arranging 4 carbon atoms and 10 hydrogen atoms in a molecule. You could figure this out by trial and error—on paper or by ball-and-stick models.

```
    H   H   H   H
    |   |   |   |
H — C — C — C — C — H
    |   |   |   |
    H   H   H   H
```

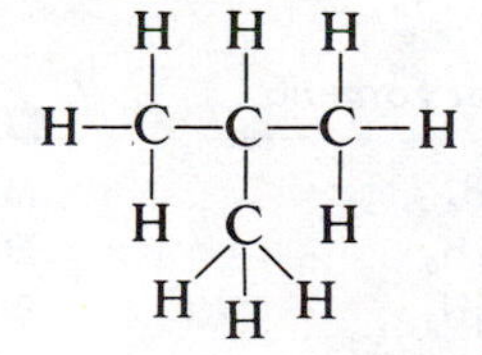

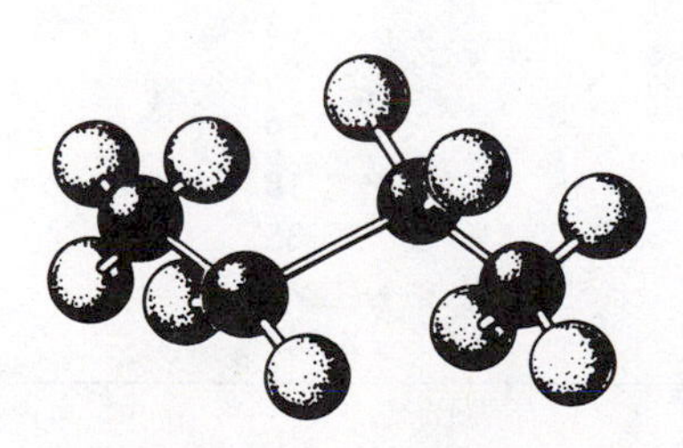

Normal butane (C_4H_{10})
(the name given to arrangement of carbon atoms in a continuous chain)

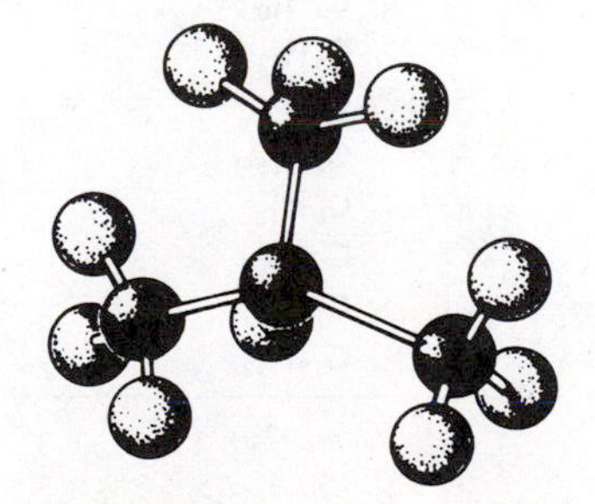

Isobutane (C_4H_{10})
(the name given to a branched chain arrangement of carbon atoms)

Molecules having the same molecular formula but different structural arrangements of the atoms are called *isomers* (Greek: same parts) and this phenomenon is called *isomerism.* Isomerism is a very common occurrence among organic compounds and contributes to the fact that there are so many different organic chemicals. Indeed, it is also of extreme importance in your body chemistry where structure and shape are of vital importance for proper functioning.

Ball and stick models for the two butane isomers. Note that the tetrahedral arrangement of the four bonds for each carbon atom requires the chains to be bent from a straight line. (Photo by Gary R. Smoot.)

If we consider the assembly of carbon atoms into growing chains past butane we find even more examples of isomers, because the more carbon atoms in the molecule, the more variety of ways they can be arranged. This again has been confirmed operationally. The number of isomers of a few of the hydrocarbons in the methane family are given below.

Molecular Formula	*Name*	*Number of Isomers*
CH_4	Methane	1
C_2H_6	Ethane	1
C_3H_8	Propane	1
C_4H_{10}	Butane	2
C_5H_{12}	Pentane	3
C_6H_{14}	Hexane	5
C_7H_{16}	Heptane	9
C_8H_{18}	Octane	18
C_9H_{20}	Nonane	35
$C_{10}H_{22}$	Decane	75
$C_{15}H_{32}$	Pentadecane	4,347

Calculations show the number of isomers for the hydrocarbon $C_{20}H_{42}$ to be 366,319. Very few of the large number of possible isomers have been isolated for the hydrocarbons past 10 carbon atoms.

Octane Number

The gasoline used in your car probably contains about a thousand different kinds of molecules. We have to get some kind of rating system for the different grades of gasoline caused by the differences in the mixtures. One common method is octane number.

You have often heard of high octane and low octane gasoline. Filling stations today provide a range of gasoline mixtures from "below regular" to "premium." These gasolines differ in octane number, which varies from about 86 for regular up to 100 for the premium grades. The gas with low octane numbers produces more "knock" or "ping" than those with high octane numbers. The pinging or knocking is caused by uncontrolled burning of the gas in the cylinders.

The mixture of gaseous gasoline and air is compressed in the car's cylinder and exploded by means of a spark plug. When the gas mixture expands and pushes against the piston, the more its volume increases, the greater the work performed. Consequently, attempts were made by automotive engineers to develop an engine with a large ratio between the volume after expansion to that before—the compression ratio. However, high compression engines knock badly with low octane gasolines. It was found by experience that the tendency toward knocking depends, among other things, on the type of gasoline molecules. Branched hydrocarbons give better performance than the continuous-chain type (the so-called normal hydrocarbons).

The octane number of a particular gasoline is based on comparison of its antiknock quality with that of a mixture of two hydrocarbons called normal heptane and isooctane.

$$CH_3CH_2CH_2CH_2CH_2CH_2CH_3$$

Normal heptane

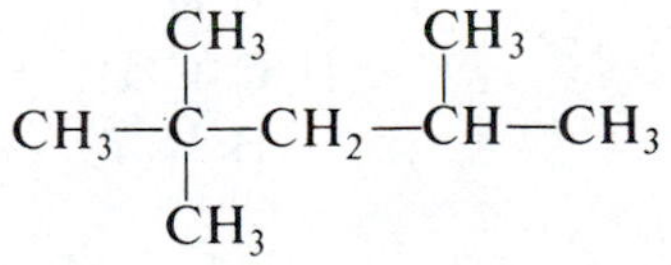

$$\begin{array}{ccccccccc} & & CH_3 & & & & CH_3 & & \\ & & | & & & & | & & \\ CH_3 & - & C & - & CH_2 & - & CH & - & CH_3 \\ & & | & & & & & & \\ & & CH_3 & & & & & & \end{array}$$

Isooctane

The continuous chain of 7 carbon atoms in the normal heptane molecule knocks severely. The highly branched compound, called arbitrarily isooctane, gives excellent no-knock properties. (Isooctane is one of the 18 possible isomers of octane, C_8H_{18}.) Mixtures of these two hydrocarbons are used as reference fuels. A gasoline's octane number is equal to the percentage of isooctane in a reference fuel having the same knock quality as the gasoline, when both are tested in a standard engine. If a gasoline knocks like a reference fuel having 94 percent isooctane, it is given an octane number of 94. This does *not* mean it contains 94 percent isooctane. It simply has knock properties like the 94 percent isooctane reference fuel. All gasolines are carefully blended products. Even though pure isooctane would provide a gasoline with an octane number of 100, it

Ball and stick models for normal heptane, on the left, and isooctane, on the right. Note the extensive branching in the isooctane molecule, accounting for its better performance in an engine. (Photo by Gary R. Smoot.)

would be practically useless in an automobile. The engine would have trouble starting without some of the lower-boiling components containing molecules of less than 8 carbon atoms.

Lead in Gasoline

One way of reducing the knocking properties of gasolines was found to be addition of a special lead compound. It is called tetraethyl lead, or TEL, and has the formula $Pb(C_2H_5)_4$, which is based on the fact that there are four C_2H_5 units to each lead atom. The C_2H_5 unit can be visualized as being derived from ethane by removal of a hydrogen atom and this accounts for the name *ethyl.*

$$\begin{array}{c} \mathrm{H\;H} \\ \mathrm{H\!:\!\overset{..}{\underset{..}{C}}\!:\!\overset{..}{\underset{..}{C}}\!:\!H} \\ \mathrm{H\;H} \end{array} \longrightarrow \begin{array}{c} \mathrm{H\;H} \\ \mathrm{H\!:\!\overset{..}{\underset{..}{C}}\!:\!\overset{..}{\underset{..}{C}}\cdot} \\ \mathrm{H\;H} \end{array} + \mathrm{H\cdot}$$

Ethane — The "ethyl" grouping — Hydrogen atom

Until recently all gasolines for automobiles contained some tetraethyl lead. The quantities used to improve knock rating are fairly small. About a thimbleful of TEL is added per gallon of gasoline on the average. The ethyl gasolines contain slightly more. Leaded gasolines are easy to identify because of red dyes which are added along with the TEL. The lead compound is volatile and can also be absorbed through the skin to cause lead poisoning. Hence a signal is given by the red dyes. (Gasoline as a part of air pollution is discussed in chapter 16.)

Gasoline Behavior and Kind of Hydrocarbons

Using conceptual know-how on gasoline behavior, organic chemists can design gasoline mixtures to operate properly under special local conditions of climate and altitude. They even adjust gas mixtures as the seasons change in a specific locality. For example, gasolines are designed for use in warm weather by making a mixture containing more molecules with larger numbers of carbon atoms. These heavier hydrocarbons are less easily changed to a gas from the liquid state than are the lighter hydrocarbons with smaller numbers of carbon atoms in the molecule. Then the high summer temperatures will not turn the gasoline mixture into a "vapor" or gas before it reaches the engine. This saves loss of gasoline by evaporation from the fuel tank. But more importantly, if too many light molecules go into the gaseous state in the fuel lines, they cause a condition called vapor lock which blocks gasoline flow and stalls the engine.

You can understand vapor lock better now that you have studied some chemistry. When gasoline vaporizes or forms a gas in the fuel lines it reduces the amount of fuel pumped to the cylinders, since vapor or gas takes up so much more space than liquid. A lot of "empty space" is being pumped instead of the concentrated fuel in liquid gasoline. The engine thus runs erratically or sputters and stops. The car will run again after it cools down and the gasoline turns to liquid.

Cold winter temperatures require the opposite adjustment. Here gasoline chemists design a mixture containing higher percentages of the lighter hydrocarbons. These molecules with fewer carbon atoms are easier to convert to a gas from the liquid state. Therefore, they make it easier to start your engine when it is very cold.

Similar adjustments are made for high and low altitudes. More of the heavier hydrocarbons are used in mountainous regions where the lighter ones would evaporate too easily, causing vapor lock and excess tank losses.

The tie-in between the operational behavior of boiling and the concept of length of carbon chains in molecules is dramatically evident from the following data. The *n* in the name column signifies the normal hydrocarbon, that is, the molecules have a continuous chain of carbon atoms, without branching.

Name	*Molecular Formula*	*Structural Formula*	*Boiling Point*
Methane	CH_4	CH_4	−162°C
Ethane	C_2H_6	CH_3CH_3	−89
Propane	C_3H_8	$CH_3CH_2CH_3$	−45
n-butane	C_4H_{10}	$CH_3(CH_2)_2CH_3$	−0.5
n-pentane	C_5H_{12}	$CH_3(CH_2)_3CH_3$	36
n-hexane	C_6H_{14}	$CH_3(CH_2)_4CH_3$	68
n-heptane	C_7H_{16}	$CH_3(CH_2)_5CH_3$	98
n-octane	C_8H_{18}	$CH_3(CH_2)_6CH_3$	125
n-nonane	C_9H_{20}	$CH_3(CH_2)_7CH_3$	151
n-decane	$C_{10}H_{22}$	$CH_3(CH_2)_8CH_3$	174

Recognizing the difference between liquid and gas, you can appreciate why the lower members of the methane family are gases at room temperature. This includes methane, CH_4; ethane, C_2H_6; propane, C_3H_8; and butane, C_4H_{10}. When the length of carbon chain reaches 5 in pentane we have more difficulty getting the molecules separated. Hence pentane is a liquid at room temperature (21°C or 70°F). In liquids the molecules slip past each other rather easily and this condition prevails with the hydrocarbons from pentane up to molecules having about 15 carbon atoms. Above $C_{16}H_{34}$ the hydrocarbons are solids at room temperature.

Where Does Gasoline Come From?

Most gasoline originates in petroleum, which is a mixture of gaseous, liquid, and solid hydrocarbons formed in the earth by decomposition of vegetable and animal matter.

No one knows exactly how petroleum was formed nor the special conditions that lead to formation of coal in some areas and petroleum in others. Scientists today generally subscribe to the organic theory of formation of these materials. We know by analysis that carbon is found in nearly all plant and animal material (salts and water being exceptions). Carbon in organisms is of course present not as the free element but in various compounds—chiefly with hydrogen, oxygen, and nitrogen.

At left, an old photo showing a typical scene in early oil fields. At right, a view showing modern oil pumping equipment.

When the primitive plants and animals died, their remains settled on bottoms of lakes, swamps, and seas. Sand and mud sediments eventually built up to provide high pressure that compacted the organic layers beneath. During time periods estimated at millions of years, various natural processes involving bacteria, heat, and pressure produced complex chemical changes of decomposition. This gave gas, oil, and coal—all now containing the same carbon atoms that were formerly part of plant and animal systems.

CARBONIZATION

Our ordinary experience will back up this kind of explanation of carbon formation. For instance, you are no doubt familiar with food that has been "burned" in pots and pans on the stove. This is not a complete combustion such as would occur if you threw food into a fire. The burned steak, overdone toast, or blackened residue left with vegetables when water is allowed to boil away are indications of how you can partially decompose organic matter by heating.

We can accomplish *carbonization* a little more carefully in the laboratory or chemical plant. Then it is called *destructive distillation.* (See Figure 8.2.) For example, we can take a few pieces of wood and put them in

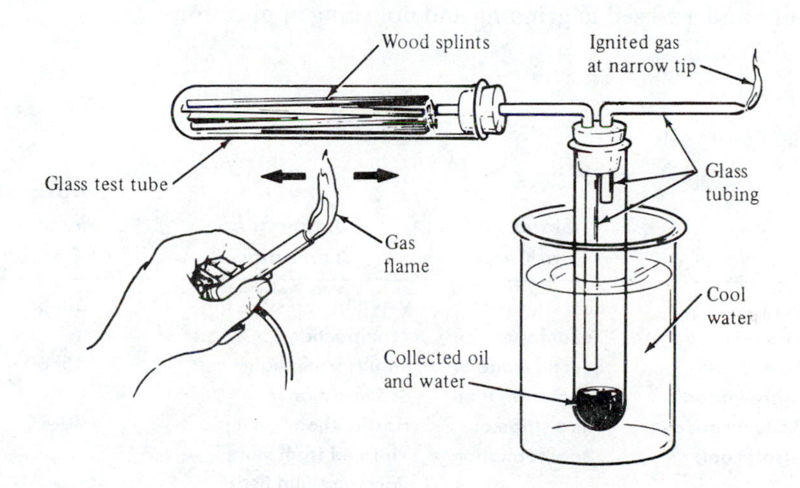

FIGURE 8.2
Experiment showing the destructive distillation of wood to give oil and gas.

a test tube with an exit for gases to escape. Heating the wood provides large quantities of gas, some of which can be condensed to a dark, oily liquid. The charred remains of the wood has the black color which we recognize as carbon. The system is set up to keep out oxygen from the air. The carbon remaining is really charcoal, a common article in your environment even if you do not burn your vegetables. You can see that the charcoal does not result from "burning" wood, as the air has been excluded. Consequently, there is no direct burning—which would convert the charcoal to carbon dioxide.

What happens in the decomposition then is apparently the breakup of the organic material in the wood, recombinations of various kinds, and eventual driving off of volatile compounds. If we heat the wood long enough the residue is chiefly carbon along with some inorganic salts which do not vaporize. This is the general nature of how petroleum and coal were probably formed in the vast laboratory of the earth's crust. It is admittedly an oversimplification.

DIAMONDS AND COAL: ATOMS OF CARBON INVOLVED

The relationship between diamonds and coal can be seen in Table 8.1. The diamond in the series of fossil fuels seems a natural culmination of the trend toward harder and harder materials formed from a basic carbon skeleton. Also note the trend toward higher and higher carbon percentages. Diamond is the hardest naturally occurring substance and it is believed by scientists to have formed when carbon was subjected to extremely high heat and pressure. Using this concept, man-made diamonds are formed in a press that creates conditions thought to exist about 250 miles beneath the earth's surface. The diamonds are the size of sand grains and are used in grinding and polishing applications.

TABLE 8.1
Types of Solid Fossil Fuels

Name	*Status in Coal Chain*	*Conditions for Formation*	*Approximate Percent of Carbon*
Peat (swamps, bogs)	First stage of coal formation	Very little pressure for compaction	50–55
Lignite (brown coal)	Second stage of coal formation	Results from slight compaction	55–60
*Bituminous coal (soft coal)	Third stage of coal formation	Harder than lignite; formed from more pressure and heat	70–80
**Anthracite (hard coal)	Fourth stage of coal formation	Harder than bituminous; formed under even higher heat and pressure	80–95
Diamond	? Fifth stage of compaction	Very hard—formed under extremely high pressure and heat conditions	100

*This is by far the chief form of solid fossil fuel; used for industrial heating and electric power plants.
**This lies deeper in the earth than bituminous; accounts for only a small fraction of coal mined.

The amazing thing about diamond to the chemist is not necessarily its brilliance when cut and polished. It is this: the property of extreme hardness is a result of the way the atoms of carbon are held in the diamond crystal. (See Figure 8.3.)

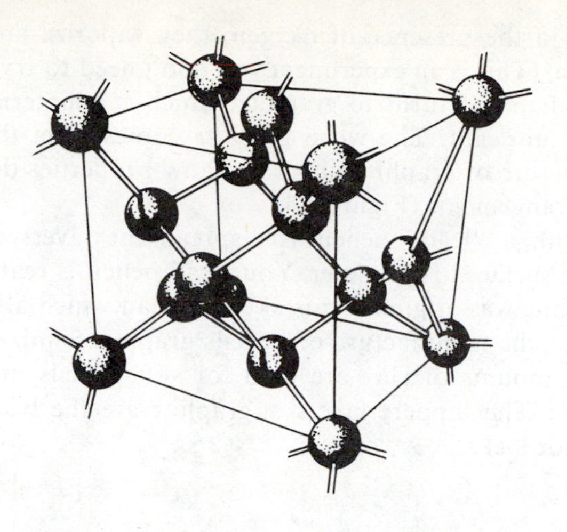

FIGURE 8.3

A diagram of the type of carbon structure in diamond. Each carbon atom is surrounded by 4 other carbon atoms in a tetrahedral interlocking arrangement. The whole crystal is thus a tightly interlinked array of carbon atoms. There are really no molecules in the structure as there are in the structure of ice. The whole crystal itself is a giant molecule. The hardness is then a result of this very rigid and strong structure where every atom is tied in tightly to the overall network.

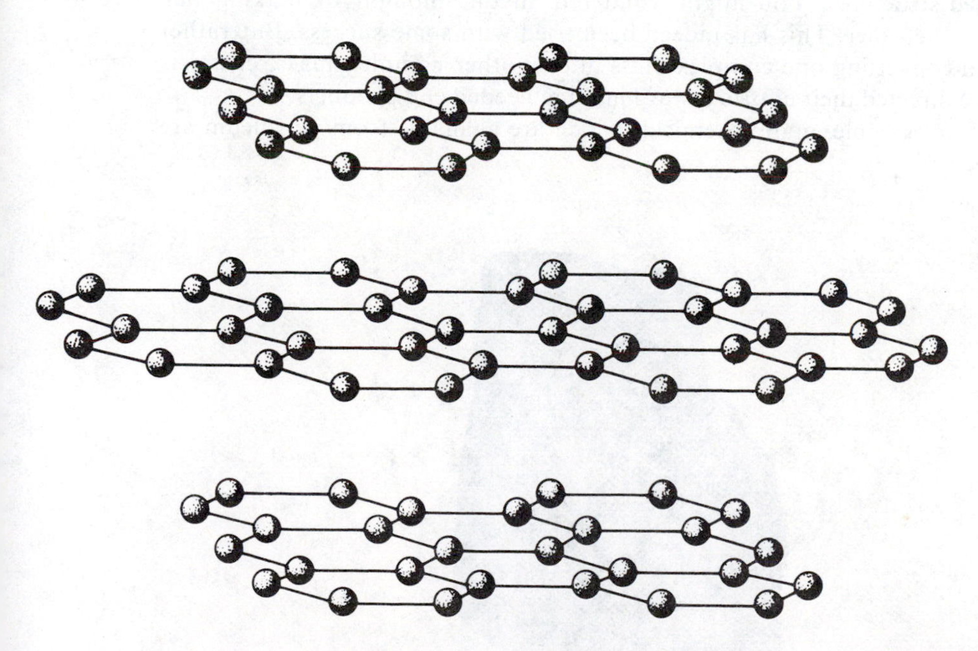

FIGURE 8.4

The arrangement of carbon atoms in graphite. Here we have layers of atoms with each layer complete in itself. Each atom is bonded to two others by single covalent bonds and to a third by a double bond, which means 2 pairs of electrons between the 2 atoms. These extra electrons in the double bonds are "delocalized" within the hexagonal structure. This is what accounts for the high electrical conductivity of graphite, compared to diamond which does not conduct electricity.

If diamonds are heated in the presence of oxygen, they vaporize and burn to give carbon dioxide. (This is an experiment you don't need to try.) If heated without oxygen, diamond turns to graphite, which is a material so soft that it is used as a lubricant. The word *graphite* comes from the Greek "to write." The structure of graphite illustrates how properties depend on internal atomic arrangements (Figure 8.4).

Whenever you write with a "lead" pencil you spread the layers of carbon atoms on the rough surface of the paper. Your lead pencil is really made from graphite. Graphite was originally mistaken for lead which also makes marks on paper. For the manufacture of pencils graphite is mixed with purified clay. Small amounts of clay are used for soft pencils and more for the harder pencils. The slippery layers of graphite are the basis for "dry lubricants" used for locks.

WHERE DO COAL AND PETROLEUM MEET?

The structure of the various kinds of coal is still not completely understood. The major part of coal structure is thought to consist of hexagonal rings which are grouped by various kinds of interlocking fusion. This is related to the structure of graphite.

Both petroleum and coal are then complex mixtures of various carbon-based structures. You might wonder if anyone thought of making one from the other. This has indeed been tried with some success. But rather than converting one complex mass into another complex mass, chemists have directed their efforts toward making needed end products.

For example, many chemicals which are available from petroleum are

Coal, or black diamond, which is useful for many other purposes besides providing heat in homes and power in electrical generating plants and manufacturing plants. (Photo by Gary R. Smoot.)

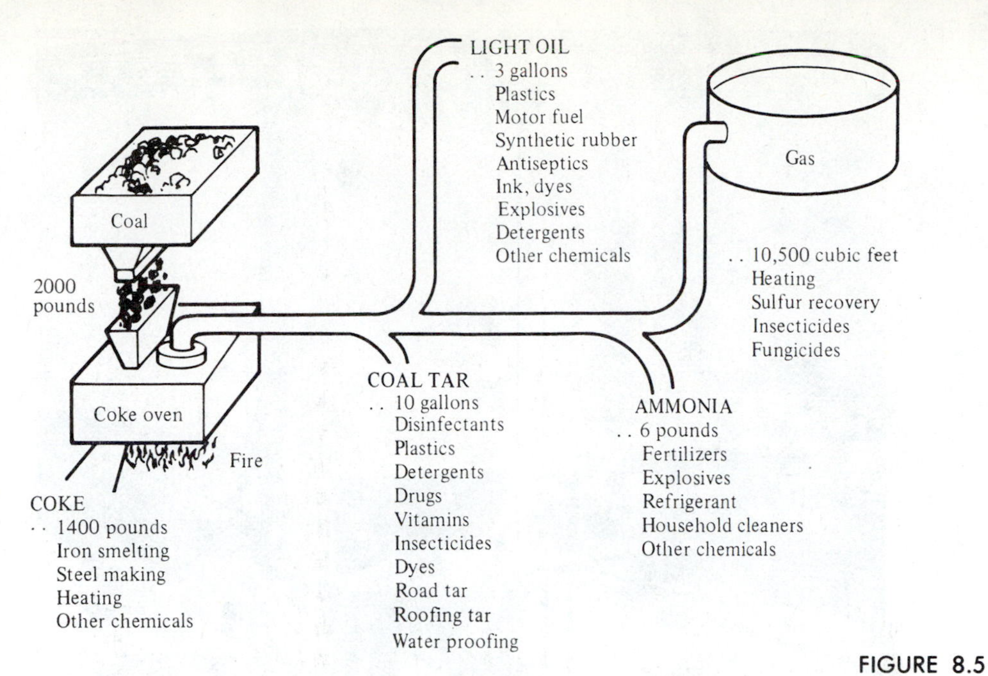

FIGURE 8.5

Schematic showing just a few of the many useful products which are obtained by heating coal in the absence of air.

also derived from coal by destructive distillation. One ton or 2000 pounds of coal yields about 1400 pounds of coke and approximately 13 gallons of coal tar and oil, along with 6 pounds of ammonia, and over 10,000 cubic feet of coal gas. (See Figure 8.5.) The U.S. production of coal tar alone now runs to more than a billion gallons per year.

The ammonia is used for household applications and fertilizers, while coal gas is used for heating. The liquid tar and oils provide hundreds of specific chemicals many of which parallel those obtained from petroleum. From these come great varieties of practical products like detergents, plastics, and drugs. Their separation resembles the procedures used in petroleum refining which we will consider below.

In addition, treatment of coal with hydrogen gas provides many other hydrocarbons like methane, ethane, propane and butane. Other conversion processes for coal involve the reaction of finely powdered coal with water to form a gas mixture. This gas mixture is then converted to a diesel fuel and into a type of gasoline.

The Energy "Crisis" of the 1970s

The energy shortage which surfaced in the mid-1970s, with scarcity of heating fuel and long lines of cars waiting for gasoline, was really not a sudden occurrence. The term "crisis" is misused here. *Crisis* comes from the Greek root meaning decision or turning point—as in a sickness where

A modern coke plant. (Courtesy Koppers Company, Inc.)

The photo on the left emphasizes our large dependence on coal. Dipper of a surface mining shovel dwarfs mechanic (left of photo) as it lifts 420,000 pounds of earth and rock from above a coal seam. The photo on the right gives visual indication that surface strip mining need not leave the earth barren and unsightly. Land at right on this West Virginia hillside was mined in 1969 and seeded in fall of 1970. The picture was taken in June, 1971. Land in left background was mined in 1955 and revegetated in 1956, fifteen years earlier. (Photos courtesy National Coal Association.)

a turning point is suddenly reached toward improvement or worsening. The energy problem has not arrived at a sudden turning point. It has been building up for many years, with an increasing population using energy in often wasteful ways. And the energy shortge promises to be with us for a long time. It will undoubtedly affect our life styles in many ways as we move into the 1990s and beyond.

There are really two major approaches to the problem. One is short range, the development of oil and synthetic fuels from coal, of which in the U.S. we have a supply estimated to last at least 300 years. Then there is the long-range problem of alternate sources of energy. Here there is really no basic energy shortage—only the need to apply concepts and technology to make energy available.

Here is a list of some research areas now being considered:

1. Solar energy. A large proportion of the sun's energy is now wasted. Our use of solar energy in space satellites has shown that this is a reliable source.
2. Geothermal energy from heat deep within the earth.
3. Wind power, like that used to operate windmills.
4. Oil recovery from shale and tar sands which are plentiful in the U.S. and Canada.
5. Gasification of coal, which has already been shown to be feasible.
6. Conversion of coal to gas and oil (see Figure 8.6).
7. Tidal energy.
8. Fuel from wastes (chapter 15).
9. Energy from nuclear fission and nuclear fusion.

General view of pilot plant for conversion of coal to synthetic crude oil and clean fuel. (Courtesy FMC Corporation.)

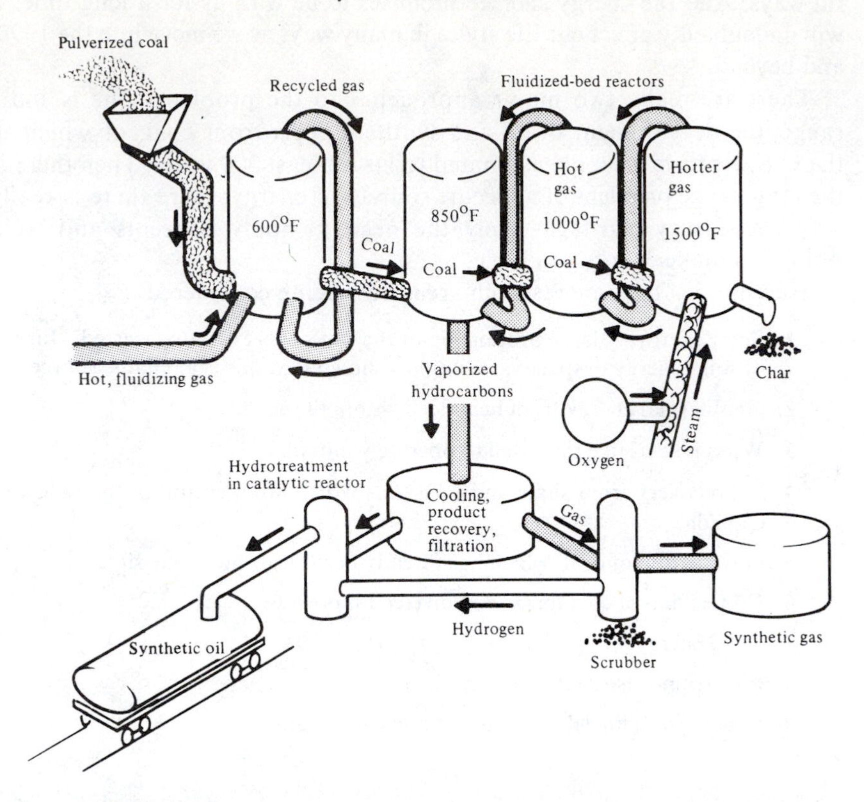

FIGURE 8.6
How one coal-to-oil plant operates. This pilot plant was developed by the FMC Corporation for the Office of Coal Research of the U.S. Department of the Interior. Original research began in 1962 and the plant was able in the 1970s to process about 50 tons of coal a day to provide raw oil and a high grade synthetic crude oil. Fine coal is passed through four successively hotter reaction vessels. The coal is heated and fluidized largely by gases recycled from the last and hottest stage. Out of the reactors comes a mixture of vaporized hydrocarbons that are cooled and separated into two streams. One is a gas which goes into storage for use. The other is a stream of oil fractions, which are hydrogenated (by hydrogen developed in the process) into crude oil. A third product, char, is removed to be used in heating boilers or generating more gas and hydrogen. Costs are still high but projection plans for the 1980s call for large scale conversion of coal to oil and gas at competitive prices.

The latter category involves delving deep into the nucleus of the atom and breaking nuclei apart or combining nuclei. (This is the realm of physics.) The energies released are of course enormous. It was the splitting or fission of nuclei which resulted in the "atomic bomb." Today the trend is toward the peaceful use of nuclear reactions, such as radioactivity involved in the treatment and diagnosis of disease, and for the production of power. The International Atomic Energy Agency reports that well over 200 nuclear power plants are operating around the world. Some nations, such as Belgium, Switzerland, and Sweden already obtain 25 percent of their electricity from nuclear power. By the year 2000, it is estimated that nuclear energy may supply about 20 percent of U.S. energy.

PETROLEUM REFINING

Daily consumption of petroleum on a worldwide basis is over 1,500,000,000 gallons. (That is over $1\frac{1}{2}$ *billion* gallons per *day!*) This makes petroleum the single most important item of international trade. The U.S. is one of the major producers of petroleum, averaging more than 3 billion barrels per year. (Each barrel contains 42 gallons.) An average well in the U.S. produces about 13 barrels or 546 gallons per day.

Yet this crude oil as it comes from the earth is a sticky, practically useless, mess. However, it is a mixture of thousands of different compounds which by proper methods of separation and chemical change provide fuels, lubricants, illuminants, solvents, asphalt road materials, roof-

Part of a modern petrochemicals plant near Ponce, Puerto Rico. Petrochemical plants produce many varieties of organic molecules used in such products as plastics, paints, cosmetics, drugs, insecticides, anti-freeze, brake fluids, fuel additives, and permanent-press clothing. (Courtesy Union Carbide Corporation.)

ing tar, fibers, synthetic rubber, detergents, paints, plastics, fertilizers, insecticides, rust preventatives, medicines, and so forth, and so on. *Petrochemicals* is the special name given to chemicals made from petroleum.

Petroleum refining covers the various processes of getting useful products from the gooey, sticky mess which is pumped out of the ground in the oil fields. Refining consists of three basic procedures: (1) separation of the crude oil into smaller "fractions"; (2) conversion by chemical reactions of the materials in the fractions into desired alternate end products; and (3) purification and removal of undesired materials.

The basic step in refining operations consists of distillation whereby the crude petroleum is heated to separate it into fractions according to boiling point. The "fractionating" or distilling apparatus is shown in Figure 8.7. The distillation process gives a more manageable variety of mixtures. The table below lists fractions obtained by distillation in the first step of the petroleum refining process. Each fraction, however, con-

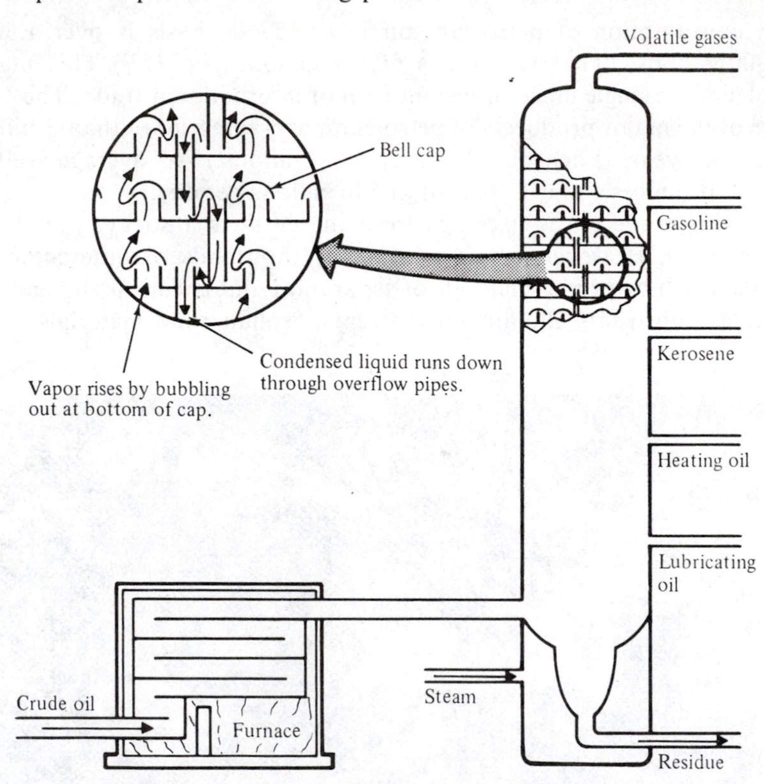

FIGURE 8.7
Simplified diagram of a fractionating tower or "still" for separating basic fractions of petroleum. The small molecules with low boiling point (gases) pass through and come off at the top. The heavier hydrocarbons which are liquids condense and come off at various points according to their boiling point. The heaviest molecules containing more carbon atoms are drawn off in the residue. California oils give asphalt residues, while Pennsylvania crudes provide paraffin types.

At left, part of a typical oil refinery showing fractionation towers, maze of connecting piping, and complex engineering structures which are required to separate the many kinds of molecules present in petroleum. At right, oil refinery area showing storage tanks and gas-collecting units.

tains large numbers of hydrocarbons, which is merely suggested in the table.

Rough Fraction	*Approximate Boiling Range in Degrees Centigrade*	*Approximate Range of Length of Carbon Chains*
Natural gas	Below 20°	1–4
Petroleum ether (solvent naphtha)	20–100°	5–7
Gasoline	70–200°	5–10, 11
Kerosene (jet fuel)	200–300°	11, 12–15
Fuel oil, diesel oil, light lube oils	300–400°	16–24
Greases	350–400°	18–22, 24
Paraffin	375–425°	20–25
Asphalt and petroleum coke	Residue	C_{24} and up

It has been found by experiment that hydrocarbons become less stable as the temperature is raised. For example, at temperatures around 500°C, chains containing 12 carbon atoms begin to break apart. The high kinetic energy which involves rapidly moving molecules causes the atoms to vibrate so much that the chains "crack" or break up. There are many possible paths and we will consider only two examples.

$$C_{12}H_{26} \xrightarrow[\Delta]{} C_{10}H_{22} + C_2H_4$$

$$C_{12}H_{26} \xrightarrow[\Delta]{} C_9H_{20} + C_3H_6$$

This process is called *cracking*. By the use of various catalysts, petroleum chemists and engineers have been able to control reactions like these to give products with desired operating characteristics.

The example above shows the conversion of a 12-carbon chain, which is a component of kerosene, into 10-carbon or 9-carbon chains, both of which are in the gasoline range. Controlled cracking is now done on a

Water or steam injection systems are now common in oil fields. This is an additional expensive operation to force the oil more completely from pockets deep in the earth. Shown here is a water injection system with a capacity of 170,000 barrels per day.

very large scale. Indeed, the gasoline obtained by "straight-run" distillation of petroleum makes up only about 10 percent of that used in the U.S. About 90 percent of the gasoline used is now made by converting larger and smaller hydrocarbon molecules into molecules of the kind and size desired in the gasoline range.

In addition to the larger quantities of gasoline available through cracking, we get many other kinds of molecules which can be used in endless varieties of ways to make everything from detergents to polyethylene bottles to plastic wrap.

The research directed toward making more gasoline from kerosene paid off in ways completely unforeseen in the beginning. (This is the nature of research.) Not only did cracking operations provide newer types of chemicals in large quantities; but other kinds of "reforming" reactions were also discovered for building up smaller molecules into larger ones. These were extensively developed in the 1930s and 1940s. Complex engineering techniques were invented for cracking and reforming. These now provide the great varieties of organic compounds which are the basis for the numerous petrochemical end products. You have certainly heard of polyethylene. It comes from a chemical C_2H_4, called ethylene, which was shown (page 182) as one of the cracked parts from the C_{12} hydrocarbon. This is the basis for a large number of common plastics (chapter 9). The molecule ethylene can be pictured structurally as:

$$\begin{array}{cc} H & H \\ | & | \\ C & = C \\ | & | \\ H & H \end{array} \quad \text{or} \quad \begin{array}{cc} H & H \\ \ddot{\ } & \ddot{\ } \\ C & :: C \\ \ddot{\ } & \ddot{\ } \\ H & H \end{array} \quad \text{or} \quad CH_2{=}CH_2$$

These all mean more than the molecular formula C_2H_4. The carbon atoms do not have the full amount of possible hydrogen atoms they could use. Ethane, CH_3CH_3, you will recall, is the second member of the methane family.

$$\begin{array}{cccc} & H & H & \\ & | & | & \\ H- & C- & C- & H \\ & | & | & \\ & H & H & \end{array}$$

Ethylene then has two of these hydrogen atoms missing. But the carbon atoms still satisfy their octets by having a double pair of electrons between them. This is called a double bond. A dash may be used to indicate a pair of electrons. Therefore two dashes are shown where carbon atoms have two covalent bonds between them.

The other cracking reaction shown above gives C_3H_6, propylene, also designated as $CH_3CH{=}CH_2$. It is used to make a plastic product called polypropylene.

THE CARBON CYCLE

Where does the gasoline go? Gasoline contains atoms of carbon and hydrogen which are for all practical purposes indestructible. This is in accord with the Law of Conservation of Mass. Consequently, the logical answer to where the gasoline with its carbon and hydrogen atoms goes is "back into the cycle." This is nature's cyclic path or circle for continual restoration of "used" materials.

You will recall our earlier discussion of the cycling of water. All of nature is full of cycles and all the atoms of the universe are themselves involved in cycles, both small and large. Some cycles are slow in turning and others rather fast. Nature is in a state of flux. "There is nothing permanent except change," said the Greek thinker Heraclitus.

When coal, oil, and petroleum products such as gasoline and plastics are "used," they eventually provide the gas carbon dioxide. The gas is an active agent in the carbon cycle until picked up by plants on land or plankton in the ocean. The details of changes involved in the chemical reactions of plants are complex. However, the overall nature of the buildup of carbon chains from CO_2 and H_2O can be easily shown in chemical equations:

$$6CO_2 + 6H_2O \xrightarrow[\text{Chlorophyll}]{\text{Sunlight}} C_6H_{12}O_6 + 6O_2$$

or the more generalized form:

$$CO_2 + H_2O \xrightarrow[\text{Chlorophyll}]{\text{Sunlight}} \text{Carbohydrates} + O_2$$

These are examples of photosynthesis—the process by which plant cells make carbohydrates from CO_2 and H_2O in the presence of chlorophyll and light. Chlorophyll is the green coloring matter of plants which is the catalyst in the process. Carbohydrates are the long-chain carbon compounds used by plants as framework of living cells (wood, leaves, stems, and so on). They are also the basis for the plant's stored food which is called starch. More details are given in chapter 10.

Looking at the diagram of the carbon cycle (Figure 8.8), you can see how important are the dynamics of plant cells. This is not only from the twofold operation of building carbon chains and releasing oxygen to the

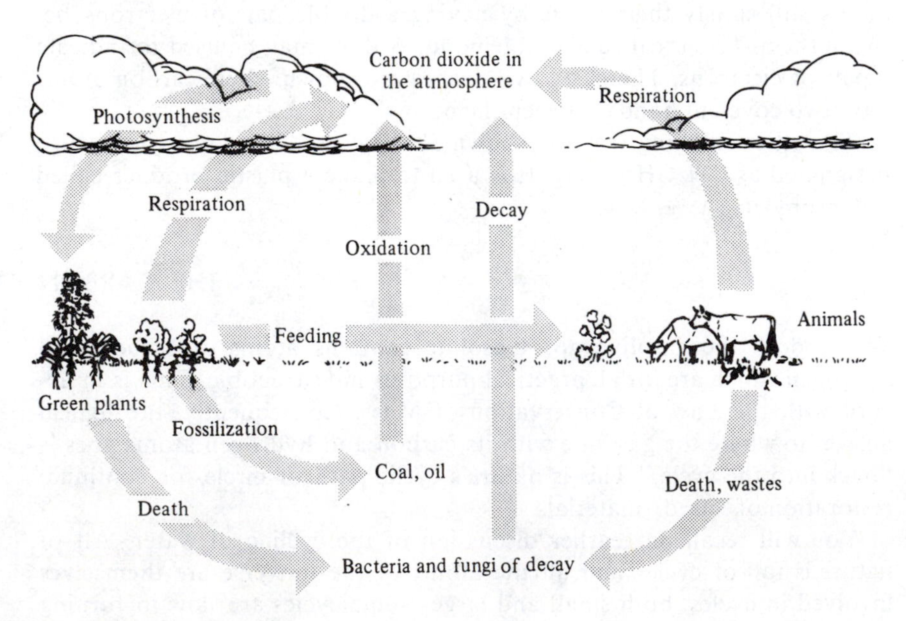

FIGURE 8.8
The carbon cycle.

atmosphere. Plants also perform the important function of absorbing radiant energy from the sun.

The energy of our bodies and the machines we use eventually depend upon capturing energy from the sun. This is a major service of plants to animals and people. Coal is sometimes called "buried sunshine" or "captured solar rays." But every time a cow eats grass a similar energy source is tapped which you may eventually use in a hamburger or a milk shake.

It has been estimated that if photosynthesis stopped abruptly, all free oxygen in the atmosphere would disappear in about 2000 years because of the disruption of the carbon cycle. Energy storage would also cease. So think before you curse the weeds in your garden. They are just displaced plants.

THE HARMONY OF LIFE'S CYCLES

The overall cycle is simple: CO_2 in the atmosphere—to plants—back to CO_2 in the atmosphere. Tied to this are many side cycles—some storing carbon atoms for millions of years. The interrelationships of these giant cycles represent ecology in the broadest sense of the word—a harmonious interdependence among all living things. Studies of nature's cycles provide a deeper recognition of this interdependence and the great responsibility we have with the atoms nature lets us handle for a time. Chemistry is especially fruitful here. It forces recognition of the natural interweaving of materials which at various times come under our temporary control. The problems are to recognize more of the pathways and to avoid imbalances. The broad picture is only now receiving proper emphasis. This will be helpful in our search for solutions to problems of pollution (chapters 15 and 16).

You can better understand now what it means when we say that the carbon in your gasoline goes back into the cycle. And thanks to the trees, shrubs, grass, and weeds, it will end up in usable form again. Maybe in another million years it will serve again for propulsion of some other type. We cannot be sure. But the carbon you burned in your engine today will still be around. It could very well have once been used in propulsion for some dinosaur. So take care of it. There may not be a dinosaur in your tank. But you can be sure that what is in your tank was very likely once in a dinosaur. He is now extinct.

PROJECTS AND EXERCISES

Experiments

1. Get a soft pencil and write a few lines on a piece of white paper. If you have a harder pencil (higher number), you can try using this too. However, the soft pencil (low number) works the best here. Then take your hand lens or

magnifying glass and examine the pencil lines carefully. What do you notice? Can you go from your operational findings to a conceptual explanation? You could not write with a diamond on paper even though it is also carbon. Does your concept base provide the explanation?

2. Get a small styrofoam ball (about one-inch diameter) of the type used in Christmas decorations. If you cannot get one you could use a piece of candy with a medium soft consistency like taffy, a gum drop, a slightly stale marshmallow, or a piece of cork cut so that its length is about equal to its width. This is to represent the central carbon atom in the compound methane. Then get four round toothpicks with the pointed ends. (The flat toothpicks are too fragile.) Or you can use nails, hat pins, or the like. Now by trial and error stick the toothpicks (or pins, etc.) into the styrofoam ball (or cork, marshmallow, etc.) so that they are equally spaced from each other. Start with one in the ball and then place the others so that there is an equal angle between each one. That is, set them in so that any two toothpicks are not any closer together or farther apart than any of the others. After you get some degree of regularity you can place your "model" down carefully on the table to note whether you have to move toothpicks a little more. Place it down on several "sides" and readjust where necessary. You are shooting for an ideal here which of course is not attainable with toothpicks. Like all models or pictures, we do not need it to be perfect. Do you now see any regular geometric shape? If you consider the ends of any three toothpicks as forming a plane, how many sides would the figure have? (Look at it on the plane of the table in several positions.) Now assume it represents the CH_4 molecule. What do the toothpicks stand for? How could you give a conceptual description of why the four "bonds" take positions as far from each other as possible?

Exercises

3. Choose one of the quotations at the beginning of this chapter. Write a brief paragraph on what it now means to you, including comments on whether its meaning has been enhanced since you first read it when starting this chapter.
4. Suppose a friend of yours says that he heard that it is less dangerous to use kerosene in an open container for cleaning metal parts than gasoline. He asks you why. Can you provide the necessary conceptual explanation of why gasoline would be more dangerous when compared to kerosene?
5. There are a couple of million organic compounds, more than the compounds of all the other elements combined. What reasons can you suggest for this great number of possible compounds?
6. There are several ways of writing the formula for normal hexane. Try to draw on paper as many ways as you can to show this compound. Which would you prefer? Why?
7. Describe gasoline in your own words, referring to its conceptual background which could be explained to someone who had not yet had a chemistry course.
8. There are three possible isomers of pentane. Draw on paper formulas to represent these three possibilities. Be sure to avoid duplications by remembering that you can turn around or turn upside down any formula drawn. There are only three possible *different* arrangements.
9. How would you explain the octane rating of gasoline in simple terms?
10. A friend of yours who did not have the advantage of a chemistry course says to you that all gasolines are the same. How would you answer him?
11. Had you ever heard of vapor lock before you read this chapter? If not, ask a

garage mechanic or longtime auto driver what it is. Then briefly describe it so that a nontechnical person would understand, relating concept to operational symptoms.

12. Describe similarities between diamond and coal. Also describe differences. Which do you think is more important?
13. Examine the carbon cycle and describe what implications it has for our modern industrial world.
14. If someone told you that study of the carbon cycle is only for use in the high schools, what would you reply?
15. A ten-year-old child asks you whether ethyl gas is named after a girl he knows called Ethel. How would you answer to explain to him a little of what is really the basis for ethyl gasoline?

SUGGESTED READING

1. Check in your favorite encyclopedia under headings such as gasoline, petroleum, coal, fuels, octane number, vapor lock. The generalized descriptions in encyclopedia articles are especially helpful in providing necessary background for understanding the complex nature of these products and the practical aspects valuable to your daily living.
2. Bolin, Bert, "The Carbon Cycle," *Scientific American,* 223, No. 3, pp. 125–132 (September 1970). Scientific American Offprint No. 1193. (Also in the Scientific American book titled *The Biosphere,* W. H. Freeman and Co., San Francisco, 1970.)
3. Hubbert, M. King, "The Energy Resources of the Earth," *Scientific American,* 224, No. 3, pp. 61–70 (September 1971).
4. Rose, D. J., "Energy Policy in the U. S.," *Scientific American,* 230, No. 1, pp. 20–29 (January 1974).
5. Perry, H., "The Gasification of Coal," *Scientific American,* 230, No. 3, pp. 19–25 (March 1974).
6. Harris, J.R., "The Rise of Coal Technology," *Scientific American,* 231, No. 2, pp. 92–97 (August 1974).

chapter 9

Giant Molecules: The Basis for Plastics

"Why," said the Dodo, "the best way to explain it is to do it."

—*LEWIS CARROLL*

See plastic nature working to this end,
The single atoms each to other tend,
Attract, attracted to, the next in place
Form'd and impell'd its neighbor to embrace.

—*POPE*

Oaks may fall when reeds stand the storm.

—*THOMAS FULLER*

A wax candle will bend when left near a source of heat, yet the branches of trees do not droop from the intense summer sun. Both the wax and the tree branch contain carbon atoms. What is the basic difference?

You may already have guessed. Longer and stronger chains! This is exactly what nature does in building fibers for plants and muscles for animals. She builds giant molecules. Following nature's example, we can try to build a few giant molecules ourselves. This is what chemists do in making the "synthetic" plastics you use every day in so many ways, from the coatings and paints on your car, your house, and your appliances to the packaging for your foods and the polyester in your clothing. In other words, nature's giant molecules are "natural" plastics. Chemists have found how to modify and use these materials, and also how to make newer long-chain materials from simple molecules. These latter are the true "synthetic" plastics—complex molecules made from simpler parts by chemical reaction. The term "artificial" is often used in connection with the synthetics and it sometimes has the connotation of something inferior. This need not be so.

THE FIRST COMMERCIAL PLASTICS

The first commercial plastics product was invented in 1869 by John Hyatt, a New York printer who was trying to win a contest for developing an ivory substitute for billiard and pool balls. The manufacturer offering the prize was being squeezed by an elephant shortage.

Hyatt used cellulose from cotton which he treated with nitric acid to form a derivative called cellulose nitrate. This was then combined with another natural product, camphor. Camphor originally came from an evergreen tree of the laurel family found chiefly in China and Japan. Hyatt's product was called celluloid and was used in combs, artificial teeth, piano key coverings, clock cases, table tennis balls, brush and mirror handles, and eyeglass frames. It didn't work too well for billiard balls and he never won the prize.

This typifies the application of a plastics material. Once its properties are determined, the same material can be used in a variety of products. Celluloid was also the film for the first moving pictures. However, the early movie houses often had unscheduled "happenings" because the cellulose nitrate caught fire very easily. In fact, smokeless gun powder is a type of cellulose nitrate.

Hyatt opened the way to further development of plastics by using a modified plant fiber material. Other chemists applied their imaginations to the problem of developing new materials. Hilaire Chardonnet in 1884 developed viscose "rayon" by treating cellulose with chemicals to make it into a fluid which was then forced through tiny holes into a hardening bath. The operation reformed the cellulose into long fibers of filaments. These are then twisted together into threads or yarn.

Later, in 1908, a Swiss chemist, Jacques Brandenberger, invented cellophane which is simply cellulose formed into a fluid, as in rayon manufacture, but then forced through a narrow slit into a chemical bath which converts the fluid into a thin sheet of recovered cellulose. Thus both rayon and cellophane are really the same natural product reformed into either threadlike, thin fibers or sheets depending on what we want the cellulose to do. The cellulose in both cases is still the same basic material that forms the main part of the cell walls of plants.

THE LONG HISTORY OF LONG CHAINS

People from the most primitive times depended on nature's giant molecular materials to supply the necessities of life: food, clothes, and shelter. Leaves, furs, leather, and textiles made from natural fibers like cotton, wool, flax and hemp have provided clothing from time immemorial.

Vegetables, meats, and grains have served as food. Leaves and wood from trees along with straw have provided various types of houses. All of these are now recognized as being made up of giant molecules.

It was not until the nineteenth century that chemists began the difficult task of analyzing these longer chain materials. Chemists gradually came to realize that the size of the molecules—that is, the large number of atoms per molecule—was the basic distinguishing characteristic of all these natural materials. The fact that cellulose and protein do not dissolve in water and resist melting—only to char on heating—suggested that possibly large molecular size was involved. This conclusion followed from the familiar behavior of smaller organic molecules where it was noted that solubility decreases and temperature requirements for melting and boiling increase with increasing chain length.

During the early days the actual making of giant molecules was chiefly an accident. Occasionally a chemist ended up with a sticky, uninvited "gunk" or "goo" in his reaction flask. These resinous products made it difficult to clean the glassware before more conventional, crystalline—and at the time more useful—organic molecules could be studied. The "gunks" were the forerunners of modern plastics but were considered misfortunes and led to many a chemist's exasperation.

By the early years of the twentieth century some intrepid souls began to look into these sticky products which were so difficult to characterize and handle. These chemists may have been spurred on by the commercial successes of products like celluloid and rayon made from modified cellulose. They were also probably influenced by the phenomenal success that followed upon the discovery in 1839 by Charles Goodyear of "vulcanization" of natural rubber.

The long chain materials were named *polymers* from the Greek for "many parts," indicating that the molecules were made up by joining many simpler units together as basic building blocks for the giant molecules. A *monomer* ("one part") identifies the basic unit itself which can serve as links for the longer chains. A *high polymer* refers to very large, or giant, molecules.

THE FIRST SYNTHETIC PLASTIC

Leo Baekeland, a Belgian chemist working in the United States, is credited with developing the first synthetic plastic. He was looking for a substitute for shellac, a natural resin secreted by tiny Indian lac bugs. It takes about 150,000 bugs six months to produce enough resin to yield one pound of shellac.

Baekeland's imagination related the type of resin produced by the bugs with that discarded as useless gunk by previous workers. And he presumed it would be easier to make such material in the chemical reactor than to have to handle so many bugs. He used a reaction between phenol (C_6H_5OH), a common disinfectant obtained from coal tar, and the acrid-smelling preservative, formaldehyde (CH_2O). This was in 1908.

Baekeland, by using high temperature and pressure, encouraged the formation of resin which previous chemists had avoided. He actually formed the two simple chemicals into a hard transparent material. This was found to have excellent electrical and mechanical properties when molded and thus became the basic material for telephones, radio and electrical control panels, knobs, and handles of pots. It was called Bakelite and inaugurated a long line of man-made polymers. This new material formed from two simple molecules was an entirely new substance not found as such in nature, whereas previous plastics like celluloid had been modifications of natural long-chain material. Modified resins based on Baekeland's initial work now find applications in such diverse products as varnishes, pump impellers, decorative laminates, plywood adhesives, cafeteria trays, buttons, helmet liners, and costume jewelry.

Baekeland's discovery of a way to control the reaction of small monomer molecules to form reproducible batches of giant molecules spurred investigations of other monomer materials.

MAKING CHAINS: CONDENSATION POLYMERS

The polymer chemist starts his experiments with an idea: to grow chains from simple molecules. You could imagine working on this problem as a thought experiment. How would you get simple molecules to add to each other so as to grow longer chains? Chemists used their knowledge of the many ways that different types of chemicals react with each other to do this. Here is one example: an acid and a base react to give water and leave a salt in solution. The reaction of hydrochloric acid and sodium hydroxide illustrate this type of acid-base neutralization:

$$HCl + NaOH \rightarrow H_2O + NaCl$$

The H^+ and OH^- thus combine because the end product water is a very stable molecule. Then if an organic molecule could be obtained with a hydrogen atom on one end which is reactive, it might be made to combine with a reactive oxygen-hydrogen grouping on another molecule to give water. But in the process the other parts of the two molecules would join together. By a repetition of such a process we could get a chain to grow. This is indeed one type of chain growing used by the chemist. Since water usually condenses out of such a reaction, the method of chain growth is in general referred to as *condensation.* The products are called *condensation polymers.*

Alcohols and Acids: How They React

We will have to digress a little first to see how chains can start growing and consider the possibility of reaction between the two simple molecules whose structural formulas are given on the following page.

$$CH_3-\overset{\displaystyle O}{\overset{\|}{C}}-OH \qquad CH_3CH_2-OH$$

Acetic acid — Ethyl alcohol

Household vinegar is 5 percent acetic acid in 95 percent water solution. Ethyl alcohol is the alcohol found in wine, beer, and whiskey. The reaction between these occurs as follows.

$$\underset{\text{Acetic acid}}{CH_3\overset{\displaystyle O}{\overset{\|}{C}}\vdots OH} + \underset{\text{Ethyl alcohol}}{H\vdots OCH_2CH_3} \rightarrow \underset{\text{Ethyl acetate}}{CH_3\overset{\displaystyle O}{\overset{\|}{C}}OCH_2CH_3} + \underset{\text{Water}}{H_2O}$$

An acid + An alcohol ⟶ An ester + Water

The product is given the special name *ester* which by definition is the reaction product of an acid with an alcohol. An alcohol can be considered to be derived from a hydrocarbon by replacing a hydrogen atom with the —OH group. For example, methyl alcohol, CH_3OH, is derived by replacement of a hydrogen atom in methane, CH_4, with the —OH group. This —OH grouping is called the alcohol functional group.

A *FUNCTIONAL GROUP* IS THE SPECIAL REACTIVE PART OF A MOLECULE.

Each functional group confers particular properties on the carbon chain and thus functions similarly in different molecules.

The carbon chain part of the molecule contributes to the overall combined properties of the particular chemical. You are probably aware that methyl alcohol, CH_3OH, is a deadly poison (sometimes called wood alcohol because it was originally obtained from wood). Ethyl alcohol CH_3CH_2OH, differs by only one CH_2 link in the chain.

You can note above another common functional group found in organic chemicals. This is the acid group, designated as

$$-\overset{\displaystyle O}{\overset{\|}{C}}-OH \quad \text{or} \quad -COOH$$

This combination of atoms provides H^+ ions which are acidic in solution. The alcohol and the acid functional groups are two of the most common groups found in organic molecules. The identification of reactive parts in molecules allows us to classify the millions of organic compounds into a small number of groups.

$$CH_3-CH_2-OH \qquad CH_3CH_2OH \qquad C_2H_5OH$$

are all equivalent and tell us much more than the simple molecular formula C_2H_6O, or the name ethyl alcohol.

$$CH_3{-}\overset{\overset{\displaystyle O}{\|}}{C}{-}O{-}H \qquad CH_3{-}\overset{\overset{\displaystyle O}{\|}}{C}{-}OH \qquad CH_3{-}COOH \qquad CH_3COOH$$

all represent the acid in vinegar which is called by the generic name acetic acid and has the unrevealing molecular formula $C_2H_4O_2$. It is sometimes even written as $HC_2H_3O_2$ to emphasize that it is an acid which gives up only *one* hydrogen atom as a hydrogen ion in water solution. From the structural formulas given above you can more readily see the reason why only one hydrogen atom is active in the acid sense. The others are tied to the carbon atoms like those in methane, and these are not acidic. Only the hydrogen atom in the organic acid functional group —COOH is active in conferring acid properties, or H^+ ion in solution.

Growing the Chain

How do chemists use this basic knowledge of alcohol and acid reactions to grow longer chains? Simply, they obtain molecules with at least 2 —OH groups and others with 2 —COOH groups. Then with the proper conditions of catalysts and heat these molecules can be made to link up by condensation at the points where the —H and —OH are connected. Water is condensed out and the adjoining pieces of each molecule join up at the places where H and OH were removed.

$$HOCH_2CH_2O{:}H \;+\; HO{:}\overset{\overset{\displaystyle O}{\|}}{C}CH_2CH_2\overset{\overset{\displaystyle O}{\|}}{C}OH \;\rightarrow$$

Etylene glycol (same chemical used in anti-freeze) + Succinic acid (obtainable from oat hulls) →

$$HOCH_2CH_2O\overset{\overset{\displaystyle O}{\|}}{C}CH_2CH_2\overset{\overset{\displaystyle O}{\|}}{C}OH \;+\; H_2O$$

Intermediate (a reactive material, not a high polymer) + Water

Here you can see the potential for further reaction since both ends of the intermediate have potentially reactive groups. Thus further reaction occurs from both ends.

$$HO\overset{\overset{\displaystyle O}{\|}}{C}CH_2CH_2\overset{\overset{\displaystyle O}{\|}}{C}{:}OH \;+\; H{:}OCH_2CH_2O\overset{\overset{\displaystyle O}{\|}}{C}CH_2CH_2\overset{\overset{\displaystyle O}{\|}}{C}{:}OH \;+\; H{:}OCH_2CH_2OH \;\rightarrow$$

$$HO\overset{\overset{\displaystyle O}{\|}}{C}CH_2CH_2\overset{\overset{\displaystyle O}{\|}}{C}OCH_2CH_2O\overset{\overset{\displaystyle O}{\|}}{C}CH_2CH_2\overset{\overset{\displaystyle O}{\|}}{C}OCH_2CH_2OH \;+\; 2H_2O$$

This reaction produces a second intermediate of a little longer chain length, still not a high polymer. However, this molecule also has reactive groups on both ends. You can guess what happens: more chain growing as more and more molecules condense through elimination of water. The example here leads to a polyester. Since the exact number of units often is not determined, we commonly designate the polymer by using the small letter *n* or *x* to indicate the extended nature of the chain. The particular polyester end product here can be designated as

$$H[-O-CH_2CH_2-O-\overset{\overset{\displaystyle O}{\|}}{C}-CH_2CH_2-\overset{\overset{\displaystyle O}{\|}}{C}-]_n OH$$

This is somewhat like a freight train whose cars always have couplings at the front and rear. If only one coupling were present on a car then we would have only two-car units and no train. By the fact of having two coupling possibilities per car (and two functional groups per molecule) we can achieve links which provide a chain of cars (or molecules). The end result is a train—or a long molecule.

Varieties of Chains by Varieties of Links

In making up chains from simpler monomers the chemist gets variety not only in the functional groups but also in the carbon chains. For example, in the polyester above, he can choose an alcohol with three carbon atoms, $HO-CH_2CH_2CH_2-OH$, instead of the one shown with two carbon atoms. Or he could use an alcohol with some other atoms substituted for the hydrogen atoms, like

$$HO-CH_2CH_2\underset{\underset{\displaystyle CH_3}{|}}{C}HCH_2-OH$$

The possibilities are endless. He could use a different acid molecule, for example,

$$HO-\overset{\overset{\displaystyle O}{\|}}{C}-CH_2CH_2CH_2-\overset{\overset{\displaystyle O}{\|}}{C}-OH$$

Chemists have used the concept that different varieties of monomers make different end-product plastics to produce the great variety of plastics available so freely today. Basic starting materials are obtained chiefly from petroleum and coal tar. Examples of plastics formed by condensation are shown in Table 9.1.

The Benzene Chain Structure

There is a very common type of natural chain structure which we have not encountered before and which you will find in many plastics. It is a

TABLE 9.1
Examples of Some Condensation Polymers

Monomers	*Polymer*	*Principal Uses*
Ethylene glycol + Terephthalic acid 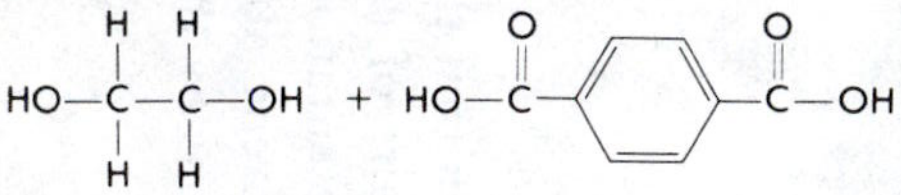	A polyester (polyethylene terephthalate) 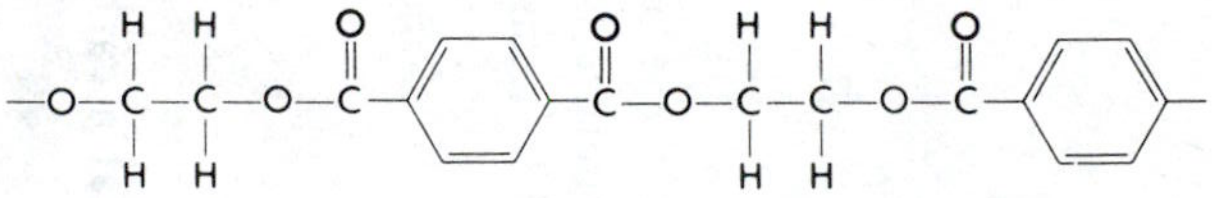	1. Films; e.g., Mylar 2. Fibers; e.g., Dacron
Dimethylsilanediol 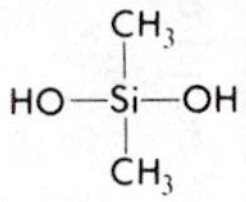	Silicone 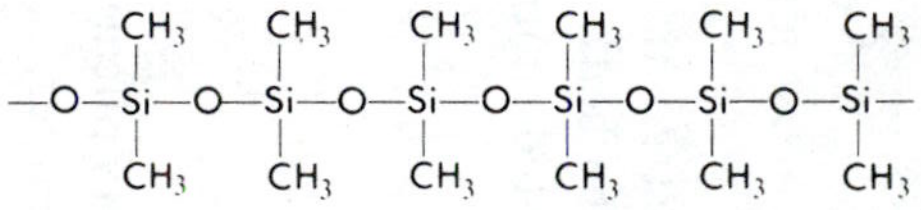	1. Temperature-resistant lubricants 2. Temperature-resistant rubber 3. Water-repellent coatings
Phenol + Formaldehyde 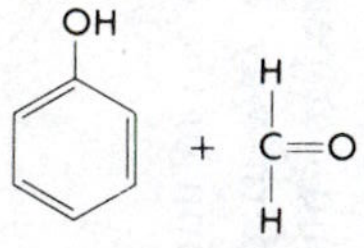	Phenol-formaldehyde resin 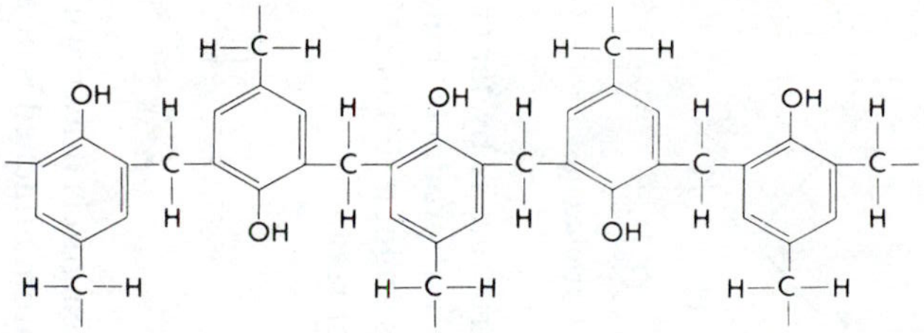	1. Reinforced molded objects; e.g., **Bakelite** 2. Varnishes 3. Lacquers 4. Adhesives
Hexamethylenediamine 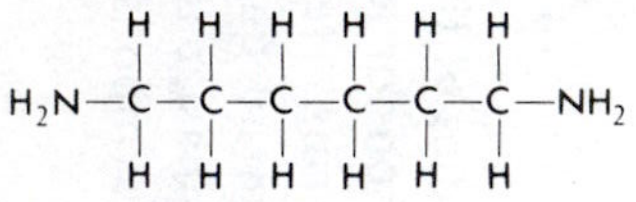 Adipic acid 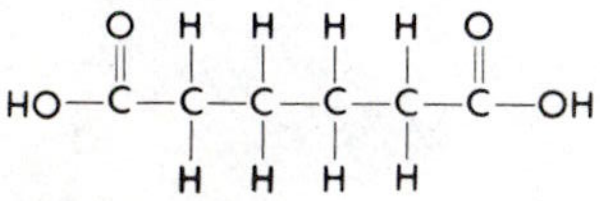	Nylon 66 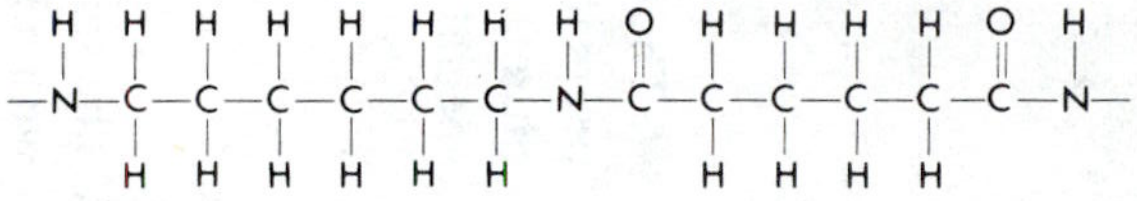	1. Fibers 2. Molded objects

"cyclic" ring structure, another variety of the chains we have considered so far. Instead of the carbon atoms sharing electrons and forming bonds in continuous chains, they are sometimes found to do so in a cyclic or ring structure. One of the structures much preferred in nature is a six-membered ring. Here we have two basic possibilities.

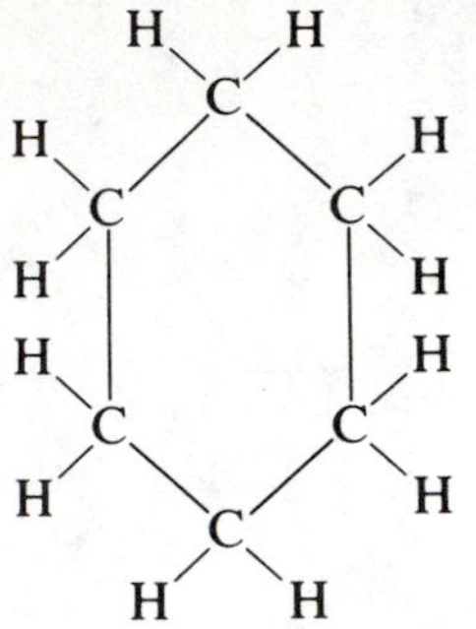

C_6H_{12}
Cyclohexane

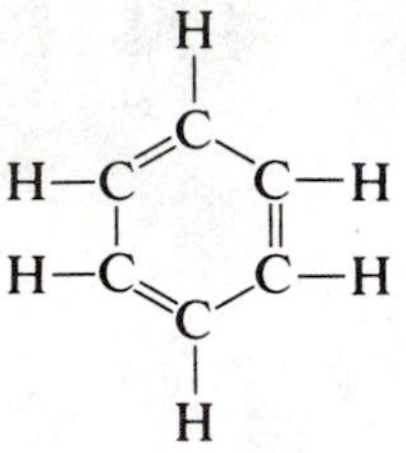

C_6H_6
Benzene

The benzene structure is by far the most common of these and it appears in many natural products. Indeed, the benzene structure is so common that chemists adopt a sort of abbreviated shorthand to designate this special six-member ring.

The simple line hexagon is drawn with the implication that a carbon atom with a hydrogen atom attached is at each corner.

You may have noted that we arbitrarily put the double bonds between certain pairs of carbon atoms. We could just as readily have placed the double bonds between the other pairs of carbon atoms.

Studies of the chemical reactions of benzene have shown that there is no reason for locating the double bond between specific pairs of carbons in the ring. Here again we find our pencil-and-paper representations are far from adequate. The modern way of representing benzene attempts to emphasize the delocalization of the double bond electrons by the following symbolism.

This standard symbol again represents a C—H bonded at each of the corner points of the hexagon. The circular representation in the middle signifies that the extra electrons are really spread equally among all the carbon atoms, or delocalized.

By convention, if a hydrogen atom at any corner is substituted with another kind of atom or group of atoms, then the atom or group is shown attached at that point. For example,

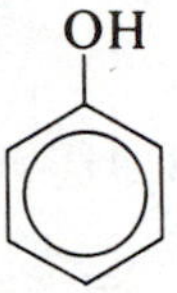

OH

is the accepted structural formula for phenol, obtained from coal tar and used in some kinds of disinfectants. Note that there are six carbons including the one at the corner where the —OH is joined. The molecular formula for phenol is C_6H_6O which does not tell us much about the type of molecule. This is the reason for structural formulas. Another way of writing the formula is C_6H_5OH.

One additional detail should be mentioned. The examples of chain growth above use two functional linking groups per molecule. You have already seen that one functional group per molecule would not result in chain growth. Maybe you wondered, "What if there are three functional groups in a molecule?" This is indeed a possibility and you might stop here for a thought experiment about the kind of product that could be formed.

You may imagine a chain forming a branch off the main chain through the third functional group. Also, there is the possibility of interlinking between chains because of the extra function available in each growing chain. In fact both of these possibilities have been found practical and have been used to great advantage to give a variety of properties in the finished polymers. Perhaps you imagine a very confused, interlocking mess of chains. This certainly would give a different overall set of properties to the finished product. We will return to this latter idea since it is a major direction followed in making certain tailor-made plastics.

MAKING CHAINS: ADDITION POLYMERS

The chemist has found another way of adding monomer to monomer to form long chains. This makes use of the extra reactivity of double bonds. Two hydrogen atoms can be removed from ethane to leave a bond network wherein the two carbon atoms share two pairs of electrons between them (chapter 8).

$$\begin{array}{ccc} \begin{array}{c} \text{H}\ \ \text{H} \\ |\ \ \ | \\ \text{H—C—C—H} \\ |\ \ \ | \\ \text{H}\ \ \text{H} \end{array} & \rightarrow & \begin{array}{c} \text{H—C=C—H} \\ |\ \ \ | \\ \text{H}\ \ \text{H} \end{array} \; + \; H_2 \end{array}$$

Ethane $\longrightarrow$ Ethylene + Hydrogen gas

If we start with ethylene, in a reverse of this process, we can add two hydrogen atoms to the "unsaturated" double bond. Each hydrogen atom will bring one electron to share with an electron from a breakup of one bond in the double bond.

$$\text{H:H} + \begin{array}{c} \text{H H} \\ \ddot{\text{C}}\text{::}\ddot{\text{C}} \\ \text{H H} \end{array} \rightarrow \begin{array}{c} \text{H H} \\ \text{H:}\ddot{\underset{..}{\text{C}}}\text{:}\ddot{\underset{..}{\text{C}}}\text{:H} \\ \text{H H} \end{array}$$

In other words, when hydrogen is added to ethylene a break occurs in the double bond joining the two carbon atoms. Each of the links in the double bond is represented by a shared pair of electrons and what happens is a shift of the two electrons in one of the links. One electron stays with one of the carbon atoms and is joined by an electron from an incoming hydrogen atom, making the third bond for that carbon atom. The other electron from the broken bond shifts to the second carbon atom where it is then joined by an electron in the second added hydrogen atom. What we have done is to link a hydrogen atom to each carbon atom.

The polymer chemist uses the same basic idea to form chains. He has developed chemicals which will break open the double bond. This is done by an active agent which has a single electron on its end. This joins with one electron in the bond being broken. Then the other electron in the broken bond swings around—ready to form another bond. But instead of adding a hydrogen atom and ending the reaction there, it merely attacks a double bond in another molecule. This is then joined to the first in what becomes a chain reaction.

An example may help to clarify the process:

$$\text{H—O}\cdot + \begin{array}{c} \text{H H} \\ |\ \ | \\ \text{C::C} \\ |\ \ | \\ \text{H H} \end{array} \rightarrow \text{H—O:}\begin{array}{c} \text{H H} \\ |\ \ | \\ \text{C:C}\cdot \\ |\ \ | \\ \text{H H} \end{array}$$

$$\text{H—O—}\begin{array}{c} \text{H H} \\ |\ \ \ | \\ \text{C—C}\cdot \\ |\ \ \ | \\ \text{H H} \end{array} + \begin{array}{c} \text{H H} \\ |\ \ | \\ \text{C::C} \\ |\ \ | \\ \text{H H} \end{array} \rightarrow \text{H—O—}\begin{array}{c} \text{H H H H} \\ |\ \ \ |\ \ |\ \ | \\ \text{C—C:C:C}\cdot \\ |\ \ \ |\ \ |\ \ | \\ \text{H H H H} \end{array}$$

This growing chain can attack other ethylene molecules and build very long chains. The process stops when single electrons on growing chains combine or meet some other source of satisfying the pair required for a stable bond.

To start the breakup of the double bond we need some kind of initiator. There are several types available but we will consider a simple one which comes from a common chemical, hydrogen peroxide, H_2O_2 or H—O—O—H. This is familiar to you as a bleaching and disinfecting agent. When heated, hydrogen peroxide can break down into two parts.

$$\text{H—O}\not\,\text{O—H} \xrightarrow[\Delta]{} \text{H—O}\cdot + \text{H—O}\cdot$$

Thus we get the combination of hydrogen and oxygen having a single, unpaired electron, H—O·. This is called a *free radical,* which is a special name for an atom or molecule having an unpaired electron. In the case above, this free radical is the initiator of attack on the double bond. The end result when we start with ethylene is the giant molecular material polyethylene, known to everyone today in varied applications from squeeze bottles to toys and transparent wrappings.

Indeed, the production figures for polyethylene are startling. In the U.S. alone the yearly production of polyethylene is in the range of 6 billion pounds. The ethylene gas comes from the petroleum industry, mostly as a by-product of various cracking and refining operations (chapter 8).

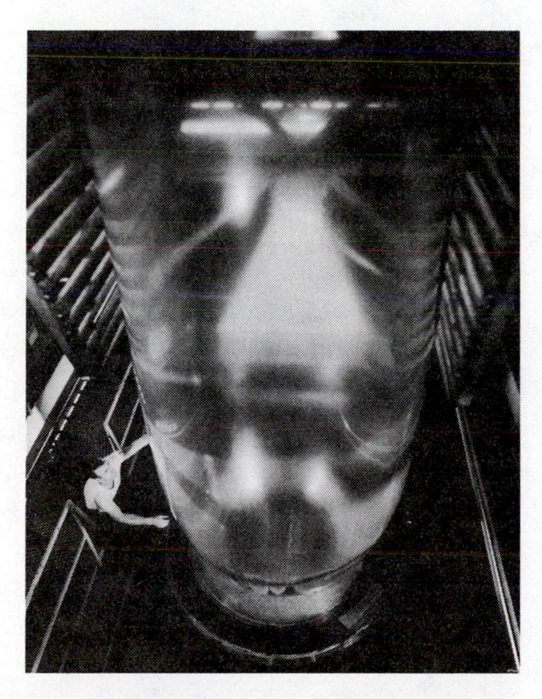

A two-story-high bubble is formed during the extrusion of polyethylene film. The film will be used to make thousands of types of plastic bags for packaging such items as foods, soft goods, and hardware. The film is also used in the building industry as a vapor and moisture barrier and in agriculture for mulching, silage covers, greenhouse glazings, pond liners, and animal shelters. (Courtesy Union Carbide Corporation.)

TABLE 9.2
Examples of Some Addition Polymers

Monomer	*Polymer*	*Principal Uses*
Ethylene $H_2C{=}CH_2$	Polyethylene $-CH_2-CH_2-CH_2-CH_2-CH_2-CH_2-CH_2-CH_2-CH_2-CH_2-CH_2-CH_2-CH_2-CH_2-$	1. Films 2. Tubing 3. Molded objects 4. Electrical insulation
Vinyl chloride $H_2C{=}CHCl$	Polyvinyl chloride $-CH_2-CHCl-CH_2-CHCl-CH_2-CHCl-CH_2-CHCl-CH_2-CHCl-CH_2-CHCl-CH_2-CHCl-$	1. Sheets 2. Phonograph records 3. Copolymer with vinyl acetate to make floor coverings, latex paints, etc.
Acrylonitrile $H_2C{=}CHCN$	Polyacrylonitrile $-CH_2-CH(CN)-CH_2-CH(CN)-CH_2-CH(CN)-CH_2-CH(CN)-CH_2-CH(CN)-CH_2-CH(CN)-CH_2-CH(CN)-$	1. Fibers; e.g., Orlon, Acrilan

TABLE 9.2 (continued)

Monomer	*Polymer*	*Principal Uses*
Tetrafluoroethylene	Polytetrafluoroethylene	1. Chemically resistant films, molded objects, electrical insulation; e.g., Teflon
Styrene	Polystyrene	1. Molded objects 2. Electrical insulation 3. Copolymer with butadiene to make rubber 4. To make ion-exchange resins on treatment with sulfuric acid
Methyl methacrylate	Polymethyl methacrylate	1. Transparent sheets, rods, tubing; e.g., Lucite, Plexiglas 2. Plastics reinforced with glass fiber

The method of growing long chains by combining monomers, without breaking out or condensing of any parts of the original monomer units, is called *addition polymerization.* As with condensation polymers, the possibility of variety is great. The chemist can get variety by using molecules containing other atoms or groups substituted in place of the hydrogen atoms in ethylene. Some common examples of representative addition polymers are given in Table 9.2.

STRUCTURE AND OPERATIONAL PROPERTIES

The symbolism shown in the chain tabulations is necessarily oversimplified. No one can really follow the exact chain growing process and it is very likely that unknown and unplanned variations occur in chain growth because of impurities in the materials used. Also, some branching may occur instead of continuous chain growth. However, polymer chemists, using the basic ideas developed above, can design "tailor-made" plastics with predetermined properties.

The simplest way of emphasizing the relationship of structure to operational behavior or properties of a particular plastic is to consider the two basic kinds of plastics materials—thermoplastics and thermosets. (See Figure 9.1.)

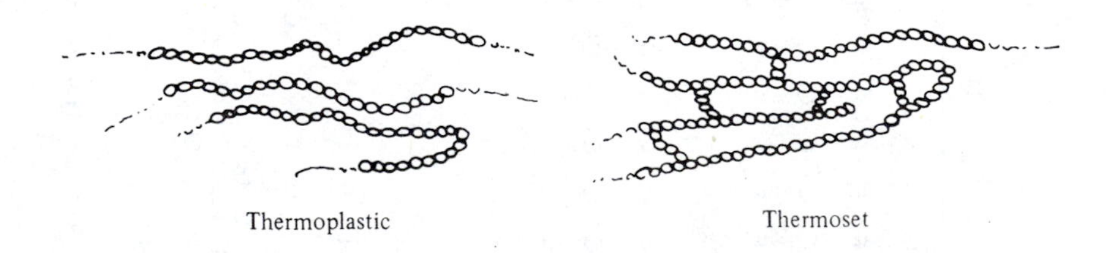

FIGURE 9.1
Symbolic representation of the two basic kinds of plastics materials. The tiny circles represent monomers. The cross links in the thermosets give extra strength to the end product. The degree of cross linking can be varied widely.

Thermoplastics

The thermoplastics are giant molecular materials which soften on heating and harden again on cooling. Thus they can be melted again and again in a similar fashion to wax used in candles. They are formed into finished products chiefly in three ways.

1. Injection molding, which consists of heating the polymer material until it is fluid and then injecting it into cooling molds which form it into the desired shape. Examples: toys, shoe heels, brush handles, mechanical pen barrels, combs, bottles.

2. Extrusion molding, which is somewhat like squeezing a tube of toothpaste. The thermoplastics are fed into heating chambers and then forced by a screw device through specially designed openings in a die. Examples: fibers, moldings, hose, drinking straws, tubing, pipe, wire insulation.

Extrusion of PVC plastic pipe for water transmission and waste treatment applications. Plastic pipe production now runs in the neighborhood of a billion pounds per year in the U.S. (Courtesy The Goodyear Tire & Rubber Company.)

3. Calendering, which involves rolling into thin films or spreading over cloth or paper used as base materials. Examples: wallet cards, wallpaper, tape, rainwear, upholstery materials.

The important operational characteristic necessary for thermoplastics is softening under heat and hardening again when cooled. In order to accomplish this we need a plastic in which the molecules can be made to flow somewhat easily by heating. But they cannot of course flow and slip past each other easily at ordinary room temperature.

Thus we need longer chains than those found in gasoline or kerosene. However, the long chain molecules must not be linked to each other or else they would not become fluid under heating. These conceptual requirements suggest to the plastics chemist monomers which can form only two links per molecule. Examples of these can be found in the Tables 9.1 and 9.2 of condensation and addition polymers. Thermoplastics formed by condensation include nylon, and polyesters such as Dacron fibers and Mylar films.

Some addition polymers which are thermoplastic are polyethylene, polystyrene, Orlon and Acrilan fibers, and polyvinyl chloride. Polyvinyl chloride is a polymer of the monomer vinyl chloride, $CH_2{=}CHCl$. The

About 2 billion feet of vinyl cellular molding is now produced per year in the U.S. Woodgrain-finished vinyl molding replaces more expensive wood trim. Reduced splitting and cracking cuts scrap loss 95 percent. (Courtesy The Goodyear Tire & Rubber Company.)

common grouping $CH_2{=}CH{-}$ is designated the *vinyl* group. Polyvinyl chloride is the most common vinyl polymer and is often referred to as PVC. It appears in large volume in vinyl floor coverings, upholstery materials, garden hose, and phonograph records.

Thermosetting Plastics

The thermosetting type of plastics, or thermosets, consists of materials that form a rigid mass on heating which cannot be remelted. This is similar to the hardening of an egg when cooked. The thermoset materials are made into finished plastics products in either of two major ways.

1. Compression molding, wherein the thermosetting polymer materials are placed in a shaped mold and the two halves pressed together under heat and pressure, somewhat like the operation of a waffle iron. The materials placed into the press are not in a fully reacted, or thermoset, form until the heat and pressure of the molding operation is applied. Familiar products here include dinnerware, ashtrays, telephone handsets and bases, wheels, and casters.
2. Molding in a laminating press, which involves coating sheets or layers of base material like paper, cloth, metal, or wood with the polymer material in a partially reacted form. Then these are layered down like a Dagwood sandwich and forced together under high heat and pressure. Common examples include structural plywood, wall paneling, tabletops, and gears.

You can appreciate the operational requirement for thermosets when you consider the exposure of cups, dishes, and dinner plates to hot foods and water. Here we need long-chain molecules which do not slip around each other when heated. The polymer chemist therefore designs not only for long chains but he also arranges to get strong links between the chains formed. The requirement of high rigidity is satisfied conceptually by merely choosing monomer molecules with more than two functional or linking groups per molecule. Such a choice insures that the monomer will form chains via two of the linkages and have a third link (or more) to join up between chains. This is called cross-linking, that is, linking across the space between chains.

A section of tree root which was anchored to a river bank. You can see here how nature used cross-linking between roots in order to provide the extra strength needed to hold the tree in the loose soil at the river's edge. Chemists use the same idea in thermoset plastics. (Photo by Gary R. Smoot.)

Thermosets can be made by either condensation or addition methods. The condensation polymer Bakelite is the classical example of a thermoset. The complex, interlocked structure of a thermoset is often considered to be really one gigantic molecule. This is similar to the diamond (chapter 8), but there is much less regularity in the thermoset plastics.

Varying Properties by Adding Chemicals

There are other ways, besides using thermosets or thermoplastic materials, of getting a variety of properties even with the same plastics material. Consider phonograph records, floor tile, pipe, imitation leather, and shower curtains. All of these are made from the same polyvinyl chloride or PVC. Here the chemist presumes that the molecular chains of the thermoplastic have to be a little looser in the shower curtain than in the phonograph record. The chemist therefore mixes the polymer with a compatible chemical which slightly loosens or lubricates the chains so that they do not provide so much rigidity to the shower curtain. A chemical added in such an application is called a plasticizer. It makes the material more plastic because it gets in between the chains.

Varying Properties by Built-in Polarity

Nylon provides another example of how the makeup and arrangement of molecular chains affect properties. We might not want the "looseness" of shower curtains in applications requiring strength like in refrigerator door latches or stocking fibers. At the same time we would hardly want extreme rigidity which would amount to brittleness. These operational requirements are obtained by a thermoplastic with provision for some degree of holding together of chains but not the extreme degree represented by thermosets. Chemists have come up with the concept of the nylons which have in their long chains the recurring group

```
   O
   ‖
 —C—N—
    |
    H
```

This is called an *amide* group. Amide groupings have an unequal distribution of charge which provides for hydrogen bonding (chapter 7).

The oxygen atom has more pulling power for electrons than carbon. Also, the nitrogen atom pulls the electron pair it shares with hydrogen closer to itself. The overall effect is a slight shift in negative charge concentration toward oxygen and away from hydrogen. The slightly negative oxygen atom in one chain is thus held to a slightly positive hydrogen atom in a neighboring chain. This is the same mechanism that we found to account for water being a liquid—the molecules are interconnected by the attraction between opposite charges. This attraction forms a sort of electrical linkage between chains in the nylons.

```
                 ¦ ← Hydrogen bond
                 O
                 ‖
etc. · · · · —C—C—N—C—C · · · · etc.
                   |
                   H
                   ¦ ← Hydrogen bond
                   O
                   ‖
etc. · · · · —C—C—C—N—C—C · · · · etc.
                     |
                     H
                     ¦ ← Hydrogen bond
```

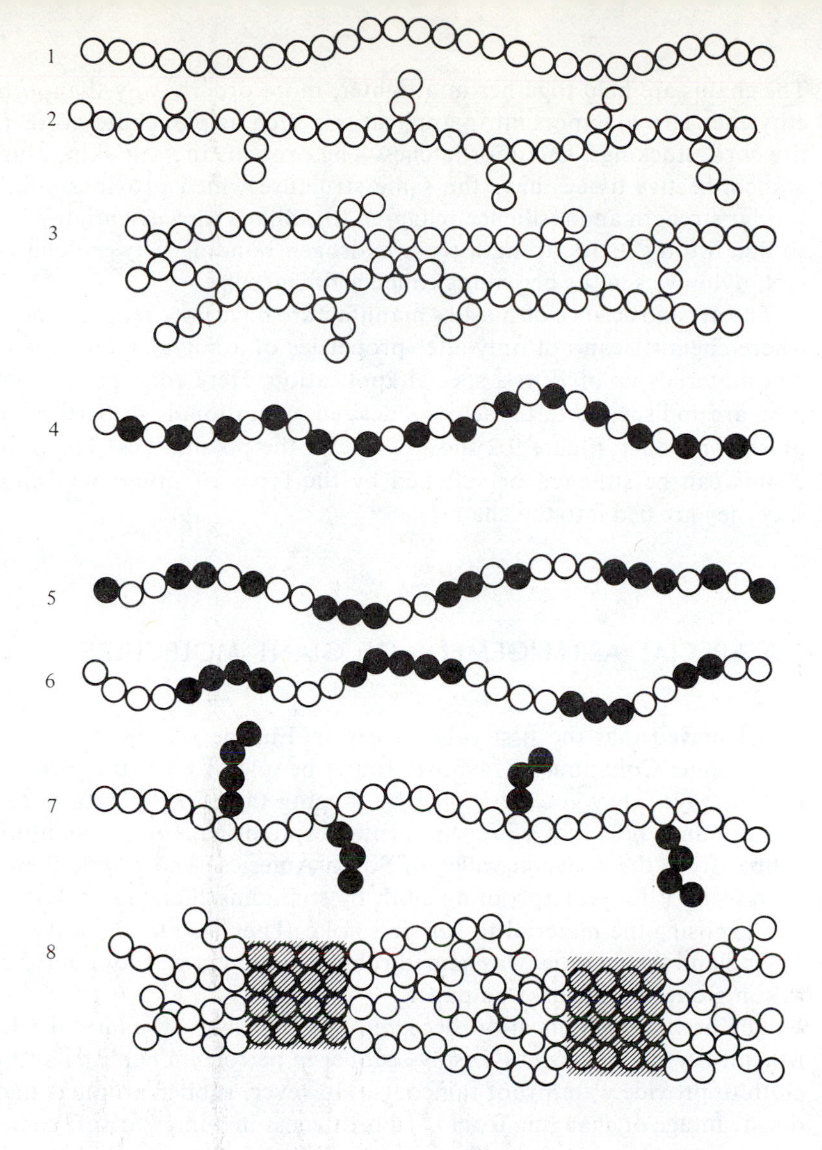

FIGURE 9.2

The polymer chemist is sometimes said to be involved in molecular architecture. Chemists now have ways of designing and building the types of chains desired in a particular end product. The symbolic chains here show some of the possibilities. Circles are used to designate a particular monomer, black one type, white another. The chains do not show the actual atoms which are the basic units for the monomers. Chain (1) is the simplest type; (2) is a simple polymer with branching; (3) shows cross-linking to give extra strength, as in thermosets; (4) is a chain with alternating monomers; in (5) the different monomers are alternating in more irregular fashion; (6) shows chains with different monomers which are grouped in "blocks"; (7) represents "graft" polymer chains with one monomer for the length direction and another in the branches; (8) is one type of chain where there is some crystallization which makes for more rigidity and greater strength.

The chains are held together in a tighter, more orderly way than in polyethylene. This is important in applications such as parachute cord, rope, tire cord, stockings, and door latches. The proteins in your skin, muscles, and connective tissues have the same structure which provides you with similar strength and resilience (chapter 10). Silk is also essentially protein so that it too gets its strength from hydrogen bonding between chains. In fact, nylon was in the beginning called artificial silk.

The art and science of plastics manufacture have now reached the stage where chemists can not only alter properties of a polymer but can make new materials tailored to a special application. Here concepts and know-how are indispensable to achieve desired operational properties in the finished product. Figure 9.2 shows some of the possibilities. The polymer chains can be stiffened or softened by the types of monomers and the way they are tied into the chain.

RUBBER: A SPECIAL ARRANGEMENT OF GIANT MOLECULES

It is believed that the first rubber seen in Europe was brought over by Christopher Columbus in 1496. He found people of Haiti using bouncing balls as toys. They got the rubber by tapping the bark of certain trees. It was not until 1731 that a French scientist brought back large quantities of rubber from the Amazon valley of South America. The people there had been using it for waterproofing cloth by spreading films on woven fabric and exposing the material to sun or smoke. They also made crude boots. In England the material was named rubber because it was found useful for rubbing out pencil marks on paper.

The first large application occurred when a Scot by name of Charles Macintosh found a way to dissolve rubber in petroleum naphtha and treat cloth to provide waterproof raincoats. However, rubber products had the disadvantage of changing from hard brittleness in winter to soft stickiness in summer. Many of the early raincoats were found completely stuck together when taken from the closet.

Charles Goodyear invented the process of "vulcanization" in 1839. He was working for years trying to eliminate the stickiness of rubber and found the solution when he accidentally spilled a mixture of crude rubber and sulfur on a hot stove. When he scraped it off and cooled it he noticed that the stickiness was gone and it sprung back when stretched.

Subsequent years saw the expansion of rubber plantations by smuggling rubber tree seeds from South America to England and thence to Southeast Asia where new plantations were started. Chemists meanwhile during the nineteenth century tried to decipher the nature of the rubber molecule. A small hint was obtained when a simple liquid, isoprene, was isolated from decomposition of crude rubber. The identification of the isoprene molecule showed it to have two double bonds.

$$\begin{array}{ccccccc} & & CH_3 & & & & \\ & & | & & & & \\ CH_2 & = & C & - & C & = & CH_2 \\ & & & & | & & \\ & & & & H & & \end{array} \qquad \begin{array}{ccccccc} & & H\;\overset{H}{|}\;H & & & & \\ H & & C & & & & H \\ | & & | & & & & | \\ C & = & C & - & C & = & C \\ | & & & & | & & | \\ H & & & & H & & H \end{array}$$

It was not until well into the twentieth century before the actual nature of the rubber structure became clear. This came about chiefly through attempts during World War II to make a synthetic rubber substitute. Chemists early recognized that synthetic rubber could not be made readily from isoprene. They therefore turned to other similar molecules; after painstaking years of effort they finally succeeded in making rubberlike materials. Examples of starting molecules these chemists used are butadiene and chloroprene.

$$\begin{array}{ccccccc} H & & H & & H & & H \\ | & & | & & | & & | \\ C & = & C & - & C & = & C \\ | & & & & & & | \\ H & & & & & & H \end{array} \qquad \begin{array}{ccccccc} H & & Cl & & H & & H \\ | & & | & & | & & | \\ C & = & C & - & C & = & C \\ | & & & & & & | \\ H & & & & & & H \end{array}$$

Butadiene　　　　Chloroprene

The synthetic rubbers were not exactly like natural rubber but each had special properties—often some special characteristic which was better than that of natural rubber. For example, some of the synthetic rubber substitutes had much better resistance to attack by oil; others had better capability of holding air in tires because of lower permeability to gas molecules.

What eventually became clear was that rubber was not a special chemical at all but a special arrangement of giant molecules. Now we know rubber is a mixed-up, tangled mass of long chain molecules resembling the random mix-up of cooked spaghetti. (See Figure 9.3.) Vulcanization consists of forming cross links at various reactive sites on the molecules. The coiled molecules are therefore responsible for the stretchability of rubbers. When the rubber is pulled out, the coils unwind somewhat, but the few cross links prevent complete dislocation. Then when the pulling is released, the long chains are pulled back by the cross links into the coiled positions they started from. This is the conceptual explanation for the operational characteristics of stretchability and resilience.

Rubber Molecules: Alternating Stretchability with Strength

Unstretched rubber has a tangled, amorphous (irregular and formless) character. By contrast, stretching aligns the chains and provides considerable strength because of a tighter, more regular arrangement of atoms in

FIGURE 9.3
Although not a perfect analogy, the mixed-up coils of spaghetti resemble the situation in rubber.

adjacent chains. The rapid change back and forth between these two states provides the special characteristics needed, for instance, in rubber tires. When the car's weight shifts to stretch the rubber side wall, extra strength is provided by alignment of adjacent chains. When the weight shifts to another area of the tire the stretching of the first part relaxes, allowing return to the resilient arrangements of randomly coiled molecular chains. Thus we get a smooth ride much improved over use of iron wheels.

Varying the Bounciness of Rubber

It is interesting to compare one of the synthetic rubbers with natural rubber. The chains can be visualized as follows.

```
       [   H  H  H           ]               [ H  H  H    ]
       [    \ | /            ]               [  \ | /     ]
       [ H    C    H    H    ]               [    C    H  ]
       [ |    |    |    |    ]               [    |    |  ]
 .... [ —C————C====C————C—  ] ....     .... [  —C————C—  ] ....
       [ |              |    ]               [    |    |  ]
       [ H              H    ]n              [    C    H  ]
                                             [  / | \     ]
                                             [ H  H  H    ]n
```

Natural rubber

Synthetic butyl rubber

The butyl rubber with two branches per monomer unit is by far the less resilient or less bouncy of these two rubbers. Knowing that rubberiness is

Blending fabric and rubber to form plies for passenger tires. (Courtesy The General Tire & Rubber Company.)

At left, tire builder applies white sidewall strip for passenger tires. (Photo courtesy The General Tire & Rubber Company.) The photo on the right shows the green, or unvulcanized, tires ready for insertion in heated mold from which finished tires are coming out. (Photo courtesy The Goodyear Tire & Rubber Company.)

caused by stretchability of coiled chains, you can visualize that branches prevent full coiling and therefore cut down on stretchability and resilience. The presence of only one branch in the natural rubber units means less entanglement and more bounce. However, the butyl rubber is tougher and is excellent for inner tubes: it prevents gas penetration and keeps tires from losing air on standing. This is understandable because the branches provide a more mixed up "jungle" of chains. These concepts are confirmed by other synthetic rubbers with no branches at all. These are even more bouncy than natural rubber.

HOW LARGE ARE THE GIANT MOLECULES?

All polymers are large molecules even though the word large is not exactly defined. In general, we say that polymers may run up to hundreds of thousands of smaller monomer units. Here are some reasons why we do not define *large* more exactly:

1. The size of molecules depends on the type of polymer—some having smaller numbers of monomer units on the average than other types. For example, ordinary nylons have about 90 monomer units to the giant molecule. Natural rubber is estimated to have between 20,000 and 100,000 monomer units in the chains.
2. Methods of synthesis for any one type of polymer do not produce all chains of exactly the same length. In other words, a simple polymer material is usually a mixture of many varying chain lengths. Part of the purification process often consists in removing chains of smaller length. Costs, technical difficulty, and impracticality preclude getting all chains of exactly the same length.
3. The thermosetting plastics do not have well-defined molecules as such, since the whole sample of plastic is really one giant, interconnected network.

The overall picture then is quite variable when it comes to specifying how many monomer units link into chains or networks. This means the numbers of atoms per molecule, while also large, is not exactly defined either. In the case of nylon (page 195) the number of atoms per long-chain molecule (of carbon, hydrogen, oxygen, and nitrogen only) runs up to about 1700 or 1800 atoms. Natural rubber may have 260,000 to 1,300,000 atoms per molecule (carbon and hydrogen only). Polyethylene can have some chains containing up to 200,000 carbon atoms along with 400,000 hydrogen atoms in the same molecular chain. The number of atoms in a thermoset molecule is not a meaningful descriptive characteristic.

CONCEPTS AND END PRODUCTS

You can see now how imagination and vision are essential in molecular design. Imagination coupled with a capability of following through on production of tailor-made materials provides fantastic possibilities. Using ideas like those we have sketched above, the polymer chemists have given us, among others, the following useful end products.

Corrosion-resistant Teflon coatings which shed both oil and water.

Coatings for cloth that provide freedom from wrinkling and eliminate the chore of ironing.

Strong yet flexible Spandex fibers for supportive clothing, bathing suits, and girdles.

Tough, heat-resistant materials for space-age applications such as heat shields on returning space capsules.

Adhesives that are stronger than steel.

Abrasion-resistant rubbers for tires.

Endless varieties of food packaging materials, including convenient cook-in-the-bag types.

Synthetic blood plasma extenders.

Foams of various types: porous or nonporous, flexible or rigid—for insulation, fire-resistance, strengthening of architectural and aircraft assemblies, seat cushions, nonrotting life preservers, and so on.

A new use of plastics is in sturdy structural polypropylene furniture frames. Because the polypropylene is tough, the frames will not shrink, swell, dry out, or creak. Stain-resistant upholstery fabric is also made from synthetic fiber. Plastic foam provides the cushion. (Courtesy Hercules Incorporated.)

The yearly production figures for plastics in the U.S. alone staggers the imagination. It has now reached the unbelievable total of about 20 billion pounds a year. Here are approximate production figures per year for some of the most common plastics:

6 billion pounds of polyethylene
3 billion pounds of polyvinyl chloride
3 billion pounds of polystyrene-type polymers
1 billion pounds of polypropylene
1 billion pounds of Bakelite-type polymers
700 million pounds of polyesters
100 million pounds of nylon

For comparison purposes, annual consumption of rubber in the U.S. runs to 6 billion pounds, of which approximately two-thirds is synthetic and one-third natural rubber.

Annual production of plastics now exceeds that of both copper and aluminum. It is anticipated that by the early 1980s it will be almost 50 billion pounds a year. On a volume basis the projected growth will, in the mid 1980s, reach approximately 1 billion cubic feet annually. This is expected to surpass the volume production even of steel and will necessarily be an important part of our waste disposal problem (chapter 15).

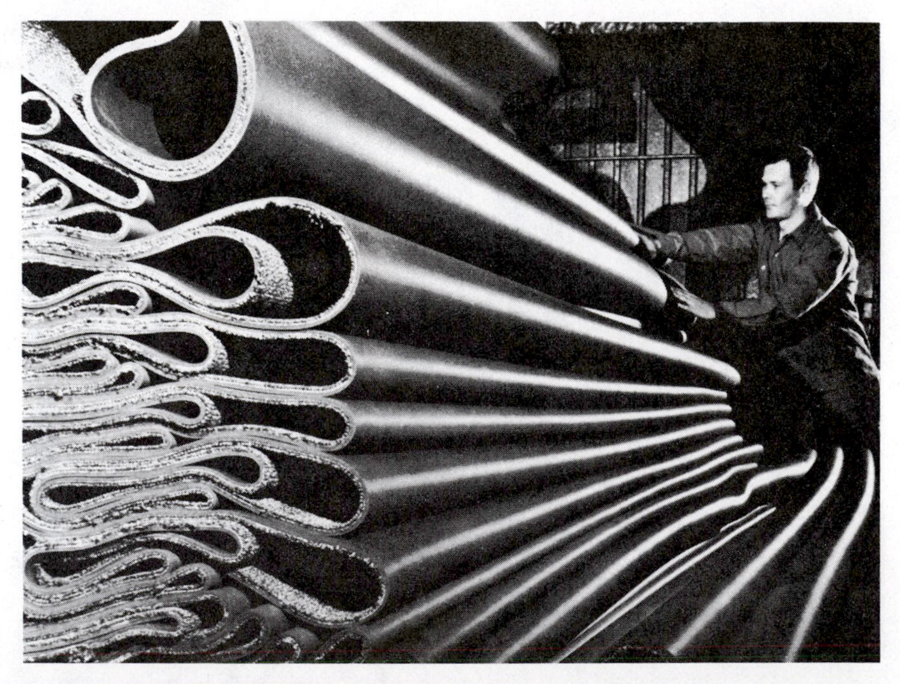

Carpeting backed with latex foam stacks up like ribbon candy in a carpet mill storage bin prior to roll up. Carpeting with foam padding already attached is growing in use because it is easier and more economical to install. (Courtesy The Goodyear Tire & Rubber Company.)

Experiments

1. This could be a thought experiment if you do not have a fireplace or cannot prepare an outdoor fire with firewood. If you have the fireplace and do it as an actual experiment, take care not to compromise safety in any way. Recognize that plastics are mostly flammable. Do not try to do the experiment indoors without a fireplace. First take several old pieces of discarded plastics products, like package material, delicatessen containers, plastic soap containers, old combs, knobs. Use small pieces. Put each into the fireplace or on the wood fire after the flames have died down and you have just placed a new log on top of a good bed of hot ashes. The idea is to place the plastic piece on the log so that you can have time to observe it under the heat before it catches fire. Can you observe behavior which would allow you to distinguish a thermoplastic material from a thermosetting type? What would you conclude as to the materials you considered for this test, that is, which are thermoplastics? Do you notice any behavior of the plastics you tested which would suggest anything about their structure? Eventually of course they catch fire. What is the result of this chemical reaction? Do not try to remove any of the materials once placed in the fire since this could be dangerous. Can you relate your experiences here with what sometimes happens when some plastic containers are placed in an automatic dishwasher? What kind of polymer should be used in "dishwasher-safe" applications?
2. Get a piece of transparent polyethylene of the kind that is used to cover clothes in the dry cleaners, or to wrap foods. Be sure it is the soft pliable packaging material, not cellophane. Cut a strip about one inch wide and about a foot long. Then holding the ends between your thumb and forefinger try stretching it slowly. Be sure to hold each end so that the whole width is anchored firmly, one end under each thumb. It is important to hold the sample evenly and to stretch very slowly, moving your hands apart smoothly. You will need to try this several times to get the feel of the experiment. If the polyethylene strip snaps, get another one. Eventually they will all snap. But the idea is to observe the changes in properties before that occurs. Describe what you observe here. Do not limit your observations to what you can see but also observe what you can feel. After you get the idea of how far the samples can stretch, try testing for rubberiness at some point before full stretch. Do this by stretching and letting go slightly to see if there is any pullback, like in a rubber band. You may want to try a few more experiments, using different width strips or a couple of strips at a time. After you have accumulated considerable operational experience with the polyethylene behavior, try to come up with some conceptual explanation of the changes and the properties noted.

Exercises

3. Choose one of the quotations opening this chapter and write a short paragraph explaining what it means now to you. Try to relate some idea about polymers to the idea expressed in the quotation.
4. Suppose someone told you he heard that eyeglass frames, ping-pong balls, and a mirror handle were made from the same plastic material. He says that he cannot believe that this is true. How would you answer him and explain such a situation?

5. Show by using formulas how an ester can be formed that will not lead to a high polymer. Next show by other formulas how the formation of an ester can lead to high polymers. Examine the labels on your clothes and try to find three specific applications of polyesters.
6. Draw a structural formula for benzene using electron dots to show the chemical bonds. Indicate how you could draw another formula for the same molecule but with the double bonds alternating in different positions. Show finally the preferred symbol for this organic ring, indicating what the symbolism means.
7. Select any addition polymer other than polyethylene from the list in Table 9.2. Then show by equations how the monomer is made into a polymer in a chain reaction.
8. Explain the basic difference between condensation and addition polymerization. Give five practical examples of common usages of polymers of each type.
9. Describe the operational characteristics of thermosetting and thermoplastic materials. Then give a brief conceptual explanation for the behavior.
10. Choose one thermoplastic and one thermosetting material and give five examples of practical products using them.
11. Look around in your environment and make a list of ten plastics products. Indicate whether you would use a thermosetting or thermoplastic polymer for each of the products. Give a reason in each case.
12. Describe briefly how interchain attraction would effect the operational characteristics of a plastics product.
13. How could polarity be introduced into polymer molecules? What would be the effect on the operational behavior or properties of the polymer formed?
14. Explain simply why rubber stretches.
15. Give a brief description of the conceptual basis for the desirable characteristics of rubber used in auto tires.
16. Describe conceptually why butyl rubber is less bouncy than natural rubber. Does butyl have any advantage over natural rubber? Give a brief conceptual explanation of this.
17. Describe or define the following terms both operationally and conceptually: rubber, nylon, plastics, natural rubber, butyl rubber, thermoplastics, thermosets.
18. Injection molding and extrusion are methods for using thermoplastic polymers. Why do you think these are not considered suitable for thermosetting resins?

SUGGESTED READING

1. Mark, H. F., ed., *Giant Molecules*, in the Life Science Library, by the editors of Life, Time Inc., 1966. A very readable and colorful book on the story of plastics.
2. Mark, H. F., "Giant Molecules," *Scientific American*, 197, No. 3, pp. 204–216 (September 1957). Scientific American Offprint No. 314.
3. Mark, H. F., "The Nature of Polymeric Materials," *Scientific American*, 217, No. 3, pp. 149–156 (September 1967).
4. Check your favorite encyclopedia for specific titles; for example, plastics, cellophane, rayon, rubber, etc.

Tell me what you eat,
and I will tell you what you are.
—BRILLAT-SAVARIN

The whole of nature is a conjugation of the verb
"to eat," in the active and the passive.
—W. R. INGE

Let onion atoms lurk within the bowl,
And, half-suspected, animate the whole.
—SYDNEY SMITH

chapter 10

Foods: The Chemicals We Eat

It was not until the end of the nineteenth century that chemists began to ask, "What kinds of molecules are in foods?" You may think it unusual that people waited so long to investigate the nature of molecules that they eat every day.

There is a basic reason, and it is essentially the same reason that we discuss giant molecules in chapter 9 of this book rather than in chapter 1. You have to understand the smaller molecules first. And scientists followed this same kind of concept building to make sense out of materials used for food. Eventually a growing body of information and understanding of smaller molecules built up so that scientists could then turn to nature's giant molecules like those in foods and cotton and leather. About three hundred years ago Newton said: "If I have seen farther than other men, it is because I stand on the shoulders of giants." When you finish this chapter you will understand more about foods than Newton ever could.

WHAT ARE FOODS?

All foods are materials which provide body nutrients. These are the chemicals for energy, growth, repair, and regulation of your body processes. In the U.S. and Canada 40 percent of foods comes from animals and 60 percent from plants. The animals in turn depend on plants so that the fundamental starting point for the food chain which ends in your kitchen is the plant.

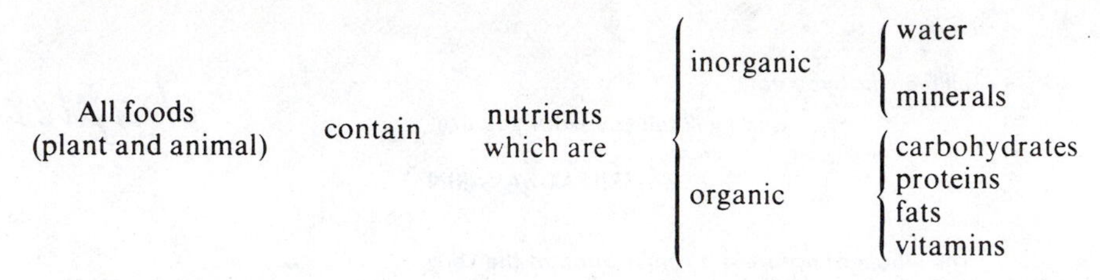

To describe foods conceptually we come down to molecules of various kinds, for example, protein molecules or fat molecules. You already know about inorganic chemicals which includes water and the minerals. Examples of minerals found in food are: sodium chloride, NaCl; potassium iodide, KI; potassium chloride, KCl; ferrous sulfate, $FeSO_4$; calcium chloride, $CaCl_2$. The usual practice in referring to the minerals in foods is to name the particular element alone without reference to how it is combined. Thus a food is said to provide the minerals calcium and iron when it may contain $CaCl_2$ and $FeSO_4$. Most foods are mixtures, but the basic nature of the fundamental building blocks that they contain is not too hard to grasp. Formulas help us here. The conceptual understanding represented by the formula aids us in thinking more profitably about foods.

CARBOHYDRATES

Most people commonly think of carbohydrates as being starch or sugar which are the major sources of energy for the body. The term also includes cellulose. All carbohydrates contain carbon, hydrogen, and oxygen. Many of the carbohydrates contain hydrogen and oxygen in the same proportion as in water: 2 hydrogen atoms to 1 oxygen atom. The structure does not really involve H_2O molecules as you will see below.

Carbohydrates with One Sugar Unit

The simplest carbohydrates are called monosaccharides (one sugar). Other molecules consist of two basic sugar units—disaccharides. The more complex carbohydrates are called polysaccharides, consisting of long chains of many sugar units. To most people sugar is the material used in the home as table sugar. This is only one of many possible sugars. A sugar in chemistry is any of a large group of carbohydrates which are soluble in water and have a sweet taste.

The monosaccharide sugar called glucose is widespread in nature and is the principal molecular type to which complex carbohydrates are reduced during digestion. Glucose is also called dextrose or blood sugar. It is the main circulating carbohydrate in plants and animals. A common representation of glucose is as a cyclic structure with the formula $C_6H_{12}O_6$.

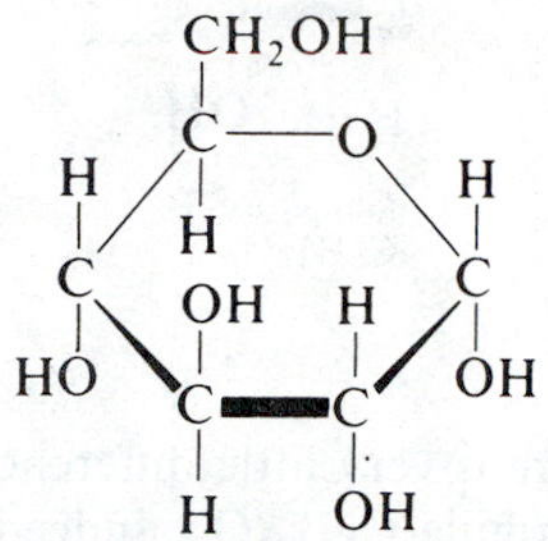

Glucose
$C_6H_{12}O_6$

This structure shows the directional placement of the groups on carbon atoms of the ring. The forward side of the hexagon has darker lines to indicate that this end is pointing out toward you. The hexagon is in a plane perpendicular to the paper with the lighter lines extending behind the plane of the paper. You are looking over the molecule. This allows you to visualize the placement of groups on the carbon atoms above or below the plane of the hexagon.

The need for structural formulas comes from the fact that there are actually many other isomers with the basic formula $C_6H_{12}O_6$. Two of the isomeric forms are named glucose, the one above being called the alpha form, the other, beta-glucose. They are shown side by side for comparison.

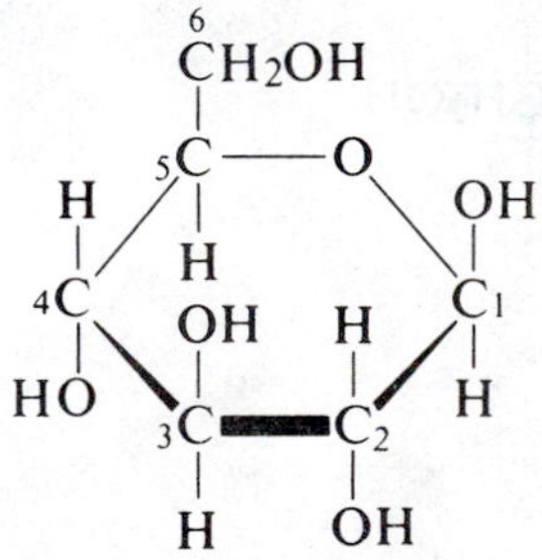

Beta-glucose

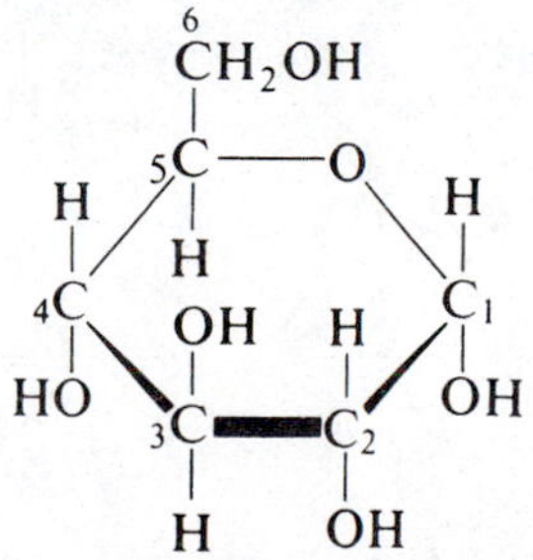

Alpha-glucose

The six carbons are numbered to allow easy reference. You can see that the beta form of glucose has a change only on the number one carbon atom where the —OH is now on top of the plane.

Another isomer which is important in nutrition is the sugar called galactose.

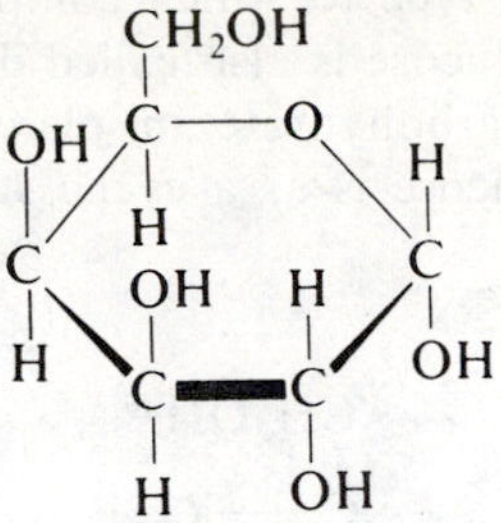

Galactose
$C_6H_{12}O_6$

You will note that there is very little difference in all of these isomers of the basic molecular formula $C_6H_{12}O_6$. Indeed, you have to look carefully at the two-dimensional representations to see any difference at all, but the subtle differences are quite critical in some of nature's operations. Starch and cellulose is one example we will look at below.

Another sugar is called fructose or fruit sugar because it is so common in fruits. It has the same formula $C_6H_{12}O_6$, and is therefore another of the numerous isomers of glucose. You will see in the new arrangement of the same atoms ($C_6H_{12}O_6$) how nature can produce an even more varied arrangement than those examples shown above. Fructose is usually pictured as a five-membered ring.

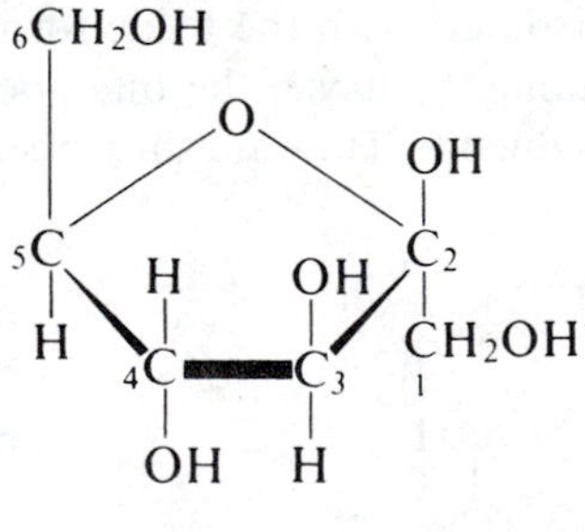

Fructose
$C_6H_{12}O_6$

Carbohydrates with Two Sugar Units

All of the sugars described above are monosaccharides, made up of a one-sugar unit. Table sugar—also called sucrose—is a disaccharide. A disaccharide occurs when two monosaccharides link together. They do this by condensation, with the loss of one molecule of water where they join together. Sucrose or table sugar is formed in many plants by joining a glucose molecule to a fructose molecule.

Glucose + Fructose → Sucrose (table sugar) ($C_{12}H_{22}O_{11}$) + Water

$$\underset{\text{Glucose}}{} + \underset{\text{Fructose}}{} \rightarrow \underset{\text{Sucrose}}{} + H_2O$$

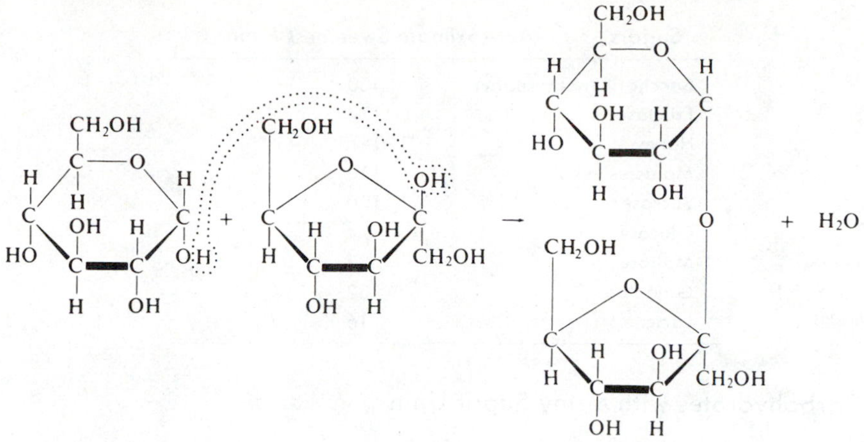

The reverse of this reaction can be caused by acids, bases, or certain enzymes which are complex proteins produced in living cells. These act as catalysts to speed up the reaction.

$$\underset{\substack{\text{Sucrose}\\\text{(table sugar)}}}{C_{12}H_{22}O_{11}} + \underset{\text{Water}}{H_2O} \xrightarrow[\substack{\text{(acid, base,}\\\text{or enzyme)}}]{\text{Catalyst}} \underset{\text{Glucose}}{C_6H_{12}O_6} + \underset{\text{Fructose}}{C_6H_{12}O_6}$$

This is sometimes called a hydrolysis reaction, which is another name for "reaction with water."

Lactose is another common disaccharide. It is found in the milk of mammals and is sometimes called milk sugar. On reaction with water it gives the two monosaccharides glucose and galactose.

$$\underset{\text{Lactose}}{C_{12}H_{22}O_{11}} + \underset{\text{Water}}{H_2O} \xrightarrow{\text{Catalyst}} \underset{\text{Glucose}}{C_6H_{12}O_6} + \underset{\text{Galactose}}{C_6H_{12}O_6}$$

A third rather common disaccharide is maltose. This sugar is available from germinated barley and is also produced by the action of malt on starch. Malt is barley or other grain soaked in water, spread till it sprouts, and then dried and aged. Maltose accounts for the special flavor of malted milk which is made from dried milk, malted barley, and wheat flour mixed with milk. Maltose gives two molecules of glucose on hydrolysis.

$$\underset{\text{Maltose}}{C_{12}H_{22}O_{11}} + \underset{\text{Water}}{H_2O} \xrightarrow{\text{Catalyst}} \underset{\text{Glucose}}{2C_6H_{12}O_6}$$

The sweetness of various sugars have not yet been measured absolutely. Sucrose is given an arbitrary value of 100 and the other compounds given comparative values.

Sugars	*Approximate Sweetness Ratings*
Saccharin (not a sugar)	400
Fructose	174
Honey	120–170
Molasses	110
Sucrose	100
Glucose	74
Maltose	33
Galactose	32
Lactose	16

Carbohydrates with Many Sugar Units

You are in daily contact with giant molecular carbohydrates or polysaccharides. Both starch and cellulose are natural polymers of glucose.

Section of a beehive showing the natural pattern of six-sided figures which make up the structure. This is an interesting pattern, especially considering that bees make honey which also has hexagonal symmetry. (Photo by Gary R. Smoot.)

Starch, which is the energy storage molecule of plants, can be digested by humans into the glucose units which serve as sources of energy. Cellulose, also a polymer of glucose, cannot. We would starve if we had only cellulose to eat. This big difference in operational behavior must have a conceptual basis.

It involves a very slight change in positioning of only one of the —OH groups on the glucose molecule. Starch has an alpha linkage where the —OH on the number one carbon atom is below the plane of the ring. Cellulose forms links through the beta form of glucose. This is a startling difference in properties caused by what you might at first think is only a minor difference in structure.

The starch type of linkage can be represented as follows.

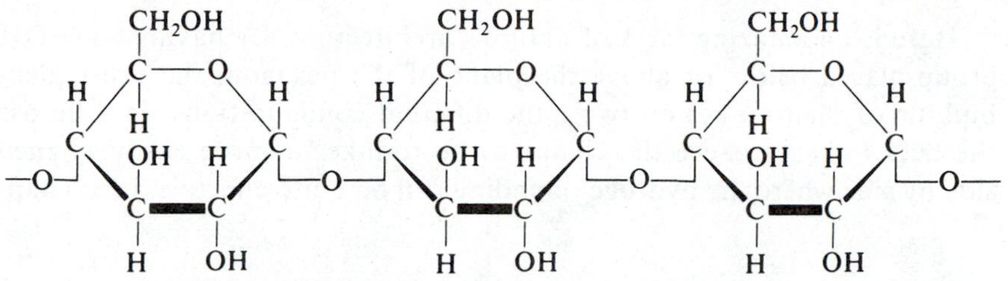

Starch is produced in plants for storage in roots, seeds, and fruits in the form of tiny granules. It does not dissolve in cold water but swells considerably in hot water, forming a paste.

More detailed examination of starch shows that it may be composed of two types of chains. One component has continuous chains running up to about 300 glucose units. There is another type of chain with branches every 25 or so glucose units in the main chain. These branched chains can have total glucose units running up to about 2000.

The animal body also stores glucose and the storage unit is called *glycogen* (which means glucose producer). This is animal starch. Glycogen is stored in the liver and muscles for use between meals as a source of energy. The carbohydrates that are not used at once for fuel, or converted to glycogen, can be converted to fat and stored. The glycogen has the same alpha linkage as plant starch but has branches at more frequent intervals. The number of glucose units can run in the range of 6000 to 30,000.

Cellulose: The Building Structure of Plants

Cellulose is the major building or supporting structure for plant cells and fibers such as wood and cotton. Plants put together glucose units in a slightly different manner from starch so that the building material will not be consumed for energy. In cellulose, glucose is combined in the beta form into long continuous chains estimated to contain between 900 and 6000 glucose units. The cellulose structure depicted below shows the upward facing —OH group (beta form) on the right of each glucose unit, rather than the opposite configuration found in starch. The bond is shown bent in order to keep the chain in the horizontal plane. Actually the bonds connecting the glucose units are straight, and the molecules of glucose move into a line off the horizontal.

Here is an amazing facet of nature's architecture. By having an —OH group placed below or above the plane of the hexagon, the subsequent buildup of chains takes on two quite different configurations. In cellulose the beta linkages cause the chains to be rodlike or more easily aligned side by side where the hydrogen bonding can be quite effective. (See chap-

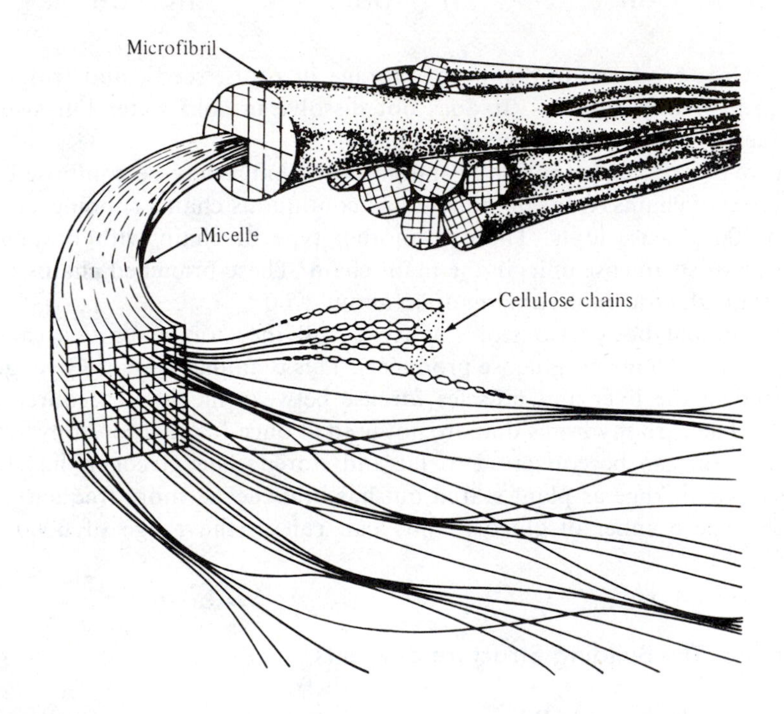

FIGURE 10.1
Simple representation of how glucose molecules form long chains of cellulose molecules, which in turn form bundles (micelles), and these in turn form even larger bundles (microfibrils) which eventually provide the structural building material for plants. The microfibrils are large enough to be seen with an electron microscope. This is nature's way of building strength which we imitate in our use of threads and wires in rope, twine, and cables. (From *Principles of Plant Physiology* by James Bonner and Arthur W. Galston. W. H. Freeman and Company. Copyright © 1952.)

ter 9, page 206.) In starch, on the other hand, the alpha linkage gives the chains a spiral configuration.

The rodlike chains in cellulose form closely packed bundles. (See Figure 10.1.) Then hydrogen bonding between neighboring chains can provide considerable strength and also hinder the penetration of water molecules. Starch molecules have less facility for formation of hydrogen bonds between chains because of the spiral arrangement. This is why starch has little strength and is easily penetrated by water molecules. In other words, it has more —OH groups which are not tied up holding chains together.

Why Does Bread Get Stale?

The presence in starch of considerable material with branched chains also contributes to its easier penetration by water. Both kinds of starch—the straight chains and the branched—are insoluble in cold water. However, in hot water the dispersion of molecules is considerable. The branched chain starch remains more easily dispersed in the water while the straight chain types are found eventually to come out, or precipitate. This is the main mechanism for the staling of bread. In other words, the straight chain starch, while not having the extreme cohesiveness of the more rigid cellulose, does have more potential for association between adjacent chains than does the branched type of starch. Since in bread the usual mixture of these is present, eventually the continuous chains of starch tend to migrate together and form a harder, stiff texture which makes the bread stale. In starched shirts this is desirable.

Different Uses of Starch

Chemists have been able to separate the two types of starch. The straight chain type is best for preparation of digestible films and is being investigated for use in packaging foods wherein the film dissolves in water during cooking. The branched type of starch is better as a thickener and in adhesives.

The chief source of starch in the United States is cornstarch, which contains about 75 percent of the branched type chain. Some cooks prefer to use this as a gravy thickener because of the high content of branched starch in comparison with ordinary flour.

Recognizing the fact that the straight chain starches have too strong a tendency for alignment, chemists thought that if they introduced groups on the chains which interfere with alignment and prevent some of the hydrogen bond formation, then they might get easier dispersion in water. They put phosphate ester groups in the chains and one of the products is now used in "instant" pudding which does not require boiling to disperse the starch and thicken.

Why Does Animal Starch Have Many Branches?

It might be interesting here to consider the question: why does animal starch, or glycogen, have more frequent branching along the main chains than plant starch? Operationally, you would expect that the more frequent branching would introduce more potential for availability in water, the reason being that the more numerous branches interfere with alignment and prevent association of chains. This is the case. So one answer to "why" might be that the branching is more numerous in order to provide more ready solubility.

This is obviously not the only answer. Further we could say that it is to provide animals with a potential for fast reaction in time of danger. Then the animal must mobilize quickly extra reserves of energy which means glucose from the chains. The branching makes glucose readily available to the water system of the blood. You can see that the question "why" may be answered on several levels. And there are always other levels beyond.

Keeping Cellulose in the Carbon Cycle

You now know one answer to another question: why can you digest starch but not cellulose? Because glucose units of each are joined into chains in different ways. Why can horses and cows break down cellulose to its glucose units? Because they have in their complex digestive system microorganisms which provide the proper enzymes. Termites also produce an enzyme which breaks wood cellulose down to glucose.

Fortunately for us there are billions of microorganisms in the world capable of breaking down cellulose to glucose. Otherwise there would be pollution of yet another sort—and on a grand scale. Cellulose is the most common organic substance on earth. Without those microorganisms there would be an accumulation on the earth of the dead debris of fallen trees, leaves, and grass which would dwarf in magnitude all our present pollution problems. There would be a break in the carbon cycle which would quickly mean an end to all life on earth. In other words, in decay there is life.

Dextrins and Other Sweet and Sticky Items

Dextrins are made by partial hydrolysis of starch. This gives mixtures of smaller length chains. They serve as adhesives, mucilages, pastes and coating for fabric and paper (called "size"). Postage stamps commonly are coated on the back with mucilage made from dextrins. Some infant formulas also contain powdered mixtures of maltose and dextrins.

Toast and the crust of bread contain considerable amounts of dextrin. This is formed from the starch by heat which breaks open some of the longer chains. The old-fashioned treatment for people recovering from stomach upset included tea and toast. This has good conceptual basis since dextrin is more easily digested into glucose than is starch.

Why do the dextrins serve so effectively as glues? The chains of glucose units have many possibilities for hydrogen bonding with the glucose chains of paper—which is chiefly cellulose. Consequently, millions upon millions of entangled chains bridge between the cellulose of the postage stamp and that of the envelope.

Corn syrup also comes from breaking down starch molecules by reaction with water. This is another controlled hydrolysis reaction which goes even further and gives a mixture of dextrins, maltose, and glucose. Such products find wide use as sweeteners in products such as candy, cake frosting, and pancake syrups.

Honey is a mixture chiefly of glucose and fructose. It is made by bees from flower nectars by reaction with a special enzyme in bee saliva. Honey tastes extra sweet because fructose is much sweeter than the other common sugars.

PROTEINS

Proteins are found in all living cells and their name indicates they are of primary importance in your body. (*Protein* comes from the Greek *proteios* which means first.) Proteins are found in muscles, nerves, hair, horns, nails, skin, hemoglobin, enzymes, and hormones. Next to water, protein is the most plentiful substance in your body. Proteins are thus building blocks of the living cell and are also involved in control of all its operations.

Chemists have been able to break down the great variety of proteins into simpler component units by heating with water and acids and have been able to isolate about twenty different basic building blocks for all the proteins, called amino acids.

$$\text{All proteins} \xrightarrow[H_2O]{\text{Catalysts}} \text{Amino acids}$$

When you eat a protein food like meat, eggs, cheese, or a part of wheat, your digestive system breaks the long chain proteins into the simpler amino acid units. Then these are available in your "amino acid pool" for rebuilding into the particular long chain protein molecules your system requires. In other words, what takes place is the reverse of the equation above.

Building Protein Chains: More Giant Molecules

All of the basic building blocks for proteins have been found to be organic acids (—COOH group on a carbon chain). In addition, they all have an $—NH_2$ or amino group on the carbon atom next to the acid COOH group, which is called the alpha carbon atom. The general formula for these units, which are called alpha amino acids, is as follows.

$$
\begin{array}{c}
\mathrm{R{-}CH{-}COOH} \\
| \\
\mathrm{NH_2}
\end{array}
$$

where the letter R stands for a hydrogen atom or a grouping of atoms. Examples are CH_3, called a "methyl" group from methane minus a hydrogen atom; or CH_3CH_2 (C_2H_5), an "ethyl" group from ethane minus a hydrogen atom (see chapter 8, page 168). The nature of the R group determines the identity and behavior of the particular amino acid. Some examples are shown in Table 10.1.

The simplest amino acid is glycine, where R = H. It can be written as

$$
\begin{array}{ccccc}
 & \mathrm{H} & & \mathrm{O} & \\
 & | & & \| & \\
\mathrm{H}{-} & \mathrm{C} & {-} & \mathrm{C} & {-}\mathrm{OH} \\
 & | & & & \\
 & \mathrm{NH_2} & & &
\end{array}
\qquad \text{or} \qquad
\begin{array}{ccccc}
 & \mathrm{H} & & \mathrm{O} & \\
 & | & & \| & \\
\mathrm{H_2N}{-} & \mathrm{C} & {-} & \mathrm{C} & {-}\mathrm{OH} \\
 & | & & & \\
 & \mathrm{H} & & &
\end{array}
$$

which are equivalent. Sometimes it is written in even more condensed form as H_2NCH_2COOH. Two molecules could link up with the elimination of a water molecule.

$$
\begin{array}{ccccc}
 & \mathrm{H} & & \mathrm{O} & \\
 & | & & \| & \\
\mathrm{H_2N}{-} & \mathrm{C} & {-} & \mathrm{C} & {-}\boxed{\mathrm{OH}} \\
 & | & & & \\
 & \mathrm{H} & & &
\end{array}
+
\begin{array}{ccccccc}
\mathrm{H} & & \mathrm{H} & & \mathrm{O} & & \\
| & & | & & \| & & \\
\mathrm{N} & {-} & \mathrm{C} & {-} & \mathrm{C} & {-}\mathrm{OH} & \\
| & & | & & & & \\
\boxed{\mathrm{H}} & & \mathrm{H} & & & &
\end{array}
\longrightarrow
$$

Glycine — Glycine

$$
\begin{array}{ccccccccccc}
 & \mathrm{H} & & \mathrm{O} & & \mathrm{H} & & \mathrm{H} & & \mathrm{O} & \\
 & | & & \| & & | & & | & & \| & \\
\mathrm{H_2N}{-} & \mathrm{C} & {-} & \mathrm{C} & {-} & \mathrm{N} & {-} & \mathrm{C} & {-} & \mathrm{C} & {-}\mathrm{OH} \\
 & | & & & & & & | & & & \\
 & \mathrm{H} & & & & & & \mathrm{H} & & &
\end{array}
+ \quad \mathrm{H_2O}
$$

The new product can link up with more glycine molecules to give us chain growth similar to that occurring with plastics (chapter 9). Indeed the same basic linkage is formed here as in nylon polymers:

$$
\begin{array}{ccccc}
 & \mathrm{H} & & \mathrm{O} & \\
 & | & & \| & \\
\mathrm{H_2N}{-} & \mathrm{C} & {-} & \mathrm{C} & {-}\boxed{\mathrm{OH}} \\
 & | & & & \\
 & \mathrm{H} & & &
\end{array}
+
\begin{array}{ccccccccccc}
\mathrm{H} & & \mathrm{H} & & \mathrm{O} & & \mathrm{H} & & \mathrm{H} & & \mathrm{O} \\
| & & | & & \| & & | & & | & & \| \\
\mathrm{N} & {-} & \mathrm{C} & {-} & \mathrm{C} & {-} & \mathrm{N} & {-} & \mathrm{C} & {-} & \mathrm{C}{-}\mathrm{OH} \\
| & & | & & & & & & | & & \\
\boxed{\mathrm{H}} & & \mathrm{H} & & & & & & \mathrm{H} & &
\end{array}
\longrightarrow
$$

Another glycine molecule — First stage reaction product from above

$$
\begin{array}{cccccccccccccccc}
 & \mathrm{H} & & \mathrm{O} & & \mathrm{H} & & \mathrm{H} & & \mathrm{O} & & \mathrm{H} & & \mathrm{H} & & \mathrm{O} \\
 & | & & \| & & | & & | & & \| & & | & & | & & \| \\
\mathrm{H_2N}{-} & \mathrm{C} & {-} & \mathrm{C} & {-} & \mathrm{N} & {-} & \mathrm{C} & {-} & \mathrm{C} & {-} & \mathrm{N} & {-} & \mathrm{C} & {-} & \mathrm{C}{-}\mathrm{OH} \\
 & | & & & & & & | & & & & & & | & & \\
 & \mathrm{H} & & & & & & \mathrm{H} & & & & & & \mathrm{H} & &
\end{array}
+ \quad \mathrm{H_2O}
$$

Second stage reaction product — Water

and so forth, and so on. This is a considerable oversimplification of the buildup of proteins but it shows the basic idea. You can imagine the variety of chains possible if units other than glycine are used. Figure 10.2 shows how the simple amino acid molecules build up in a particular protein molecule.

The many proteins that make up the variety of our body tissues use only approximately twenty different amino acids. All of the twenty different amino acids are not absolutely required in the diet because the body manufactures all but eight of them. The eight which we humans cannot make in the body are called essential amino acids. Animals are dependent on plants as basic sources of amino acids for proteins.

You can see from Table 10.1 how variety is possible in the molecular arrangement of five different kinds of atoms: carbon, hydrogen, oxygen, nitrogen, and sulfur. Further complexity of the protein molecules comes from different arrangements of the twenty amino acids.

Consider that if you had three letters, A, B, and C, you could arrange them in six different ways: ABC, ACB, BAC, BCA, CBA, and CAB. If you had ten letters—or ten amino acids—you could arrange them in 3,628,800 different ways! Imagine the staggering variety of ways you can arrange twenty amino acids in protein molecules. Nature chooses special sequences of amino acids for special tasks such as building smooth muscle tissue, hair, or the protein in eggs. Then she arranges the protein chains in even more complex ways suggested by Figures 10.3, 10.4, and 10.5. It is interesting that we eat meats and eggs and other protein foods to get the breakdown products of the amino acid units for our own rebuilding.

Scanning electron micrograph of sliced ends of human hair. (Courtesy Emil Bernstein, Gillette Company Research Institute; published in *Science*, 173, cover, July 16, 1971.)

TABLE 10.1
Some of the Twenty Amino Acids which Are the Building Blocks for Proteins

Molecular Structure	*Name*
$H{-}CHCOOH$ NH_2	Glycine
CH_3 $CHCHCOOH$ CH_3 NH_2	Valine
CH_3 $CHCH_2CHCOOH$ CH_3 NH_2	Leucine
CH_3 $CHCHCOOH$ C_2H_5 NH_2	Isoleucine
OH $CH_3{-}CHCHCOOH$ NH_2	Threonine
$-CH_2CHCOOH$ NH_2	Phenylalanine
$HO-$ $-CH_2CHCOOH$ NH_2	Tyrosine
$HS{-}CH_2CHCOOH$ NH_2	Cysteine
$H_2N{-}CH_2CH_2CH_2CH_2CHCOOH$ NH_2	Lysine
$COOH$ $CH_2CH_2CHCOOH$ NH_2	Glutamic acid

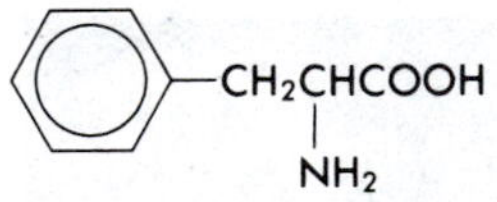

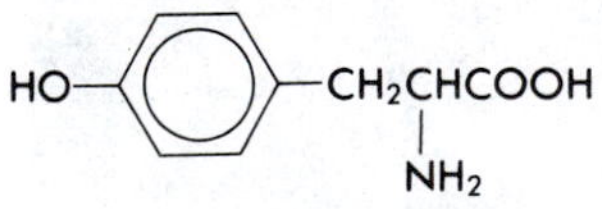

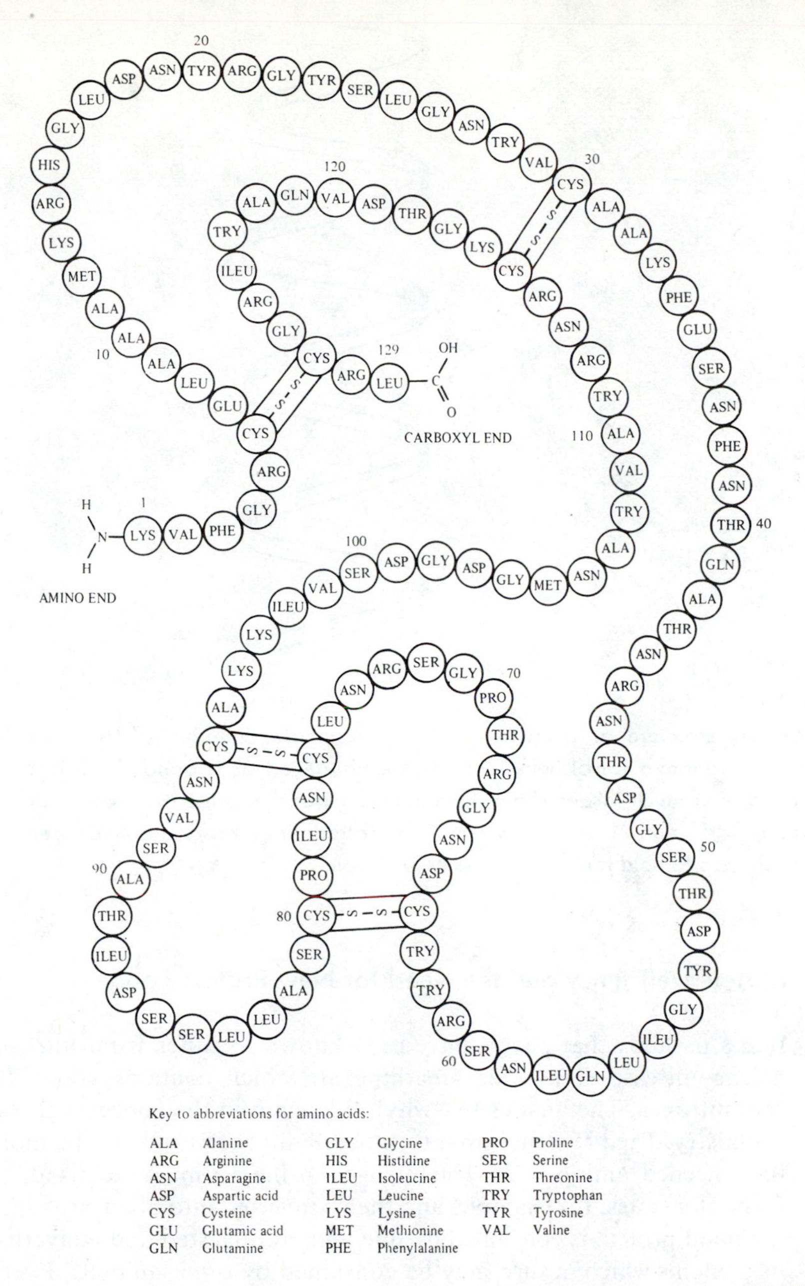

FIGURE 10.2
Two-dimensional model of the primary protein structure of a small protein, the enzyme lysozyme. Primary structure refers only to the sequence of amino acids in the polymer chain. This molecule contains 129 amino acid subunits with a total of 1,950 atoms. The abbreviated symbols refer to specific amino acids. Cross links in the chain are made through sulfur atoms. (From "The Three-Dimensional Structure of an Enzyme Molecule" by David C. Phillips. Copyright © November 1966 by Scientific American, Inc. All rights reserved.)

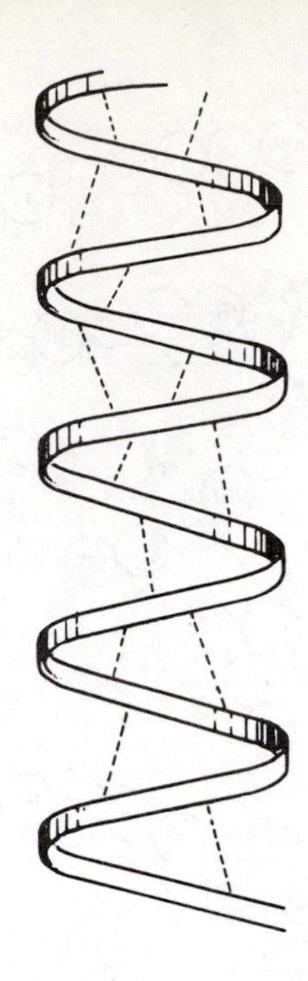

FIGURE 10.3
The secondary structure of proteins involves even more complexity. Here the chains are coiled into a spiral helix form. This is visualized as a band which is a small section of a much larger chain. In reality the band represents a chain of amino acids with many atoms involved. The dotted lines represent hydrogen bonds which help to hold the chain in the helix shape.

Protein Deficiency and the Need for New Protein Foods

It is a paradox that people have been known to suffer from nitrogen deficiency disease even while breathing air which contains about 78 percent nitrogen. The answer to "why" lies again in the conceptual realm of chemistry. The problem is that this "free" nitrogen is not in the molecular form needed, amino acids. The nitrogen in the air must be "fixed," in the form of nitrates, for example, and then converted into plant protein. Later the plant protein is consumed by one animal, digested and converted into its proteins which in turn may be consumed by other animals. Eventually the amino acids reach you.

The devastating effects of lack of protein in the diet can be clearly seen in the disease called kwashiorkor. This is a severe form of protein-calorie malnutrition which affects many children in underdeveloped nations of the world. It is really a stage of starvation, and its symptoms are distended stomach, skin lesions, rashes, and retarded growth.

The word "kwashiorkor" comes from the language of one of the principal tribes of Ghana although there is some disagreement on its mean-

FIGURE 10.4
A representation of the structure of hair, fingernail, muscle, and related fibrous proteins. The protein molecules have the helix structure and each molecule is here represented as a rod with circular cross section. These fibrous proteins contain seven-stranded cables, consisting of a central helix and six others twisted about it. (From *College Chemistry,* Third Edition, by Linus Pauling. W. H. Freeman and Company. Copyright © 1964.)

ing. One meaning is "red boy" which refers to the red patches or bleeding lesions on the body and the red bands which may appear in the hair. Another meaning is "disease of the older child when the baby is born," relating to the fact that the disease often occurs in the first child when it is weaned after a second is born.

Kwashiorkor was thought at first to be caused by a diet high in starch. Recent work has shown that the lack of protein balance is the cause. This happens in many tropical countries where the diet is low in protein and rich in starch foods such as yams, corn, or maize.

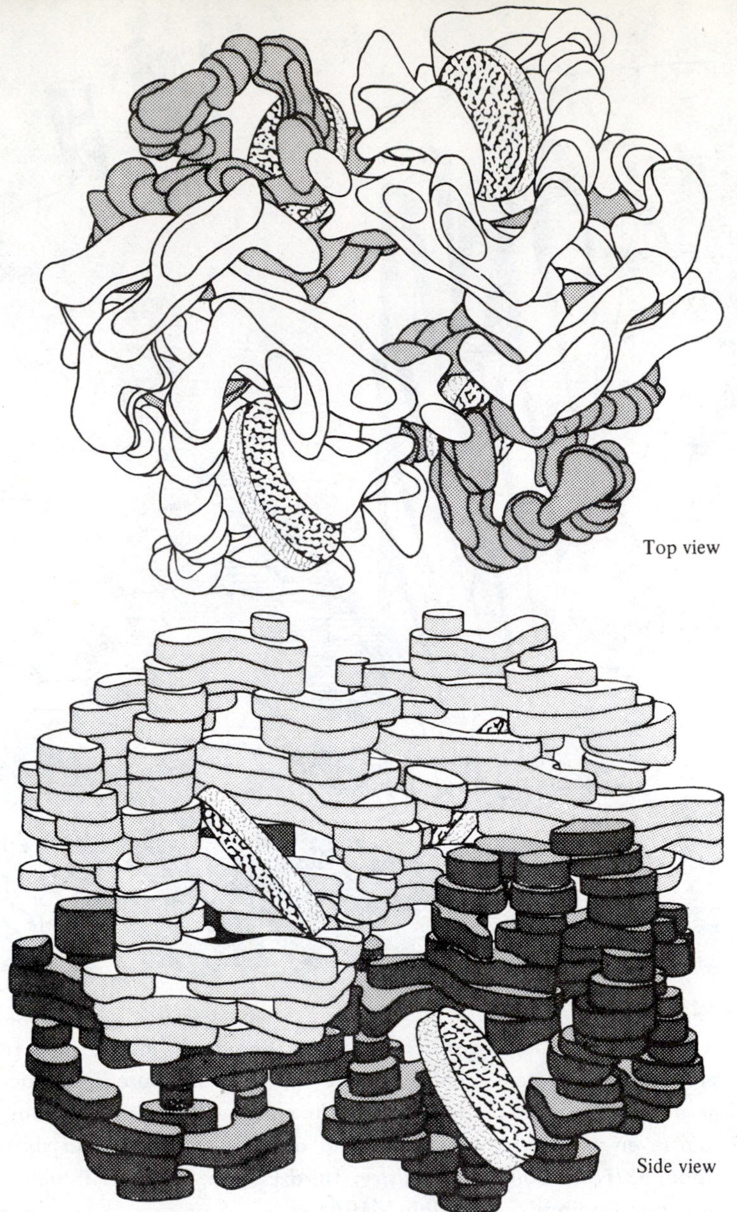

FIGURE 10.5
Even more complexities have been discovered in proteins wherein molecules fold and coil into unusual shapes related to the special purposes they serve. Dr. M. F. Perutz developed the above representation of the hemoglobin molecule, the main component of red blood cells which carry oxygen from lungs to tissues. A single red blood cell contains about 280 million molecules of hemoglobin. Each hemoglobin molecule contains about 10,000 atoms of hydrogen, carbon, nitrogen, oxygen, and sulfur, plus four atoms of iron. The round disk shapes shown are the heme units containing iron. Individual atoms are not shown. The dark and light sections identify subunits in the molecule. (From "The Hemoglobin Molecule" by M. F. Perutz. Copyright © November 1964 by Scientific American, Inc. All rights reserved.)

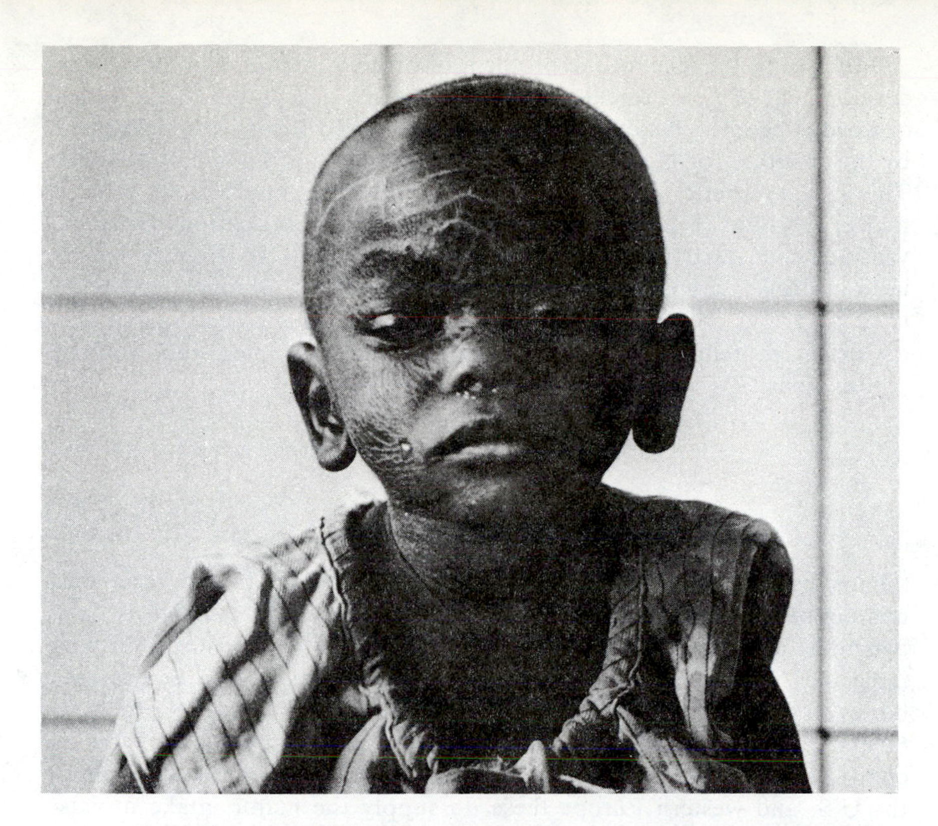

Photo showing effects of protein-calorie malnutrition, or kwashiorkor. World Health Organization teams are now working in undeveloped countries to treat cases of malnutrition and prevent deficiencies by introducing changes in foods and food habits. (WHO photo by Paul Almasy.)

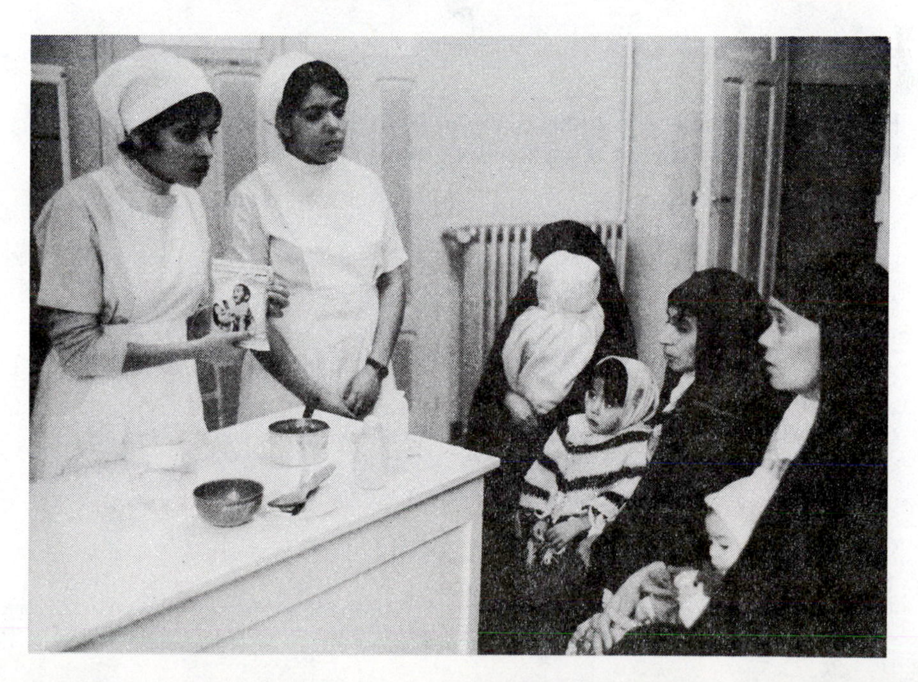

At a mother and child health center in Algeria, mothers learn about the value of proper nutrition for the child and watch the preparation of one of the newly de-

Much hope for relief of the problem of protein malnutrition lies in the development of new protein-rich foods in forms acceptable to the tastes of various peoples. An example is a mixture called Incaparina developed by the Institute for Nutrition in Central America and Panama (INCAP). This is a synthetic vegetable protein-rich mixture made from ground maize, sorghum (a tall cereal grass resembling corn), cottonseed flour, yeast, calcium carbonate, lysine, and vitamin A. The Incaparina can be used in drinks, soups, and puddings. Other synthetic products are being tried in various areas of the world. Raw materials being investigated that are rich in protein include soybeans, peanut meal, sesame, sunflower and cotton seeds.

FATS

Almost every food has some quantity of fats. Most fruits and vegetables contain between 0.1 and 1 percent of fat. A few fruits are high in fats, avocados and olives containing approximately 20 percent of fat-type molecules. Whole grain cereals range in fat content from 1 percent for barley to about 7 percent for oatmeal. Nuts are rich in fats, running up to 60 and 70 percent. However, you think of most of your intake of fats as coming from animal products—meat, milk, and eggs. For most people in the U.S. and western Europe these do supply the major intake of fats. Fats are also often added to foods as shortening in cakes and cookies, or as flavoring or enriching agents like butter and oils in vegetables and puddings.

Fats as Esters

Fats can be visualized as esters (chapter 9) formed from the natural substance glycerol which has three alcohol groups in one molecule.

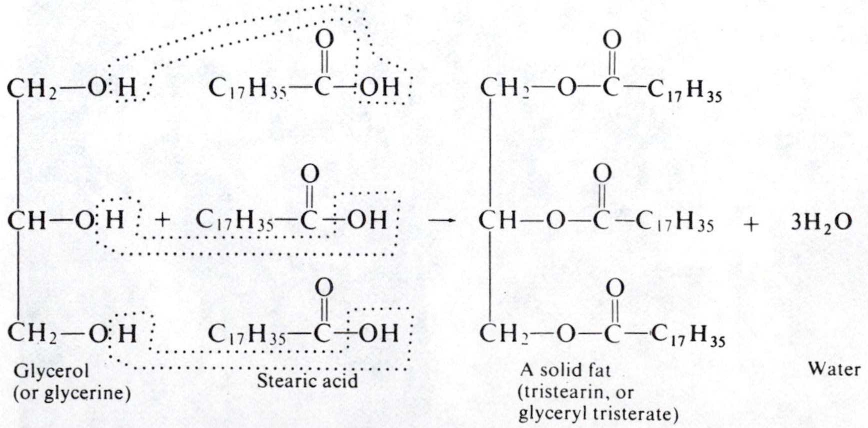

$$
\begin{array}{l} CH_2-O-H \\ | \\ CH-O-H \\ | \\ CH_2-O-H \end{array}
\;+\;
\begin{array}{l} C_{17}H_{35}-\overset{\overset{O}{\|}}{C}-OH \\ \\ C_{17}H_{35}-\overset{\overset{O}{\|}}{C}-OH \\ \\ C_{17}H_{35}-\overset{\overset{O}{\|}}{C}-OH \end{array}
\;\longrightarrow\;
\begin{array}{l} CH_2-O-\overset{\overset{O}{\|}}{C}-C_{17}H_{35} \\ | \\ CH-O-\overset{\overset{O}{\|}}{C}-C_{17}H_{35} \\ | \\ CH_2-O-\overset{\overset{O}{\|}}{C}-C_{17}H_{35} \end{array}
\;+\; 3H_2O
$$

Glycerol (or glycerine) — Stearic acid — A solid fat (tristearin, or glyceryl tristerate) — Water

Here we have a typical fat. It is an ester formed from three acid molecules reacting with the three alcohol groups in the glycerol molecule. The general formula for a fat is:

$$
\begin{array}{l}
CH_2{-}O{-}\overset{\displaystyle O}{\overset{\|}{C}}{-}R \\
| \\
CH{-}O{-}\overset{\displaystyle O}{\overset{\|}{C}}{-}R' \\
| \\
CH_2{-}O{-}\overset{\displaystyle O}{\overset{\|}{C}}{-}R''
\end{array}
\quad \text{or} \quad
\begin{array}{l}
R{-}COOCH_2 \\
\qquad\quad | \\
R'{-}COOCH \\
\qquad\quad | \\
R''{-}COOCH_2
\end{array}
$$

Natural fats are mixtures of esters, having variety in the three long chain acids which combine with a single glycerol molecule. These long chain acids (like $C_{17}H_{35}COOH$) are called fatty acids from their source in fats.

Some fats like lard and butter are solid fats, which means they melt at temperatures above room temperature. Other fats are liquids at room temperature and are usually called oils. Peanut, olive, cottonseed, and soybean oils are common liquid fats. What is the difference between these and the solid fats like lard and butter? The answer lies in the nature of the long chains. The solid fats have a predominance of chains which are saturated whereas the liquid fats, or oils, have more unsaturation in the chains.

Unsaturation in Fats

Molecules like ethylene, $CH_2{=}CH_2$, illustrate that carbon atoms can form double bonds instead of having full saturation with hydrogen atoms (chapter 9). This is what provides for liquidity in the oils. They have one or more double bonds or unsaturation in the side chains. The unsaturation then comes originally from the acid which is condensed with glycerol. An example will make the picture clearer.

$$
\begin{array}{l}
CH_2{-}O{-}\overset{\displaystyle O}{\overset{\|}{C}}{-}(CH_2)_7\overset{\displaystyle H}{\overset{|}{C}}{=}\overset{\displaystyle H}{\overset{|}{C}}(CH_2)_7CH_3 \\
| \\
CH{-}O{-}\overset{\displaystyle O}{\overset{\|}{C}}{-}(CH_2)_7\overset{\displaystyle H}{\overset{|}{C}}{=}\overset{\displaystyle H}{\overset{|}{C}}(CH_2)_7CH_3 \\
| \\
CH_2{-}O{-}\overset{\displaystyle O}{\overset{\|}{C}}{-}(CH_2)_7\overset{\displaystyle H}{\overset{|}{C}}{=}\overset{\displaystyle H}{\overset{|}{C}}(CH_2)_7CH_3
\end{array}
$$

This is the formula for triolein, a glycerol ester or fat which is a liquid at room temperature. It is formed from oleic acid and glycerol.

In nature we also find side chains with two and three unsaturation groups. Examples are shown on the following page.

Oleic	$C_{17}H_{33}COOH$	$CH_3(CH_2)_7CH{=}CH(CH_2)_7COOH$
Linoleic	$C_{17}H_{31}COOH$	$CH_3(CH_2)_4CH{=}CHCH_2CH{=}CH(CH_2)_7COOH$
Linolenic	$C_{17}H_{29}COOH$	$CH_3CH_2CH{=}CHCH_2CH{=}CHCH_2CH{=}CH(CH_2)_7COOH$

Now you can see where the term polyunsaturated comes from: it refers to a fat with extra unsaturation. In certain cases doctors have recommended cutting down on saturated fats in order to lower blood levels of a product called cholesterol (chapter 14).

Why Olive Oil Is a Liquid and Butter Is a Solid

You might wonder how the presence of unsaturation makes such a drastic change in properties of the fats. A simple conceptual explanation applies. The double bond introduces a restriction of freedom for the sections in the chain. You might expect that the two carbon atoms on either side of a double bond are much less free to move and rotate than those joined by only one bond. This restriction of movement contributes in general to a more random or mixed-up condition for the molecules.

You may remember from Kinetic-Molecular Theory that molecules in liquids are less regularly aligned than those in solids. Thus anything which contributes to a difficulty of alignment will provide a greater tendency toward the liquid state. The linkage at the double bond restricts chain movement.

It may help to compare stearic acid from lard with oleic acid from olive oil.

Acid		*Melting Point*	*Condition at Room Temperature*
$C_{17}H_{35}COOH$	stearic acid	69°C	solid
$C_{17}H_{33}COOH$	oleic acid	14°C	liquid

The only difference between these two acids is that the oleic acid has two hydrogen atoms missing in its chain. This gives one double bond. The double bond provides enough built-in randomness by restriction of the free movement of molecular chains that oleic acid is a liquid at room temperature (24°C). Stearic acid molecules, on the other hand, can more easily align with each other and set up as a solid because of unrestricted rotation at all carbon bonds. Consequently, stearic acid will not melt until given additional kinetic energy by raising its temperature to 69°C. The same condition carries over to the glycerol ester compounds or fats. Thus you can see that olive oil, which has a large portion of unsaturates, is a liquid. Lard and butter, with a higher content of saturated molecules, are solids.

Salad, cooking, and frying oils are derived from various vegetable sources like cottonseed, corn, soybean, olive, and peanut oils.

Once the conceptual basis for structure of various fats is known it is possible to convert one to another. "Oleo" or margarine is made from liquid oils by partial hydrogenation. This process adds extra hydrogen atoms to convert some of the unsaturated bonds to saturated. Shortenings (Crisco, Spry) are made in a similar fashion by partial hydrogenation of vegetable oils like those from cottonseeds and soybeans.

Hydrogenation is now accomplished routinely by industry in converting liquid vegetable oils to solid fats, and food chemists are able to formulate margarine with a high proportion of unsaturated molecules. This is what the advertising on polyunsaturates is all about.

Fats: Good or Bad?

Young animals fed on diets containing only saturated fats develop a fat deficiency disease. Certain unsaturated acids are therefore essential in the diet.

While it was once believed that the body could make all the necessary unsaturated fats by removal of hydrogen from saturated chains, we now know that this is false. Here we have a situation similar to that of proteins. Quality is more important than quantity.

Fats are of course useful as fuels for energy. They serve as an important reserve food supply in the event of an emergency. The body reserves of glycogen can last only a few hours. However, the body normally contains enough reserve fat to sustain life for about five weeks providing water is available. The reason that fat represents a much more efficient energy storage system than the carbohydrate glycogen is because the energy obtained from a given quantity of fat is more than twice that from the same amount of carbohydrate. Both these fuels are burned completely to provide carbon dioxide, CO_2, and water, H_2O, as end products. Carbohydrates have considerable oxygen already in the molecule compared with fats. Therefore, more energy comes from the conversion (by burning or oxidation) of fats to CO_2 and H_2O than by the conversion of carbohydrates. In other words, the carbon and hydrogen in carbohydrates are already—figuratively at least—"on the way" toward being "oxidized." Therefore we get less extensive reaction with oxygen and less heat out of their burning. (See Table 10.2.)

Everyone knows that the body stores fat in various places such as beneath the skin and around the abdominal area. A more important depot for fat is around vital organs to provide insulation against sudden temperature changes and protection against sudden jolts and bumps as

TABLE 10.2
Relative Energy Values from Foods*

Carbohydrate	4 Calories per gram
Protein	4 Calories per gram
Fat	9 Calories per gram

*These are approximate values and depend slightly on digestibility. A good comparison is one teaspoon of sugar (5 grams) gives 20 Calories whereas 5 grams of fat provides 45 Calories. Tables of energy values for specific food items are found in books on nutrition and diet.

the body moves. Fat deposited in various areas of the body is called depot fat. At one time it was thought that fat provided insulation like an ordinary blanket; now the analogy is to an electric blanket. Fat is penetrated deeply by a network of capillaries. The cells in the fat depots continually generate heat which helps to keep the temperature of vital organs fairly constant—a tremendous advantage in preserving life. However, excess fat can be a considerable problem. The heart is put under extra strain, not only in moving the extra weight around, but also in pumping blood through the widely extended network of capillaries in the enlarged fat depots.

Fats are noted for giving a feeling of satisfaction after a meal. All the physiological reasons why certain foods "stick to the ribs" or give a longer feeling of fullness are not exactly known. However, fat is known to retard the rate at which food leaves the stomach. This thus contributes a feeling of fullness and delays the onset of hunger.

Use Chemical Laws to Keep Thin

The body's efficient fat storage system is often a handicap to those who want to "keep in shape." Not only will fat deposit in the body from eating excess fatty food. If carbohydrates are eaten in excess of overall body energy needs, they can be broken down and made into fats. Reducing diets merely recognize the Law of Conservation of Mass as a conceptual fundamental. If you don't want to put on fat, keep your total intake of foods about even with your energy needs, always with a proper balance of varieties of foods so as to maintain the right amount of essential nutrients.

The amount of energy produced by various types of food is often expressed in calories. The term used in diet measurements is called the large calorie and is equal to 1000 of the calories the chemist refers to as the basic unit. The chemist's small calorie is the heat needed to raise the temperature of 1 gram of water 1°C. It is usually abbreviated cal.—with a small letter c. (You may remember that a gram in mass or weight is about one-fifth that of a nickel.) The large calorie, or dietary calorie, is equal to 1000 small calories. It is abbreviated Cal.—with a capital C. This larger unit is a more practical unit for diet discussions. It represents a large amount of heat since water takes in so much energy to raise its temperature (chapter 7). A dietary calorie is the amount of heat needed to raise the temperature of 1000 grams of water one degree centigrade or to raise a pound four degrees fahrenheit.

Lecithin: A Marriage of Fat and Phosphate

Among the other known compounds present in some fats are the lecithins. The word is from the Greek *lekithos* for egg yolk, which is a good source. Lecithins are distributed throughout the body but are found especially in nerve cells and brain tissue. They are closely related to fats. A typical lecithin can be represented with the following structural formula:

$$
\begin{array}{l}
C_{17}H_{33}COOCH_2 \\
C_{17}H_{31}COOCH \\
\quad\quad\quad\quad O \\
(CH_3)_3\overset{(+)}{N}CH_2CH_2O\underset{O^{(-)}}{P}OCH_2
\end{array}
$$

The important additional element here is phosphorus. In the lecithins a phosphate group, (PO_4^{-3}), has been joined to one of the side chains, which also contains nitrogen, giving a sort of "inner salt" compound. This is somewhat like the salt ammonium chloride, NH_4Cl, where the ammonium group, NH_4^+, is positive and the chloride ion, Cl^-, is negative. Considerable quantities of lecithins are found especially in oils from seeds. Lecithin prepared chiefly from soybeans is used as an additive emulsifying agent in many foods, margarine, for example. Lecithin is the natural emulsifier in milk that helps keep the butter fat and water from separating out. You have probably seen the name also on the labels of other food and cosmetic products.

Why the Cookie Crumbles

Shortening refers to butter, margarine, lard, or other fat used to make pastry, cake, or cookies rich and crumbly. The original derivation of the term is probably through the idea of shortness in length, and the "short" in shortening means "brittle" or "not holding together." Thus cakes and cookies crumble into short pieces more easily than bread because they contain more shortening. The percentage of fat in cakes, pastry, and cookies runs in the neighborhood of 15 percent whereas bread averages only 1 percent.

We can come up with a reasonable conceptual explanation for the crumbly nature of cookies and cakes. You recognize all fats and oils as being greasy in feel. And you know that bread is more chewey than cake. Fat molecules provide lubrication to the flour ingredients starch and gluten (which is the predominantly protein part of flour). When you bite a piece of cake the extra interspersed fat molecules allow starch and gluten molecules to slip away from each other easily as the walls of the tiny gas chambers crumble. (The gas is air or CO_2 purposely introduced into the cake to provide lightness and easier digestibility.)

There are of course further ways to confirm these ideas. And they can be pleasant experiences. Compare the texture of ordinary cake with that of sponge or angel cake. These are much less crumbly and more chewy. Why? The recipes for sponge and angel cakes call for no shortening at all—just large amounts of egg white, and the protein of egg white provides much less lubrication than the fats.

Chemical Concepts and Sweeter Cakes

When sugar is incorporated into a cake batter it is held in close proximity to the overall mixture of starch, gluten, and residual water by hydrogen bonding. The amount of sugar which can be added is limited because addition of too much water and sugar molecules—which are not long chain types—weakens the cake and could cause it to fall during baking or cooling. Here the long chain starch and gluten molecules of the flour have to carry the major load structurally. They do this by hydrogen bonding to other smaller molecules, holding everything in a conglomerate togetherness.

Conceptually it was figured out that if molecules of fatlike material with less than the full content of fatty acid groups were used, there would be more free —OH groups to bind water and sugar into the overall structure. This additional hydrogen bonding would provide less tendency for the cake to collapse. Special shortenings were then developed by adding to regular shortening molecules like those shown below.

$$\begin{array}{r} C_{17}H_{35}COOCH_2 \\ | \\ HOCH \\ | \\ HOCH_2 \end{array} \qquad \begin{array}{r} C_{15}H_{31}COOCH_2 \\ | \\ C_{17}H_{33}COOCH \\ | \\ HOCH_2 \end{array}$$

A "monoglyceride" (*one* fatty acid joined to glycerol)

A "diglyceride" (*two* fatty acids joined to glycerol)

You will note that there are still some free —OH groups left on the glycerol chains compared to the usual fat which has three long fatty acids attached in each molecule. Compare these structural formulas with those of fats on page 237.

These combination molecules worked as planned. The long chain part was compatible with the regular fat or shortening molecules, and the added —OH groupings provided additional hydrogen bonding to help hold and entangle the added sugar and water. This prevented the structural weakness which had previously been found to cause cake to fall when extra sugar was added. The success of this conceptual plan allows preparation of cake mixes with up to 40 percent more sugar. Is this good?

VITAMINS

Vitamins are complex organic substances essential for health and growth. An average person eats about 1000 pounds of food in a year. Yet the total intake of all vitamins comes to only about one and a half ounces. This is why vitamins were first discovered negatively—by vitamin deficiency diseases.

The development of conceptual knowledge about vitamins has been a slow and painful process, even though people from ancient times undoubtedly noticed that certain foods were beneficial to health. The British early recognized the value of citrus fruits and juices for use on long voyages to prevent scurvy, a disease characterized by bleeding gums, weakness, and prostration. Sailors who ate chiefly bread and salt meat commonly suffered from the disease. Limes and lime juice as a regular part of the diet eliminated scurvy on ships. (This is the origin of the name "limeys" for British sailors.) We now know limes contain necessary vitamin C.

The Concept of Deficiency Disease Developed Slowly

The history of thiamine (Vitamin B_1) is typical of how slowly the concept of vitamins developed. Thiamine is involved in beriberi, which is a degenerative disease of the nervous system. *Beriberi* comes from the language of Ceylon and means "weakness" or "I cannot." It emphasizes the paralysis of muscles and swelling of the extremities which occurred frequently in those parts of Asia where polished rice was a principal item of diet. The disease was once widespread and initial attempts to find the cause concentrated on looking for a particular microbe. This was logical since the prevalent conceptual explanation for disease was associated with bacteria.

A Dutch medical officer in Sumatra, Christiaan Eijkmann (1858–1930), was investigating the disease in prison camps and he happened to observe accidentally that chickens fed on leftover polished rice got sick with symptoms similar to those of people with beriberi. He wondered: Could the cheap *un*polished rice—the chickens' usual fare—make the difference? To try his hypothesis out he fed them brown unpolished rice. They got well again!

Eijkmann got the Nobel Prize in recognition of his development of a valuable experimental method of studying dietary diseases. He had been able to produce and cure a deficiency disease in fowls and show that the disease was the same as human beriberi.

Other people continued his work and tested the effect of different rice on fowl and on prisoner volunteers. The eventual conclusion was that some unknown antiberiberi factor was present in the rice polishings or hulls (outer coatings). This was in 1901 and the beginning of conceptual thinking about a deficiency disease.

Chemists tried to isolate the factor in rice polishings and a few years later a biochemist noticed that an amine was involved (contains the $—NH_2$ group). Hence the word vitamine was coined, meaning "vital" amine. This became shortened to vitamin as a general term when it was later recognized that vitamins are not necessarily amines. Indeed, they have no structural family features in common like carbohydrates, proteins, or fats.

After many years of patient work the antiberiberi factor was eventually isolated (1926) and the identical chemical synthesized (1936). The history of each of the known vitamins is a long and intriguing search from initial operational descriptions of disease all the way to the difficult identity of the structure of molecules which cure or prevent it.

Operational ....→ to ....→ Conceptual ....→?

Deficiency disease recognized	Certain foods implicated in prevention or cure of the disease	Isolation of certain food factors in complex mixture	Purification and isolation of particular organic molecule	Determination of structure of molecule	Making of synthetic samples of the same molecule which on testing cures and prevents the disease

Like all refinements in conceptual development there is room for even further conceptualizing. For example, how does the identified molecule cure the disease? Biochemists have developed some information for certain of the vitamins. However, most of our understanding of the detailed chemical mechanisms involved is still in its infancy. We recognize vitamins as catalysts essential for regulating body processes. And, while over twenty-five vitamins have been identified, research people believe that there are still others.

Conceptual Descriptions of a Few Vitamins

Vitamin A

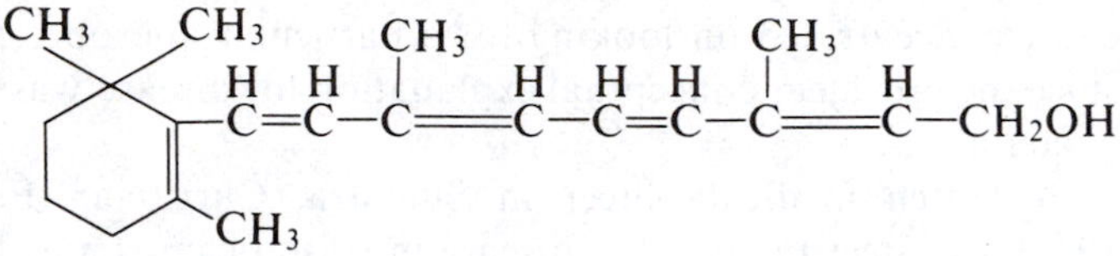

CH_3 CH_3 CH_3 CH_3
H H H H H H
$-C{=}C{-}C{=}C{-}C{=}C{-}C{=}C{-}CH_2OH$
CH_3

Vitamin B1
(Thiamine)

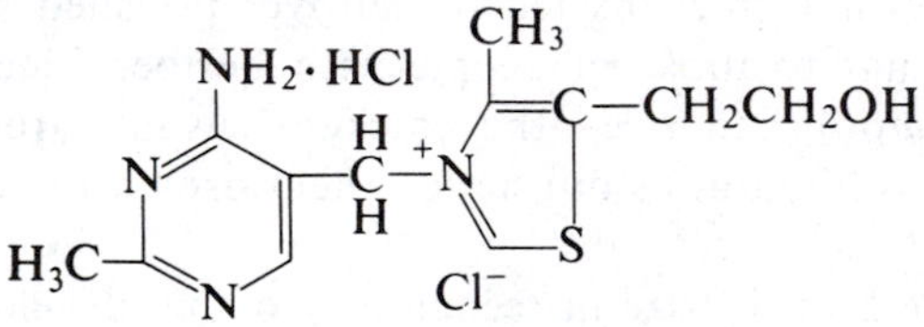

CH_3
$NH_2 \cdot HCl$
$C{-}CH_2CH_2OH$
N H $^+$N
C
H
H_3C N S
Cl^-

Vitamin B2
(Riboflavin)

OH OH OH
$CH_2{-}CH{-}CH{-}CH{-}CH_2{-}OH$
CH_3 N N O
N—H
CH_3 N
O

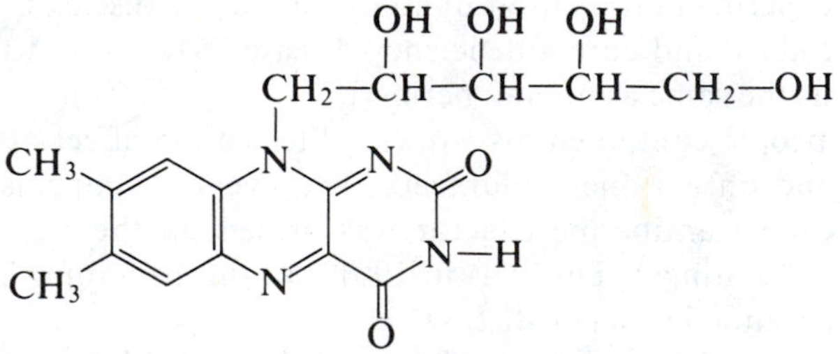

Niacin
(another
B vitamin)

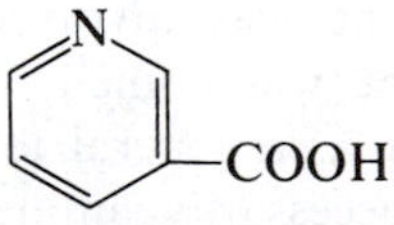

N
COOH

Vitamin C

OH OH
H H
$C{-}C{-}OH$
O O
O H
H

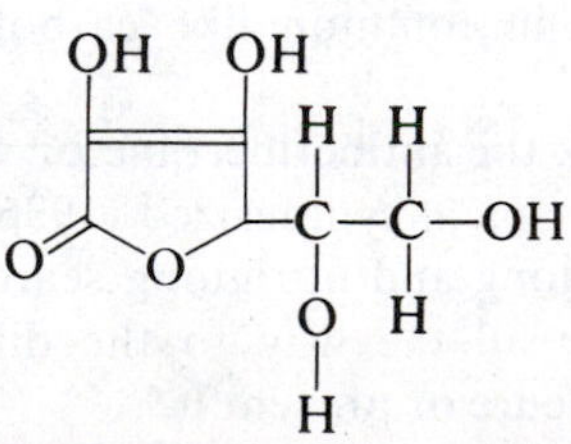

Vitamin D_2

CH_3 CH_3 CH CH CH CH_3 CH CH CH CH_3 CH_3 CH_2 HO

Operational Descriptions of a Few Vitamins

We must obtain most of our vitamins from outside sources even though a few can be made to some degree by the body (such as Vitamin D by sunlight on the skin). Certain fat-soluble vitamins like A and D are stored in your tissues so that you need not replenish them daily. Others, like Vitamin C and thiamine, cannot be stored and therefore must be consumed daily. The six vitamins believed to be most critical in the North American diet are described in Table 10.3.

Photo taken in Tanzania showing a severe case of acute pellagra, caused by lack of niacin, a B vitamin. (WHO photo.)

TABLE 10.3
Description of Vitamins

Vitamin	*Use*	*Major Source*	*Deficiency*
A	Growth; good vision, healthy skin and hair.	Milk, butter, margarine, eggs. The body also makes it from foods containing carotene, e.g., leafy green and yellow vegetables like spinach, carrots and chili peppers.	Eye diseases; night blindness.
B_1 Thiamine	Proper functioning of heart and nervous system.	Enriched bread, enriched cereals, fish, lean meat, pork, milk, liver, oranges.	Early deficiency signs include loss of appetite, insomnia, constipation, and irritability. Eventually, beriberi.
B_2 Riboflavin	Healthy skin, building and maintaining body tissue.	Eggs, enriched bread and cereals, leafy green vegetables, lean meats, milk.	Inflammation of tongue; skin lesions.
Niacin (also a B vitamin)	Converting food to energy. Health of nervous system.	Lean meats, enriched bread and cereals, eggs, peanuts.	Pellagra (Latin: rough skin)
C Ascorbic acid	Healthy teeth, gums and bones. Strong body cells and blood vessels.	Citrus fruits, berries, cantaloupe, tomatoes, cabbage, green vegetables, peppers, new potatoes.	Scurvy
D	Strong teeth and bones. Helps utilization of calcium and phosphorus.	Vitamin D fortified milk, salmon, tuna, egg yolk.	Rickets

You will notice that Vitamin E is not included in the table because it is readily available in many common foods. Vitamin E is a fat-soluble vitamin (like A and D). The presence of Vitamin E in foods has been known for decades. Its chief function is to act as an antioxidant. This means it inhibits the combination of a substance with oxygen and thus acts as a preservative. The amount of Vitamin E needed by most people is apparently supplied by the average, well-balanced diet. Many common foods contain the vitamin and it is present in large quantities in leafy vegetables, whole grain foods, and vegetable oils.

Many extravagant claims have been made in recent years for Vitamin E, but the Food and Drug Administration scientists have as yet found no support for adding Vitamin E supplements to the average diet. Further research on this vitamin, as well as others, is undoubtedly needed.

Saving Vitamins

Much of the natural vitamin content in foods can be lost because of poor cooking habits. Some common examples follow.

Part of the water-soluble vitamins in foods may be lost by discarding meat juices and water used for cooking vegetables. For example, vitamin C and all the B vitamins are soluble in water. Nutritionists recommend avoiding prolonged cooking, using as little water as possible, and recycling the water and juices into food preparations like gravy and soup.

Thiamine can be lost through oxidation, a loss that is further increased by heat and exposure to an alkaline medium. Thus the practice of adding baking soda to vegetables to improve their green color is very unwise. Cooking vegetables in covered pots has a firm conceptual basis. It prevents excessive loss of volatile flavor components and prevents escape of water molecules, thus facilitating cooking in a smaller amount of water. It also cuts down on access of oxygen and shortens cooking time, keeping vitamin destruction to a minimum.

The Vitamin A value of fruits and vegetables is directly proportional to the intensity of color, that is, the deeper the green or yellow, the higher the Vitamin A value. Discarding deep-colored leaves can thus result in lower Vitamin A availability. In fact the leafy parts of turnip greens, broccoli, and kale have much more Vitamin A value than the whiter stems. Consequently, nutritionists recommend trimming away the fibrous stems to make the nutritious parts more acceptable for eating. Leaf lettuce and the outer green leaves of head lettuce have higher calcium, iron, and Vitamin A value than the white leaves.

Vegetables keep all their nutrients best when stored at high humidity to prevent withering. They should never be stored uncovered in the open part of the refrigerator.

Some vegetables even give you a nice operational signal that they should be cooked and eaten when freshly picked. Green peas and corn contain sugar which after harvesting is quickly turned to storage starch. This change is catalyzed by enzymes which are destroyed by heating. Thus vegetables which are blanched right after picking and then quick frozen are often tastier than so-called fresh vegetables which stand around after picking or are shipped too long a distance.

Boiling vegetables like carrots and potatoes in their skins retains more vitamins and minerals than cooking these vegetables pared and cut up where the molecules can migrate more easily into the boiling water. Baking potatoes whole in their skins also conserves nutritive value.

Cooking vegetables for reheating at later meals may save time but this saving is at the expense of nutrients. Vitamin C losses in cooked vegetables increases with length of time they are held in the refrigerator. After two or three days only one-third of Vitamin C is left. Orange juice retains Vitamin C content well even when kept several days in the refrigerator. Berries, however, lose Vitamin C quickly if capped or bruised. Here the greater access of oxygen is probably involved. Rice should not be washed because it is enriched with iron, thiamine, riboflavin and niacin, all of which dissolve in water. Cooking with just the right amount of water so as to avoid draining will also prevent losses of these added vitamins.

Starting vegetables in boiling water rather than cold quickly destroys enzymes which contribute to vitamin loss. It also allows shorter overall cooking time.

FOOD ADDITIVES

The practice of adding chemicals to food probably began when people first learned to preserve meat by smoking or salting it. Written records show that spices were used to flavor bread as far back as 3500 B.C.

The Industrial Revolution brought the growth of large concentrations of population in cities. Consequently, more and more people came to depend on food grown and stored by others. This started the large-scale use of chemicals for many purposes connected with the monumental job of feeding efficiently and healthfully so many people who no longer live "on the farm." Today we use extremely large numbers of additives to enhance the wholesomeness, attractiveness, convenience, and nutritional value of foods. The total quantity used in the U.S. now comes to about 3 pounds per person per year.

Some of the purposes to which additives are put are: (1) improving nutritional value, (2) flavoring, (3) coloring, (4) bleaching, (5) antifoaming, (6) glazing, (7) moisturizing, (8) sweetening (nutritive), (9) sweetening (nonnutritive for diet control), (10) preserving, (11) stabilizing, (12) emulsifying, (13) thickening, (14) foaming, (15) neutralizing, (16) acidifying, (17) firming, (18) anticaking, (19) conditioning, (20) propelling, and (21) antistaling. Table 10.4 gives a few examples of formulas for reference purposes only.

Iodized Salt: What Good Is It?

The first intentional nutritional supplement added to foods was the mineral potassium iodide (KI) which was added to table salt beginning in the early 1920s ("iodized" salt). Before that time many people, especially

TABLE 10.4

Formulas for a Few Intentional Food Additives

Name	*Formula*	*Purpose*
Potassium iodide	KI	Goiter Prevention
BHT Butylated hydroxytoluene	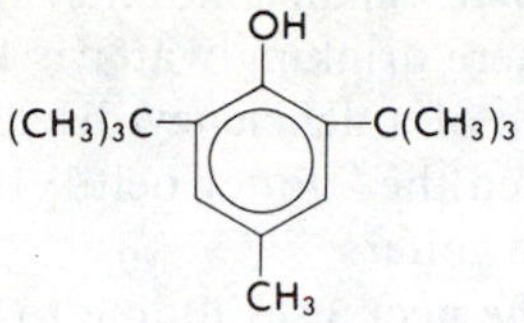	Antioxidant
Propyl gallate	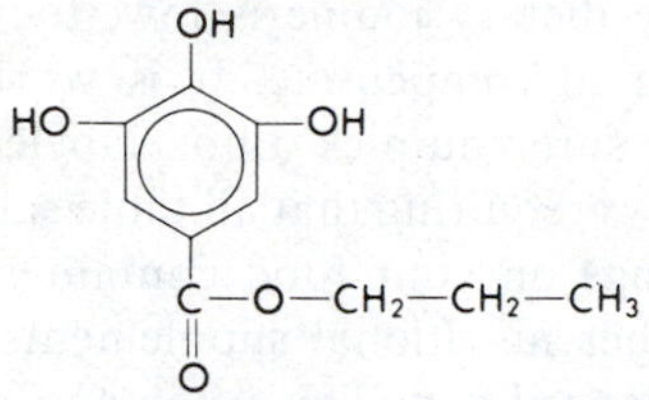	Antioxidant
Saccharin	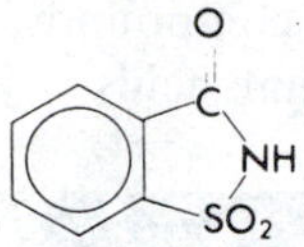	Non-nutritive sweetener
Sodium propionate	CH_3CH_2COONa	Mold and fungus inhibitor
Lactic acid	$CH_3CHOHCOOH$	Mold and fungus inhibitor
Sorbic acid	$CH_3CH{=}CHCH{=}CHCOOH$	Mold inhibitor
Sodium sorbate	$CH_3CH{=}CHCH{=}CHCOONa$	Mold inhibitor
Citric acid	COOH | CH_2 | HOCCOOH | CH_2 | COOH	Sequestrant, tart flavor, acidity
Tartaric acid	COOH | HCOH | HCOH | COOH	Sequestrant, acidity
Potassium acid tartrate (Cream of Tartar)	COOK | HCOH | HCOH | COOH	Acidity
Calcium phosphate	$Ca_3(PO_4)_2$	Prevents caking
Phosphoric acid	H_3PO_4	Acidity
Cyclamate (Sodium)	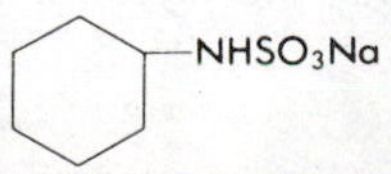	Non-nutritive sweetener

women, had goiter, a condition characterized by a large and somewhat bulbous neck.

Goiter is caused by a lack of iodine in the diet. Although large amounts of iodine are toxic, small amounts are essential. Goiter often occurs in inland areas where drinking water is lacking in iodine content. It is thus another example of a deficiency disease. The Great Lakes area of the U.S. was once called the "goiter belt." In the early 1900s even the fish in Lake Michigan had goiters.

The thyroid gland in the neck uses iodine to make a certain regulating hormone. When the dietary iodine is low, the body reacts by enlarging the thyroid in order to compensate. It is worthwhile to be alert when buying salt to make sure you pick a box labeled "iodized." Canada and Switzerland have laws requiring that all table salt be iodized.

Look at the listings on your food containers. You will recognize the large amount of other nutritional supplements being added in order to give you a more balanced diet. This often covers chemicals like vitamins, minerals such as iron and calcium compounds, and even proteins to provide a higher balance of needed amino acids.

Villagers in Paraguay, some suffering from goiter for which the remedy is iodized salt. In numerous countries, financial difficulties and ethical objections have prevented the addition of potassium iodide to salt. More recently, however, iodized salt is beginning to find acceptance in even the most remote villages. (WHO photo by Paul Almasy.)

All Kinds of Preservatives

There are many different types of preservatives. Each type is suited for a particular product or effective against a particular organism or unwanted chemical change. Some of those in longest use are sugar, salt, and vinegar (acetic acid).

If you have ever tasted rancid butter or stale potato chips you know about the storage problem with fats. The foul flavor means the fat has gone rancid. Rancidity generally involves reaction of the fat with oxygen from the air. Farm women knew this long ago and stored lard in crocks in a cool place with as small a surface exposed to the air as possible.

Many years ago it was noticed that vegetable fats had a resistance to oxidation and deteriorated much less rapidly than animal fats. Investigation of this operational advantage showed the presence of natural antioxidants. Antioxidants are chemicals which preferentially react with the oxygen and thus prevent or delay the onset of rancidity. Examples of some of the synthetic antioxidants which inhibit oxidation are: butylated hydroxytoluene (BHT), butylated hydroxyanisole (BHA), and propyl gallate. You will see statements like "BHA added to prevent spoilage" on many food containers, such as for dry cereals, cooking oils, shortening, and bread crumb batter, to name a few.

Preservatives in bread are useful for inhibiting mold growth. Examples are sodium and calcium propionate, lactic acid, and monocalcium phosphate. Sorbic acid and sodium and potassium sorbates are used to slow mold growth in cheeses. Some antimold chemicals are naturally present in foods, for example, benzoic acid in cranberries and propionic acid in Swiss cheese.

Certain kinds of chemicals are called sequestrants because they "sequester" or "tie up" offending constitutents involved in spoilage. Chemicals of this type are often used to prevent physical or chemical changes which affect color, flavor, texture, or appearance. Typical examples of sequestrants are those commonly used in dairy products such as the sodium, calcium, and potassium salts of citric, tartaric and phosphoric acids.

pH and Leavening

Acidity or alkalinity is important in many classes of processed foods. For example, acids cause leavening in baked goods, give taste to soft drinks, and provide proper pH in churning cream for butter. Some common pH control agents are citric acid, phosphoric acid, sodium bicarbonate ($NaHCO_3$), calcium carbonate ($CaCO_3$), tartaric acid, and potassium acid tartrate (cream of tartar).

Special strains of yeast are added to bread in order to provide carbon dioxide gas which lightens or leavens the bread. Other chemical agents are considered in chapter 13.

Flavors: Lots of Variety

Flavors are natural to foods. The differences between foods are related to a delicate balance of mixtures of chemicals which affect both your taste and smell. The exact mechanism by which they work is still largely unknown. Many of these natural flavoring chemicals have now been identified. Once such conceptual information is obtained it is possible to make synthetic flavoring agents—i.e., to make the same kind of molecules.

One of the practical problems in making synthetic flavors is that nature prefers quite a mixture—sometimes with slight proportions of many trace

TABLE 10.5
Some Basic Flavor Molecules

Name	*Formula*	*Used in Artificial Flavors for*
Vanillin	Benzene ring with OH, OCH_3 and CHO substituents	Vanilla
Citral	$(CH_3)_2C{=}CHCH_2CH_2C(CH_3){=}CHCHO$	Lemon
Octyl acetate	$CH_3COOC_8H_{17}$	Orange
Ethyl butyrate	$C_3H_7COOC_2H_5$	Pineapple
Benzaldehyde	Benzene ring with CHO	Almond, cherry coconut
Cinnamaldehyde	Benzene ring –CH=CHCHO	Cinnamon, cola
Methyl salicylate	Benzene ring with $-C(=O)-O-CH_3$ and –OH	Wintergreen, walnut, root beer
Menthol	Cyclohexane ring with CH_3, –OH and $CH(CH_3)_2$	Mint

Terpene

flavor agents. More than one hundred different compounds have been identified in coffee aroma. Often we do not exactly duplicate nature's delicate mixtures simply because the research costs would make the food product too expensive.

There are in current use over 1100 different flavoring substances, including 750 synthetic or imitation flavorings. Examples of a few flavor molecules are shown in Table 10.5.

Flavor Intensifiers: MSG

For hundreds of years Chinese and Japanese cooks used special materials like seaweed and the shavings of dried fish to enhance the flavor of diets which consisted largely of rice, soybeans, and vegetables. Oriental cooks also used chicken and meat extracts as well as soy sauce, obtained from fermented soybeans, to improve both flavor and color of foods.

Eventually a Japanese chemist identified the flavor principle of seaweed as monosodium glutamate, called MSG for short. It is also the chief active ingredient in soy sauce. The first Japanese patent for preparation of MSG as enhancer of flavor was granted in 1908. Japan and China began large-scale production of MSG and it was also adopted by the canning industry in other countries primarily for soups, tuna, meat pies, and stews.

MSG is obtained by fermentation or hydrolysis of any number of vegetable proteins like wheat, corn, sugar beets, and soybeans. It is the sodium salt of the naturally occurring amino acid called glutamic acid. The compound can thus be represented with a hydrogen from one acid group replaced with a sodium ion.

$$\begin{array}{l} \text{COONa} \\ | \\ \text{CH}_2\text{CH}_2\text{CHCOOH} \\ \qquad\qquad\;\; | \\ \qquad\qquad\;\; \text{NH}_2 \end{array}$$

The action of MSG has been said by one scientist to be "like turning up the volume on your hi-fi." It intensifies flavor of protein foods and has no great effect in chiefly carbohydrate foods like candies, fruits, and cereals. Like many other additives in food use, there is not a definite agreement on exactly how it works.

MSG has a long history of testing and usage. However, in May 1969 a researcher reported that brain damage resulted when high doses of MSG were injected under the skin of two- to ten-day-old mice. The newspaper reports disturbed many mothers because MSG was used in many baby foods. Because of the alarm and uncertainty, baby food producers voluntarily stopped using MSG. A study by the National Academy of Sciences in July 1970 indicated that the risk of using MSG in baby foods was extremely small but that its use was unnecessary, as babies probably did not find the food more attractive although mothers who sampled it did. There was no evidence that reasonable use of MSG is hazardous to older children or adults, except to a few persons who might be unusually sensitive to it. The findings pointed out that the flavor-enhancing property of MSG is definitely beneficial to the general consumer.

The public, however, cut down considerably on home use of MSG. It is interesting that the story about the damage to baby mice was never, in the minds of many people, superseded by the later findings about its safety. Nor do most people know of its long history of safety testing and use and the fact that glutamic acid is one of the twenty important amino acid building blocks for proteins which the body needs. MSG has never been officially banned in foods and it is still used commercially in many foods containing meat and fish. The MSG experience is often cited as an example of the possible scare effect of newspaper stories on the behavior of the general public.

The Flavor Business: Always Something New

A recent development of the food industry has been the encapsulation of food flavors. Examples are instant powdered breakfast drinks like Tang and Start. Here the attempt to achieve natural flavor involves fixing the orange oil in gum arabic, which is a juice from certain tropical acacia trees. (It is also used in candies and mucilage.) The delicate orange flavor is thus stabilized from oxygen attack. The vegetable gum also prevents evaporation by loss of the molecules into the air during processing or preparation. If you have ever tasted a can of liquid synthetic orange drink you recognize the problem and the advantage of the newer powders using the encapsulation idea. Encapsulation is also being researched in meat substitutes to allow for timed release of flavors so that they will not be completely leached out during the initial chewing.

THE ABUSE OF FOOD ADDITIVES

With all the possibilities for additives in your food (including nonintentional ones like bacteria, dust, and hair), you certainly could ask, Who is going to make sure the wrong things don't go in? This big job is handled chiefly by the federal Food and Drug Administration (FDA) of the U.S. Department of Health, Education, and Welfare.

During the later years of the nineteenth century many abuses appeared involving food additives. In 1883 Harvey W. Wiley (1844–1930), a former professor of chemistry at Purdue University, became chief chemist in the U.S. Department of Agriculture. He launched a vigorous campaign against food and drug adulteration. As a result mainly of his pioneering revelation of food abuses, the Pure Food and Drugs Act of 1906 was passed. A Meat Inspection Act was passed at the same time. The shocking disclosures of unsanitary conditions in meat packing plants, use of poisonous preservatives and dyes in foods, and cure-all claims for worthless and dangerous medicines led to the enactment of these laws. Many changes have been made in the food and drug laws. Major updatings were made by passage of the Federal Food, Drug, and Cosmetic Act of 1938 and a special Food Additives Amendment in 1958. Changes are continually being made in regulations under these laws in order to keep up with the latest research findings.

Briefly, the law now requires a manufacturer to make extensive animal feeding tests on any new additive and submit results to the FDA to prove the material safe for intended use. If FDA scientists are satisfied, they issue a regulation permitting use. This specifies the amount which may be used and any other conditions necessary to protect the public health. One of the provisions of the 1958 amendment specifies that no additive may be permitted if the tests show that it produces cancer when fed in any amounts to man or animal.

A Few Examples of FDA Actions

The experience of FDA scientists and inspectors in properly carrying out their watchdog activities for the American consumer occupy volumes. The FDA people are continually checking the wholesomeness of foods, drugs, cosmetics and therapeutic devices. A few details will show the nature of some of the operations involving foods.

1949—The FDA prohibited use of nitrogen trichloride for a bleach and accelerator in the aging of flour. This chemical had been used for over twenty-five years and no evidence was actually obtained that treated flour was harmful to humans. However, dogs fed large quantities of nitrogen trichloride developed running fits or canine hysteria and the milling industry voluntarily stopped using it.

1950—The FDA prohibited further sale of an artificial sweetener called Dulcin after it was shown to produce cancer in the livers of rats.

1954—FDA banned coumarin and its natural source, the tonka bean, as a food additive. This chemical had been widely used for over seventy-five years in synthetic vanilla flavors and in chocolate confectionery products. It was found to cause liver damage to dogs and rats.

1956—The FDA withdrew its approval of FD&C Orange No. 1 (Food, Drug & Cosmetic designation for colors used in foods and cosmetics), FD&C Orange No. 2, and FD&C Red No. 32. The color FD&C Orange No. 1 had been deemed safe as previously used but a particular candy manufacturer loaded Halloween candies with it to get a particular deep orange color, and many children suffered severe gastrointestinal illness as a result.

1959—Diethylstilbesterol (DES) was banned for use in poultry. This is a synthetic female sex hormone which had been used for about ten years on young male chickens to produce artificial caponization and thus faster fattening capability. The DES was implanted at the base of the skull so that any residue could be thrown away with the head at killing time. But more sensitive tests showed small residues in skin and liver, and the ban was imposed because large quantities were found to cause cancer in rare cases.

1959—The FDA condemned a large stock of cranberries because some berries had been found to be contaminated with a weed killer (aminotriazole). The weed killer had been approved by the Department of Agriculture for use in cranberry fields after picking the crop. Some producers used it incorrectly before harvest and these berries got mixed in with many others in a cranberry packing plant. FDA scientists were aware that it causes cancer in laboratory animals and required no residue in finished cranberries. The enforced destruction of a large part of the cranberry crop just before Thanksgiving made newspaper headlines. It is often referred to as the big "cranberry scare" of 1959.

1960—The FDA banned safrole, a flavoring material extracted from the root bark of the sassafras tree. For many years this had been the chief flavoring ingredient in root beer. Through cooperative, long-term animal feeding studies carried out by the food industry, the FDA learned that safrole could produce cancer in the livers of rats. Sassafras leaves and the root bark can still be used in foods but they must now be free of safrole.

1966—Cobalt salts (like cobalt chloride or cobalt sulfate) were officially banned for use as beer additives. They had been used in Europe since about 1956 to improve the stability of beer foam. They were first used in the U.S. in 1963 and were implicated in the unexplained deaths in early 1966 of twenty people in Quebec, Canada, and eighteen in Omaha. All had been heavy beer drinkers.

1969—The nonnutritive sweetener cyclamate was removed from the GRAS category (GRAS = generally recognized as safe). GRAS substances are compounds, such as salt, pepper, baking powder, and citric acid, which were exempted from the safety provisions of the 1958 Food Additives Amendment because of their long history of safety in food uses. At the time of the Food Additives Amendment the provision was made for setting up such a GRAS list so as not to disrupt completely the consumption of foods. Because of the large number of materials like salt and pepper which had been used by generations of people, it was physically impossible and impractical to require assurances of their testing in laboratory animals, each test taking up to several years. In the GRAS list were other chemicals like the cyclamates which had previously been tested quite extensively and used in foods for many years. The removal of cyclamate from the GRAS list and the eventual ban against its use in foods were the results of more recent tests. In these tests, carried out by the manufacturer, the sweetener was found to cause bladder cancer when fed to rats in large amounts over long periods of time. Consequently the FDA was required by law to prohibit its use. Some people feel that cyclamates are safe, based on new studies. In 1974 the FDA approved a new artificial sweetener, aspartame (Equa), which is 180 times as sweet as table sugar. Aspartame is a condensation product of two amino acids, aspartic acid and phenylalanine. It has a similar caloric value to that of sugar but smaller amounts are used to provide the same sweetening effect.

1972—The synthetic hormone DES was banned in cattle feed. It had been used for many years without any evidence of trouble. Then even more sensitive testing methods showed tiny amounts in animal livers. The 1958 cancer amendment required a ban because DES can cause cancer under some circumstances. The ban stirred controversy because the price of meat had to rise considerably. Cattle receiving DES gain on the average 15 percent more than control animals but consume about 11 percent less feed for every pound of gain. About 80 percent of U.S. beef cattle were fattened on rations containing DES until it was banned in 1972.

1974—FDA rules for labeling became effective in a new program to provide consumers with better nutrition information on most foods.

It is worthwhile to note here that chemicals (like safrole and coumarin) that occur naturally in foods may be harmful also. In fact, many chemicals that occur in plants can be toxic.

The FDA and Food Labels

The new FDA regulations on labeling of food products are the result of several years of studies and open hearings. The new program has essentially two aims: (1) to improve nutritional information on food labels and prohibit false claims, and (2) to make information on food labels more meaningful. This will be done in part by setting standards for identifying essential vitamins and minerals. The program will also replace the outmoded system of measuring nutritional intake. The new measurement system is based on recommended daily allowances (RDA) of vital nutrients. The old system was based on minimum daily requirements (MDR). The U.S. RDA thus specifies the amounts of proteins, vitamins, and minerals that a person should eat every day to keep healthy.

The regulations set a standard of identity for dietary supplements of vitamins and minerals. There is a clear distinction now between ordinary foods, special dietary foods intended for diet supplementation, and drugs intended for the treatment of diseases. If a product contains less than 50 percent of the U.S. RDA, it is not a dietary supplement and only nutrition labeling is pertinent. If it contains 50 percent to 150 percent of the U.S. RDA, it is a dietary supplement and must meet certain standards. If it exceeds 150 percent of the RDA, then it cannot be sold as a food or dietary supplement, but must be labeled and marketed as a drug. Figure 10.6 gives examples of labeling recommended by the FDA.

The food law officially defines standard food items. Some of the foods for which standards have been set are bread, flours, macaroni products, margarine, evaporated milk, cheeses, tomato ketchup, canned fruits and vegetables, ice cream, and carbonated beverages.

Enriched bread and flour are thus defined standard foods. In the 1930s the alarming incidence of the nutritional deficiency diseases pellagra and beriberi and the disclosure that more than one-third of the U.S. population had poor diets prompted medical, public health, and nutrition leaders to plan corrective measures. Because of the universal use of bread and cereal products, these were chosen for an enrichment program. Now products like bread, flour, and rice, if labeled "enriched," must meet federal standards. A majority of states have subsequently passed mandatory enrichment laws. Even though some states do not have bread enrichment laws, it is estimated that about 80 to 90 percent of all white bread and white family flour (flour sold in food stores) is now enriched. This

NUTRITION INFORMATION
(PER SERVING)
SERVING SIZE = 1 OZ.
SERVINGS PER CONTAINER = 12

CALORIES	110
PROTEIN	2 GRAMS
CARBOHYDRATE	24 GRAMS
FAT	0 GRAM

PERCENTAGE OF U.S. RECOMMENDED DAILY ALLOWANCES (U.S. RDA)*

PROTEIN	2
THIAMIN	8
NIACIN	2

*Contains less than 2 percent of U.S. RDA for Vitamin A, Vitamin C, Riboflavin, Calcium and iron.

This is the minimum information that must appear on a nutrition label.

NUTRITION INFORMATION
(PER SERVING)
SERVING SIZE = 8 OZ.
SERVINGS PER CONTAINER = 1

CALORIES	560	FAT (PERCENT OF CALORIES 53%)	33 GM
PROTEIN	23 GM	POLYUNSATURATED*	2 GM
CARBOHYDRATE	43 GM	SATURATED	9 GM
		CHOLESTEROL* (20 MG/100 GM)	40 MG
		SODIUM (365 MG/100 GM)	830 MG

PERCENTAGE OF U.S. RECOMMENDED DAILY ALLOWANCES (U.S. RDA)

PROTEIN	35	RIBOFLAVIN	15
VITAMIN A	35	NIACIN	25
VITAMIN C (ASCORBIC ACID)	10	CALCIUM	2
THIAMIN (VITAMIN B_1)	15	IRON	25

*Information on fat and cholesterol content is provided for individuals who, on the advice of a physician, are modifying their total dietary intake of fat and cholesterol.

A label may include optional listings for cholesterol, fats, and sodium.

FIGURE 10.6
Examples of labeling recommended by the FDA.

means that thiamine, riboflavin, niacin, and iron are added. Enrichment is especially valuable in white flour where the milling process removes a large portion of the mineral and vitamin content.

WHAT IS THE FUTURE FOR FOODS?

The food picture today is one of gradual and constant change. You may not notice the great variety of new food products and processes which are being made available by food research. The typical large supermarket today carries over 8000 different food items. This compares with only about 1500 items thirty years ago. The number and diversity of products are steadily climbing and estimates point to 12,000 items on big supermarket shelves before 1990.

Here are a few of the chemical areas which will contribute to new foods:

1. Freeze drying as a method of food preservation. In freeze drying the food is quickly frozen and then placed under a high vacuum with controlled input of heat. The molecules in the ice crystals then go directly from solid to gas (sublime). This leaves behind a porous product from which up to 99 percent of free water has been removed. Since water is necessary for spoilage organisms to multiply, the food is automatically preserved if sealed in a moisture-proof container. Originally a specialty item for campers and backpackers who want to travel light, freeze-dried products are now expanding rapidly into home use. Examples are freeze-dried coffee, beef stew, fruit, and cottage cheese.

As shown in the photo at left, part of the manufacturing process in the General Mills vegetable protein foods plant at Cedar Rapids, Iowa, resembles the process for spinning rayon or nylon. Here, liquid protein from soybeans has been fed into a spinning machine and emerges as bands of tiny white fibers. To these bands of vegetable protein fibers, color and flavor will be added. Then they are cut into various shapes and sizes for end use applications. In the photo on the right is a typical entree made from textured soybean protein. Granules with a flavor like beef can be served in loaf, patty, and ball forms. (General Mills photos.)

2. Simulated foods, of which nondairy substitutes for cream and simulated fruit juices are the most common. A growing impetus for research in simulated foods is the population boom. Examples of other items used today are meat substitutes, made basically from spun soybean protein fibers to simulate bacon, roast beef, ham, fried chicken, and pork sausage. These also contain a whole gamut of food additives—flavors, colors, thickeners, vitamins, emulsifiers, preservatives, anticaking compounds, and humidifiers. Such products are finding a growing market with vegetarians and heart-disease patients concerned about intake of animal fats.
3. Laboratory synthesis of basic food components such as amino acids for proteins. This is an exciting development for the world food problem. The addition of synthetic amino acid lysine to bread and flour products can raise the usability of wheat protein from about a half to two-thirds. Adding the amino acid threonine could make grain proteins almost as fully usable as those of meat and milk. The addition of amino acids to grain products where they have been found to be low would thus make the enriched product about equal to meat from the nutritional standpoint.

Foods are obviously important materials in an operational sense. We all need them for growth, energy, health, and our very existence. However, in this chapter you have seen how much more important is a conceptual understanding of the molecules that make up foods. Synthetic foods are still not looked on favorably because many people do not yet fully realize that all materials are chemical, even the purest food we eat. We need concepts both for proper selection of a balanced diet as well as for planning on new foods. The experience of learning is basically the experience of relating concepts and operations. The learning here can lead to better foods and to better ways of growing foods (chapter 11).

PROJECTS AND EXERCISES

Experiments

1. This will show you how mayonnaise is formed. You will need a small clear glass or bowl and a rotary beater. Place about 1/3 cup of salad oil or cooking oil in the glass. Then add about half this amount of vinegar, 1/6 cup. Beat with a rotary beater for about a minute. Then let the glass stand so that you can observe it. Look for indications of separation. After about one or two minutes you should see definite results. What happens? Where are the oil and the water located? (Vinegar is a 5 percent solution of acetic acid, CH_3COOH, in water.) Next carefully crack an egg shell with a knife and separate the yolk from the white. Add the egg yolk to the glass and beat with a rotary beater again for about a minute. Then let it stand where you can look at it occasionally over the next hour or so. Especially observe it for the first few minutes in order to compare behavior of the mixture with that before you added egg yolk. What does this indicate about the egg yolk? Can you come up with a conceptual explanation? You might want to check back in this chapter on the formula of lecithin and remember that the name is from the Greek for egg yolk. Next you should check a cook book for a recipe to make mayonnaise. You will notice that there are many variations of flavorings added besides oil and vinegar, and that the proportion of water components is much less than in the experiment above. This would make thorough mixing even easier. This experiment uses a ratio of about 2:1 for oil:water whereas in mayonnaise the ratio may run close to 6:1. The increase in the water component (vinegar) is to enable you to see the layering more readily. Mayonnaise is essentially a water and oil emulsion, stabilized by egg yolk.
2. This experiment will require a few drops of food colors, some white paper towels, some tenacious transparent wrap like Saran-wrap, and the tallest clear glasses available. You will need only a tiny drop of as many varieties of food colors as you can locate, so perhaps you may borrow from a neighbor what you don't have in your kitchen. Cut strips about 2 inches wide from the white paper towels. These should extend the full length of the tall clear glass. Now put a very tiny drop on the paper towel strip about 3/4 of an inch up from one end. Most food colors are in plastic squeeze tubes so that you can carefully squeeze out a small drop. You will get practice by repeating this and even a big drop will work. Then put about 1/4 inch of water in the glass and carefully lower the strip in with the spot end nearest the bottom. Your paper strip may be stiff enough to stand by itself in the glass, leaning against the side. If the paper towel strip is not stiff you can attach the top end over the edge of the glass. Put the plastic wrap tightly over the top, stretching it so that no water molecules can escape. You may want to put a rubber band around the outside to make sure that the system is tightly closed. This keeps the inside of the glass saturated with water molecules so as to prevent the towel from drying out during the experiment. Now observe carefully the be-

havior of the food color during the next fifteen to thirty minutes. After a practice first run you may be able to put two strips in the same glass. It is important to try several colors. What do you observe? Does this indicate that the colors are mixtures? What conceptual explanation can you offer for what happens here? (If you want to experiment further you can try other solvents like rubbing alcohol, or gin or vodka if you can afford it.) The general idea is that different molecules move through the cellulose fibers at different rates. Some can migrate faster because of their particular molecular structure and the relationship of this to the structure of the cellulose. It is really a race between molecules to see which one gets to the top of the towel first. If you think this is kid's stuff, you are wrong. The method you have just used is called *chromatography*, which means color writing or color picturing. This simple technique has now been developed into a tool of the research chemist for the detection, identification, and preparation of various chemicals. Indeed, many of the findings which were casually mentioned in this chapter as facts were discovered or confirmed in some way by chromatography. It is used daily in such diverse and important applications as disease diagnosis and crime detection. In hospital labs it has become a rapid method of finding minute components in body fluids which indicate the presence of disease.

3. This experiment involves baking a couple of cakes, or at least checking the recipes if you think you cannot bake. Baking the cake is best since this makes for a rewarding experiment. Do not use the prepared cake mixes. They already have almost every ingredient added and you won't learn much from just mixing water and the prepared mix. Go to a cookbook for the recipes that start from flour. You may want to make one cake this week and another next week. Pick a recipe for the usual type of cake and another for a sponge cake. Pay particular attention to the ingredients that are called for. Then when you eat the cakes make notes on the characteristics of bitability, crumbling, chewability, and so on. After you have eaten a piece of both kinds, give a conceptual explanation of some of the differences in characteristics, tying it in with the ingredients which went into the recipes.

Exercises

4. Choose one of the quotations at the beginning of this chapter. Write a brief paragraph on what it means to you now and whether its meaning has changed since you first read it.
5. Glucose has six carbon atoms as does hexane, which is a component of gasoline. What reason can you suggest to explain conceptually why glucose is soluble in water but hexane does not appreciably dissolve?
6. Give a brief description of what you mean by sugar. Is your present knowledge more sophisticated than your former understanding?
7. Someone tells you that he heard that starch and cellulose are both made of glucose. How would you explain the difference?
8. What would you answer to the question: Why does animal starch or glycogen have more branches than plant starch?
9. Compare table sugar or sucrose with corn syrup. Consider both the conceptual background you now have and also what a cook would think of the two products.
10. Explain what proteins are and give some examples of common food sources. What is the essential difference between these and carbohydrates?

11. Explain how conceptual understanding can lead to the development of a new food product to alleviate world hunger.
12. Distinguish between a disease caused by bacteria and a deficiency disease. Give a few examples of each. Can you explain why deficiency diseases were not at first easily recognized? People tended in the early years to automatically favor the bacteria or germ theory for disease. Can you guess any of the reasons?
13. Consider the statement: "All fats are esters but not all esters are fats." Explain what this means in terms of your present understanding of these chemical types. You can use formulas if you like.
14. Someone tells you that he heard that both lard and soybean oil are the same kind of chemical. He wants you to explain this and also why one is solid and the other a liquid.
15. Write a brief paragraph explaining the difference between margarine and butter. Did you think they were essentially the same before reading this chapter?
16. Do you think we could run into vitamin deficiencies in the United States among the general consumer population? Why or why not?
17. Someone tells you that all food additives are bad and should be banned by law. How could you answer this?
18. Do you think MSG should have been banned? Give reasons why or why not.
19. Carbohydrates have sometimes been referred to as "protein sparers." The terminology comes from the fact that proteins when eaten without carbohydrates will be largely converted to oxidation products in order to provide energy. This means that the important amino acid units of the protein will be burned for fuel instead of being used for repair and building of body cells. Consequently, nutritionists highly recommend eating proteins and carbohydrates together so that the carbohydrate will be burned for energy and thus "spare" the protein for more important construction work. Think of some of your present eating habits and give a few examples where you follow this recommendation. (You may never have heard the conceptual reason for your practice which is passed down from generation to generation.)
20. It is also a good general habit to eat fats along with carbohydrates. This is because if fats are eaten predominantly alone they will be incompletely burned to give intermediate products which cause headache, nausea, and loss of appetite. Think of your eating experience and give a few examples where you by habit eat fats and carbohydrates together. Have you ever had the experience of eating too much fat without accompanying carbohydrate?

SUGGESTED READING

1. There are many topics in this chapter that you could profitably check into further by reference to various encyclopedias. Suggested headings may include the following: nutrition, foods, fats, carbohydrates, proteins, vitamins, disease, calories, fruits, vegetables.
2. **Mitchell, H. S., Rynbergen, H. J., Anderson, L., and Dibble, M. V., *Cooper's Nutrition in Health and Disease, 16 Edition,* J. B. Lippincott Co., Philadelphia, 1976. This is a classical text in nutritional science. It gives a comprehensive review of the principles of nutrition as they apply both to normal persons and individuals suffering various illnesses. It is a very good reference source providing nutritional information on a variety of food topics.**

3. Lowenberg, M. E., Todhunter, E. N., Wilson, E. D., Feeney, M. C., and Savage, J. R., *Food and Man,* John Wiley & Sons, Inc., New York, 1968. This is a good readable book on the historical, sociological, medical and political aspects of world-wide nutrition problems. These are aspects which deserve everyone's attention and are not emphasized in this book only because of necessary space and time limitation.
4. Gillie, R. B., "Endemic Goiter," *Scientific American,* 224, No. 6, pp. 92–101 (June 1971).
5. Leverton, R., *Food Becomes You,* Dolphin Books, Doubleday & Co., Garden City, New York, 1961.
6. Williams, R. J., *Nutrition in a Nutshell,* Dolphin Books, Doubleday & Co., Garden City, New York, 1962.
7. U. S. Department of Agriculture, *Food,* Yearbook of Agriculture 1959, U. S. Department of Agriculture, Washington, D. C., 1959. Somewhat dated but still good reading in many general areas.
8. U. S. Department of Agriculture, *Food for Us All,* Yearbook of Agriculture 1969, U. S. Department of Agriculture, Washington, D. C., 1969.
9. Scrimshaw, N. S., "Food," *Scientific American,* 209, No. 3, pp. 72–80 (September 1963).
10. Pirie, N. W., "Orthodox and Unorthodox Methods of Meeting World Food Needs," *Scientific American,* 216, No. 2, pp. 27–35 (February 1967).
11. Brown, L. R., "Human Food Production as a Process in the Biosphere," *Scientific American,* 223, No. 3, pp. 160–170 (September 1970).
12. Kermode, G. O., "Food Additives," *Scientific American,* 226, No. 3, pp. 15–21 (March 1972).
13. Fernstrom, J. D., and Wurtman, R. J., "Nutrition and the Brain," *Scientific American,* 230, No. 2, pp. 84–91 (February 1974).
14. Hall, F. K., "Wood Pulp," *Scientific American,* 230, No. 4, pp. 52–62 (April 1974).
15. Dovring, F., "Soybeans," *Scientific American,* 230, No. 2, pp. 14–21 (February 1974).
16. Hanawalt, P. C., and Haynes, R. H., eds., *The Chemical Basis of Life, An Introduction to Molecular and Cell Biology,* W. H. Freeman and Company, San Francisco, 1973. This is one of a series of collections of *Scientific American* articles bridging the chemical and biological sciences. Molecular biology is the fascinating area wherein life scientists use so much basic organic chemistry to explain biological activity, reproduction, expression of genetic information, and regulation of cellular activity.
17. Stroud, R. M., "A Family of Protein-Cutting Proteins," *Scientific American,* 231, No. 1, pp. 74–88 (July 1974).

chapter 11

Chemicals on the Farm

That which the palmerworm left, the locust has eaten; and that which the locust left, the beetle has eaten; and that which the beetle left, the mildew has destroyed.

—JOEL (Ch. 1, v. 4)

If insects want our crops they help themselves to them. If they wish the blood of our domestic animals, they pump it out of the veins of our cattle and our horses at their leisure and under our very eyes. If they choose to take up their abode with us we cannot wholly keep them out of the houses we live in. We cannot even protect our very persons from their annoying and pestiferous attacks. And since the world began, we have never yet exterminated—we probably never shall exterminate—so much as a single insect species.

—S. A. FORBES

It is the microbes who will have the last word.

—PASTEUR

The growing variety and quantity of food products in developed countries of the world has been a phenomenon of our times, especially when contrasted with severe food shortages in nonindustrialized areas. All foods begin with the farmer and his nurturing of plants. The efficiency of farming has progressed in the U.S. so that approximately 60 persons are now suplied food by one farm worker. This compares with 4 in 1800 and 7 in 1900. (See Figure 11.1.)

What is the reason for this astounding agricultural growth? The answer involves

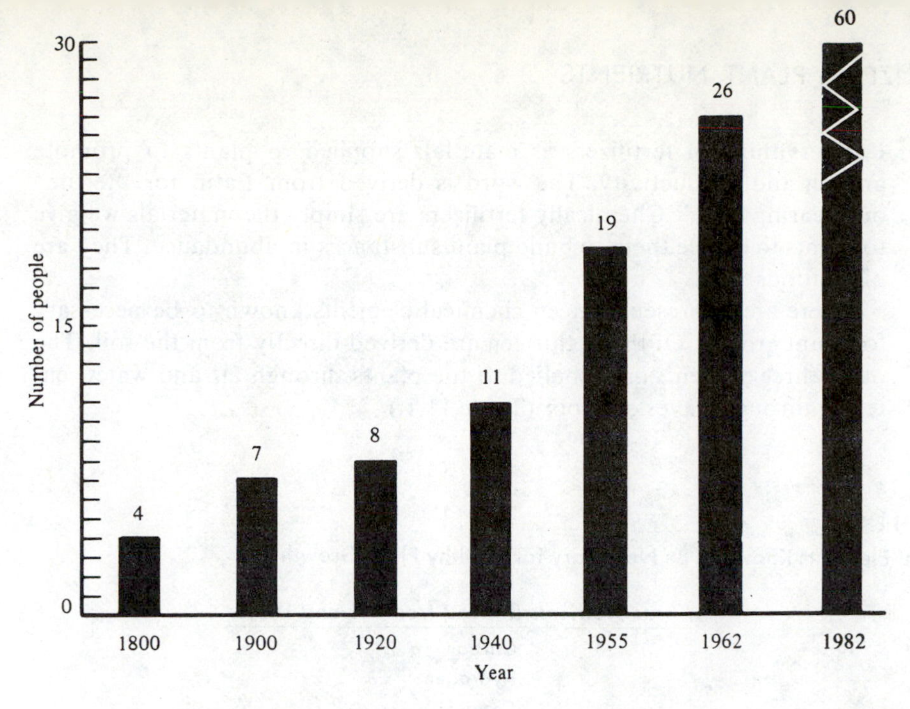

FIGURE 11.1
People provided for by one farm worker.

motivational, sociological, and political aspects which are beyond the scope of this book. However, the contributions of science and technology lie in three major areas: chemicals; machinery for plowing, planting, weeding, irrigation, pumping, harvesting, threshing, cleaning, and so forth; and breeding for disease-resistant plants and animals, hybrid varieties of plants, high-yield food staples such as wheat, rice and corn, and high-protein grains.

The availability of foods in large quantity from fewer and fewer farm workers is the result of a successful cooperative blend of the chemical, mechanical, and biological arts along with the natural human drive for understanding and improvement. Chemistry represents a large part of the picture. Chemicals are even involved in the mechanical and biological areas. Applications of chemistry to machines include special steels, oils, plastic foams, insulation, paints and fuels. Biological breeding improvements depend heavily on chemistry, from fertilizers and insecticides to chemical analysis of the high-protein end products. This chapter, however, will discuss only the most direct applications of chemicals to farming.

A few areas where chemicals directly help farmers are the following:

Fertilizer
Soil Conditioners
Pesticides, insecticides, fungicides, rodenticides
Plastic mulches
Growth regulators
Weed killers
Sprouting controls
Ripening agents
Disinfectants
Paints
Petroleum products
Plastic pipe
Chemicals preventing premature fruit drop
Antibiotics and other additives in livestock feed
Sex attractants

FERTILIZERS: PLANT NUTRIENTS

Farmers think of fertilizers as materials supplied to plants to promote growth and productivity. The word is derived from Latin for "fertile" or "bearing fruit." Chemically fertilizers are simply the materials we give to plants to enable them to build plant substances in abundance. They are plant foods.

There are at present sixteen chemical elements known to be necessary for plant growth. Of these, thirteen are derived directly from the soil. The other three elements are supplied to the plants through air and water, entering through leaves or roots. (Table 11.1.)

TABLE 11.1
Nutrient Elements Known to Be Necessary for Healthy Plant Growth

Three Nutrients Derived from Air and Water

Carbon
Hydrogen
Oxygen

Thirteen Nutrients Derived from the Soil

Three Primary Plant Nutrients
(required in very large quantities)

Nitrogen
Phosphorus
Potassium

Three Secondary Plant Nutrients
(required in fairly substantial quantities but less than primary)

Calcium
Magnesium
Sulfur

Seven Micro-Nutrients
(needed in only very small quantities)

Boron
Copper
Iron
Manganese
Zinc
Molybdenum
Chlorine

There is at present some feeling among plant physiologists that sodium, vanadium, and cobalt are very likely also required by plants. Other elements probably are also beneficial but our research information is not yet definite enough to add them to the list of sixteen essential elements.

Plant Deficiency Disease

Since the nutrients listed in Table 11.1 are essential for proper plant growth and production, you would expect that plants would show deficiency diseases just like people and animals. This is indeed the case. Farmers and agricultural researchers recognize many symptoms of deficiency and have been able to tie operational symptoms to a lack of certain elements. Table 11.2 lists a few examples.

TABLE 11.2
Some Essential Plant Nutrients and Deficiency Symptoms

Essential Element	*Operational Indications of Deficiency*
Nitrogen	Yellowing and shriveling of leaves; lack of dark green color.
Phosphorus	Slow starting growth and initial root formation; difficulty in blooming and seed formation.
Potassium	Lack of disease resistance, for example, to fungus blights; lack of plumpness in grain and seed; lowered quality of fruit and tuber growth.
Calcium	Slow formation and growth of early root hair; decreased calcium content of food and feed crops; lowered grain and seed production.
Magnesium	Lack of green color because magnesium is essential in chlorophyll molecule; poor uptake of other plant foods and lowering of sugar formation.
Iron	Lowered production of green chlorophyll.
Zinc	Lowered chlorophyll production and growth.

At the present stage of our collective knowledge we are able to tie in a particular nutrient with some characteristic of plant growth, as we are with our own body nutrients. However, detailed conceptual explanations of the complex processes which the plants use for making food and fiber are far from fully understood.

Fertilizers Are Now Essential

The fertilizer industry is vital for agricultural production in all developed nations of the world. The use of natural and artificial fertilizer materials in many countries stands in the way of widespread starvation. This certainly includes the U.S., where annual production of synthetic ammonia for fertilizer now runs in the neighborhood of 15 million tons. This compares with a figure of only 5 million tons in 1960. And this is only part of the picture—the nitrogen component. Total world fertilizer consumption was less than 10 million tons annually at the end of World War II. Now it is

Cane field on the island of Kauai at the planting site of the first commercial cane crop in Hawaii in 1835. Fields in this area have been cultivated continuously since that time. Because of fertilization and agricultural technology, these fields are still producing record harvests. (Courtesy C and H Sugar Company.)

in excess of 100 million tons.

Bringing unused land into cultivation is not as promising for the food production needs of growing world populations as the expansion of science and technology for commercial fertilizer production. People often think that there is unlimited land available, but there are many obstacles to expanding land usage. For example:

1. The world land area capable of producing crops is quite limited. (Figure 11.2.) At present only about 10 percent of land surface is under cultivation, with pasture land occupying another 17 percent and forests 29 percent. The remaining 44 percent of land is considered unusable for various reasons like cold climate, rocky or sandy soil, or usage for buildings and roads.
2. Vast new irrigation and drainage projects have to be developed to reclaim usable land.
3. Much time and capital investment are required for deforestation, leveling and preparing land sites for particular farming needs.
4. In the end even this new land will need conditioning and replacement of plant nutrients.
5. Much of the potentially available land is now in forests. It is questionable whether displacement of these forest lands to crops would not provide an unpredictable upset to nature's vital balances.

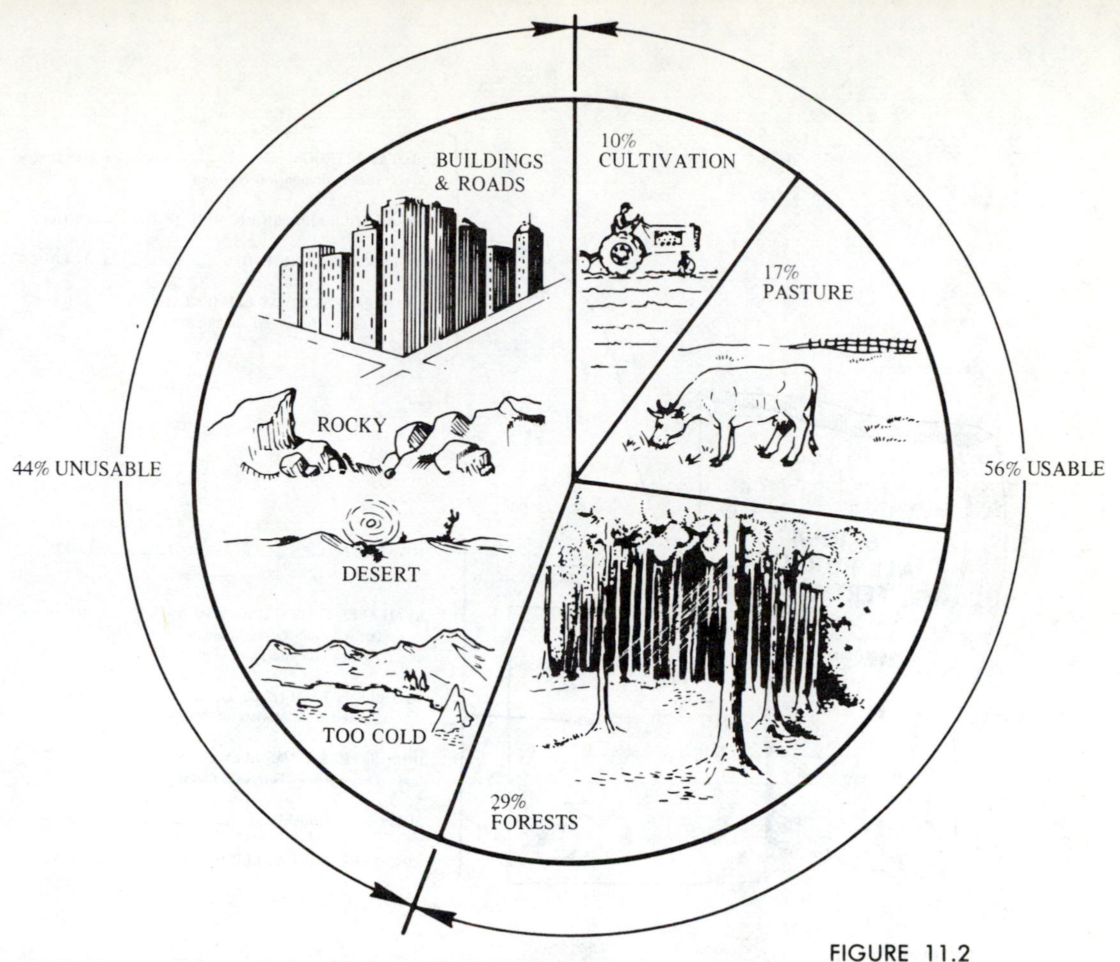

FIGURE 11.2
Approximate distribution of world land area.

What Is a 12-9-6 Fertilizer?

The answer to increased food needs then necessarily depends upon fertilizer. You may have observed boxes or bags of fertilizer used around the home and farm. There is usually a set of three figures like 6-10-4 in a significant place on fertilizer bags. These numbers indicate the percentages of the primary plant nutrients—nitrogen, phosphorus, and potassium, respectively—which plants really "gobble up" in large quantities (along with carbon, hydrogen and oxygen which are available from air and water). The percentages of these major nutrients, N, P, and K, are expressed as percent N, percent P_2O_5, and percent K_2O. The method of expressing phosphorus and potassium as oxides goes back to the early days of fertilizer analysis. A fertilizer bag labeled 12-9-6 would thus have 12 percent nitrogen, phosphorus expressed as 9 percent P_2O_5, and a potassium content of 6 percent K_2O.

Other nutrients required in smaller quantities are also added, usually

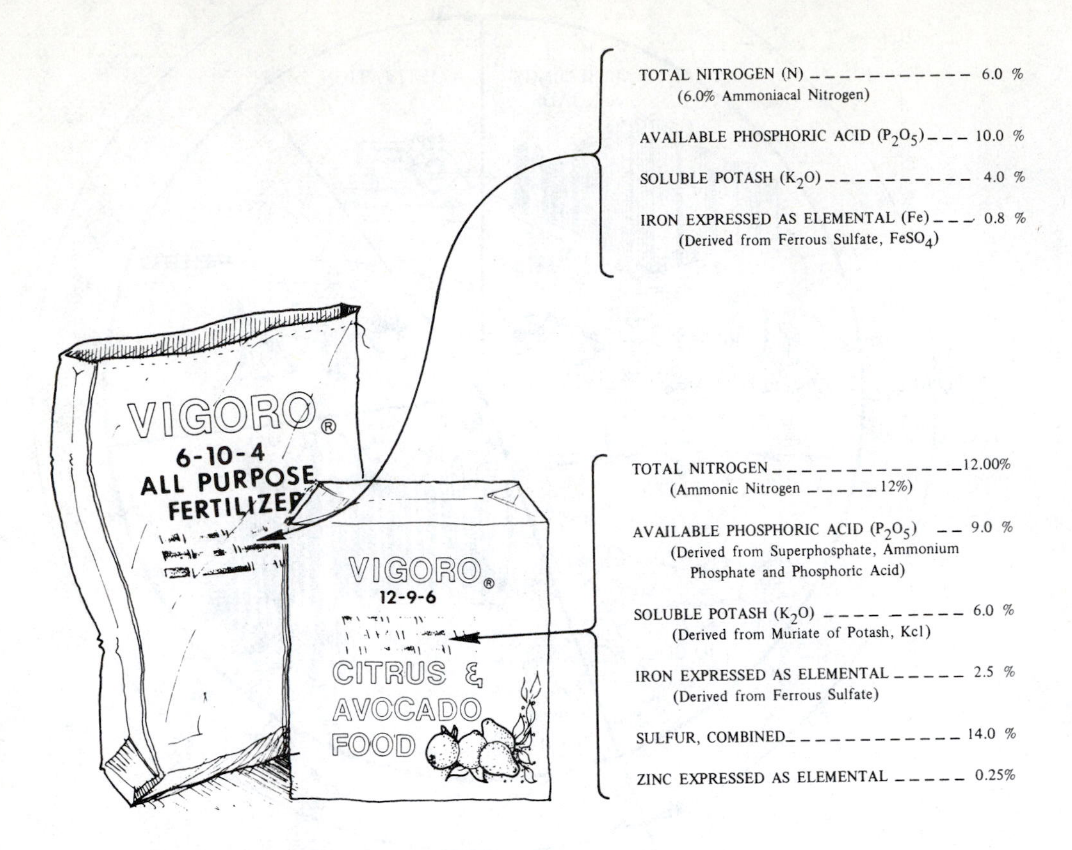

FIGURE 11.3
Details on nutrient labeling for two fertilizer containers. (Courtesy of Swift Chemical Company.)

selectively for certain crops. These are separately listed on fertilizer containers. Figure 11.3 gives two examples.

Plants as Complex Chemical Factories

From your knowledge of foods you can recognize that plants are really complex chemical factories. They have to build the chains of their leaves, fruits, and operating apparatus from simpler chemicals. Fertilizer provides the atoms in the form plants can use or in forms which soil microbes can convert for plant needs. This usually involves getting the necessary nutrient atoms into ionic forms which can be absorbed by the roots from water solution. Nitrogen is very plentiful in the air as N_2 but only a few plants (like clover, alfalfa, and some legumes) can use it in that form. Major factors in progress using fertilizers over the years have been the conceptual recognition of the plant's need for certain chemical elements (see Table 11.1) and the ability to make necessary elements available for plant growth.

The first use of fertilizers probably goes back to early times when people noticed that the grass was always greener where carcasses and droppings from animals were deposited. In medieval Europe farmers recognized the value of growing clover for later ploughing into the soil. The first settlers to North America noticed that the Indians obtained greater yields by placing dead fish in the soil when planting seeds. In various ways then the use of organic fertilizers became common. These included manures, compost, meat and fish wastes, blood and ground bones. Gradually over the years scientists began to develop conceptual understanding of fertilizer use. Even after the nitrogen content of manures was recognized, however, there was a strong feeling that organic forms of nitrogen were the only ones which plants could really thrive on. The pressure for understanding grew as did the need for more and more organic fertilizer. The Industrial Revolution and increasing concentrations of population added urgency.

One of the reasons for the slow recognition of the value of inorganic nitrogen compounds was the basic lack of understanding of how the plants use the nutrients of organic fertilizers, and how nitrogen in the air is involved as a source of nitrogen compounds that plants use.

The Nitrogen Cycle

One representation of nature's vast and complex nitrogen cycle is shown in Figure 11.4. You can see how the atoms of nitrogen move around in cyclic paths and change from inorganic to organic and back again depending on environmental conditions. There is really no fundamental distinction which makes the nitrogen atom in nitrate ions (NO_3^-) or ammonium compounds (like NH_4Cl) any different from that in amino acids. Some of the cyclic paths take hundreds, or thousands, or even millions of years. But the nitrogen atoms merely wait their turn to participate in the various branches of the cycle. They move along, never being worn out or used up. They are never "tagged" for inorganic or organic usage.

Which are more important, the nitrates or the amino acids in proteins? They both contain nitrogen atoms which previously have served in compounds of the opposite kind. The most important nitrogen atoms to you are those in the amino acids making up vital body proteins. However, even while you think about this some nitrogen atoms are being prepared for leaving your system, and will soon rejoin the inorganic ammonium compounds in the soil, nitrogen in the air, and nitrates in the soil.

The History of Nitrogen Fertilizer

The name *ammonia* comes from the salt formerly called sal ammoniac, now ammonium chloride NH_4Cl, which was first brought to Europe from Egypt. It was originally prepared from the soot or deposit obtained from

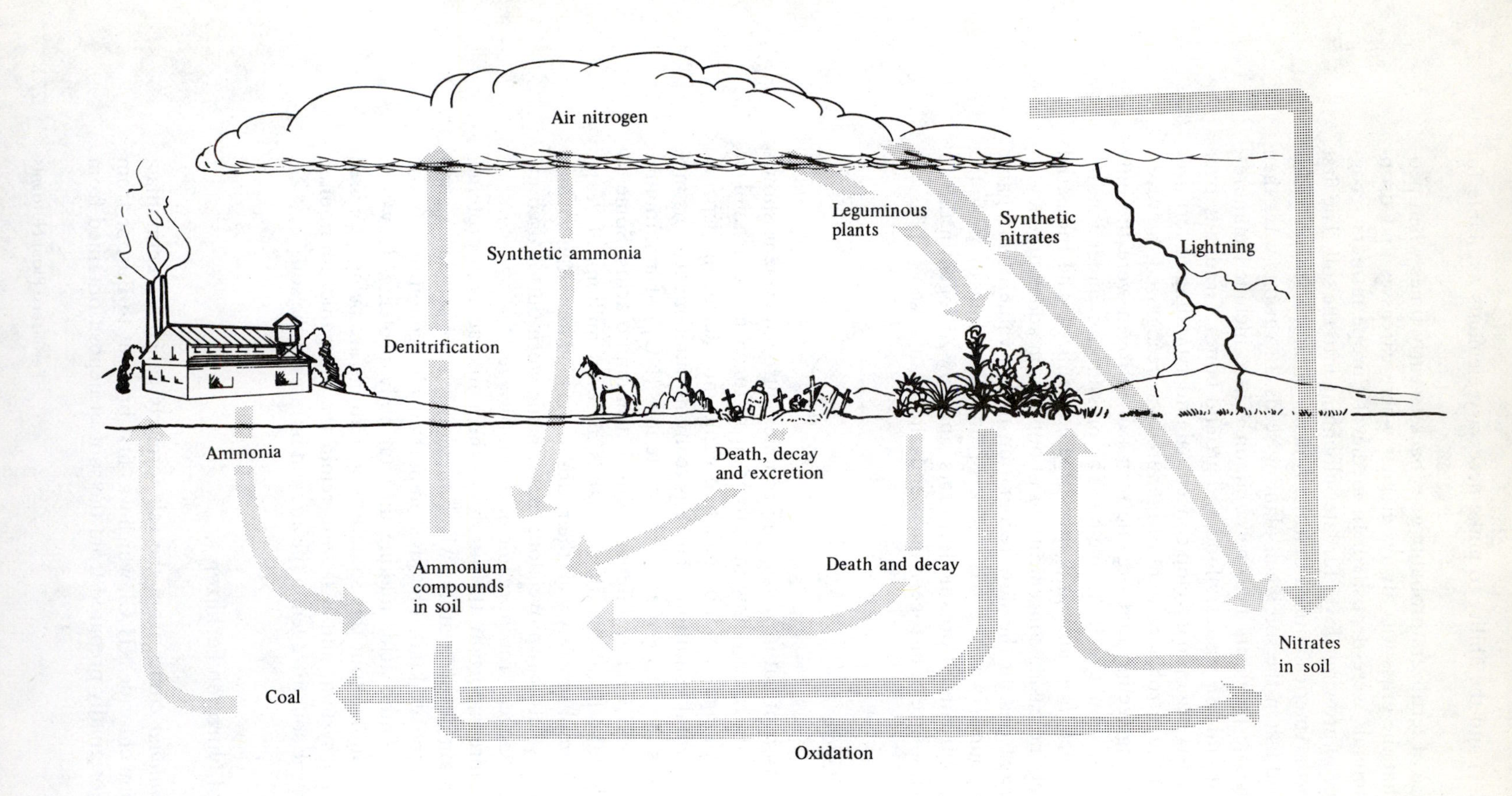

FIGURE 11.4
The nitrogen cycle.

burning camel's dung at the shrine of Jupiter Ammon in Libya. Hence the term sal ammoniac or salt of Ammon. Later in 1774 Joseph Priestley—who is given credit also for discovery of oxygen—obtained ammonia gas by treating sal ammoniac with lime.

$$2NH_4Cl + Ca(OH)_2 \xrightarrow{\Delta} CaCl_2 + 2NH_3 + 2H_2O$$

Priestley did not know these formulas and called the gas *alkaline air* because it had an alkaline reaction in water. Around 1800, chemists finally established the formula of ammonia as NH_3.

Toward the end of the eighteenth century, coal gas was first developed as an illuminating gas for overhead lighting fixtures. It is a mixture chiefly of hydrogen, carbon monoxide, methane, and other hydrocarbons. Coal gas is obtained when coal is heated in the absence of air (chapter 8). One of the byproducts from manufacture of coal gas, besides coal tar, is ammonia and ammonium salts. During the nineteenth century these were accumulating at the same time that people began to develop a primitive conceptual understanding of the nitrogen cycle and the possible plant use of inorganic nitrogen compounds like NH_4Cl and $NaNO_3$.

By the early 1900s extensive deposits of $NaNO_3$ were being mined in Chile and exported for use in chemicals and explosives. The logical next step was to try inorganic salts from Chile and coal gas operations for nitrogen fertilizer. They worked. This essentially set the stage for the major breakthrough which was an attempt to provide nitrogen compounds from the inexhaustible supply in the air (which is about 78 percent nitrogen).

In the later 1800s it had been shown that certain plants like peas and beans produced rounded lumps on their roots where bacteria live in what is called a *symbiotic* arrangement (Greek: living with) with the plants. They act to convert nitrogen of the air into a form available as food for the plants on which the bacteria live.

After frustrating attempts by many workers, a German chemist, Fritz Haber, finally developed the special conditions necessary to make ammonia from nitrogen of air and hydrogen (Figure 11.5).

$$N_2 + 3H_2 \xrightarrow[\text{Pressure}]{\text{Catalyst}} 2NH_3$$

The significance for food production of this simple reaction cannot be overemphasized. It represented a direct shortcut to ammonium compounds. For example, ammonia reacts with nitric acid to form ammonium nitrate.

$$NH_3 + HNO_3 \rightarrow NH_4NO_3$$

Any process for causing the free nitrogen in the air, N_2, to form compounds is said to "fix" the nitrogen. The Haber process represented an

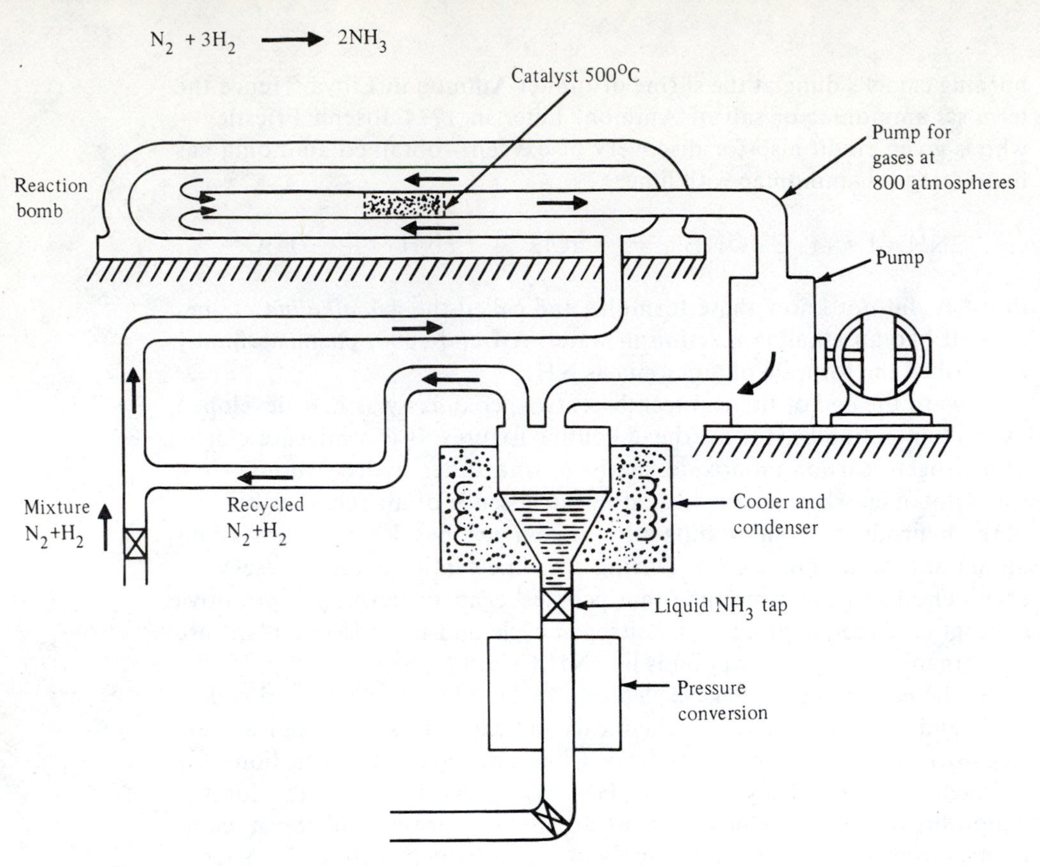

FIGURE 11.5
Haber process for making ammonia by fixation of nitrogen from the air.

important synthetic method for fixing nitrogen by forming NH_3 molecules. These then can be made available to plants.

Look again at the diagram of the nitrogen cycle in Figure 11.4. The ammonium compounds available from the Haber process are the same ammonium compounds formed slowly in soils by death and decay of plants. They are molecules identical also to those obtained in soils by excretion and decay of animals. They are also the same as those obtained by tapping into the nitrogen cycle at the fossilized plant stage represented by coal.

With the availability of ammonia in large quantities, processes were developed for oxidation of NH_3 to nitric acid. This in turn provides nitrates in unlimited quantities independent of natural nitrate deposits like those in Chile.

It is interesting to note that the process of manufacturing ammonia was invented shortly before World War I broke out in 1914. The problems involved in large-scale production were overcome by the Germans, who were desperate to provide nitrogen compounds for explosives when supplies of Chilean nitrate were cut off by the British blockade. Thus the process was originally instrumental in considerably prolonging the war. Yet today it provides millions of tons of nitrogen fertilizer which in turn contributes in no small measure to feeding and keeping large populations alive. Are nitrogen compounds and chemical technology good or bad? It depends.

Phosphorus: From Bones and Rocks

The history of the use of phosphorus in fertilizer has also followed a slow development. Phosphorus was first discovered about 1669 by a process involving distillation of a mixture of sand and the solid residue from evaporated urine. About 1769 the Swedish chemist Johan Gahn showed that calcium phosphate occurs in bones. Shortly thereafter the element phosphorus was prepared from ground bones heated with sand.

However, the knowledge that bones contained phosphorus remained academic for some time. Ground bones or bone meal or bone manure was used as fertilizer for many years before its nutrient value as a carrier of phosphorus became more thoroughly understood. This refinement came about toward the middle of the nineteenth century. At that time the German chemist Justus von Liebig (1803–1873) began work in the study of soils and fertilizer.

He showed that bones treated with sulfuric acid would make phosphorus more easily assimilated by plants. This laid the groundwork for expansion of fertilizer usage based on bones and also other sources of phosphates. These sources include materials shown by analysis to be high in phosphate content, like slag from production of iron and steel, and phosphate rock from areas like Algeria, South Carolina, and Florida. Phosphate rock contains various phosphate minerals, shown simply as $Ca_3(PO_4)_2$.

The origin of sedimentary phosphate rock deposits is still the subject of hypothesis since they are generally presumed to have been laid down in oceans beginning around 480 million years ago over a time span of about 400 million years. It is likely that the phosphate used in fertilizer and reaching you via plant foods once served also as a part of the bones or control mechanism of a dinosaur. (The energy release or driving force for muscle contraction is controlled by compounds containing phosphorus; thus the term "control mechanism.") So the phosphorus atoms of a dinosaur are very likely in your bones and blood. This is even closer than the dinosaur in your gas tank!

The Phosphorus Cycle

The world's phosphorus atoms thus are all involved in a complex cycling process somewhat similar to that of nitrogen. You represent a vital reservoir of phosphorus atoms since your bones and teeth contain about 60 percent $Ca_3(PO_4)_2$ and even your muscles and nerves depend heavily on phosphate compounds. Like other animals, you are dependent on plants for replenishment of phosphate which you continually return to nature's giant cycling networks. Waste decomposition products from nerve and muscle tissues are continually carried by your blood to the kidneys where they are excreted in urine, chiefly as sodium ammonium phosphate.

You can see how much easier it is to provide the quantity of phosphorus needed by plants once we recognize conceptually what they need. Formerly, bone manure was thought necessary. But now we realize that the key is phosphate—not bone per se, and there is a need to develop new sources of phosphate supply. It has been reported that even human bones from the battlefields of Europe were once recovered and ground for use in plant foods because of the dire needs for fertilizer materials. Now we go more directly to slag and phosphate deposits. Even sewage might very soon become, of necessity, a vital source for phosphate. The days are long past when mankind could depend on just manure and bones.

At present there is no known natural process for returning phosphorus to the earth, such as through lightning and bacteria as in the case of nitrogen. For this reason, a vast global experiment is now being carried out by removing phosphorus from beds of rock for fertilizer. Scientists estimate that over 3 million tons of phosphorus in various compounds washes into the sea annually. Much of this eventually settles in sediments and lies unavailable for practical return to the cyclic needs of plants and animals. This may have been the origin of much of the rock we now use to extract phosphorus. But at the present rate of usage we will not be able to wait another couple of hundred million years to recover it. Reserves of phosphorus are thus much more limited than nitrogen. Some scientists believe that phosphorus compounds may be a major item to be mined from the sea floors in the not too distant future.

Potassium: The Third Primary Plant Nutrient

The first use of potassium in fertilizer also extends backward to very early times. Potash was the name for the material (now known to be chiefly potassium carbonate) obtained from ashes left after wood and plants were burned. This ash was apparently concentrated in round fire containers or pots in the early days; this is the derivation of the name potassium.

The value of ashes as aids to plant growth was probably recognized in the same incidental way as the value of manures and bones. It is interesting that plants require potassium and not sodium, which is a close family member of the alkali metals. The higher absorption of potassium in soils by plants is thought by some scientists to account for the preferential concentration of sodium in rain runoff into the oceans. Thus the oceans contain a large proportion of sodium salts and only very small amounts of potassium salts.

Potassium is a very prevalent element in the environment, representing 2.4 percent of the earth's crust, or seventh in relative abundance order of elements (chapter 3). However, many minerals have potassium so tied up as not to be suitable for fertilizer. The nutrient is now usually recovered from sources of KCl deposits either as solid or, in some deep mines, by solution mining where water is pumped into the earth to dissolve the salts. The solutions are then pumped out and evaporated to provide KCl. The salt is used in blending with nitrogen and phosphorus fertilizer material.

Application of Fertilizer

Everyone who has had a garden or farm experience knows that the best growth is not provided for by dumping fertilizer on the land. A few examples are given below to show how the form of fertilizer materials and the conditions of application are of vital importance.

1. The pH of soils is often a controlling factor for proper assimilation of nutrients by plants. Some nutrients may be present in a soil but because of acidity or alkalinity may not be in proper ionic form for absorption through the plant root system. For example, nitrogen, phosphorus, potassium, calcium, magnesium, and molybdenum are all much less available if the soil is too strongly acid. Salts of these are more adequately absorbed in soils that are very slightly acid or very slightly basic, that is, from pH around 6.5 to 7.5.
2. Given the proper pH conditions, there is still a considerable difference in solubility of various fertilizer compounds. Consequently, it is necessary to choose compounds with a balance between ready availability which might provide rapid leaching and loss from the soil on the one hand, and the slow release of nutrients over a longer time period on the other. For instance, nitrogen present in nitrate (NO_3^-) form is rather quickly leached whereas nitrogen based on ammonia (NH_3, or NH_4^+ compounds) is more readily attached to soil structure and held for gradual release to the plants. For this reason, and because ammonia gas (NH_3) represents a much more concentrated and efficient medium for application of nitrogen, it is often used directly. It can be injected below the surface of moist soil where it is quickly absorbed by the ionic soil structure. Anhydrous ammonia contains 82 percent nitrogen compared to the nitrogen content of barnyard manure which contains only 3 to 5 percent nitrogen. Liquid ammonia solutions are also used for injection. These can be made at the site of application by adding ammonia from tanks to water and thereby saving considerably in fertilizer transportation costs.
3. Large increases in crop yields are obtained by the first application of fertilizer, the yield sometimes having a value up to ten times the cost of fertilizer. However, proportional yield increases are not obtained by continuing increases of fertilizers. Optimum quantities are recognized by all experienced farmers and gardeners. Overfertilization can be uneconomical. But, more importantly, it can result in damage to plants.
4. Micronutrients are needed in such small amounts that addition to fertilizers is often not required since the soil may have sufficient supply. Extreme care is required in adding micronutrients since the balance is very delicate and an oversupply which is toxic to plants can easily be obtained.

Plant analysis is now becoming more common. This consists of taking samples of the growing plants and analyzing them chemically in order to determine the kinds, amounts, frequency, and methods of fertilization for efficient crop production. This is often done in cooperation with experts from the various agricultural extension services of universities and the U.S. Department of Agriculture. A good plant analysis program has been said to be as simple as reading the gasoline gauge in your automobile and adding more gas when the reading is low. In the case of plant analysis the conceptual basis is just a little more refined. You get readings for nitrogen, phosphorus, potassium, sulfur, and so on. These readings are then compared with critical nutrient concentrations which have been determined in carefully controlled experimental studies of the particular plant. Then you know which nutrients are running low so that proper fertilization can be applied to get maximum crop yields.

WHAT ARE PESTICIDES?

Pesticides are chemicals selected to kill or in some way inhibit the plant and animal competitors which interfere with our health, comfort, or production of foods and fibers for clothing or shelter. These chemicals can be broadly categorized according to the plant or animal group toward which they are aimed (the target species). Then we get a breakdown which, broadly speaking, includes the following types of pesticides: (1) insecticides; (2) fungicides (to control mildews, rusts, and molds); (3) miticides (to kill ticks and mites); (4) nematocides (to control nematodes, which are slender threadlike worms, such as hookworms and pinworms); (5) rodenticides (to kill rodents like rats and gophers); and (6) herbicides (to control or kill weeds).

How Many Insects Are There?

No one knows exactly the actual number of the different kinds of insects. Latest estimates place the number of *species* of insects between one and three million. This is far more than that of all other plant and animal species combined. Note that this is merely the number of *different kinds* of insects! Then within each species there are uncountable millions upon millions of individual insects. The total number of insects in the world is a staggering figure which can only be roughly guessed at. One estimate places it at around a billion billion. This far outnumbers the number of people on the whole earth, which is only a little over 4½ billion.

Only a Small Fraction of Insects Are Harmful

Fortunately for us, most of the vast numbers of insects are not harmful. Indeed, in some cases they even help plants, animals, and people. Consider the value of bees in pollinating flowers, vegetables, and fruit trees, and providing honey. More important, but seldom recognized, is the fact that many beneficial insects prey upon others which destroy crops and animals. A common example is the ladybug which devours aphids. Other insects provide food for fish and birds. Estimates place the number of destructive and disease-carrying insects at only 0.1 percent of the total population.

But the Damage They Cause Is Fantastic

In the U.S. alone estimates indicate that there are nearly 100,000 different kinds of insects, mites, and other flying or crawling pests. About 600 of these are major insect pests causing an annual U.S. crop and livestock loss in excess of $4 billion. Here are some annual agricultural losses in the U.S.

Insects	$4 billion
Fungal disease	$3 billion
Weeds	$5 billion
Rats, mice and other rodents	$2 billion
Nematodes	$1 billion

Total direct losses run to approximately $15 billion per year or in the neighborhood of one-fourth to one-third of total crop value. To combat these losses, more than 60,000 chemical formulations are currently registered with the U.S. government for sale to the farmer. These represent

Plant pathologist examines apple seedling for scab disease. The healthy plant on the right was treated with a new insecticide which proved highly effective for preventing this disease. (Courtesy E. I. du Pont de Nemours & Co.)

the total of all manufacturers' registered brands and are based on approximately 900 different chemicals.

You might wonder at the staggering losses in spite of all the efforts at control. The answer lies partly in the tremendous reproductive capacity of pests. Curtis Sabrosky, an entomologist with the U.S. Department of Agriculture, estimated that a pair of houseflies reproducing without any controls at all could theoretically produce 191,000,000,000,000,000,000 new flies in a season starting in May and running only until September of the same year (1.91×10^{20}). It is estimated that these could cover the earth to a depth of about 45 feet! Even without such large numbers, flies are of course very formidable germ carriers. Typhoid, dysentery, diarrhea, and other digestive troubles are transmitted by flies who live in filth and contaminate our foods and utensils. (See Figure 11.6.)

Glenn Herrick, a Cornell University entomologist, found that in New York State the cabbage aphid averaged 41 young per female and 16 generations between March 31 and October 2. He calculated that one female could provide in one season a posterity numbering 1,560,000,000,000,000,-000,000,000 (1.56×10^{24}) if all lived. Herrick weighed cabbage aphids and obtained an average weight of about one milligram or 1/1000 of a gram for each aphid. Then he calculated that the theoretical maximum of one female aphid's descendants, if all lived to the end of one season, would weigh more than the total weight of all the people on earth!

Obviously flies have natural enemies besides humans on the end of a fly swatter. Fortunately for us nature exercises considerable control over pests. The tools of nature include predators, parasites, and disease.

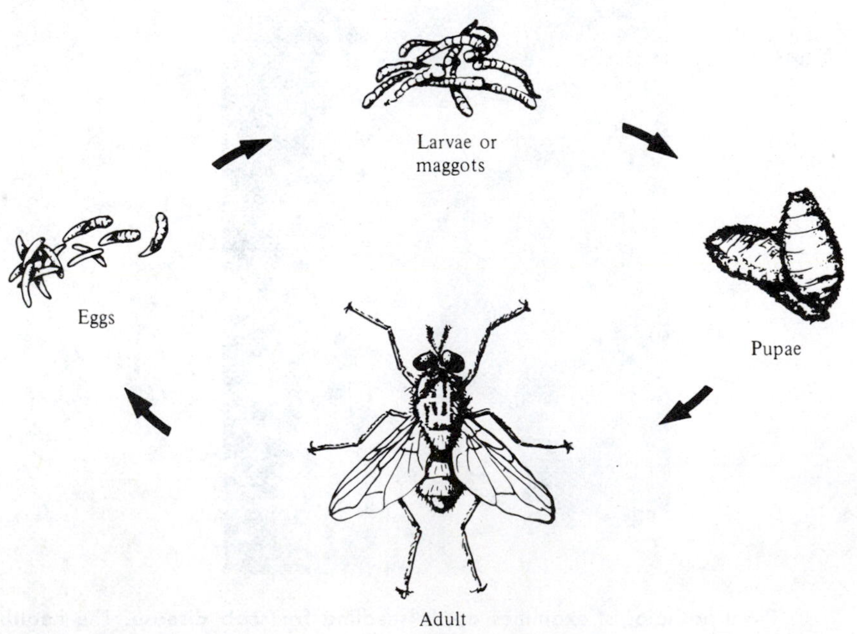

FIGURE 11.6
Stages in life cycle of the house fly.

Casual Insect Control: Pesticides before 1800

Attempts to use chemicals for control of pests goes back thousands of years. The Greek Homer, about 1000 B.C., described the "pest-averting" properties of burning sulfur. This is still used today as a fungicide and in powdered form against certain pests such as red spider mites. The Roman statesman Cato recorded, around 200 B.C., that fumes from burning sulfur and asphalt would kill insects infesting trees. Other written sources indicate that insects were considerably pesty in ancient times. Many odd materials were tried in order to control them. Some of the concoctions were chosen because they were obnoxious to people—the assumption being that they would also be repellent to insect pests—such as camel's urine and burning dung. Unfortunately there was no connection and some of the odd mixtures used were even beneficial or attractive to some types of pests.

There is evidence that the Chinese used arsenic compounds to kill pests in the sixteenth century. Many materials of botanical origin, like tobacco water and certain powdered flowers, were tried and found helpful for some kinds of insect control.

Insecticides From Flowers

The pyrethrins are today very common in many household insecticides. *Pyrethrum* refers both to various chrysanthemums and also to the insecticide made of the powdered flower heads of certain plants of the chrysanthemum family. The word derives from the Greek for fire, probably because of the showy red and lilac colors of many of these flowers. The chemical constituents or active molecular principles of pyrethrum are referred to as pyrethrins.

The use of pyrethrum goes back to the early 1800s when the method of making the powdered insecticide was developed in various Asiatic countries. It was kept a closely guarded secret for many years and flea and louse powders were sold at exorbitant prices. In the mid 1800s the secret was obtained and production spread to Japan, Kenya, Brazil, and the Congo. Finally in the late 1940s chemists identified the active agents in the pyrethrum extracts. One of these is shown for reference in Figure 11.7 to emphasize the extremely complex job involved in the detective work of identification.

One advantage of knowing a formula is the possibility of making compounds of a similar nature. The commercial product called allethrin was made after fifteen years of research by scientists of the U.S. Department of Agriculture. You can see it is very similar to the natural pyrethrin. The pyrethrins are probably the most widely used ingredients in household insecticides. Their special values are very low toxicity to mammals and fast paralytic or "knockdown" action on insects.

You can check a label for household insecticides and will likely find pyrethrins as a major component. The active insecticide ingredients in aerosol spray cans represents only a small fraction, something in the

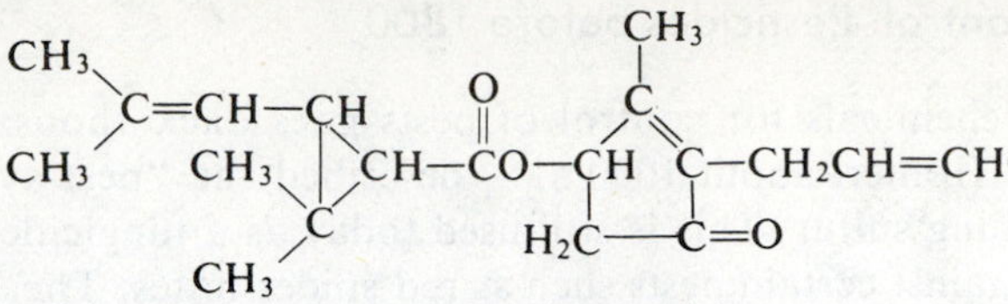

CH_3
CH_3 $\quad$ O
C=CH—CH $\quad$ ‖ $\quad$ C
CH_3 $\quad$ CH_3 CH—CO—CH $\quad$ C—CH_2CH=CHCH=CH_2
C $\quad$ H_2C—C=O
CH_3

One of the pyrethrins

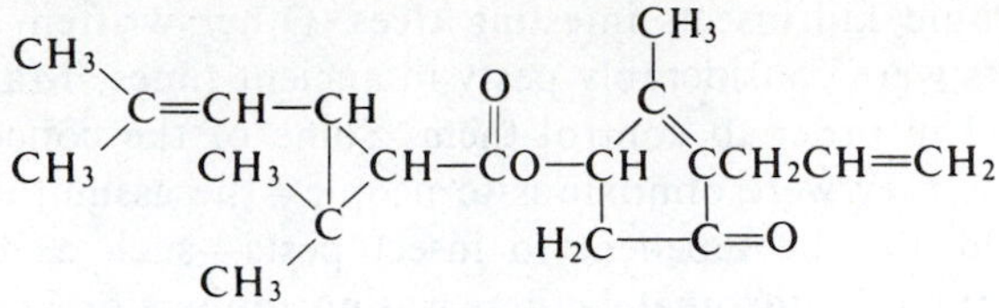

CH_3
CH_3 $\quad$ O
C=CH—CH $\quad$ ‖ $\quad$ C
CH_3 $\quad$ CH_3 CH—CO—CH $\quad$ CCH_2CH=CH_2
C $\quad$ H_2C—C=O
CH_3

Allethrin, a synthetic pyrethrin

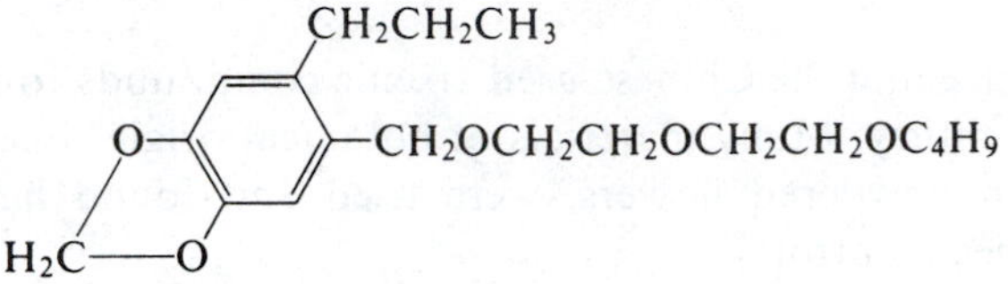

$CH_2CH_2CH_3$
O— $\quad$ —$CH_2OCH_2CH_2OCH_2CH_2OC_4H_9$
H_2C—O

Piperonyl butoxide

FIGURE 11.7
Some chemicals often used in household insecticides.

neighborhood of 2 to 2.5 percent. If you look carefully you will find that pyrethrins represent only a small percentage of that—being present to the extent of only about 0.04 to 0.25 percent. There is usually five to ten times as much of the additive piperonyl butoxide, and also a large proportion of petroleum distillates.

The piperonyl butoxide is one of a large number of chemicals which are called synergists or activators. *Synergism* (Greek: work together) refers to a united action by two or more agents which gives a greater overall effect than the total of the agents acting separately. This happens often with drugs. In the case of insecticides like the pyrethrins, the addition of certain chemicals—which by themselves are not spectacular insect killers—provide a combined toxicity which is amazingly high. This allows much lower percentages of the expensive pyrethrins.

Because of their relative safety, the pyrethrin-based insecticides find wide use in homes against flies, mosquitoes, ants, bed bugs, clothes moths, cockroaches, and silverfish. They are also often used on farms as livestock spray, in flour mills and grain storage warehouses, and as dusts and sprays for control of aphids and chewing insects on truck crops shortly before harvesting.

The First Generation Insecticides

It was in the latter part of the nineteenth century that people began to try special chemicals for insect control. Farmers had long suffered terrific

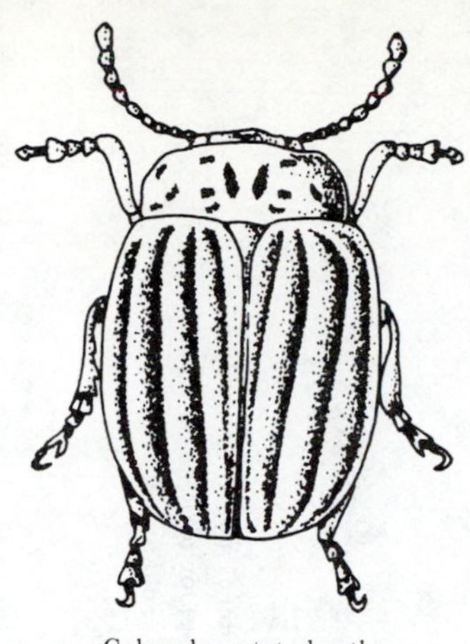

FIGURE 11.8
Colorado potato beetle.

losses to the ravages of insect hordes such that whole crops were often lost. In the 1860s a particularly bad problem arose when U.S. potato fields were infested by an insect that came to be called the Colorado potato beetle (Figure 11.8).

The beetle had been first discovered in 1824 on the eastern slopes of the Rocky Mountains where it fed on weeds. It was not until the 1850s that it surfaced as a dangerous pest. Then the pioneers moving westward after the gold rush brought the beetle a potential new food, the potato. As soon as the cultivation of potatoes began, the beetles switched from weeds to potato plants.

Slowly but relentlessly, the Colorado potato beetles moved eastward. Whole potato crops were destroyed in some areas. The beetle was recorded in Nebraska in 1859, in Illinois in 1864, in Ohio in 1869, and reached the east coast potato fields in 1874.

Spraying was unknown in those days and the losses were enormous. Then about 1865 the compound paris green, a paint pigment, was found to poison the beetles. It is a compound in which the major toxic agent is arsenic. The use of an arsenic compound caused considerable controversy in those early days—some fearing that the potatoes would be poisoned by application to the potato vines. However, the fears did not materialize and the beetle menace was eventually controlled through arsenic poisoning.

The success against the beetle started a rapid development of other inorganic insecticide formulations based on varied compounds of sulfur, copper, zinc, lead, and mercury, for example. In 1892 a compound of lead and arsenic called lead arsenate was developed for use in control of the gypsy moth (Figure 11.9). This compound could be applied in greater concentration than paris green and was found to provide protection against a wide range of insects that chew leaves.

TABLE 11.3

A Few Early (First Generation) Insecticides

Name	*Formula*	*Purpose*
Hydrogen cyanide (Hydrocyanic acid)	HCN	Fumigation of grain, greenhouses, ships, flour mills, warehouses.
Ethylene dichloride	CH_2Cl-CH_2Cl	Fumigation of grains or soils.
Ethylene dibromide	CH_2Br-CH_2Br	Fumigation of grains or soils.
Para-dichlorobenzene	$Cl-C_6H_4-Cl$ (para)	Domestic use against clothes moths and carpet beetles. Not too toxic at low concentrations. Sold as mothballs or moth flakes.
Naphthalene	$C_{10}H_8$ (two fused benzene rings)	Domestic use against clothes moths and carpet beetles. Not too toxic at low concentrations. Sold as mothballs or moth flakes.
Methyl bromide (Bromomethane)	CH_3Br	Fumigation in flour mills, freight cars, ships, warehouses,and residences. Also useful as soil fumigant, for instance in treating nursery stock infested with insects against which quarantines have been set.
Chloropicrin	CCl_3NO_2	Soil fumigation. Was also used in mixture with other gases during World War I where it was known as "vomiting gas."

TABLE 11.3 (continued)

Name	*Formula*	*Purpose*
Nicotine (From tobacco leaves)	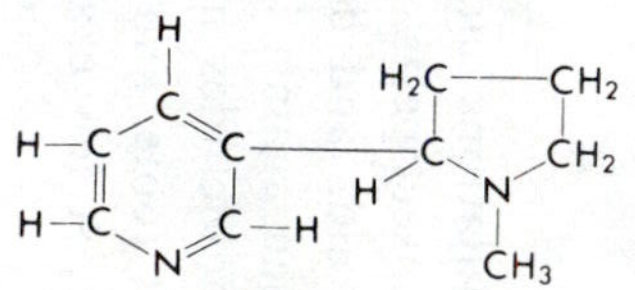 [structural formula: H, C, H—C, C, H—C, N, C—H, H₂C, CH₂, C, H, N, CH₂, CH₃]	Used in liquid formulations and also as a fumigant in greenhouses. First used in 1770 by burning tobacco in a closed space. A very strong poison.
Bordeaux mixture	Mixture of $Cu(OH)_2$ and $CaSO_4$	Fungicide.
Sodium fluoride	NaF	Roach powder.
Paris green (copper acetoarsenite)	$Cu(C_2H_3O_2)_2 \cdot 3Cu(AsO_2)_2$	Paint pigment found very effective against Colorado potato beetle.
Lead arsenate (several compounds go under this name)	$PbHAsO_4$ (acid lead arsenate) $Pb_4(PbOH)(AsO_4)_3$ (basic lead arsenate)	Control of chewing insects.
Rotenone (obtained from roots of certain plants like the derris, an East Indian and African plant of the pea family)	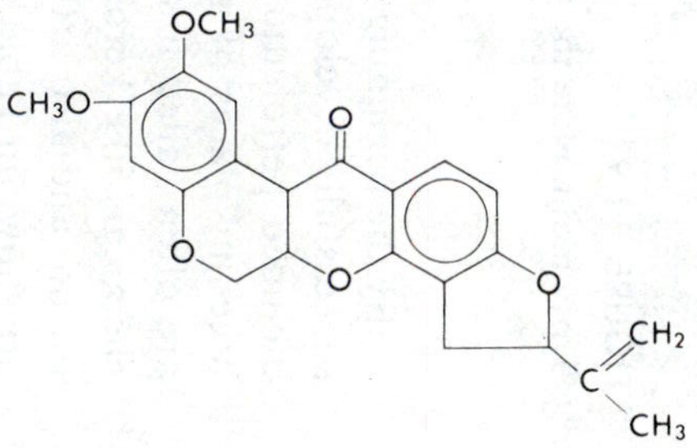 [structural formula: OCH_3, CH_3O, O, O, O, O, CH_2, C, CH_3]	"Natural" insecticide useful against aphids, caterpillars, and fleas and lice on animals.
Pyrethrins (obtained from flowers of various plants of the chrysanthemum family)		Details given in separate listing.

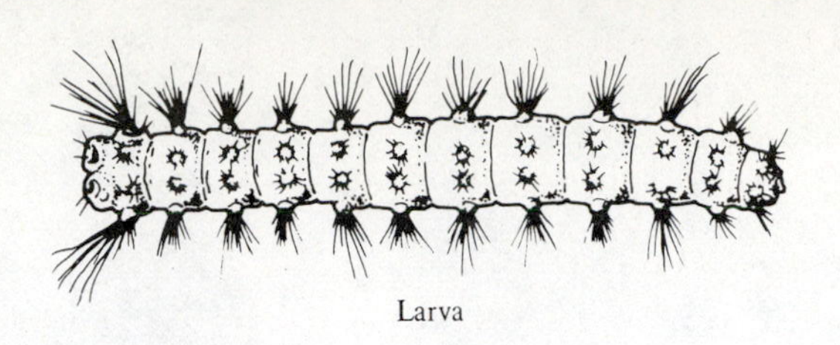

FIGURE 11.9
Gypsy moth. Note the feathered nature of the male's antenna.

By the beginning of the twentieth century many formulations had been successfully developed for use against a whole array of insect pests. These included petroleum fractions like kerosene to coat ponds and destroy breeding places of mosquitoes. The early chemicals used against insects are often called first generation insecticides. The term includes natural chemicals like kerosene, nicotine, rotenone from plant roots, and pyrethrum, and also artificial compounds like lead arsenate. A few examples are shown in Table 11.3.

The Intensified War on Insects: 1942

The second generation insecticides were ushered in with the introduction of DDT in 1942. Its development was spurred by the need during World War II to control insect populations in tropical climates and thus reduce the incidence of diseases like yellow fever, typhus, and malaria. (DDT is short for the generic name *d*ichloro*d*iphenyl*t*richloroethane.) DDT was effective with a large variety of insects and is therefore called a broad spectrum insecticide. The insects very quickly went into what has been called the "DDT's." Sensory organs and nervous system are affected so that insects go into convulsive fits, buzz around aimlessly for a while, and eventually die.

DDT was not the first synthetic organic insecticide but its success was so great as to shift emphasis to organics away from inorganics like lead, arsenic, and copper compounds. Thus a veritable insecticide revolution began, with DDT the forerunner of a whole host of second generation insecticides. Several of the more important ones are shown in Table 11.4.

DDT: Its Dramatic Advantages

The value of DDT lay in both its phenomenal success against so many different kinds of insects and its great residual effect. This means it could kill insects crawling over the residues on previously treated surfaces.

DDT first became known to the general public in February 1944 when the U.S. Army used it to stop an epidemic of typhus fever in Naples. A DDT powder formulation was dusted on a million of the inhabitants to destroy body lice which carry the disease. Within three weeks the typhus was contained. This was an amazing proof of effectiveness never before seen in medical history.

The use of persistent chemicals represented by the chlorinated hydrocarbons has been instrumental in control of many diseases carried by insects such as malaria, viral encephalitis, cholera, Rocky Mountain fever, and tularemia. It is estimated that the lives of at least 10 million people were saved in the years following 1942 when the persistent chlorinated insecticides were extensively used. In addition, hundreds of millions of illnesses were avoided.

DDT: Disadvantages and Danger

Yet the great effectiveness of DDT should have carried a warning. Its very advantages entailed some potential risks. For example, its persistence pointed toward possible problems of buildup in the environment (a major problem which is considered in chapter 15). In addition, other difficulties arose relating to its wide-range effectiveness. The continual use of a broad spectrum insecticide like DDT may result in killing off of large numbers of beneficial insect predators and parasites. A twofold effect may follow. (1) A new pest which was not formerly doing any damage and which is not affected by the insecticide may begin to build up to dangerous and damaging levels. This is called a secondary outbreak. (2) The original target species of insect may begin to develop resistance or immunity to the insecticide by genetic changes over many generations. Then released from natural predator and parasite control, there may be a resurgence of infestation which will not be controllable easily by the original insecticide.

Such a resurgence can lead to a grower using excessive amounts of the insecticide to which the pest has developed resistance. This in turn can contribute to a vicious circle of more environmental pollution and more killing of beneficial insects. A better course in recent years has been to

TABLE 11.4
A Few Chlorinated Hydrocarbons of the Second Generation Insecticides

Name	*Formula*
DDT	
Lindane (benzene hexachloride)	
Chlordane	
Aldrin	
Dieldrin (pronounced deel-drin)	

switch to another insecticide to which the target insect is not resistant. But this too can lead to dangerous side effects if not carefully controlled. DDT is now banned in the United States but it is still used in some parts of the world.

Malaria, Mosquitoes, DDT, and Defeat

Malaria is still one of the most serious world diseases in terms of numbers of people affected. It was also, prior to the 1940s, a very prevalent disease in the U.S., especially in the southern states. The following table tells an interesting story of how this mosquito-borne disease declined in the U.S.

Year	*Reported Cases of Malaria*	*Deaths from Malaria*
1940	78,129	1,442
1945	62,763	443
1950	2,184	76
1955	522	18
1960	72	7

The 1940–1960 period saw an extensive program of DDT use by spraying in houses along with antilarval drainage operations and enforcement of laws designed to eliminate breeding sites. During the peak years of the campaign over 1,300,000 homes were treated annually with a total yearly DDT usage greater than 650 tons. This amounted to more than a pound per house.

The anticlimax is, however, of considerable concern to public health people. The Anopheles mosquito, which is the malaria carrier, has now become resistant to DDT. Also the malaria protozoon which the mosquito transmits has developed resistance to some of the modern antimalarial drugs, thus necessitating a continuing search for replacements.

Organic Phosphate Insecticides

Another group of insecticides in the second generation are the organophosphorus compounds. (See Figure 11.10.) These contain a phosphate or related grouping in the molecules and are sometimes called organic phosphates since they are small carbon chain compounds with the phosphate as part of the molecular structure. Many thousands of candidates for insecticides were investigated and only a few survived the hurdles, including studies of toxicity to humans and persistence of residue on plants. Some of the organophosphorus compounds are extremely toxic and include nerve gases; these have been eliminated as insecticides. But special care still must be exercised. The organophosphate insecticides have caused fatalities by unsafe handling procedures or premature re-entry of workers into sprayed fields. Most of the organic phosphate insecticides are not extremely persistent like the chlorinated hydrocarbons.

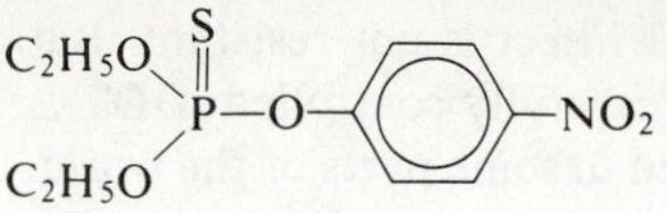

$(C_2H_5O)_2P(=S)\text{—O—}C_6H_4\text{—}NO_2$

Parathion

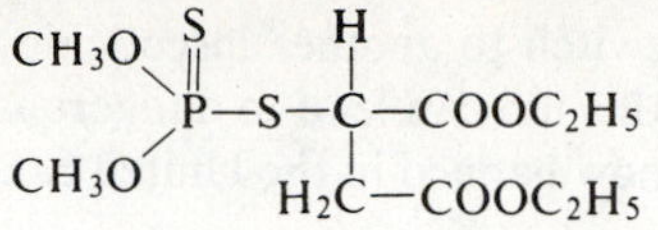

$(CH_3O)_2P(=S)\text{—S—}CH(COOC_2H_5)\text{—}CH_2COOC_2H_5$

Malathion

$(CH_3O)_2P(=S)\text{—O—}C_6H_4\text{—}NO_2$

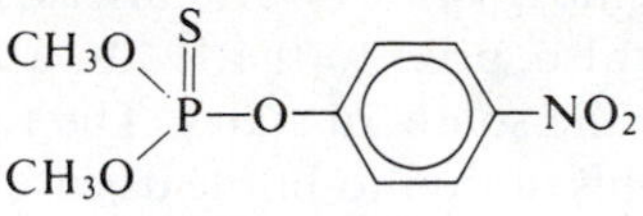

Methyl parathion

FIGURE 11.10
A few organic phosphate insecticides.

One of the interesting features of some of the organophosphorus insecticides is their systemic activity. *Systemic insecticides* are those which penetrate the plant and move within the circulatory system so that insects are killed by feeding. Some inorganic systemics were known since 1936 when soil high in content of the element selenium had been observed to kill aphids on the wheat grown in such soil. Following this observation, selenium compounds were developed as systemics. Usage was limited to ornamental flowers like carnations because of residual toxicity to animals if used in food crops.

The application of systemics may be through seeds, leaves, bark, trunk, roots, or stems. The insecticide then travels to all parts of the plant. Selective use of organic phosphates makes them applicable to many varieties of crops—especially as seed treatment prior to planting. This includes crops such as wheat, cotton, alfalfa, vegetables, peanuts, and potatoes.

A Possibility for Third Generation Insecticides

An exciting recent concept for insect control is "juvenile hormone." This is a chemical secreted by an insect at certain stages of its life cycle. The name comes from experimental evidence that it suppresses adult characteristics and promotes juvenile stages of insect growth. Final metamorphosis to the adult stage occurs in the absence of juvenile hormone (abbreviated JH in the following discussion).

The hormone is secreted by a special gland at certain stages to control the extreme changes an insect goes through from the crawling, wormlike larva (caterpillar) to the quite different structure of the adult. If the hormone is provided at stages where it should be absent, then abnormal development can be forced, or the treated specimens can be made to grow too fast or too slowly so as to emerge at the wrong season or to die prematurely. Research has been concentrating on questions like the following: How can JH be extracted and identified? What is the molecular structure of JH? How can synthetic JH be made which is selective to only detrimental insect pests? Could other chemicals similar to JH ("hormone mimics") be made which would be more stable and practical for insect control?

In 1966, researchers isolated JH from certain types of moths and identified the active component. Since that time hundreds of hormone mimics have been synthesized and screened. Several are under active testing today. One is being tested especially for mosquito control since many mosquitoes have developed resistance to standard insecticides. (See Figure 11.11.)

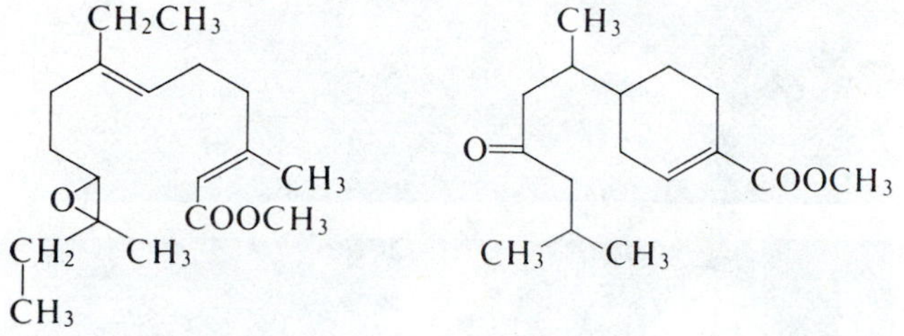

CH_2CH_3 CH_3 O $COOCH_3$ CH_2 CH_3 CH_3

Juvenile hormone from certain type of moth

CH_3 O $COOCH_3$ CH_3 CH_3

Compound with JH activity isolated from balsam fir

(The carbon and hydrogen atoms in the chains are not shown.)

FIGURE 11.11
A few chemicals isolated in juvenile hormone research.

Much additional work has to be done to find out more fully the potential and safety of the JH-type compounds, including studies of toxicity, carcinogenicity (cancer-forming capability), and teratogenicity (malformation of organisms).

One interesting finding in JH research was that a Canada balsam fir tree produced a compound which has a strong JH effect on a certain type of insect, the linden bug. It is thought that the balsam fir developed the particular compound over many years as defense against insects. Consequently, some scientists feel that analysis of the complex mixtures of chemicals made by trees and plants may provide a source for materials to counteract certain insect pests.

So far JH has been found to have no effect on forms of life other than insects. In addition, since it is secreted as a control chemical by insects themselves, some people have suggested that insects will hardly be able to develop resistance to it without committing suicide. Other scientists, while optimistic about the immediate possibilities of JH, are less optimistic that it represents the last word in control since insects at some stages of their life normally inactivate or cut down on JH in order to proceed to the adult stage.

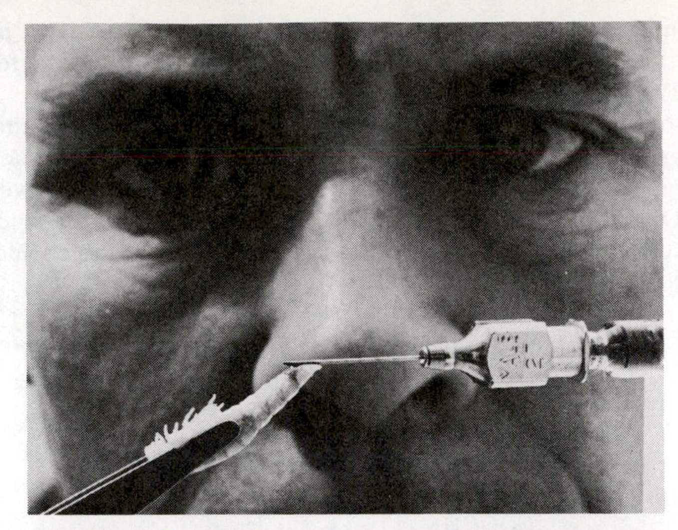

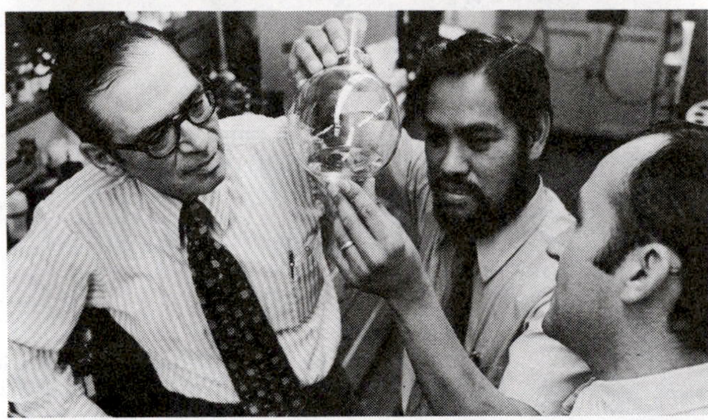

At top, USDA entomologist Robert Redfern applies one-tenth of a billionth of a gram of a new JH mimic to the abdominal end of a yellow mealworm pupa. In middle photo, a fraction of a new hormone mimic (JH-25) is discussed by the chemists who developed it, Morton Beroza, Rafael Sarmiento, and Terrence McGovern. Bottom photo shows normal pupa of the yellow mealworm at right and a normal adult on the left. In the middle is a JH-25-affected "monster" adult where the head and thorax are those of an adult but the abdominal end retains the pupal appearance. The half-pupa, half-adult died without feeding or reproducing. (USDA photos.)

It is difficult to draw a definite line between chemical control measures for pests and what are called "biological" methods. Currently much emphasis is being directed toward what is predominantly a biological approach to insect control. In many cases this is used in combination with chemicals or some form of chemical alteration. Four areas presently receiving attention will be considered briefly below: (1) growing and dissemination of enemies of harmful insects, (2) propagation of insect diseases, (3) sterilization of insects, and (4) sex attractants.

Lions and Tigers of the Insect World

The use of insect eaters—predators and parasites—encourages a basic part of nature in the insect world—the continual wars and fights-to-the-death going on among the billions of insects on the earth. There is no doubt that this continual conflict is much more effective in the way of keeping insect populations under control than any of the synthetic approaches we have so far invented.

You are familiar with a few common beneficial insects. An example is the dragonfly (or "darning needle") which is considered an excellent balancing agent in ponds and lakes. This is a beautiful insect as you know if you have ever seen it darting in daredevil aviation fashion over lakes and streams. Dragonflies can achieve speeds of 60 mph and catch and eat prey, like flies, mosquitoes and wasps, while on the wing. They are especially noted for having a terrific appetite for mosquitoes, both larvae and adults. Another example is the lady beetle (also called a lady bug or lady bird) which most people easily recognize and know as a great benefactor because of its ravenous appetite for scale insects and aphids. (See Figure 11.12.)

Use of predators and parasites, the "lions and tigers" which devour myriads of insects, depends on detailed study of the habits of various insect species, identification of the value of beneficial ones, and encouragement by breeding and release of those which contribute to destruction of the pests. This approach has been highly successful in certain instances and it is a very specialized field of study for biologists and entomologists.

The classical case for successful biological control involves the so-called vedalia beetle which was imported into the U.S. from Australia to control the cottony-cushion scale, a small insect that feeds on sap of leaves and twigs of citrus trees. This pest was first discovered in California in 1872. Within fifteen years it had spread throughout the citrus-growing area. Many orchards had their complete fruit crop wiped out and many trees were killed. The situation was desperate when entomology people from the U.S. Department of Agriculture determined that the scale had originated in Australia where it was apparently kept under control by some native predator. A USDA entomologist went to Australia where he

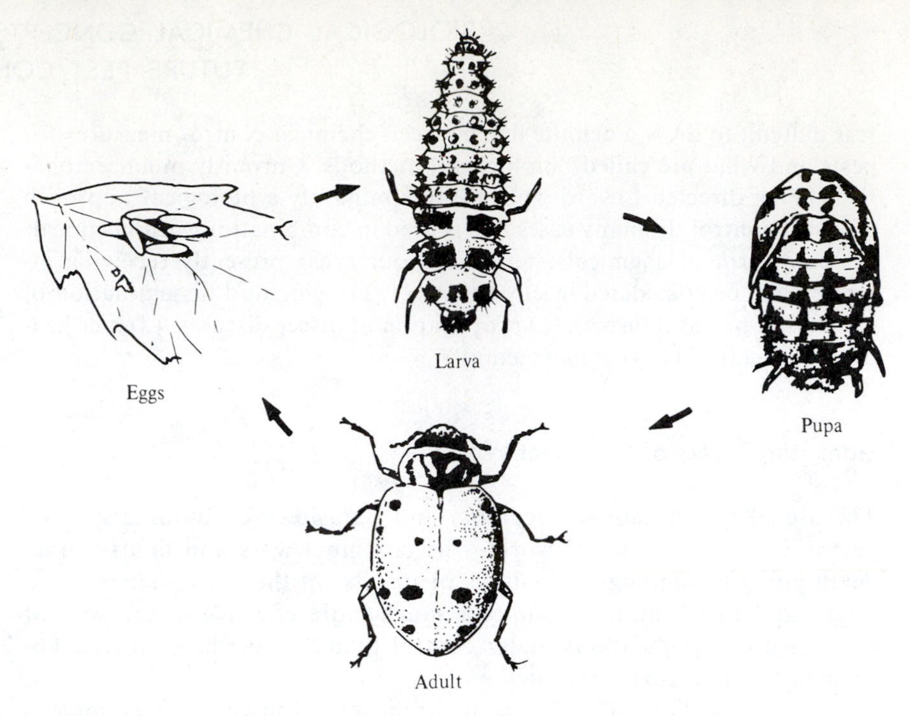

FIGURE 11.12
Stages in life cycle of a lady beetle. Both adults and larvae feed on aphids.

was able to find a parasite fly and also the vedalia beetle which fed on the eggs and larvae of the scale. He shipped specimens of both back to the U.S. where the beetles proved to be the major control mechanism. A total of only 514 beetles were shipped but they thrived on the heavily infested orange trees. Within two years the cottony-cushion scale was under complete control throughout the citrus-growing area. The scale has never since been a problem.

The vedalias are still grown in certain insectaries for release when the occasion demands. They have also been successfully used in Hawaii and New Zealand.

The success of the beetle imports from Australia spurred further efforts in this kind of biological control. More than 600 species of insect parasites and predators have now been used in the U.S. alone. More than one hundred are now established and it is claimed that they control over 20 of our most serious insect pests. Many insectaries are in the business of breeding beneficial insects which cannot be easily established. These are released in fields and forests for seasonal outbreaks of pests.

Insect Disease Can Help

Ever since it was discovered that insects, like humans, are susceptible to various diseases, the possibilities of using this knowledge as a control method has excited entomologists. The infectious agents that cause insect diseases are of the same major types as those that cause disease in other animals—bacteria, viruses, fungi, and protozoa. However, most of the microorganisms that result in insect diseases are harmless to plants and other animals. Bacteria, viruses, and fungi have been used successfully against insects. In some cases they are applied manually as a powder. Aircraft are also used to spray agents of insect disease.

Sterility Stops the Screw-Worm Fly

It is possible to sterilize insects by means of X-rays or other forms of high energy radiation. This method eliminated screw-worm flies on the island of Curacao off the coast of South America and is also used in the southeast United States. This fly was the cause of extreme misery and millions of dollars of damage by attacks on horses, cattle, hogs, and goats.

The adult fly is about twice as large as a house fly and lays eggs on the skin of animals near a wound or infected body opening caused by bites of other insects. The eggs hatch in about 10 to 20 hours and the small maggots or larvae tear their way into the flesh. In the process they secrete a toxin which prevents healing, thus contributing to reinfestation by other flies. The sickened animal usually dies if not treated. The larvae become full grown in from 4 to 10 days when they drop to the ground and remain there as pupae for about 3 to 14 days. Then they become flying adults ready to spread the infestation. A generation averages only 3 weeks, and there may be 8 to 10 generations in one summer.

The screw worm is controlled by raising millions of flies, sterilizing the males by radiation, and then releasing them in the area of infestation. Since the female mates only once, the large number of sterile insects preempt the field and prevent reproduction.

The sterilization method involves problems including the expense of breeding and sterilizing so many insects and the isolation of a treated area from invading, nonsterile insects.

Sex Attractants

This is an extremely interesting subject both to the entomologist and also to the insects. Certain animals like insects produce chemical odors which have an effect on others. The word *pheromone* refers to any chemical secreted by an animal to influence the behavior of other animals of the same species. Interesting studies have indicated that chemical secretions are involved in insect behavior in trail setting, alarm signals, gathering of workers, and sexual attraction. The latter pheromones are called sex attractants.

Individual sex attractants for quite a few species of insects have now been isolated, identified conceptually by molecular structure, and synthesized. Examples include gypsy moth, silkworm moth, pink bollworm moth, cabbage looper, army worm, American cockroach, black carpet beetle, and boll weevil.

This work involves much patience and the most sophisticated chemical tools of analysis. For example, when the sex attractant of the female gypsy moth was identified, researchers at the U.S. Department of Agriculture had to use over 500,000 female moths to obtain a minute sample of about 20 milligrams of the sex attractant (1 milligram being 1/1000th of a gram or 1/5000ths of the weight of a nickel). In other words, each moth provided less than a millionth of its body weight of the active sex attractant, named "gyplure." Once it was identified specifically as

```
   H H H H H H H H H H H H H H H H
   | | | | | | | | | | | | | | | |
H—C—C—C—C—C—C—C—C—C=C—C—C—C—C—C—C—OH
   | | | | | | | |     | | | | | |
   H H H H H H O H     H H H H H H
               |
             O=C
               |
             H—C—H
               |
               H
```

Gyplure

synthetic material could be made. This was tested and showed the same attracting power as the natural material.

The sex attractant chemicals are astounding in their physiological potency. It has been estimated that the amount of attractant produced by one female gypsy moth would theoretically be capable of exciting approximately a billion male moths if spread with maximum efficiency, which fortunately for us it never is. It is also effective more than two miles downwind. The sex attractants are especially valuable as controls because it is not likely that insects will develop a resistance to them.

Evaluating Sex Attraction

One of the difficulties involved in use of attractants is evaluation of the effectiveness of various natural and synthetic chemicals. A tentative beginning was made in this direction during work with the sex attractant of the female silkworm moth. A minute sample of only 5 milligrams was isolated from the abdominal tips of more than 313,000 moths. This pure attractant was then evaluated by a bioassay method, that is, using living insects for measurement of potency. Samples were made more and more dilute until the smallest amount necessary to affect 10 out of 20 males was determined. This minimum quantity was called the sex attractant unit. The procedure involved holding a glass rod dipped in the dilute solution for one second before the antenna of male silkworm moths. A positive reaction is a male fluttering its wings in a characteristic manner. The

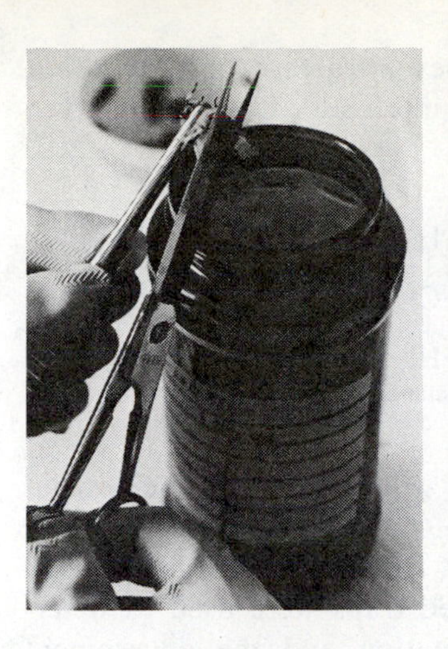

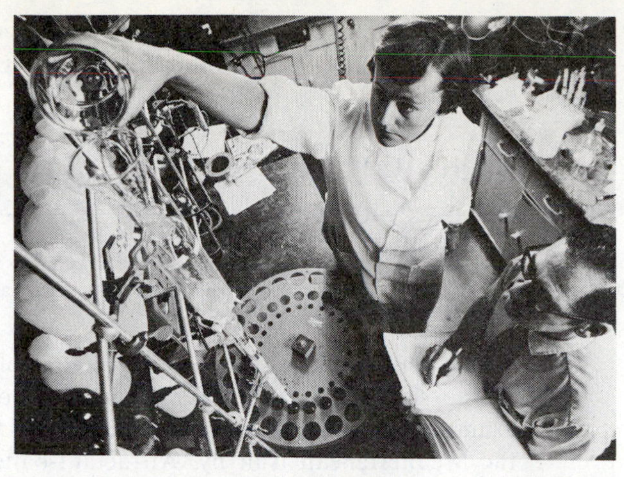

At left, tail sections are being clipped from female adult gypsy moths by the researcher to make extracts of the natual sex lure. At right, Morton Beroza and Barbara Bierl, chemists with the USDA Agricultural Research Service, conduct chromatographic separation of an active gypsy moth sex attractant, a step in the development of the synthetic lure for trapping the gypsy moths. (USDA photos by Larry Rana.)

At left, Daniel Cook, USDA researcher, sets up rack for testing synthetic gypsy moth sex attractants. At right, male gypsy moth swings and writhes when exposed to synthetic female sex attractant. (USDA photos by Larry Rana.)

silkworm moth extract had an estimated sex attractant unit of about 0.0000005 grams. The formula for silkworm moth sex attractant is given below.

```
   H  H  H  H  H  H  H  H  H  H  H  H  H  H  H  H
   |  |  |  |  |  |  |  |  |  |  |  |  |  |  |  |
H—C—C—C—C=C—C=C—C—C—C—C—C—C—C—C—C—OH
   |  |  |              |  |  |  |  |  |  |  |  |
   H  H  H              H  H  H  H  H  H  H  H  H
```

Silkworm moth sex attractant

It is unlikely that this one type of control will replace insecticides but the possibilities as a complementary control measure are exciting. For example, in 1956 a million acres of Florida fruit trees were infested by the Mediterranean fruit fly. Attractant-baited traps were used to determine the location and density of the infestation and the government people were able to eradicate the flies by spraying with an insecticide-bait formulation.

Other potential uses of sex attractants are the following:

1. Wider use of insecticides in traps baited with sex attractants may cut down on the use of insecticides over large crop areas, thus reducing possible pollution and the killing of large numbers of beneficial insects. This is possible because sex attractants will lure only members of a particular species.
2. Possible use of chemical sterilants in combination with sex attractants so as to sterilize insects in the field and thus eliminate the very high cost of breeding, sterilizing, and distributing sterile insects over the areas infested.
3. Use of what is called sex pheromone masking. This consists of permeating the atmosphere with the sex attractant chemical and confusing the insects so that they cannot find each other.
4. Luring females to radiation-sterilized males or males to sterilized females.

WHAT IS THE ANSWER TO PESTS?

You can see that there are more possibilities for pest control than using chemicals to kill the offending pest. If we took the extreme measure of banning all pesticides as an answer to the problems of pesticide use, we would inevitably have famine and widespread disease. And we cannot adopt the other extreme of uncontrolled use and continued multiplication in quantities applied. Scientists today almost unanimously agree that an integrated or whole view needs to be taken of the pest problem. In the past, because of lack of knowledge and various urgencies, procedures have been used without looking at the complex web of interrelations involved.

The newest approach to pest control has been called integrated control or pest management. Integrated pest control combines and correlates all the concepts we have developed over the years. It puts the whole gamut of control procedures—biological, chemical, and mechanical—together into an integrated system. It recognizes the complex interrelationships among pests, beneficial organisms, people, and the environment.

Integrated control naturally requires a very sophisticated knowledge of all the factors involved. Pest managers are specialists in a sense but also require broad background in many fields, such as pest biology, ecology, crop culture, chemistry, toxicology, and residue analysis. Many states are now licensing pest control advisors after certifying their expertise through examinations.

The newer approach to pest control requires expanded research into all aspects of the environment—into new biological, chemical and crop culture methods; and into developing more integrated conceptual and operational understanding of agricultural ecosystems, particularly the complex interactions of pests, parasites, predators, host plants, climate, and chemical fertilizer treatment. There should incidentally be exciting employment opportunities for a new type of farmer's aide—the pest management specialist.

Weed Control Chemicals: Herbicides

Weeds are one of the most expensive farm "pests." Consequently, agricultural researchers have been interested in developing better weed control methods than mowing, tillage, burning, and crop rotation.

In the early days farmers used salt and ashes to kill all vegetation along roadsides. Some farmer noticed that the copper salts used on grain to control fungus diseases also killed certain broad-leaved weeds. This was the beginning of selective weed control which began to develop in the 1940s. Then USDA scientists discovered that an organic chemical, 2,4-D, could kill weeds systemically. The inorganic compounds previously used, like copper, iron, and arsenic compounds, killed the weeds by scorching the parts they touched.

Following the success of this systemic herbicide, hundreds of other organics were evaluated for effectiveness and safety. Now dozens of herbicides are available commercially to the farmer and home gardener. (See Figure 11.13.)

Herbicides have had a great influence on farm practices. Here are a

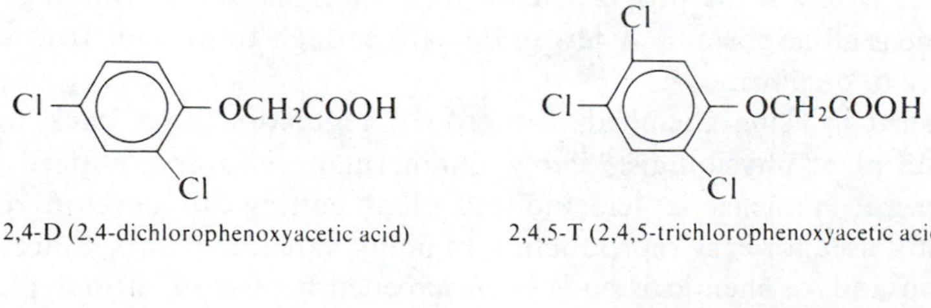

Cl — (benzene ring) — OCH_2COOH, with Cl

2,4-D (2,4-dichlorophenoxyacetic acid)

Cl — (benzene ring) — OCH_2COOH, with Cl, Cl

2,4,5-T (2,4,5-trichlorophenoxyacetic acid)

FIGURE 11.13
Examples of chemicals for weed control.

few of the advantages which have resulted from use of chemical weed killers:

1. Substantial reduction in manpower for producing corn, cotton, grapes, citrus, vegetables, and seed crops. In cereal grain crops which are sown broadcast, herbicides are often the only method of killing weeds in the growing crop.
2. Large increases in crop yields. For example a cornfield treated with a weed killer at planting time yielded 106 bushels per acre compared with a control field which was not treated and yielded only 45 bushels per acre.
3. Elimination of the problem of weed seeds being harvested along with grain.
4. Improvement of the efficiency of mechanical harvesting, cutting down on wasted manpower, and thus providing savings in food costs.
5. Improvement of the operation of waterways for irrigation and drainage.

The safety features of weed killers have apparently been less troublesome than those associated with other pesticides. Some herbicides break down in the soil and become harmless before crop seeds are planted; some persist for a long time and are used in nonagriculture areas. When correctly used they leave no residues in plants, soils, or water which might be harmful to people or other animals. They were extensively used in Vietnam as defoliants to open jungle trails and reduce problems with sniper and guerilla hideouts.

Plant Growth Regulators

Certain chemicals have been found to stimulate or retard the growth of plants, or of various plant parts. It is interesting that the use of 2,4-D as a weed killer developed out of studies on how it regulates growth. It can be used to improve "fruit-set" on citrus trees and to retard the drop of Winesap apples. Here it is used in very small quantities.

In fruit growing areas thousands of acres of apples and pears are sprayed from the air with "stop-drop" chemicals, often reducing losses from 20 percent to less than 6 percent. Because the chemicals practically eliminate premature fruit drop, they can be abused. If growers delay harvesting too long, fruit can become overripe on the tree. Also apples sprayed improperly may cling so tightly that they may be damaged on picking. The use of the proper chemical in the right concentration at the right time is all important. A few grams are enough to prevent fruit drop in an acre of orchard.

The start in using chemicals for growth regulation goes back to an American plant physiologist, Percy Zimmerman, who first noticed that the chemical indoleacetic acid induced plant cuttings to develop roots. This is still used for easy reproduction of many valuable plants. Since that time thousands of chemicals have been screened for use in various phases of plant regulation. (See Table 11.5.) Here are a few of the results:

1. The buttons on the end of lemons can be made to remain attached, thus preventing rotting by disease or fungi during storage.
2. Chemicals can be used along highways as a "chemical lawn mower" because they retard grass growth without killing it.
3. A special group of growth regulators is called the "gibberellins," derivatives of gibberellic acid. Some of these compounds cause stem elongation, while others promote early flowering, change branching, and set the fruit. Thompson seedless grapes, for example, exhibit larger fruit and substantially greater yields on elongated stems.
4. Small amounts of 2,4,5-T, which was mentioned previously as an effective weed killer, can be applied to cut plant stems causing them to form roots readily.
5. Chemicals that retard growth can be used for dwarfing plants like azalea and chrysanthemum so that they take up less space and require less pruning.
6. Chemicals that remove leaves can aid in harvesting. For example, defoliation before mechanical picking keeps leaves out of harvested cotton. Defoliation also lowers insect population, reduces deterioration of fiber and seed, and speeds up harvesting.

TABLE 11.5

Some Conceptual Description of Some Chemicals Used for Control of Plant Growth

Name	*Formula*
Indoleacetic acid	$-CH_2COOH$; N; H
Naphthaleneacetic acid	CH_2COOH
Gibberellic acid	O; OH; OH; C; CH_3; O; COOH; CH_2
2, 4-D	Cl—; OCH_2COOH; Cl

SAFETY PRECAUTIONS AND LEGAL SAFEGUARDS

Safety testing of agricultural chemicals to make certain that they can be used without hazard to health is a continuing process. There has now been developed a fairly comprehensive system of registration and controls.

Agricultural chemicals must be registered with the USDA in order to be shipped legally in interstate commerce. The manufacturer has to provide results of tests on the effectiveness of the chemical for the particular purpose for which it is being sold and also toxicity tests on mammals. The USDA scientists must be satisfied that the particular material is not hazardous when applied for the particular purpose and according to the label instructions provided. Then the USDA grants the registration.

If the intention is for use on food crops, however, there is a special additional requirement for experimental test data on the amount and nature of residues. If the tests show conclusively that there is no residue on crops, then the registration is given USDA approval. If there is a residue, special clearance has to be obtained from the Food and Drug Administration of the U.S. Department of Health, Education, and Welfare.

The FDA is responsible for determining and enforcing tolerances or the permitted concentration of a chemical on foods. Here the manufacturer has to submit to FDA scientists reports on extensive tests of acute and chronic toxicity using several species of test animals. Tolerances are usually set at 1 percent of the lowest level at which harmful effects were observed on the most sensitive species of test animal. If a chemical causes cancer in test animals, regardless of the amount used, then a zero tolerance is required by the 1958 amendment to the Food, Drug, and Cosmetic Act. The FDA inspectors regularly examine samples of food in interstate and overseas commerce, and they may seize shipments found to contain chemicals in excess of tolerances.

There are no recorded instances of seizure of crops following use of pesticides in accordance with recommendations on the label. In the case of the "cranberry scare" mentioned in the last chapter there was gross disregard for the timing of the chemical application.

TECHNICAL PROGRESS MAY BRING BROADER PROBLEMS

The application of modern science and technology to agriculture in the developing nations of the world has in recent years produced an enormous increase in crop yields. This has been called the Green Revolution. Here the rapid spread of modern agricultural techniques, particularly in Asia, has provided unparalleled expansion especially in food grains like wheat, rice, and corn, doubling and sometimes tripling production. The progress has been achieved by use of fertilizer, pesticides, irrigation, modern equipment, and new high-yield varieties of grain. Traditional food importers like the Phillipines and Pakistan have become self-sufficient and now export some foods.

The spectacular success has paradoxically brought many new problems in areas other than science: these include education, political science, economics, morality, sociology, psychology, and international relations. We cannot consider these in detail but the situation is worth mentioning because it shows the complexity of side issues which are necessarily involved when science is applied to problem-solving. A few of the potential difficulties generated by the Green Revolution are the following:

1. The new seed varieties are so attractive to farmers because of high yield that they may be adopted on larger and larger tracts of land. If they then should become susceptible to disease there is the likelihood of massive crop losses. This has happened in some cases. Previously, small farms that planted a variety of strains had built-in protection since not all varieties are equally susceptible. This is similar to the problem we ran into in the U.S. when large specialized farming replaced the small, all-crop farm. That is when we began to need pesticides on a large scale.
2. The agricultural credit arrangements in developing countries have to be developed to help the small farmer. The initial investment in the new technology can be up to ten times that required for traditional methods. The yield is so great that the net return is also considerably increased but this is not available to the small-tract farmer unless he can get started. Consequently, in some areas the Green Revolution has greatly accentuated social problems. Large farmers tend to get richer and the individual small farmer is displaced.
3. There are tremendous dislocations in storage and marketing because of increased output. The small peasant farmer had consumed up to 80 percent of his crop under traditional methods. Doubling his yield does not lead to doubling his consumption and hence transportation, storage, and marketing now become problems. In one area of India a very large increase in crop yield resulted in makeshift storage in schools and open fields where much of the advantage gained by technology was lost due to deterioration and attack by pests.
4. When countries which formerly were importers of food become self-sufficient, what is the impact on the former food exporters? The whole system of economic interdependence is changed.
5. Great changes are required in educational practices in order to provide properly trained manpower. Farmers have to develop expertise in crop timing, fertilizing, watering, and use of pesticides. Also preparations have to be made for absorption of surplus manpower released from agriculture which involves complex educational changes, possibly increased industrialization. Political consequences here are also considerable.

These are just some of the possibilities which may need to be considered as the developing nations spread the Green Revolution. It has already lifted the spectre of famine in Asia and has begun to lift the burdens of rural poverty in many areas. However, it has been so spectacularly successful that many have been caught unawares. There is no doubt that the many problems it has raised can be solved. And the solutions will probably come by similar conceptual approaches that went into much of the progress in the first place.

THE FUTURE

From the brief coverage in this chapter you can see how the conceptual approach—even if not called by that name—has significantly contributed to the progress of agriculture. Sometimes primitive, sometimes refined—it is behind all the efforts to keep up with rapidly expanding world food demands.

The case for agricultural chemicals and agricultural research in keeping a balance between growing population and potential starvation cannot be overstated. Farmers are in the vitally important business of providing foods. Chemistry or conceptual understanding of what foods are, how they are formed, and how they are most efficiently produced will play a large part in solving the agricultural problems of the future. Here are a few examples of new approaches to food production:

1. Changes of eating habits in the developed countries to substitute plant protein for animal protein. A basic limiting factor for animal protein is of course the protein from plants grown on grazing land. But land is limited, and present trends of land use reflecting increased urbanization and leisure seem to indicate further limits to the supply of feed grains. In addition, the efficiency of protein production through animals is considerably lower than the potential via plants. The changes here will probably be gradual and will obviously have to involve thorough conceptual analysis of amino acid needs and supply sources. Mention is made in chapter 10 of synthetic meats from soybean protein and the improvement of plant protein by supplementation with needed amino acids. Other possibilities lie in breeding plant hybrids providing higher protein yields.
2. Further development of microorganisms or microbes as sources of food. Microorganisms are plants and animals so small that they can be seen only through a microscope. Important examples of these organisms are bacteria, fungi (mushroom), yeasts, and protozoa. We have already considerable practical experience in limited food usage of yeasts and mushrooms. Yeasts have been used by mankind for centuries in bread- and winemaking. In recent years yeasts have been investigated as direct food sources.

   Certain yeasts high in protein and vitamins were developed for food by the Germans during World War II. The yeasts were grown on hardwood waste residues from wood pulp production, thus serving a double purpose. Successful growth of various yeasts has been achieved not only with the sugars contained in wood pulp waste but also with residues from various fermentation processes like beer and cheese manufacture, and also from grain, cane sugar, and molasses. Yeast is also high in the minerals phosphorus, potassium, and calcium. By analysis and blending with other sources of amino acids in which yeast is deficient, a more balanced product is obtainable. Potential additive uses include products such as flour and breads, baby foods, candy, cheese, soups, and cereals.

   The advantages of microorganisms include very rapid growth and much simpler requirements for microbe food sources. A chicken doubles its weight in about a month whereas a yeast cell doubles in about two hours.

Yeasts and other fungi are also very fast in comparison with plants in converting nutrients into cell structures, as you know if you have ever seen a mushroom form overnight in grass or moist forest land. The logistics problem of feeding microbe crops would require whole new technologies. But if we could use sewage waste as food for growing microbe crops, two problems might be solved.

3. Harvesting insects for food. Peoples in various parts of the world have long used insects as a source of food, including toasted termites, roasted and ground red ants, roasted grasshoppers, fried caterpillars, dried flies, and fried bees and beetles. These are not standard agricultural products yet. But some people have suggested that we have not given enough thought to insects as foods. It would certainly solve some of the insecticide problems and might even serve as a possibility for upgrading sewage.
4. Cultivation of the oceans as sources of food. This sounds exciting, and it is, but it will not be easy. It will involve complex studies to keep ecological balances and prevent potential problems like pollution, depletion of oxygen, overharvesting without conservation, and lack of international cooperation.
5. New and improved methods of storage and handling for foods. This would cut down on the tremendous waste by deterioration which negates much of the effort expended at growing the crops in the first place.
6. A more perfect union between nutrition and agriculture. Our progress over the years indicates a trend toward a system where enough knowledge may be developed to provide a balance of all the chemical raw materials for crops and livestock with the end products they are capable of supplying. This would involve a large-scale conceptual coordination of the natural needs for food and health of all the living things in the ecosystem: people, plants, animals, and microorganisms.

PROJECTS AND EXERCISES

Experiments

1. This is a thought experiment. Even though insects have been attacking crops since man first began cultivation, it is only within the last 100 or 125 years that pests have become such a critical problem. Before that time agriculture was based on small units where great varieties of crops were grown. By the mid-nineteenth century, larger and larger areas were devoted to growing a single crop. (For example, note the terms corn belt, cotton belt, bread basket for wheat growing.) Think about this and imagine an experiment where you had a farm in 1850 and started to add land and convert from growing many crops to growing more and more of the same crop, eventually specializing in a single crop grown on a very large scale. What pest problems would you likely run into? Would they change from those you had previously when you grew many varieties of produce? Would rotation of crops help control pests? Would it be practical for you now to go back to the small farm with a variety of crops? Give reasons for all your answers.
2. This is an experiment with fruit flies and can only be done if they are available. These are the small flies which congregate around fallen fruit and trash

cans containing discarded food items. You will need a clear bottle, some cotton, a paper towel and some old fruit (bananas, peaches, plums, apples). Find a place where flies are active. Place about a one-inch lining of paper toweling in the bottom of your bottle. Then cut up a few pieces of fallen fruit and insert this so that it forms a layer on the towel. Decayed fruit is okay to use. It is all right if some sticks to the side. Place the bottle where the flies are plentiful. After about an hour come back and see if you have got many flies in the bottle. If not, wait till you have a few dozen in there. Then quickly insert a *loose* cotton plug in the bottle. This should be loose enough so that air can get in but not so loose that flies can get out. It should extend about one inch into the neck of the bottle. Next remove the bottle to a convenient place where you can observe it for a few weeks. Make a tabular listing of day and date in one column and then a wide column for your daily observations. You can summarize briefly since a lot of changes will occur. You can run this experiment for quite a while. Be sure not to discard it too soon, and do not get squeamish when the fruit gets moldy or decayed. Just be patient and see what happens. When you are ready to terminate the experiment, you can try a control method, such as putting a tight cap on the bottle or inserting an insecticide. What are your general conclusions? Did your control method involve chemicals?

Exercises

3. Choose one of the quotations opening this chapter and write a short paragraph explaining how it strikes you now. Try to find some relation between the ideas of this chapter and the idea expressed in the quotation.
4. Studies have been made of the amount of nutrients that are removed from the soil by growing crops and grazing animals. For example a ton of wheat takes 40 pounds of nitrogen, 8 pounds of phosphorus, and 9 pounds of potassium. A ton of cattle grown by grazing represents a loss to the soil of about 54 pounds of nitrogen, 15 pounds of phosphorus, 3 pounds of potassium, and 26 pounds of calcium. What does this mean conceptually? In other words explain briefly what the mechanism is. What does this imply for fertilizer? What methods could be used to keep the necessary nutrients in the cycle? Include at least one natural method.
5. Take one of the major plant nutrients and relate how progress occurred from first operational recognition up to modern conceptual views of taking care of plant needs for this particular nutrient.
6. Look carefully at the nitrogen cycle. Then imagine one atom of nitrogen in the nitrogen of the air. Trace its path conceptually into one of the subcycles and describe the changes as it moves throughout the cycle until it emerges again back in the atmosphere as free nitrogen gas. You can use a large graphic outline with information written in on arrows if you like.
7. A few examples of the potential for sex attractants in control of insects were mentioned in this chapter. Think up another possible use you would consider worthwhile for insect control using sex attractants.
8. Below are listed several items. Consider *each separately* and decide whether it is an insect attractant. Give a reason why or why not *in each case* using reference to your past observations.
   (1) decaying organic matter
   (2) sugar in an open bowl
   (3) flowers
   (4) a lighted electric bulb

9. The procedure today in evaluating chemicals is to weigh risks and dangers versus benefits in order to consider whether the benefits justify the risks involved. What do you think of this approach? What are the difficulties? Compare the risks and benefits of DDT usage against malaria in the 1940s. Would the risk now be evaluated the same as in 1945? Explain.
10. Think about the question: Could insects develop resistance to juvenile hormone? Consider this very carefully, realizing that there are two quite different points of view today among scientists. Then decide for yourself one way or the other. Give the reasoning for your choice.
11. Someone said that "microbe crops and livestock" may be a part of our future. Comment on this and give your ideas on the possibilities. Include some conceptual distinctions so that your answer is not just a bald opinion.
12. A bright ten-year-old said she heard that you could possibly get food from wood waste products. How would you explain the situation? Would the food really be made of wood?
13. Pick out two areas in the listing of future developments in agriculture which you feel are most important. Using your imagination, what can you suggest as possibilities in these areas?
14. What do you think of the statement made by one entomologist, also interested in the food supply problem, who said, "If you can't beat 'em, eat 'em."
15. Do you think eating insects could ever be accepted in our society? If not, how do you account for many people who do eat them? If yes, which ones do you think will be accepted first? No matter which way you answered above, now try to tie in conceptually why or how insect-eating could relate to nutritional needs.

SUGGESTED READING

1. Check in your favorite encyclopedia under headings such as insect, insecticide, larva, pupa, fly, moth, fertilizer, farm, etc., etc. In other words start in the encyclopedias if you want further information on a particular topic which interests you.
2. United States Department of Agriculture, *Insects,* Yearbook of Agriculture 1952, USDA, Washington, D.C., 1952. The date here makes this an old reference but it is very readable and contains a wealth of information assembled by many experts in the field. Much of this is still not dated.
3. United States Department of Agriculture, *Plant Diseases,* Yearbook of Agriculture 1953, USDA, Washington, D.C., 1953. Here again hundreds of specialists provide readable information on the vast variety of plant diseases.
4. United States Department of Agriculture, *After A Hundred Years,* Yearbook 1962, USDA, Washington, D.C., 1962. Another easy-to-read reference which provides interesting background information on agriculture. Subjects are very varied and include plants, conservation, forests, animals, insects and economics.
5. United States Department of Agriculture, *Protecting Our Food,* Yearbook 1966, USDA, Washington, D.C., 1966. Another of the excellent USDA yearbooks providing authoritative information on many of the subjects of this chapter. Articles are short and complete in themselves as in all yearbooks making browsing an easy and pleasurable activity.
6. Pratt, C. J., "Chemical Fertilizers," *Scientific American,* 212, No. 6, pp.

62–72 (June 1965). Scientific American Offprint No. 328, W. H. Freeman and Co., San Francisco, 1965.

7. McMillen, W., *Bugs or People,* Appleton-Century, New York, 1965.
8. Williams, C. M., "Third Generation Pesticides," *Scientific American,* 217, No. 1, pp. 13–17 (July 1967), Scientific American Offprint No. 1078, W. H. Freeman and Co., San Francisco, 1967. A very interesting account of the experiments with juvenile hormone by one of the pioneers who has contributed much to our knowledge of the field.
9. Delwiche, C. C., "The Nitrogen Cycle," *Scientific American,* 223, No. 3, pp. 136–146 (September 1970).
10. Epstein, E., "Roots," *Scientific American,* 228, No. 5, pp. 48–58 (May 1973).
11. Waterhouse, D. F., "The Biological Control of Dung," *Scientific American,* 230, No. 4, pp. 100–108 (April 1974). A very interesting story of one type of biological control of both insects and pollution.
12. If you are interested in local plant and livestock problems including suggestions for control measures, be sure to consult your local Farm Advisor and the Agricultural Extension Office of your state university. They have available pamphlets and brochures giving up-to-date information on various agricultural problems.
13. Schneider, D., "The Sex-Attractant Receptor of Moths," *Scientific American,* 231, No. 1, pp. 28–35 (July 1974).
14. Revelle, R., "Food and Population," *Scientific American,* 231, No. 3, pp. 160–170 (September 1974).

chapter 12

Technology: Using Materials

We are as much gainers by finding a new property in the old earth as by acquiring a new planet.

—R. W. EMERSON

Nothing fails like success.

—CHESTERTON

He who accounts all things easy will have many difficulties.

—LAO-TSZE

Future shock refers to the mental shock of having the future sneak up on us too fast. Undoubtedly, technological advances, while providing powerful and effective tools, have their share in future shock. To be aware of the possibilities of technological change, to understand the conceptual bases of technology and to integrate this conceptual understanding in your total understanding of your environment, will help you grow toward the future, not be shocked by it.

The proper assessment and direction of technology can only come from a basic conceptual realization of what it's all about. This means a wide-ranging effort at education for everyone—not just the specialist, because people are going to have to make the choices. And judgments can no longer be made simply on operational terms. Proper preparation for understanding the changing world and how technological advances introduce new sets of problems should allow you as an individual to live with future shock and even possibly to enjoy it.

Research is at the very heart of all science. Dictionary definitions say it is a careful hunting for facts or truth about a subject. You know that curiosity is at the very center of all science. *Research* then describes the formalized seeking for understanding. It represents the process of inquiry and investigation. Research sometimes results in material improvement, sometimes not.

The chemical industry is heavily involved in research. This includes the research of the past hundreds of years, which forms the conceptual base, and also the active research of today. About sixty years ago, only 3 patents out of every 100 issued by the U.S. Patent Office were in the chemical field. Today, chemical patents run approximately 20 percent of each year's total. The practical dividends for you include more than 500 new and improved chemical products which go on sale in the U.S. each year.

Types of Research

Research is of two kinds: basic and applied. Basic or fundamental research is any planned search for unknown facts and principles without regard to practical or commercial objectives. You have already seen many examples such as the structure of atoms; the nature of fundamental particles like electrons, protons, neutrons; hydrogen bonding in water; and the composition of petroleum.

Applied research is an investigation made for a specific practical objective. It is a search for truth in an attempt to achieve a new process or product. The basic research of today is the foundation of tomorrow's applied research.

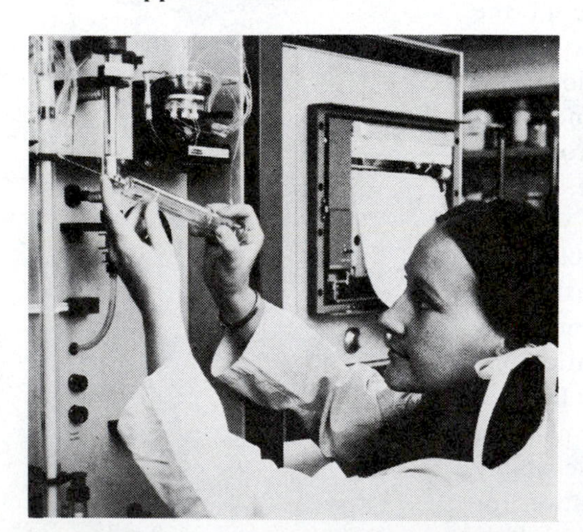

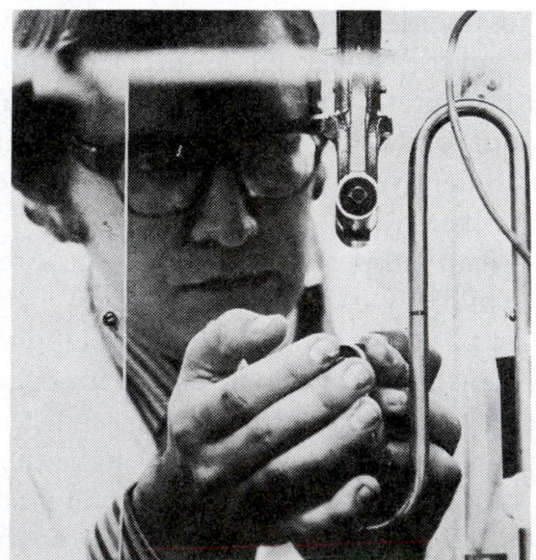

Two photos of applied research. Chemist on the left uses an amino acid analyzer to determine the protein composition of color-treated hair. Chemist on the right places a strand of hair in a tensile testing instrument to determine strength and elasticity after treatment with hair products. (Courtesy Clairol.)

Applied research also concerns itself with what is called product development. This is an organized attempt to find suitable uses for new materials, to improve old products, and to find new uses for existing products. The word development is often used in the common phrase "research and development." This term refers to the combined efforts of an industrial company to investigate and bring to fruition a new product.

Successful laboratory research—like the discovery of polyester fabrics or the development of the color TV tube—is usually a complex team effort. The initial laboratory work may be started as an investigation of a basic conception in someone's mind. It may involve both basic and applied research aspects. However, a long succession of development steps must follow to bring an eventual product to the consumer. This evolutionary process may require, for example, the work of organic, inorganic, and analytical chemists; biologists and biochemists; toxicologists; chemical and mechanical engineers; patent attorneys; experts in pilot-plant design and full-scale production; and specialists in market research, packaging, transportation, advertising, accounting, sales, and customer service. It sounds complicated, and it is. But the major theme which carries it all off is a team effort to bring a concept to application. Most successful research in the chemical industry is thus accomplished only after intensive efforts by many people and considerable expense. It generally takes from one to three years but sometimes up to seven years from the time a project is conceived to the time a chemical product reaches the market.

There is nothing inevitable about the progress of the system. Most projects never reach fruition. For each success story there may be a hundred or a thousand failures. A successful industrial chemical firm is one which gets enough hits to pay for the misses. A few examples will show that along with good concepts must go a considerable supply of patience.

1. Development of a new marketable insecticide costs an average of $2.4 million. The process may require screening of 4000 to 8000 compounds and use approximately 100,000 man-hours of research and testing.
2. Basic research on nylon started in 1927. But it was not until 1939, after spending about $27 million for research and development, that it was possible to get into commercial production.
3. More than 5000 pain-relieving compounds have been developed, but less than 50 could be put into general use.

Charles Kettering, who invented the self-starter for automobiles, once said that science is 99 percent error. This is in reference to the wrong guesses which are discarded. But they are useful in that they lead to improved guesses.

Applied Research Products Are All Around You

Practically everything you handle has some connection with applied research as the following examples will illustrate.

Pesticides.

Food products, including the many additives listed in chapter 10.

Polymers and plastics products running into the hundreds of thousands.

Synthetic textiles and fabrics like polyester, nylon, and rayon.

Aerosol products in push-button, propellant cans.

Paints and coating formulations.

Alloy metal products.

Detergents and cleaning agents.

Adhesives and glues.

Drugs.

WHAT IS TECHNOLOGY?

Briefly, technology is applied science. Technology encompasses all activities of people which are directed toward satisfaction of needs by alteration of materials found in the environment. In other words, it involves techniques or skills of converting, combining, or shaping matter to serve

Technology reached a high degree of complexity in manufacture of space vehicles. Shown here is the most powerful rocket engine developed in the U.S. It was used in a cluster of five to provide 7.5 million pounds of thrust in the first stage of the Apollo/Saturn V moon launch vehicle. (Courtesy Rocketdyne Division, Rockwell International Corp.)

human purpose. The word *technology* (and *technique*) is derived from the Greek *techne* meaning art, or skill, or craft.

Technology has a long history. When a person first learned to tie a rock on the end of a stick and use it for an ax, this was technology in a primitive form. Other examples are spears, bowls cut from rocks, plows, and rope made from grass. Thus technology's basic products are tools to cope with the material environment.

The definition of technology as "applied science" has to be taken broadly. Technology predates formal science. However, if you consider that the basic capability of mankind is forming ideas, then the application of these ideas is the beginning of technology. Einstein says: "The whole of science is nothing more than a refinement of everyday thinking." And the application of ideas to making things is technology.

Perfection of scientific theory is not necessary for a useful technology. What is required is the use of concepts, however primitive and unrefined. In other words, the caveman thinks that a rock on the end of a stick will be an improvement over the old way of just holding the rock in his hand. He does not need to know what a rock is, how cellulose forms long chains so that it can be used in rope, how gravity is involved, and so on. None of this understanding is required in building concepts that lead to an ax or a hammer. However, thinking is required. And man is the only animal to invent and use tools and machines to make hard jobs easier.

Effects of Technology

The products of technology are endless. Every day human ingenuity and creativity turn out new inventions. These range from color TV to new detergents, drugs, jet aircraft, communications satellites, farm tractors, artificial lighting, and TV dinners.

It is important to emphasize the several aspects of technology. One is people affecting technology. Another is technology altering the physical universe. A great problem of our time is coordinating the work of technology with proper respect for the raw materials that nature provides. A third aspect is the effect of the new tools on people, individually and in groups.

Until recently little thought was given to effects of technology on various sectors outside of manufacturing. Today there tends to be considerable overemphasis on one aspect—pollution, an important problem but one that very likely will be controlled by advances in technology itself. The most neglected area has been the enormous influence that technology exercises on the political and social aspects of our national life. Only recently have sociologists begun to recognize that technology is a dominant influence in social change. Think of the social effects of the invention of the wheel, the printing press, and the telephone.

Following are some examples of chemical technology and the major changes they have brought.

Technology	*Influence*
Invention of practical steam engine by James Watt (1736–1819)	Brought about industrial revolution, large factories, concentration of population in cities, etc.
Gunpowder	Changed the whole nature of warfare.
Gasoline and the automobile	Influenced outward sprawl of cities, growth of suburbs, increased mobility of population, aggravated pollution, etc.
Agricultural chemicals	Tremendously increased productivity of the farmer, allowing for growth of urban areas and eventually the population explosion.
Drugs, vaccines, and medicines	Provided longer life span under more healthful conditions.

TECHNOLOGY REQUIRES MATERIALS

The needs for materials were not extensive during the early years of human existence. As technology developed from its first primitive forms, more specialized materials were required. For example, the early hunting people needed a certain kind of stone for axes and certain kinds of plants for spears. The use of fire increased the need for wood as fuel.

Gold was probably one of the earliest metals to be used by man. It occurs free in nature and is found in the gravels of many rivers. However, gold was rare and did not see wide use except in decorations and primitive jewelry. Copper was probably the first metal which people used widely. Metallic copper is native to many localities and early people found that it could be easily beaten or hammered into tools, weapons, utensils, and ornaments. The Copper Age followed the Stone Age and copper was thus known in prehistoric times—even before written records.

Copper is estimated as being in use as early as 8000 B.C. by people in the region of present-day Iraq. It was not until much later, about 3800 B.C., that someone found out how to make an alloy of copper and tin metal which was much harder and more durable than copper. This alloy is called bronze. The Bronze Age is often used to describe the period in the development of a people or region when bronze, rather than the earlier stone and copper, was used as the major material for tools and weapons. The term Bronze Age does not necessarily refer to a particular time period. Some regions had their Bronze Age earlier than others. In some places it was long and in others short. Some people skipped the Bronze Age entirely. In the region of the Mediterranean Sea, people continued using chiefly bronze until around 1100 B.C. Then their Iron Age started. Of course, the use of bronze did not cease. Shields and swords made of bronze were instrumental in the building of the Roman Empire. We use considerable quantities today, for example, in bells, statues, and door pulls.

Another copper alloy is called brass. This consists of copper mixed with zinc. Estimates place its discovery some time between 1000 B.C. and 600 B.C. Brass is a common metal today in such applications as locks, doorknobs, keys, clocks, watches, and precision instruments.

COPPER: THE METAL MORE USEFUL THAN GOLD

The process of smelting copper from ores was discovered around 3500 B.C., helping to meet the great demand for copper-based tools and utensils. People used up native copper deposits in civilized areas rather quickly. In early times copper was recovered from only the highest-grade ores. By the late 1800s, however, the rapid expansion of copper applications for wire in telephones, telegraph, and electric lighting required use of lower-grade ores. Fortunately, by this time the conceptual base was established to allow recovery of copper even where present only to the extent of about 10 pounds in a ton of ore (2000 lbs).

The recovery of copper from concentrated ore requires a low temperature in comparison to that required for iron. This is probably a major reason copper came into wide use much earlier than iron. The initial discovery of methods for recovering copper from chemical compounds happened long before the corresponding theory was developed. It probably was a chance event which some observant person noticed and then developed by further investigation.

Today almost 6 million tons of copper are recovered from ores each year throughout the world and the technology has become very elaborate. The U.S. mines about one-sixth of the world's copper, more than half of which comes from Arizona.

Five Major Steps in Copper Recovery

The basic process is outlined briefly below and is shown in Figure 12.1:

1. Mining. This is chiefly the open-pit type wherein large power shovels or other machines remove the ore from wide steps which can extend deep into the earth. The enormous size of the mining operations is emphasized in the photograph on page 318.

2. Crushing and Milling. The ore is crushed and then run into ball mills, which are huge rotating cylinders half filled with iron balls. The balls grind the ore, which is slurried with water, into a very fine dust. The particles are so small they pass through a screen with 10,000 openings per square inch.

3. Flotation. The copper-bearing ore is concentrated during this process. Surface-active chemicals and oil are mixed with the powdered ore by means of propeller agitators and air jets. This produces a frothy mixture. The powdered copper compounds are coated with oily material and stick to the bubbles which rise to the top. This concentrate contains from 15 to 33 percent copper. The waste materials consist of clay, sand, and

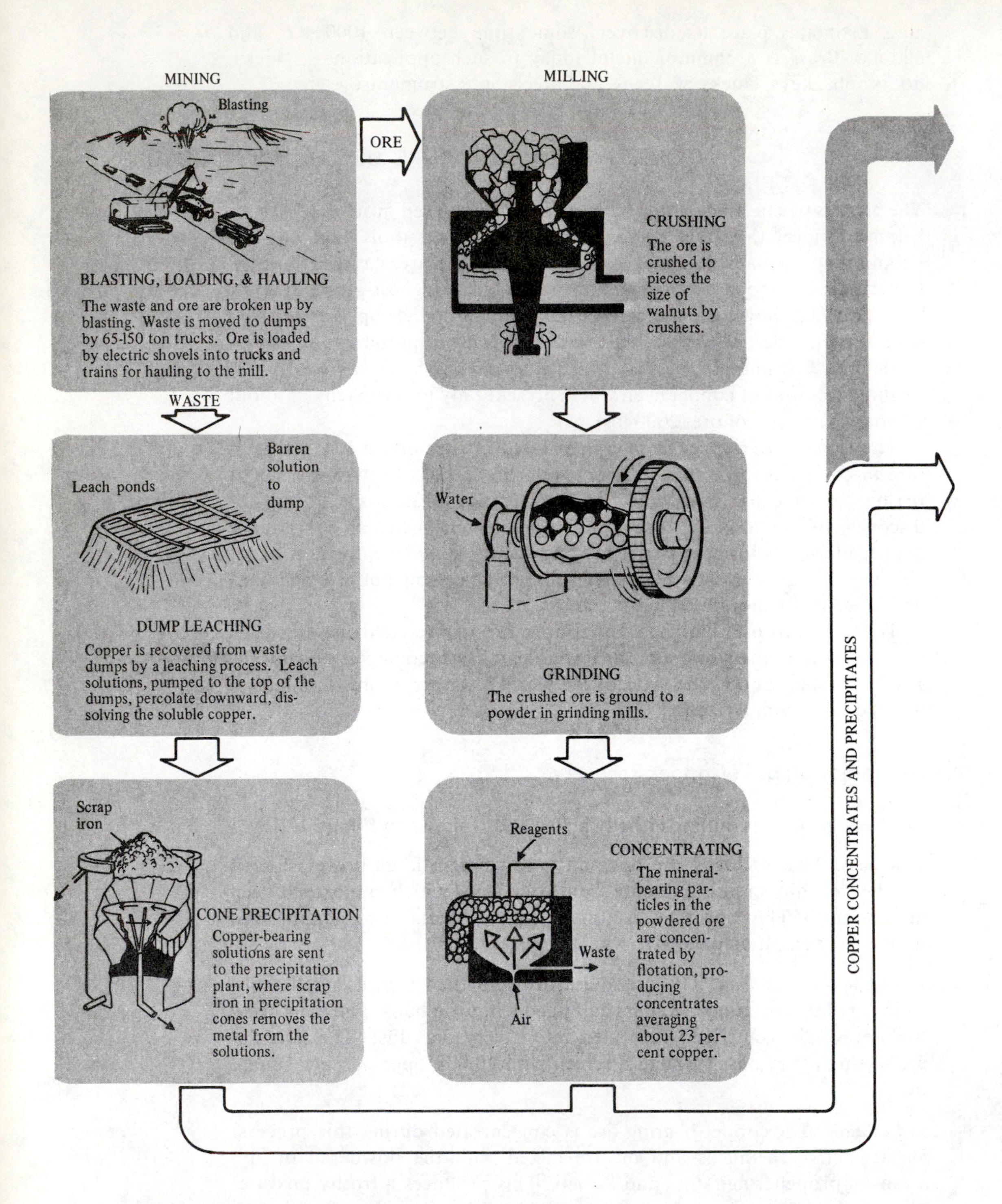

MINING
MILLING
Blasting
ORE
BLASTING, LOADING, & HAULING
The waste and ore are broken up by blasting. Waste is moved to dumps by 65-150 ton trucks. Ore is loaded by electric shovels into trucks and trains for hauling to the mill.
CRUSHING
The ore is crushed to pieces the size of walnuts by crushers.
WASTE
Barren solution to dump
Leach ponds
DUMP LEACHING
Copper is recovered from waste dumps by a leaching process. Leach solutions, pumped to the top of the dumps, percolate downward, dissolving the soluble copper.
Water
GRINDING
The crushed ore is ground to a powder in grinding mills.
Scrap iron
CONE PRECIPITATION
Copper-bearing solutions are sent to the precipitation plant, where scrap iron in precipitation cones removes the metal from the solutions.
Reagents
CONCENTRATING
The mineral-bearing particles in the powdered ore are concentrated by flotation, producing concentrates averaging about 23 percent copper.
Waste
Air
COPPER CONCENTRATES AND PRECIPITATES

FIGURE 12.1
Basic steps—copper sulfide ore to finished product. (Adapted courtesy of Kennecott Copper Corporation.)

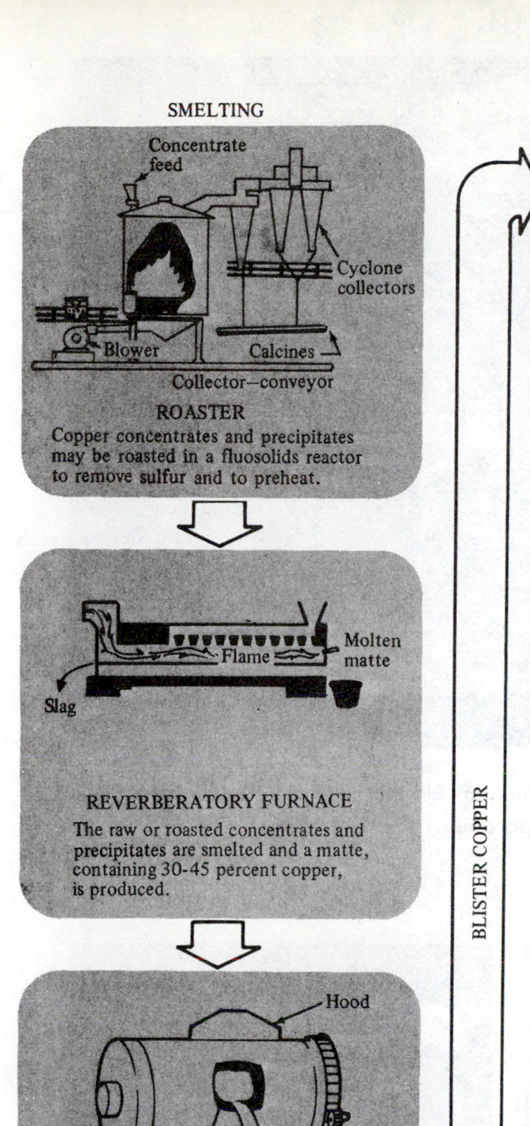
SMELTING
Concentrate feed
Cyclone collectors
Blower
Calcines
Collector—conveyor
ROASTER
Copper concentrates and precipitates may be roasted in a fluosolids reactor to remove sulfur and to preheat.
Flame
Molten matte
Slag
REVERBERATORY FURNACE
The raw or roasted concentrates and precipitates are smelted and a matte, containing 30-45 percent copper, is produced.
Hood
Air
Slag
Blister copper
CONVERTER
The matte is converted into blister copper with a purity of about 99 percent.
BLISTER COPPER

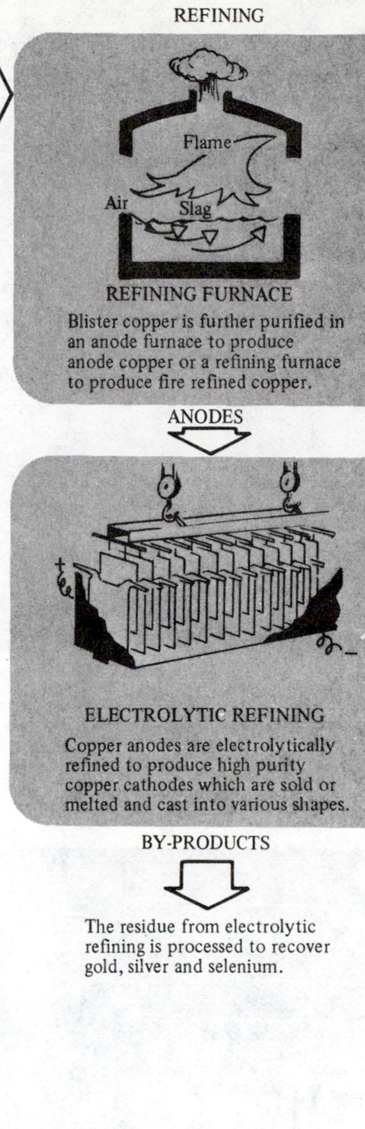
REFINING
Flame
Air
Slag
REFINING FURNACE
Blister copper is further purified in an anode furnace to produce anode copper or a refining furnace to produce fire refined copper.
ANODES
ELECTROLYTIC REFINING
Copper anodes are electrolytically refined to produce high purity copper cathodes which are sold or melted and cast into various shapes.
BY-PRODUCTS
The residue from electrolytic refining is processed to recover gold, silver and selenium.

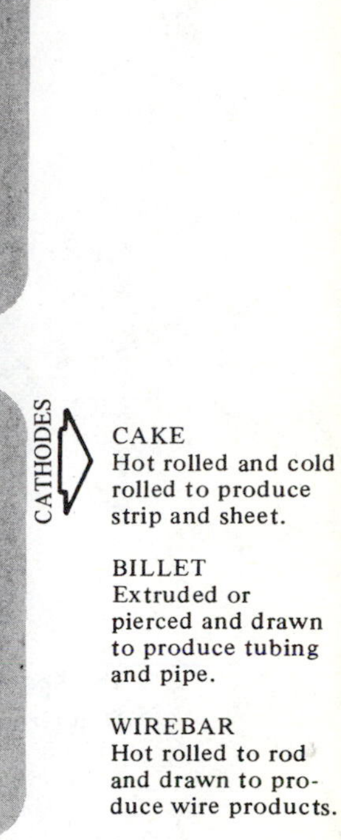
CASTING
INGOTS
Melted for brass products.
CATHODES
CAKE
Hot rolled and cold rolled to produce strip and sheet.
BILLET
Extruded or pierced and drawn to produce tubing and pipe.
WIREBAR
Hot rolled to rod and drawn to produce wire products.
ROD
Drawn to produce wire products.

The Kennecott Copper Corporation open-pit mine at Bingham, Utah. (Photo by Don Green, courtesy of Kennecott Copper Corporation.)

At left, copper flotation showing frothy surface layer which concentrates the copper-bearing materials. At right, molten copper matte is transferred in 50-ton ladles to a converter (the world's largest) at the Kennecott Copper Corporation Nevada Mines Division. (Photos by Don Green, courtesy of Kennecott Copper Corporation.)

other rock debris. These are wet by the water through hydrogen bonds. They therefore sink to the bottom. (See photograph, page 318.)

4. Smelting. This consists of the recovery of impure copper by a process which involves melting. Concentrated copper ore is first run into a furnace where flames and heat drive off some impurities as gas. Other impurities rise to the surface and are skimmed off as slag. The lower layer of copper compounds is called copper matte and is run in molten form into the converter. Here blowers force air through the molten mass, and sand (SiO_2) is added. The sand combines with impurities like iron and floats to the top as slag. The copper melt is run off and cast in molds. It is called blister copper because escaping sulfur dioxide gas causes the surface to form blisters as it cools. It contains from 97.5 to 99.5 percent copper. This is still not pure enough for most applications. The impurities make it brittle and also reduce its conductivity below that required in electrical wiring. (See photograph, page 318.)

5. Refining. The molten copper is held in a furnace and workers dump in pine logs. These produce a boiling, bubbling cauldron as they decompose and burn. In the process they remove most of the remaining impurity, which is chiefly oxygen. The resulting copper is then 99.9 percent pure. However, copper for electrical applications must be purer than 99.9 percent. To obtain this high level, the electrolytic refining process is used. Here the blister copper is cast into large plates which are used as electrodes in a bath carrying an electric current. By the proper arrangement of the direct current source, very pure copper is plated out on a starter sheet of copper. (See photograph below.)

Electrolytic refining of copper is carried out on a very large scale, as shown in this photo of the Utah refinery of the Kennecott Copper Corporation. (Photo by Don Green, courtesy of Kennecott Copper Corporation.)

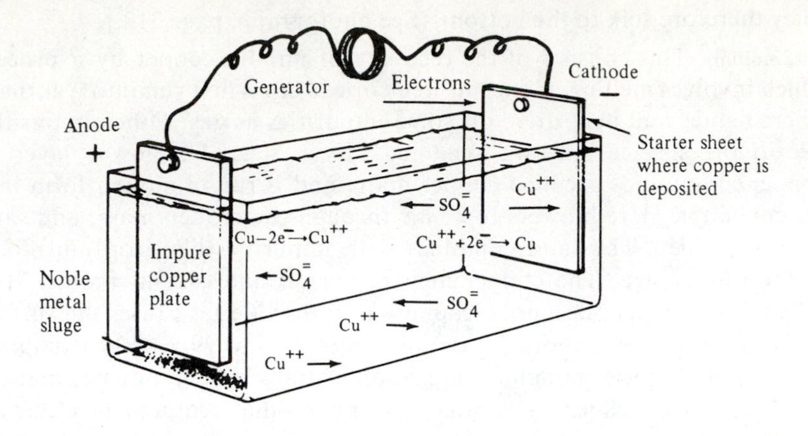

FIGURE 12.2
The refining of copper by use of electric current.

Electro-refining of Copper, or Copper-plating

The process of electro-refining of copper is as follows. (See Figure 12.2.) Copper atoms are removed from the impure plates (+) and transported as Cu^{++} ions to thin starter sheets (−). The plates to be purified are hung in a copper sulfate solution and connected to a positive pole of a direct current (D.C.) source. Then the negative pole, where the D.C. source pumps in electrons, is connected to the thin starter sheet. This electrode is called the *cathode* from the Greek for "down-way" because of the fact that electrons are added, or go down, at this point. This negative sheet with its buildup of electrons thus attracts positively charged copper ions in the solution (Cu^{++}). Those receive two electrons each and are deposited as free copper. This builds up a layer of very pure copper—up to 99.98 percent.

Meanwhile the plate of impure copper is "eaten up" as the copper leaves its surface. This electrode is called the *anode* from the Greek for "up way" because electrons are removed, or go up, from this point. The electrons left behind by each atom of copper ($Cu \rightarrow Cu^{++} + 2e$) are pumped by the D.C. source to the other pole where they are used to deposit copper atoms from ions ($Cu^{++} + 2e \rightarrow Cu$). Impurities—mainly the elements gold, silver, selenium, and tellurium—sink to the bottom where they are recovered. The bath liquid is also removed from time to time to recover metals like iron, zinc, and nickel, which build up in it.

What is involved here is copper-plating; that is, copper metal is plated out on the negative electrode (or cathode). The same process is carried out to coat other articles like iron and steel to prevent corrosion. You are familiar with chromium-plated articles like auto bumpers and plumbing fixtures. For these items a thin coating of copper is plated first, followed by the chromium, as the adherence of chromium is much better over copper.

Leaching: An Alternate Recovery System

Some copper ores do not separate in the flotation process. Then an alternate recovery system called leaching is used. In this process water containing sulfuric acid (H_2SO_4) circulates over the ore. The acid dissolves the copper compounds or leaches out the copper in the form of copper sulfate, $CuSO_4$. Then the copper solution passes into vessels containing suspended pieces of iron metal. The copper deposits on the iron while iron atoms go into solution. Copper is removed by scraping and new pieces of iron are added as needed. The iron simply replaces the copper:

$$Fe + CuSO_4 \rightarrow FeSO_4 + Cu\downarrow$$

The copper thus obtained is called precipitate copper because it precipitates as a fine copper powder when the copper bearing solutions react chemically with the iron. It is a low-purity product running about 60 to 90 percent pure, which is smelted and refined in the usual way. The iron solutions formed are used as sources of iron.

Concepts behind Copper Recovery from Ores

It is interesting here to note how technology can be aided by refinement of scientific ideas. The earliest practices of recovery of metals were all due to accidental discoveries. Metals like copper, tin, silver, gold, mercury, and iron were all known to the ancients. Indeed, the practice of metal recovery from ores (metallurgy) was considered an ancient practice by the Hebrews at the time the Old Testament was written. As higher levels of concepts developed, the older practices of metallurgy were refined. This was indeed the case in the whole history of technology.

The example of copper concentration by two radically different methods is a case in point. Some copper ores separate with the flotation process. Others do not. Why the difference? The answer lies in the kinds of compounds in which the copper is found. The ores containing copper as sulfide, such as CuS, Cu_2S, and $CuFeS_2$, are easily concentrated in the flotation process.

This is because flotation depends on the presence of two different types of compounds. First there must be those compounds which form hydrogen bonds easily. These contain oxygen and can form links with the hydrogen atoms in water. They can then be surrounded by water and sink to the bottom in a flotation process. Example: limestone, $CaCO_3$. Secondly, there are the compounds like Cu_2S which are said to be hydrophobic; that is, they "fear" water. These are not wet by water because they do not form hydrogen bonds. They are more compatible with the oily phase and float off in the frothy foam.

What about the other ores of copper which we now recognize as containing compounds like $Cu(OH)_2$ or $CuCO_3$? If you think about this you will see why they cannot be concentrated by the flotation process. They are like the compounds which sink by wetting with water because of formation of hydrogen bonds with water. The OH^- and $CO_3^=$ groups allow these to be wet similar to compounds like $CaCO_3$, $CaSiO_2$, and Mg_2SiO_4, which we want to separate out. Then everything sinks together. Conceptual analysis shows you why leaching with sulfuric acid is a suitable method. Leaching even gives a signal which was recognized in the early days, since copper in solution turns blue. Thus an ore containing $Cu(OH)_2$ would react readily with H_2SO_4 as follows:

$Cu(OH)_2$	+	H_2SO_4	$\rightarrow$	$CuSO_4$ + $2H_2O$
As solid in complex ore mixture		In water solution as "leaching" agent		In solution as deep blue (i.e., "leached" in this form)

The formation of the blue color would be a visual operational confirmation of our idea that copper is dissolving or leaching out of this particular kind of ore. When copper is present as sulfide, however, the compound is one of the most insoluble known. Here again the refined concept, even if not fully applied to the original trial-and-error guesses, is extremely effective in explaining why certain practices work, suggesting when they are suitable, and providing clues as to how they might be modified and improved.

Uses of Copper

Your car contains on the average 50 pounds of copper. This includes electrical wiring, radiator core, and undercoating for chrome-plated bumpers. The automobile industry alone takes 10 pounds out of every 100 pounds of copper used in the U.S. Homes, factories, and other buildings take another 15 pounds. If you have ever made repairs in or built a house, you have seen copper in wiring, plumbing fixtures, door locks, hinges, and gutters. The Empire State Building in New York City contains over a million pounds of copper and brass. About half of the total copper used goes into motors, generators, transformers, and transmission lines. Among miscellaneous applications are fungicides, pigments, solid state circuits in radio and TV, and metal for currency.

The penny is obviously predominantly copper. It is not so obvious that the nickel contains about 75 percent copper. Gold and silver coins even contain about 10 percent copper and most gold and silver jewelry also contains considerable copper to provide hardness.

Even the body cannot do without copper, although large amounts are toxic. Depending on your size, you contain between 100 and 150 milligrams of copper in various compounds. These are necessary in your blood system even though no copper is present in the hemoglobin. Greatest concentrations are in the liver and bones. Copper is also one of the necessary micronutrients (chapter 11).

Another of the common metals in use today which has a long history is iron. The Iron Age around the Mediterranean began about 3100 years ago in the area now occupied by Turkey. The advantage of iron was its cheapness, because its ores were abundant and widespread. Occasional use of iron had been made even earlier, as far back as 3000 B.C. during the Bronze Age. In these exceptional cases iron was beaten or hammered from meteors where the free metal itself was literally provided out of the blue.

The use of iron spread rapidly once methods of "winning" it from ores were available. The major difficulty here is that temperatures required are very high. Eventually, people found out how to make what we call steel, described below. The Iron Age has never really ended.

Iron from Iron Compounds

Very likely the first winning of iron occurred when lumps of iron ore were accidentally placed with the stones in fire pits where meats were roasted for a feast. Then some observant cook may have noticed that silvery iron metal was freed from the ore when the wood fire was maintained for a long period. Later, during the planned process of winning iron, someone noticed that the higher temperatures obtained when the wind was blowing gave better results. This was the primitive technology which led eventually to our modern blast furnace.

A blast furnace may be from ten to fifteen stories high and more than 30 feet wide at the base. (See Figure 12.3.) The name is derived from the blast of hot air which huge blowers push into the bottom of the furnace. The air supplies the oxygen to burn the coke which then provides the heat to melt the ore. Complex chemical reactions occur with the raw materials and provide iron and slag at the bottom. The hot air blasting into the furnace through nozzles or holes keeps the coke burning at a very high temperature (in the range of 1500–1600°C or 2700–2800°F). The materials melt down and make room for new raw material fed in at the top. The iron, which is formed at a point about half way down the furnace, trickles down to a collection space at the bottom. The slag, which contains impurities, floats on top of the molten iron.

The chief ores of iron are the oxides or compounds which can be converted into oxides prior to smelting. Consequently, the formula Fe_2O_3 can be used as representing iron ore.

The recipe for smelting iron has remained the same down through the ages. The ingredients are ore (Fe_2O_3), limestone ($CaCO_3$), and coke (C) which go in at the top of the furnace, and air which is blasted up from the bottom.

Pig Iron

The iron produced in blast furnaces is often called "pig iron." The name derives from the practice of pouring the hot iron into small sand molds

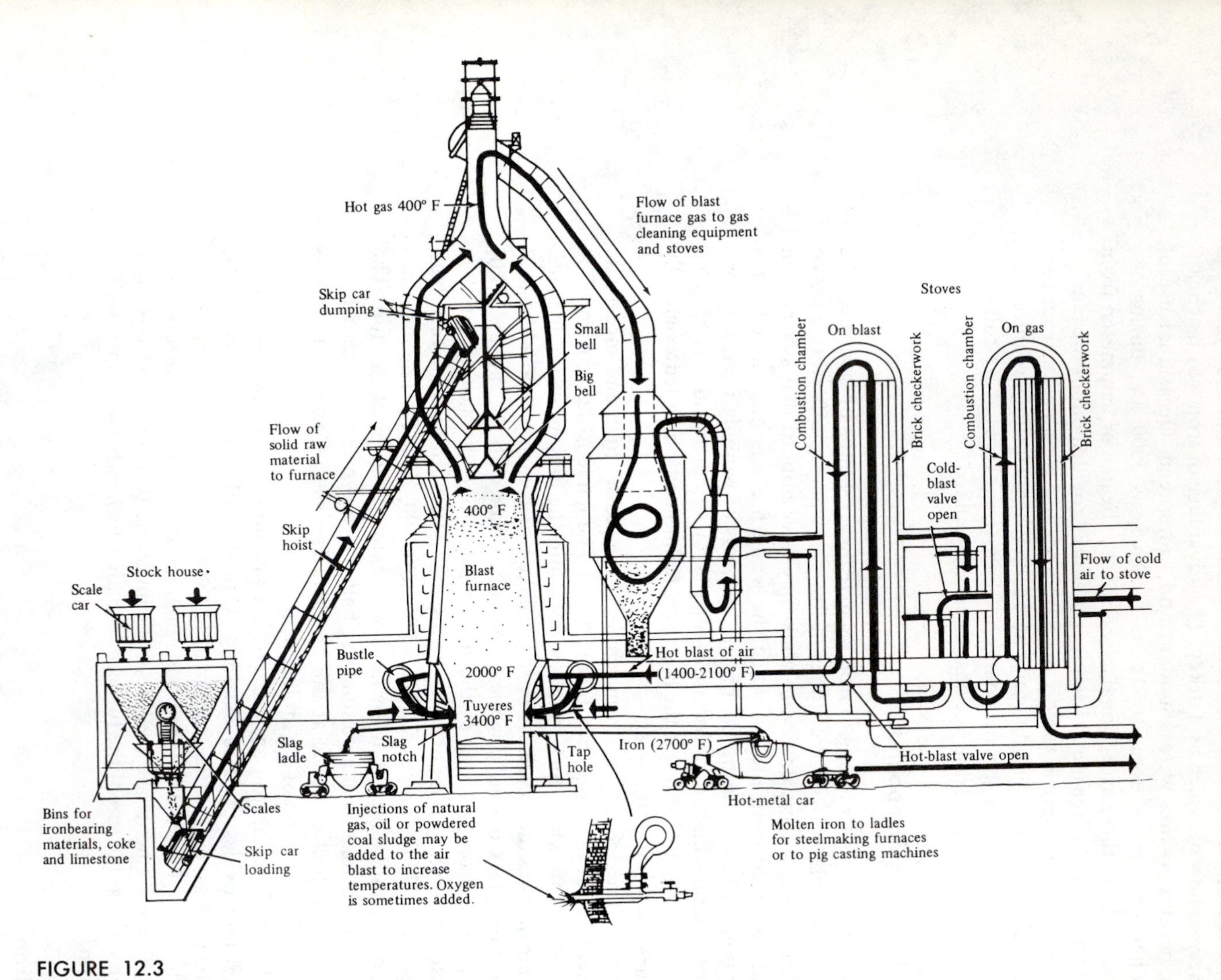

FIGURE 12.3
Operations of a blast furnace in changing ore to iron metal. (Courtesy of the American Iron and Steel Institute.)

arranged around a main channel. The iron shapes reminded people of a sow and her piglets, and ever since then, cast iron at this stage has been called pig iron.

Today most of the liquid iron tapped from the blast furnaces is carried molten to nearby steel-making plants. It is cast in pigs only if necessary to ship it long distances.

Pig iron can be used for some applications, but it was early discovered that it is very brittle. Pig iron is only about 95 percent iron, containing a large amount of carbon (3 to 4 percent) and smaller amounts of silicon, phosphorus, sulfur, and other elements.

When pig iron is used directly for castings it is called cast iron. Foundries shape the iron by pouring liquid iron into molds. You are familiar with some of these molded forms such as fire hydrants, auto engine blocks, pipes, radiators, and fire gratings. Cast iron has limited uses because it is brittle and cannot be formed into shapes by rolling or hammering.

What is Steel?

Scientists very early learned that if the carbon content of pig iron was lowered to less than 1 percent, the product would be stronger and much less brittle. This is the origin of steel, which is an alloy or mixture of iron with small amounts of carbon and other elements. The origin of *steel* is believed to be from the old Teutonic word meaning firm or rigid. Steel-making thus basically consists of removing the excess of carbon and other undesirable impurities and adding other desired elements in exact and controlled amounts.

The three basic methods of making steel today are listed below.

Openhearth Furnace. (See Figure 12.4.) The name comes from the saucer-shaped depression or hearth where the molten iron is open to the flames which blow across it. The furnace is charged with scrap iron, limestone, and some iron ore, along with molten pig iron. The scrap and ore provide oxygen content which results in removal of impurities. The carbon is oxidized to carbon monoxide or carbon dioxide which goes off in the gas stream. Other impurities are trapped in the slag by the limestone, forming various products like $CaSiO_3$. Each furnace is about the size of a two-story house. The raw materials are loaded by skilled steelworkers through openings on one side. When the heating and purification is completed, the steel is tapped from the opposite side where the floor is one story lower. Tests are taken of the steel at various stages in order to provide control of its chemical composition. Eventually the necessary elements are added according to the particular steel recipe. A furnace can make between 100 and 300 tons of steel in a typical batch, which takes between five and eight hours.

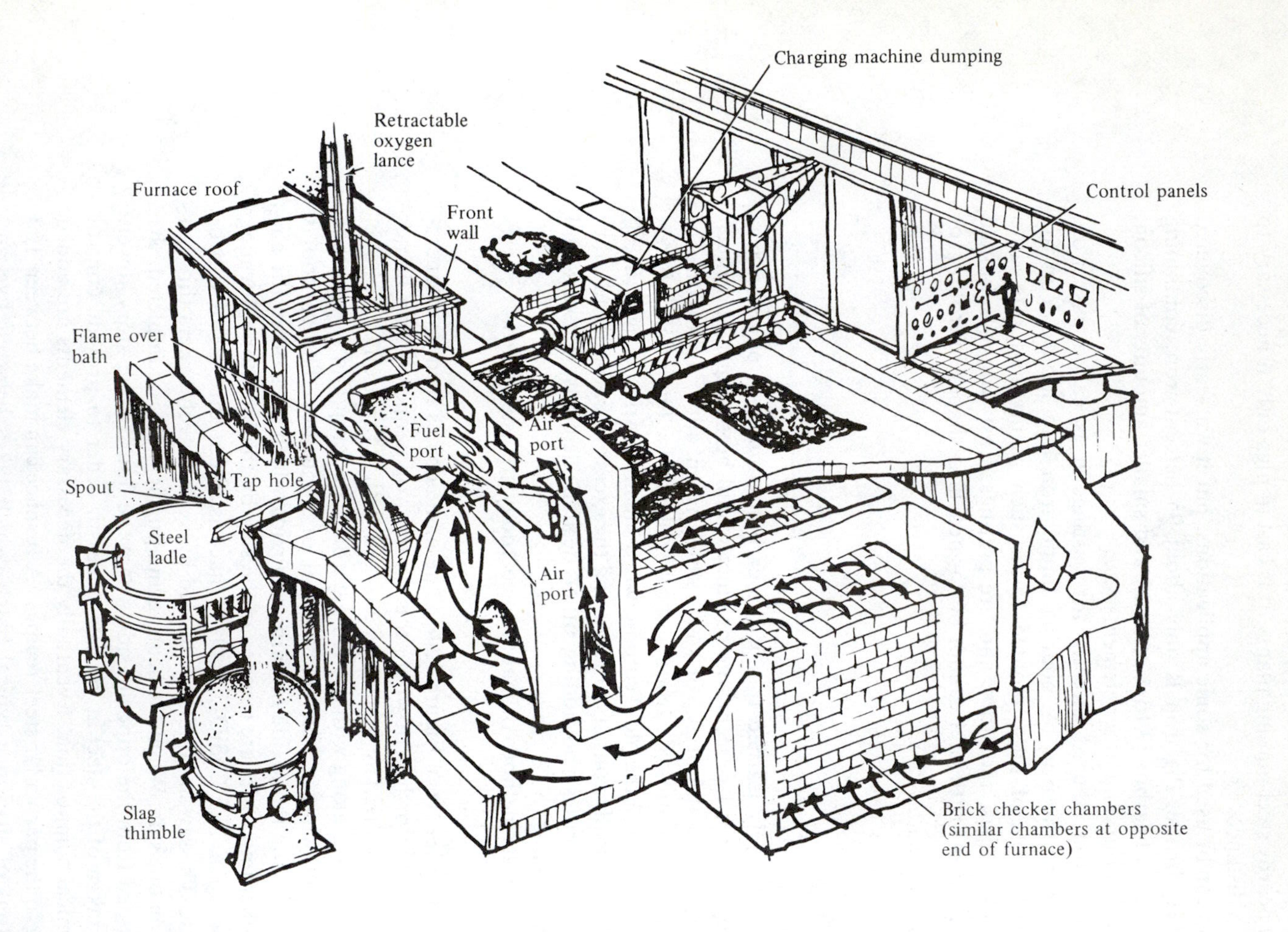

FIGURE 12.4
Operations of a modern open hearth furnace for making steel. (Courtesy of the American Iron and Steel Institute.)

Ladle of molten iron is positioned by crane for pouring into open hearth furnace. (Courtesy American Iron & Steel Institute.)

Electric Furnace. (See Figure 12.5.) This is a large round shell, lined, as are all the furnaces, with heat-resistant brick. It uses a high current of electricity which forms arcs within the furnace and thus provides the high temperature to melt the steel. The raw materials are chiefly scrap steel and various alloy additives according to the particular steel recipe. The electric furnace permits precise adjustments of conditions and is used to make the steels which have very special requirements. A typical electric furnace can produce between 60 and 90 tons of steel a day.

Basic Oxygen Process. (See Figure 12.6.) This is the newest of the methods of steel-making and is now taking over as the leading method in terms of tonnage output. A basic oxygen furnace can produce up to 300 tons of steel in less than an hour. The method uses a pear shaped, brick-lined, steel container which can be mechanically tipped for pouring out a tap. As with all steel-making methods, scrap steel is an important raw material. It is loaded first and molten pig iron added to melt the scrap. Then a special lance blows oxygen gas into the molten mass. This burns away carbon and other impurities, the carbon exiting as carbon monoxide or carbon dioxide gas. No fuel is used. Added limestone forms flux with other impurities, which floats on top and is removed in the usual way. Alloying elements are added as required for the particular steel recipe.

Steel companies turn out sheets, bars, pipes, girders, reinforcing rods, wires, and tubes. Each of these are made with various compositions and heat treatments. There are literally thousands of different kinds and forms of steel, which is much more valuable in your life than gold or silver.

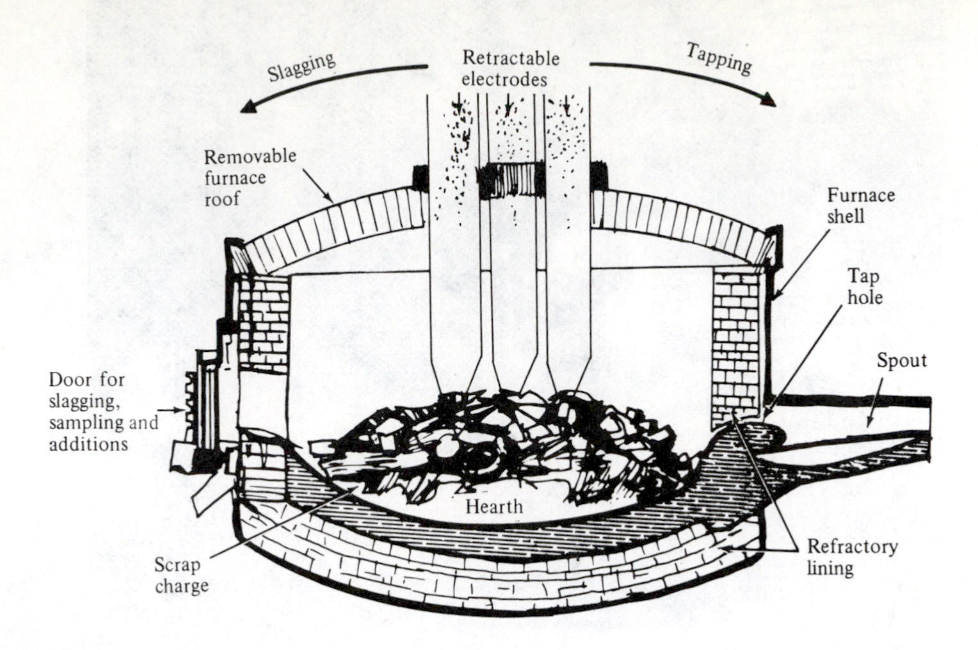

FIGURE 12.5
Cross-section of electric arc furnace showing the three consumable electrodes which pass current through the solid scrap charge and melt it. The entire roof swings aside for charging, and the furnace tilts left and right for slagging and pouring. (Courtesy of the American Iron and Steel Institute.)

Left photo: With the basic oxygen furnace tilted, a charging box is elevated and tipped to dump a load of scrap steel into the furnace. Molten pig iron will be added from a ladle. Right photo: When oxygen blowing is completed and analysis of the batch shows the proper chemical content, the furnace is tilted and steel is poured into a ladle. (Photos courtesy American Iron & Steel Institute.)

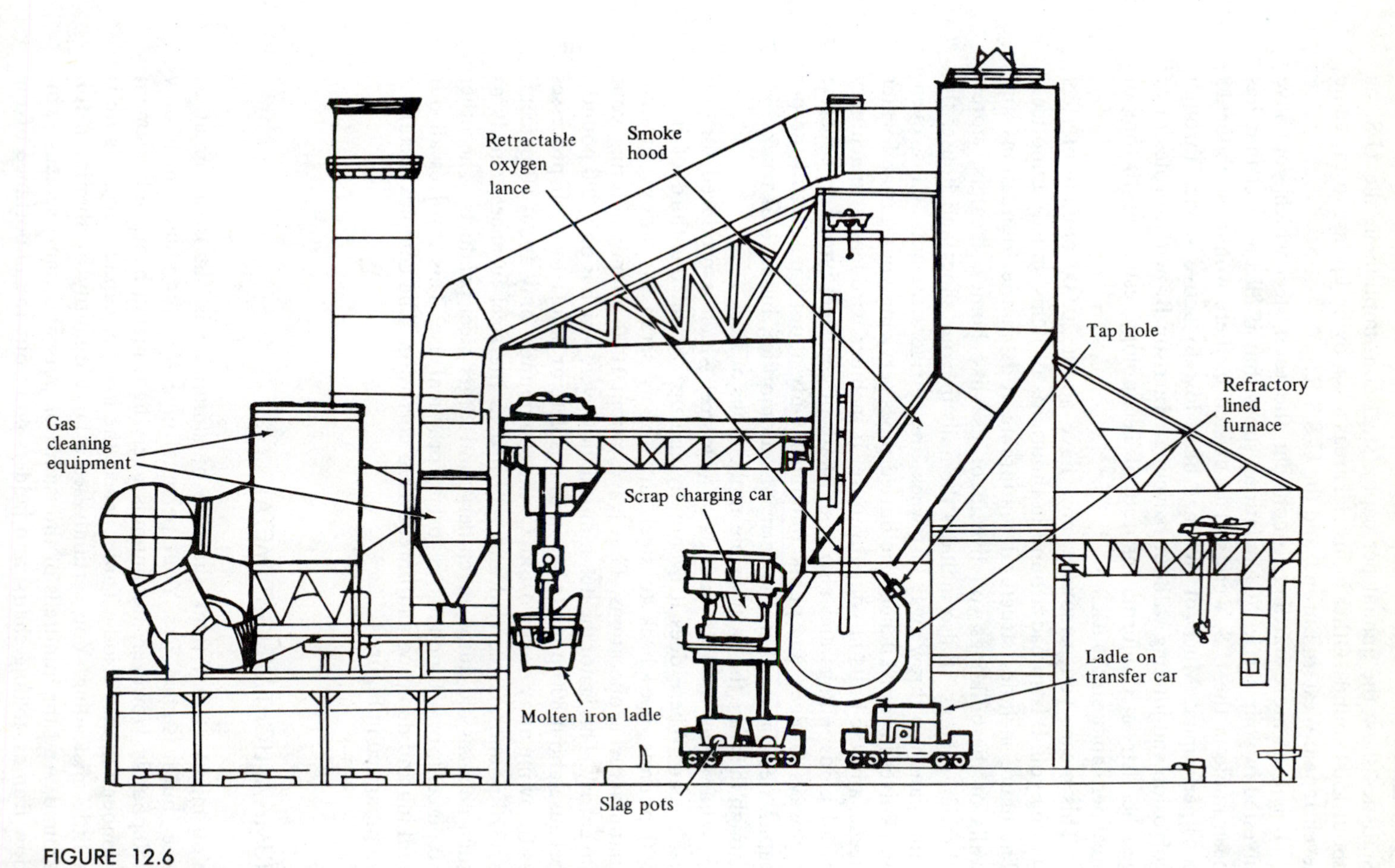

FIGURE 12.6
Schematic of a basic oxygen furnace shows the facilities needed to charge scrap and molten iron into the vessel and receive the steel after the oxygen blowing process is complete. (Courtesy of the American Iron and Steel Institute.)

How Much Steel Is Consumed Each Year?

Statistics show the annual per capita steel consumption in the U.S. as about 1500 pounds. Other industrial areas are moving up fast to the same level: Japan, Western Europe, and the U.S.S.R.

It is even more startling to consider the total steel which you have invested in your service. Calculations show that if all the steel that has been made in the U.S. were still in existence, there would be about 15 tons per capita. Much of this has been lost to active service through corrosion and junking without recycling. Harrison Brown, a professor of geochemistry at California Institute of Technology, estimates 10 tons of steel per capita still in use.

This means you personally now have about 20,000 pounds of steel serving you. It is spread around in thousands of items, mostly unnoticed. The figure includes structural reinforcing rods in the foundations and walls of the buildings you live in or use like homes, theaters, stores, libraries, schools. It also includes the nails in your shoes and house, the "tin" cans you threw away last week, the refrigerator, pots, pans, oven, and knives in your kitchen, the buried pipes servicing your gas and water needs, and the bolts in your furniture. It includes wire fences, machinery of all kinds, pilings and also your share of the steel buried in concrete highways, bridges, and public buildings. Only about 8 percent of the per capita steel in use is in the form of automobiles, buses, and trucks, although these are the most visible applications of steel.

What about the 1500 pounds of new steel which is produced for you each year? Where does it eventually end up? About one-third of it, or 500 pounds, goes back to steel furnaces as scrap in various product manufacturing operations. That leaves about 1000 pounds entering your inventory. This is not the net addition, however, which is only 200 pounds, because about 800 pounds becomes obsolete or is lost by such processes as throwing away or corrosion. Of the 800 pounds, only about 40 percent, or 320 pounds, is recycled and returned to the steel furnaces, chiefly as such items as old autos, machinery, and larger steel products. The other 60 percent, or 480 pounds, is lost to practical use. This involves all our junking and dumping and unrecovered corrosion. The complex picture is summarized in Figure 12.7.

ALUMINUM: A LIGHT-WEIGHT METAL

You might at first think that use of aluminum is almost as great as steel, but it is much less. Your personal inventory for aluminum is only 250 pounds. However, aluminum is a very light metal, being only element number 9 in the Periodic Table; hence, the low per capita weight figure is a little misleading. You certainly see a large quantity of aluminum foil around. And the hundreds of thousands of pots and pans, screens, window frames, folding chairs, and ladders point out the importance of the

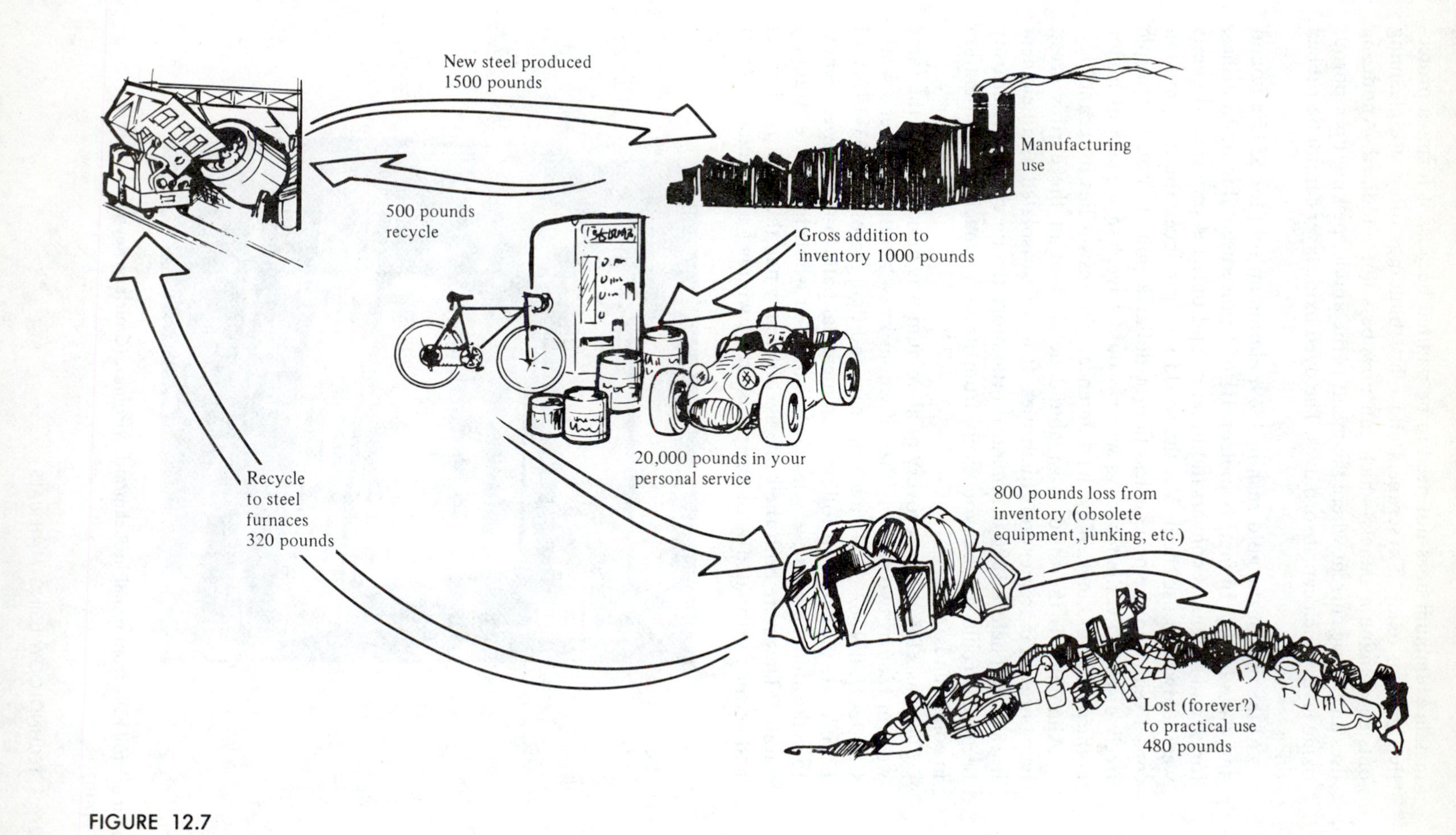

FIGURE 12.7
U.S. per capita steel inventory changes in one year.

metal. Aluminum does not need to be painted because it forms a protective layer of oxide. Skyscrapers and office buildings often use gleaming outside aluminum panels which can now be color anodized. Anodizing uses an electric current where the metal is the anode, picking up a porous oxide layer which can absorb dyes. The dyed oxide layer cannot be rubbed off.

At the time of the gold rush in 1849, aluminum sold for $545 a pound and was a luxury. In fact, Napoleon III of France used aluminum dishes, forks, and spoons as conversation pieces, and only his very special guests were allowed to use them. Yet the world now produces about 10 million tons a year, and the price is less than a dollar a pound. You see people freely discard TV dinner trays which would have had a place of honor in the palaces of kings. Why the difference? The answer lies in technology.

Aluminum is the third most abundant element in the earth's crust, being present in tremendous quantities. It was first isolated as an element in 1825. But before 1886 it remained little more than a chemical curiosity because it could not be won easily from its very tenacious combination in compounds.

The basis for cheap recovery of aluminum was discovered in the same week in 1886 by two men, neither of whom knew about the other's work. Charles Hall (1863–1914), an American, and Paul Heroult (1863–1914), of France, are the two who followed an identical path to the answer of how to make aluminum an everyday chemical. Their whole lives were strangely parallel. They were both the same age, fifteen, when they first read about and became fascinated with the problem of winning aluminum from its

Aluminum foil rolls from a mill in Richmond, Virginia. (Courtesy Reynolds Metals Company.)

chief ore, which is a mixture of hydrates of aluminum oxide. This is called bauxite and is often, for simplicity, designated Al_2O_3. The bauxite ore may run between 50 to 70 percent Al_2O_3 and about 25 to 30 percent water, with varying amounts of impurities like sand and iron oxide.

Both Hall and Heroult used electric current to force electrons into Al^{+3} ions. The Al^{+3} ions were obtained by a neat trick of getting the Al_2O_3 into ionic form in a previously melted mineral called cryolite. At the time cryolite could be obtained only in Greenland, where it was called "ice stone" by Eskimos because it had the appearance of frozen snow. It has the formula Na_3AlF_6.

The recovery of aluminum required a method of getting Al_2O_3 into solution at a practical temperature. The Al_2O_3 by itself melts only at about 2000°C (3630°F) and, consequently, it would be very difficult to find a container to hold it. Iron melts at about 1535°C (2795°F). The cryolite melts at 1006°C (1843°F) which then permits the use of an iron shell for containing the liquid. The essentials of what is called the Hall-Heroult electrolytic cell are shown in Figure 12.8.

The cryolite is not used up in the process. Aluminum ions (Al^{+3}) are attracted to the negative electrode. They are each given three electrons:

$$Al^{+++} + 3e \rightarrow Al$$

The ions thus become liquid aluminum metal. From time to time it is tapped off and more Al_2O_3 is added to the molten mass. (Aluminum melts at 600°C or 1220°F.)

Meanwhile at the other electrode (+) the following type of reaction can occur:

$$C + 2O^{-2} \rightarrow CO_2 + 4e^-$$

You can see here how the simple concepts of ions, electrons, and melting points can be essential to the success of technology in making important materials available for everyday use.

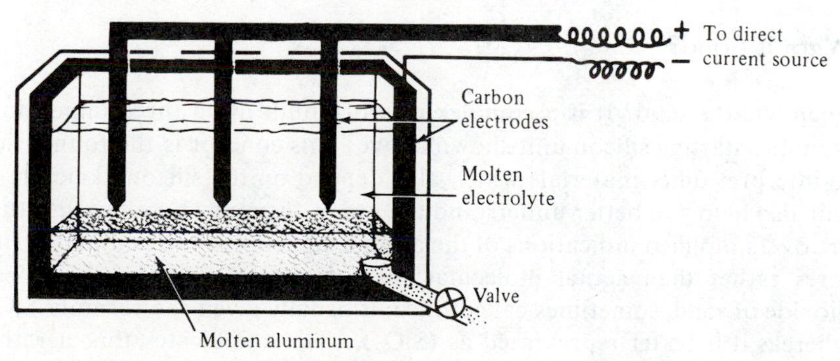

FIGURE 12.8
The Hall-Heroult electrolytic cell which made aluminum recovery practical.

POTTERY, BRICKS, AND GLASS: ALL BASED ON SILICON CHAINS

If you visit a museum, you will find pottery and procelain from China and Egypt going back thousands of years. The various mixtures required were often highly guarded secrets developed from experience and passed on from generation to generation. Nothing in the inorganic world is more complex structurally than these materials which we use every day—chinaware, bricks, and concrete.

It is interesting that bricks, pottery, and glass are all made up of silicon-oxygen compounds. These are, in turn, the most abundant of all the compounds in the earth's crust. They are the backbone of most minerals and rocks.

The chemistry of all of these is best understood conceptually if we start with sand because the chain structure in sand sets a pattern which we find in many related materials. Sand is often given the formula SiO_2, but it does not exist as free SiO_2 molecules. There is a giant network similar to that present in the diamond (see chapter 8).

In other words, silicon, being the second member of the carbon family, behaves in a similar fashion to carbon. However, whereas carbon forms chains by joining to other carbon atoms, silicon generally joins in chains through oxygen atoms between each silicon atom. Thus we get

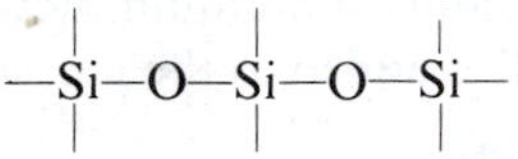

$$-\overset{|}{\underset{|}{Si}}-O-\overset{|}{\underset{|}{Si}}-O-\overset{|}{\underset{|}{Si}}-$$

You can see that each silicon atom has a possibility for joining to two other atoms in addition to the two chemical bonds shown within the chain. The electron dot notation for silicon gives an indication of its basic chemistry.

$$\cdot\dot{\underset{\cdot}{Si}}\cdot$$

Silicon, like carbon, shares four pairs of electrons and forms four bonds to each atom. This is the basis of brick, concrete, pottery and glass.

What is Sand?

Then what is sand? It is a complex network built up in three dimensions from the starting silicon unit shown above. This concept is the foundation to interpret other materials which also depend on the silicon skeleton. It will also help you better understand the use of formulas here because they are oversimplified indications of the atomic makeup of the complex structures rather than actual molecular formulas. For example, the silicon dioxide of sand, sometimes called silica, is usually given the formula SiO_2, whereas it is better represented as $(SiO_2)_x$. The *x* indicates that a rather large polymeric arrangement is really involved. This is different from the indication of carbon dioxide as CO_2. In the latter case separate molecules of CO_2 do exist and you know this is a common gas. However, the grains of sand on the beach or in your backyard or in your concrete walk are giant molecules containing billions upon billions of silicon atoms, each joined to others through oxygen atoms.

Silicon dioxide exists in nature in a rather pure form as quartz. Sand, agate, opal, amethyst, flint, and onyx are other examples of SiO_2 with traces of impurities. These are all hard, brittle materials because of extensive three-dimensional structures similar to those in diamonds.

Most of the solids in the earth's crust then are built up around the basic skeleton

$$\begin{array}{ccccccccc} & & | & & & & | & & \\ & & \mathrm{O} & & & & \mathrm{O} & & \\ & & | & & & & | & & \\ - & \mathrm{O} - & \mathrm{Si} & - & \mathrm{O} & - & \mathrm{Si} & - \mathrm{O} & - \\ & & | & & & & | & & \\ & & \mathrm{O} & & & & \mathrm{O} & & \\ & & | & & & & | & & \end{array}$$

The SiO_4 grouping is sometimes referred to as a silicate. A few of the important silicate minerals are listed in Table 12.1.

TABLE 12.1

A Few Representative Minerals Containing Silicon

Name	*Relative Atom Ratios*
Quartz	SiO_2
Garnet	$Ca_3Al_2(SiO_4)_3$
Zircon	$ZrSiO_4$
Emerald (beryl)	$Be_3Al_2Si_6O_{18}$
Asbestos	$Mg_3Si_2O_5(OH)_4$
Mica	$K_2Al_2(AlSi_3O_{10})(OH)_2$
Talc	$Mg_3(Si_4O_{10})(OH)_2$
Feldspar	$KAlSi_3O_8$ (orthoclase, a common feldspar example and a most abundant mineral)
Kaolinite	$Al_2Si_2O_5(OH)_4$ (the chief component of clay)
Zeolite	$Na_2(Al_2Si_3O_{10}) \cdot 2H_2O$ (used in some water softener applications)

Note: There are other varieties of some of the minerals shown.

There are many variations even of a particular type of substance so that there is really no *single* formula for say asbestos, or mica, or feldspar. These complex materials can be said to be derivatives of sand. The basic silicon-oxygen structure is given fantastic variety by attaching other atoms to the oxygen, or even by substituting other atoms for the silicon or oxygen to give many thousands of additional minerals. The molecules in asbestos form long chains and this accounts for the stringy nature. Mica, on the other hand, contains a structure where the silicon framework stretches out in layers, somewhat modeled after the situation we found in graphite. Hence it slices easily into sheets.

Pottery, Porcelain, Bricks, and Tile

You can see why the technology of pottery, brick, and glass are still based chiefly on art developed over the years. The little we know about the structures of the complex products is due to modern techniques, using X-rays and electron microscopes. Early potters and brick makers never knew the mechanism behind their work. Even though today we know very little of the details, we can understand basically what they are doing. They are rearranging silicon chains in various ways, and they are taking advantage of the possibility of joining adjacent chains through linkages involving other materials like water. Thus they can go through the plastic stages and then, by firing or curing, set up the materials in rigid form. The overall process is very similar conceptually to cross-linking in plastics manufacture.

Ceramic materials commonly include pottery, porcelain, bricks, and tile. The word *ceramics* is derived from the Greek for potter's clay. Pottery-making is considered one of the fine arts even though it would also come under the heading of technology. Ceramic materials are shaped or molded at room temperature, chiefly from clay, sand, and feldspar, and then permanently hardened by firing or heating.

The Chinese perfected the art to such a degree that the products they introduced were called porcelain, or chinaware, or china for short. Porcelain represents one of the highest forms of the potter's art. In other words, all chinaware is pottery, but not all pottery is china.

What makes chinaware distinctive is its whiteness and ability to let light through—its translucency. The basic reason for its special properties is the pure form of clay the Chinese used. There have been intriguing stories revolving around attempts to get the secrets and duplicate the special qualities of china.

The commonest kind of building brick is red. The color comes from iron compounds in the clay. Bricks for use in furnace wall lining are called firebricks and have a high percentage of SiO_2.

Although the basic art of ceramics has been unchanged for hundreds of years, there have been new developments based chiefly on growth in conceptual background. A few examples are: magnetic ceramics, which contain iron compounds and are finding use as memory agents in computers; electrical and thermal insulator ceramics based on "alumina" (Al_2O_3) and zirconia (ZrO_2) instead of silica; and special glass-ceramics.

Glass or Glasses?

From the preceding discussion, you may have begun to realize that there are many kinds of glass. In fact there are hundreds of thousands of different kinds. But their manufacture depends very heavily on the technology developed as art over thousands of years.

The basic raw materials for glass are cheap. They are chiefly sand, limestone, and sodium carbonate. Chemically these are SiO_2, $CaCO_3$,

Skilled glass blower makes a special piece of glassware needed in chemical research. (Courtesy Abbott Laboratories.)

and Na_2CO_3. The process of making a glass consists of heating these raw materials in a furnace fired by gas or oil. Carbon dioxide is given off as the sodium and calcium components melt and begin to link up in the long silicon oxygen chains. The temperature is raised to about 1500°C (2730°F) at which the glass is very fluid. After the gas is removed, the molten glass is ready for shaping, blowing, molding, drawing, or extruding.

The glaze on pottery, and also the enamel or porcelain on steel kitchen pots, sinks, and bathtubs are glass materials containing ground chemicals of various kinds like titanium dioxide to make the glass opaque. Other coloring agents are added as desired.

Glass-making today represents a beautiful blend of art and technology. Glass is made with various substitutes for some of the original major components as well as the addition of minor ingredients to provide special properties. Other variations can be obtained by the treatment of the glass after forming. Some examples of the variety are listed below.

1. Lead glass uses lead oxide for the calcium carbonate. It is more expensive but has valuable optical properties so that it is often used for the finest tableware and art objects.
2. Special shockproof glass does not expand and crack with sudden changes in temperature. One composition uses boric oxide along with sand as a major ingredient and small amounts of aluminum oxide and sodium carbonate. You know of it under the trade names Pyrex or Kimax, where it is very useful in kitchen bowls. Other applications include baking ware,

Bowl made of 96 percent silica glass, sitting on block of ice, shows resistance to extreme heat shock as molten metal is added. (Corning Glass Works.)

sealed-beam headlights, glass pipe lines in chemical plants, and the giant 200-inch mirror at the Mount Palomar Observatory.

3. Still another type of heat-resistant glass is made of 96 percent silica. It is sold under the Vycor trade name and is used for chemical laboratory utensils, windows for space ships, and in the nose cones of guided missiles.
4. Coloring of glass is an art that goes back to the early Egyptians, who recognized that brilliant colors could be obtained by adding impurities. They are needed in only very small quantities: the addition of one part of nickel oxide to 50,000 parts of glass will give strong tints ranging from yellow to violet, depending on the type of glass; cobalt oxide will give a deep blue color when one pound is added to 10,000 pounds of glass.
5. Many specialty glasses are obtained by heat treatment. All glass objects are annealed, which removes stress and strain set up in formation. Annealing consists of reheating the glass and then cooling it gradually. Other special heat or chemical treatments provide what is called tempered safety glass. It is much stronger than ordinary glass and finds application in optical safety glasses and glass doors. It is hard to break even with a hammer. When it does break, it turns into a mass of small, dull-edged, relatively harmless pieces.
6. Laminated safety glass is common for automobile windshields. It is a "sandwich" made by the adhesion of two layers of glass to a tough, transparent plastic sheet. When the glass breaks by an impact, the broken pieces are held to the sticky interlayer.
7. Fiber glass was known to the Egyptians in rather coarse form before the

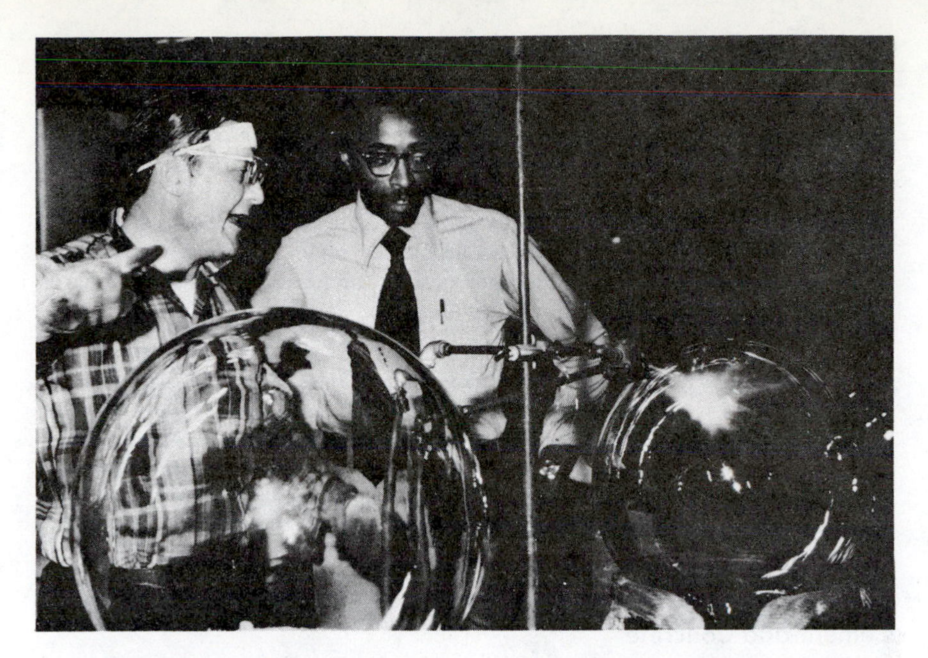

Spheres for undersea exploration being heat-treated. These spheres actually increase in strength under pressure of lower ocean depths. (Corning Glass Works.)

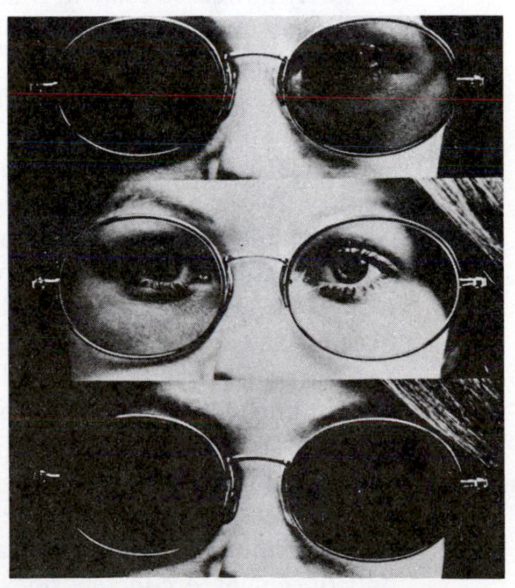

As shown at left, sheet of glass-ceramic is used for an extremely practical smooth-top cooking unit. At right, sunglasses made from photochromic glass darken and clear automatically in response to sunlight. The glasses contain silver halides which undergo reversible changes in form on exposure to UV light, one form absorbing visible light. (Corning Glass Works.)

time of Christ. Practical methods of making it commercially were developed in the 1930s, and now it appears in hundreds of products like helmets, auto bodies, fishing poles, and surfboards. It is an excellent reinforcing agent for plastics. Fiber glass can be formed by spinning out fibers through tiny holes at the bottom of a furnace. Up to 95 miles of fiber can be obtained from an ordinary-sized marble a little over one half inch in diameter. The fibers can be drawn as fine as one-twentieth of the thickness of a human hair. The tiny fibers can be broken off in the skin of your hand and cause considerable irritation, as some amateur molders of surfboards and dune-buggies have found out.

8. Another interesting development is called glass-ceramics. The process was invented by S. D. Stookey at the Corning Glass Works. He and the researchers at Corning found that addition of certain chemicals like metallic silver and sodium fluoride would cause very tiny crystal areas in molten glass. The effect of the billions of tiny crystal areas is to prevent cracks from starting. Glass ceramics are harder than steel yet lighter than aluminum. Household cookingware is sold under the trade names Corning Ware and Pyroceram.

Cement and Concrete

Cement is another of the materials in which rocklike strength is a reflection of interlocking silicon-oxygen bonds. It is made by heating a mixture of powdered limestone and clay in a rotary kiln to a high temperature. The product is small fused lumps or "clinkers" about the size of marbles. These are ground with a small amount of gypsum ($CaSO_4 \cdot 2H_2O$) to give the finely powdered cement. Blast furnace slag may be substituted for part of the clay. You can see from Table 12.1, presented earlier, that clay is chiefly kaolinite, $Al_2Si_2O_5(OH)_4$. Thus the waste slag obtained by removal of silicon-containing materials from iron ore can be put to very effective use.

Cement was first made in England in the early part of the nineteenth century. Because of its similarity to a natural rock from the Isle of Portland, the fine gray cement was named Portland cement. The term *portland cement* is still used today to refer to the type which hardens on contact with water. This distinguishes it from other cements based on rubber or plastics.

Concrete is a mixture of portland cement with sand and crushed stone, slag, or gravel, all made into a paste with water. This is the vital building material of our time. On setting it cures to a rigid solid which serves many uses in foundations, walls, floors, highways, airplane runways, bridges, dams, pipes, and so forth. Sometimes a reference is made incorrectly to a "cement sidewalk." But a sidewalk would be much weaker if it were made only of cement. A concrete sidewalk is extremely strong, since the cement helps in bonding or "cementing" all the particles of sand, rock, slag, and gravel together into a conglomerate mass of great strength. Even more strength can be added by steel reinforcing rods.

There is no simple equation for the curing of concrete. There is no doubt, however, about the basic concept. It is that countless bonds are

Boring from a concrete foundation showing graded size of stones and sand used in the complex mixture. (Photo by Gary R. Smoot.)

formed between the particles in the concrete mix. The four-way bonding of silicon through oxygen is the heart of the complex interlocked network just as it is in ceramics. The role of water is vital. Indeed, one of the functions of the prior heating to high temperature is to drive out water from side chains. This prepares for the later addition of water where bridges can then be formed between chains. The aging process is believed to continue over many years even though the initial set which occurs over the first day or so is sufficient to supply rigidity. Newly poured concrete is kept moist by sprinkling in order to provide the water molecules which are necessary for bonding.

ORGANIC VERSUS INORGANIC MATERIALS

You have seen a few examples of how technology provides materials required by modern civilization. Before considering the problems of technology, you might pause to consider the nature of the production of organic versus inorganic materials. This is important in judging problems of pollution and technology.

Organic materials like wood, cotton, food, leather, and paper are capable of self-renewal. This means they go back into the various cycles of nature and eventually can be made available again for you and other people. The Law of Conservation of Mass is at work here. Matter is not destroyed—just changed in form.

Inorganic materials like metals, stone, and concrete are not self-renewable in the same sense. They do not renew themselves on a practical time scale of human usefulness. In other words, the rotting fencepost or the burned leaves or the discarded wooden chair are likely to return their atoms to active cycling much more quickly than the cracked and dumped concrete sidewalk or the refrigerator buried in the dump. Carbon dioxide from oxidation or burning is picked up the very same day by trees to make more wood for whatever use we wish to put it to. Meanwhile the ceramic coating on the dumped refrigerator door is not likely ever to decorate a kitchen or to be a part of a flower vase. The silicon, oxygen, and iron atoms do not normally find themselves being picked up by the winds for distribution to plants and trees which so effectively recycle much of our organic wastes.

The fact that we are unearthing so much of the inorganic materials like stone and metals creates a twofold problem. First, there is the problem of conservation of limited supplies. Secondly, since there is no natural recycling or self-renewing process, the growing piles of waste inorganic materials become a mounting problem.

TECHNOLOGY INVOLVES MANY PROBLEMS

In thinking about how technology helps to solve some problems, you recognize that it also brings others. The need for a balanced conceptual outlook at environment is so vital today that it would be an oversight to fail to mention several problem areas related to technology. This is of course a vast subject, but several problem areas can be identified. You should note that the pollution problem, which you hear about almost exclusively, is only one aspect of the changes introduced by advancing technology. Here are some of the broad technology problems.

Widening of the gap between rich and poor nations. Studies by the United Nations indicate that the gap between the developed and underdeveloped nations is steadily widening. The development of technology, in contributing to industrialization, has accelerated the economic growth of the countries inhabited by people of European origin. One of the major problems of our time is bridging the gulf between the world of affluence and that of poverty.

Growing need for financing more complex technologies. For example, the expenditure of resources involved in research and development for proposed controlled thermonuclear fusion is enormous. The spiraling demand for electric power emphasizes the need for new sources of energy other than fossil fuels.

Multiplication of automation problems. Technology inevitably leads to automation which entails both monotonous repetition of the tasks of reproducing a single part of a product as well as decreases in the numbers of people employed. The latter is often called technological unemployment. While it may be true that technology also produces new jobs, these are of a different kind, usually involve highly skilled workers, and still leave the difficult problem of retraining the people displaced.

Complexity of technology is emphasized by this control room in a large petrochemicals plant using computerized control systems. (Courtesy Union Carbide Corporation.)

Dehumanizing loss of artistic skills. Technology is often damned as a destroyer of artistic skills. In the old days a carpenter would make a whole chair or table, a weaver a piece of cloth or tapestry, a shoemaker a pair of shoes. Today technology assigns a workman to do part of a job like putting door handles on autos, installing the outside trim, emptying tank cars of chemicals, putting fertilizer into bags, or inspecting electronic parts.

Extreme growth in complexity and quantity of information required. Increasing complexity and proliferation of information is such that an individual person will be unable to grasp and understand the meaning of all the scientific data accumulated. This could, if not taken in stride and with a somewhat philosophical outlook, turn a person against technology. However, computer science promises some help by storage of information in memory banks and associated retrieval systems.

The possibility of the loss of individual freedoms through technological control. This area has been of considerable interest in recent years, especially to social psychologists and students of the newer effects of electronic media like movies and TV. The effects of the communications media on social change has been a phenomenon of our times, and the media depend very heavily on complex technologies. There are growing fears that technology may provide a base to control human personal and social development. There is the additional potential threat implicit in sophisticated information-gathering systems and devices for wiretapping.

Growing needs for precious raw materials. New technology advances require more and more materials, both organic and inorganic. Even the early use of fire introduced a great need for firewood which at first was no problem. However, after the introduction of blast furnaces for smelting of iron, the consumption of charcoal practically wiped out the trees in England.

Pollution and degeneration of the natural environment. Expanding technology invariably involves the by-product of pollution which must be controlled in order to maintain an ecological balance in nature (chapters 15 and 16).

Loss of spiritual values. To some people technology represents an overemphasis on the material in place of the spiritual. There is an urgent need to provide meaning and moral and ethical values to living so that people do not lose a sense of balance and become overwhelmed by what appears to them an overmechanized society. It would surely be the supreme paradox if technology provided only boredom and meaninglessness. In earlier times people had to work so hard just to survive that the arduous work itself gave purpose to life. With the new leisure brought by technological progress, the purpose of life must be sought elsewhere.

EVALUATION OF TECHNOLOGY

How does one evaluate technology? This is another problem of technology. Technically, the term used is *technology assessment.* In other words technology, which is a human development, needs human thought and judgment as to how it is applied. And this is not easy.

Technology per se has no morality. It is not itself either good or bad. When the automobile was first invented it was praised to the skies. Now it is often damned. The chief reason is pollution, to which the auto is a major contributor. However, there is the equally devastating effect of death and injury on the highways. In the U.S. alone more than 50,000 people are killed every year in highway accidents. And hundreds of thousands of others suffer injuries as a result of the technologies of autos, highways, and gasoline. There are more people killed each year on U.S. highways than the total of American deaths in ten years of the Vietnam war. This was hardly the intention of those who pushed the mass production of autos and trucks.

Or take the airplane. When Wilbur Wright and his brother Orville first got off the ground they certainly had no idea of skyjackers holding hundreds of innocent people in jeopardy with guns and bombs.

Here is a more down-to-earth example. Consider the three-hole paper punch. It is a convenient device to punch three holes in paper so that it can be more easily and neatly inserted in a three-ring binder. But what about the thoughtless person who just dumps the tiny paper circles on the floor when the punch is full? That is certainly an abuse of technology. Building custodians probably often wish the three-hole punch was never invented.

These are all side effects of technology. Yet we do not necessarily need highly refined technology to accomplish evil ends. Rocks were probably used to injure others many times before technology evolved the knife. But technology does provide many more refined methods of adversely affecting others. The examples above are oversimplifications, but the principle is clear. Often you hear such statements as, "Man must establish proper goals for technology" or "Mankind must use technology properly to control the environment." These can be misleading because people then figure some man-in-the-abstract or mankind will take care of things. Who actually must finally evaluate or assess technology? It is you and hundreds of millions of other people.

A special panel was convened by the National Academy of Sciences to study the problem of the assessment of technology. These experts in various disciplines recognized that even for them it was difficult to come up with answers. They pointed to the need for more information on potential side effects of technology and development of a way to quantitatively measure benefits against risks. Society today is not equipped to handle the conflicting interests involved. In the end their recommendations included one for setting up federal mechanisms for evaluating the broad social consequences of advancing or holding back various technological developments.

It might be well to end this chapter with the realization that we do not by any means know all the answers to the problems of research and technology. But both scientists and people in general are now looking carefully at the influence of a technical advancement on the whole of mankind and the environment. This is an important attitude to develop.

PROJECTS AND EXERCISES

Experiments

1. This is a research experiment on a very simple matter—the distribution of heads and tails in a group of coins. It is not related to any subject covered in this chapter. You will need some kind of covered container, possibly a box from the local supermarket, and about 50 coins. The box should be of such a size as to permit you to easily count the numbers of heads and tails. Start by placing all coins in the box with heads up. Then close the top, so as not to allow coins to fly out, and shake vigorously. Then count the number of heads and tails and shake them up again. Start a short tabulation like this, filling in all the columns.

Shaking Trial	*Number of Heads*	*Number of Tails*
0	50	0
1		
2		
3, etc.		

Do not purposely make the coins all heads except at the start. Run the experiment for a little while to see if you can get the 50 coins to turn up all heads or all tails. Do you think this is likely to happen? Is it possible? What kind of distribution of heads and tails do you find after running for say 15 or 20 trials? Now try the same experiment with just 10 coins, starting with all heads again and keeping a tabular record of the count. Do you think it would be more likely or more probable to get all heads or all tails here? Next try with just 4 coins. What's the probability of getting all 4 either heads or tails? Try with 2 coins if you have not been successful with 4. What do you think would be a probable distribution if you took 100 coins? Would 1000 coins give you a more likely even distribution (that is, 50 percent heads, 50 percent tails) than 4? In other words, what can you predict from the result of this simple experiment? You may think this is a very unimportant finding. But behind your small research lies a large meaning. It has to do with a concept in science called *entropy,* which means that nature normally prefers a state of disorder to one of order. The entropy or degree of disorder in the universe is said to be increasing in spite of the fact that in some isolated areas things may get more orderly.

2. This experiment will involve considerable imagination on your part. Also you can pick a research project which interests you particularly. Many of the operations that you and other people perform by habit are actually based on some definite reasoning. The purpose of the experiment is to briefly investigate a particular procedure, process, or behavior to determine its basis. Here are a few examples to start you thinking, although you may choose any topic you wish, not necessarily one of these.
   (1) The refrigerator as a dehydration device. We had an early experiment in this area which was a sort of research project to demonstrate once and for all that you should not leave items like lettuce uncovered in the refrigerator if you want them to remain fresh and crisp. (See Experiment 2, chapter 2.) You could expand on this in a more finished way as a research project, although it is probably too "old-hat" for you now.
   (2) Research on the best procedure for wiping up kitchen counter spills. This is a very practical area for cooks. Does a hot dishcloth work better than a cold one? A soapy one better than plain water? Anyone who ever cleaned up in a kitchen knows intuitively how best to wipe up old spills, but research on a small scale might be a more convincing back-up for established cleaning habits.
   (3) Does soaking clothes before washing make washing easier? Or more efficient? This could involve comparison tests with duplicate soiled articles, or clothes half soaked and half not. (Normal soaking is done in warm water and detergent or soap.)
   (4) A related research project could investigate the best way to wash dishes. Is it an advantage to wash dishes right after use? Or does it work just as easily if you pile up dishes from breakfast and wash them at night? You may already have carried out research here without calling it by that name. At any rate, a controlled research experiment might be enlightening. We will consider soap and detergents in the next chapter.

Exercises

3. Choose one of the quotations opening this chapter and write a short paragraph explaining what it means to you now, based on some insights developed in this chapter.

4. Take a look around you and pick out ten specific products which are the result of applied chemical research. For each one indicate some of the concepts and procedures which could have been involved in the research.
5. In copper-plating or electro-refining of copper, what the technologist is really doing is transferring copper atoms from one electrode to another. Draw a simple sketch showing the conceptual back-up for this statement. Then, using your now rather sophisticated conceptual background, explain what happens. Do you see here how the atomic theory model is essential in making sense out of the operational process?
6. Look around your environment and find ten applications for copper. List these in a column and then next to each indicate which is most likely: (a) recovery through recycling or (b) loss through junking. Could the materials which today are sent to dumps with mixed-up trash, garbage, and so on, be more likely recovered in the future? Would conceptual understanding help? How?
7. Consider the problem presented by the inorganic materials which you use every day in providing necessary, or at least desired, conveniences. Suggest three materials of this kind which could be recycled even though now they are normally dumped. How would you go about setting up a practical system?
8. Think about the way conceptual background plays a vital part in directing human efforts involved in conservation of inorganic materials and recycling of obsolete products. Illustrate three cases by practical examples. Show briefly how concept was involved in each case.
9. In this chapter we have discussed several examples of technology influencing society. Can you think of any historical cases of society influencing technology? Describe a few instances and indicate the effects involved.
10. List ten areas of your home life which are considerably influenced by technology, indicating briefly the inventions, products, or materials involved.
11. How would you suggest nations should proceed to lessen the gulf between developed and underdeveloped peoples? Include some ways that technology could help.
12. Select one of the problem areas listed on pages 342–344 as related to advancing technology. Indicate ways which could be followed to minimize or eliminate the problem.

SUGGESTED READING

1. This is a chapter where it would be especially valuable to consult encyclopedia articles for further information on topics of special interest to you. You will be able to find more extensive illustrations and pictures than can be incorporated in this text. Examples of topics worth checking: iron and steel, copper, aluminum, cement and concrete.
2. Briggs, A., "Technology and Economic Development," *Scientific American,* 209, No. 3, pp. 52–61 (September 1963). This is the lead article in a whole issue devoted to various aspects of technology and development. Some of the major articles which are pertinent to this chapter are listed below. Others concern especially the economics of developing nations. You could browse in these if interested.
3. Schurr, S. H., "Energy," *Scientific American,* 209, No. 3, pp. 110–122 (September 1963).

4. Feiss, J. W., "Minerals," *Scientific American,* 209, No. 3, pp. 128–136 (September 1963).
5. Brooks, H., and Bowers, R., "The Assessment of Technology," *Scientific American,* 222, No. 2, pp. 13–21 (February 1970). Also Scientific American Offprint No. 332, W. H. Freeman and Co., San Francisco, 1970.
6. Brown, H., "Human Materials Production as a Process in the Biosphere," *Scientific American,* 223, No. 3, pp. 194–208 (September 1970). Also Scientific American Offprint No. 1198, W. H. Freeman and Co., San Francisco, 1970. This is a very good overall summary of the materials usage by modern industrial society. Interesting comparative data on various materials is provided along with historical perspectives.
7. Shiers, G., "Ferdinand Braun and the Cathode Ray Tube," *Scientific American,* 230, No. 3, pp. 92–101 (March 1974). An interesting history of how the technology of the TV tube evolved.
8. Myrdal, G., "The Transfer of Technology to Underdeveloped Countries," *Scientific American,* 231, No. 3, pp. 172–182 (September 1974).

Great flame follows a tiny spark.

—DANTE ALIGHIERI

A man's house is his castle.

—ANONYMOUS

Beauty is something wonderful and strange that the artist fashions out of the chaos of the world in the torment of his soul.

—W. SOMERSET MAUGHAM

chapter 13

Chemicals in the Home

Looking over the chemicals in the home is much like looking at a kaleidoscope. Chemical usage at home represents variety as well as continually changing patterns. These changes come from the introduction of newer materials into home use and also because the very use of materials involves changes in the materials. Look, for example, at a box of matches with the regular pattern of colored coatings on each and every match. Then observe the change required to get the fire started as you strike the match. A sort of chemical kaleidoscope.

When you casually strike the match on the box coating you may not be aware that some mind similar to your own, but of earlier vintage, sat down with a problem. That person, whose name is lost in the anonymity of time, did not like the idea of rubbing sticks together or striking stones to get a spark for starting a fire. The application of concepts resulted in an invention—creative and practical. Then other minds and hands refined the technology so that matches are now produced in such enormous numbers that they are very cheap to buy.

But too often the cheapness or easy availability of things produces negative attitudes in the user: (1) We tend to take common articles of daily use for granted with no appreciation for the work of human minds and human hands which went into their development. (2) We may lose respect for the materials themselves. These materials are amazingly combined to give us service and make life easier, which is their true service to humans and the only proper way in which man subdues or controls nature. We let nature work for us if we know the ways in which she operates best. Remember Francis Bacon's statement, "Nature to be commanded must be obeyed."

It is worthwhile when beginning a survey of some of the many services of matter to human needs in the home to recall the balance of Bacon's statement. The rest of this chapter will be devoted to details of the operations of things—atoms and molecules directed in the service of people.

CHEMISTRY IN MATCHES

Matches, which are now consumed in the U.S. at about 10 per day on the average per person, are a relatively recent invention. Charles Dickens (1812–1870) in the mid-1800s said that it took a person about half an hour to start a fire by striking the hard, gray, siliceous stone called flint against steel.

In the first attempts to produce chemical fire-making systems, the element phosphorus played an important part, as it was early recognized that phosphorus easily catches fire. *Phosphorus* is derived from the Greek for "I bear light." A few ancestors of the match were:

1. Paper strips tipped with yellow phosphorus. These were sealed in a glass tube until ready to expose to the air which caused them to ignite spontaneously.
2. A French invention called the Instantaneous Light Box. Wooden sticks were coated first with sulfur and then a mixture of potassium chlorate and sugar. The sticks were dipped into a bottle containing asbestos moistened with sulfuric acid. The acid reacted with the potassium chlorate ($KClO_3$) to produce a rapid exothermic reaction which supplied oxygen for burning the complex mixture of fuel. These were very popular in the nineteenth century and were safer than the early phosphorus-based matches.
3. Various types of wooden matches using white or yellow phosphorus. This is an extremely toxic substance and thousands of people were crippled or killed by exposure. The danger was not only to the match-makers, but to people who breathed the fumes while using the matches. Others died by suicide or murder using the match tips.

Finally, in the early part of the twentieth century new mixtures were perfected which incorporated phosphorus in a safer form, for example, phosphorus sulfide, P_4S_3.

The various match formulations now all involve:

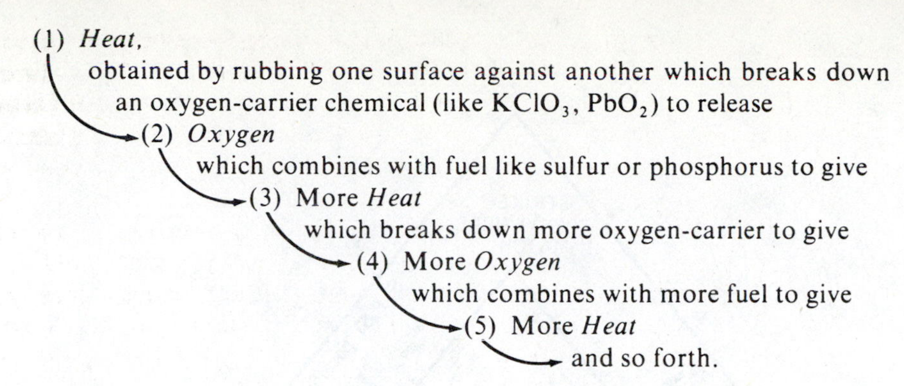

(1) *Heat,*
obtained by rubbing one surface against another which breaks down an oxygen-carrier chemical (like $KClO_3$, PbO_2) to release
(2) *Oxygen*
which combines with fuel like sulfur or phosphorus to give
(3) More *Heat*
which breaks down more oxygen-carrier to give
(4) More *Oxygen*
which combines with more fuel to give
(5) More *Heat*
and so forth.

The result is a rapid build-up of heat such that the ignition temperature is soon reached and the match is in flames. The combinations of chemicals used can be a little complicated, but the underlying action is simple. Figures 13.1 and 13.2 show common match formulations and how a match works, respectively.

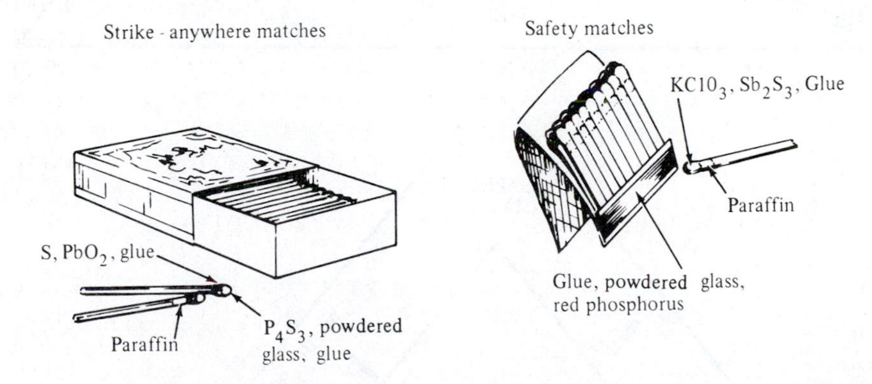

FIGURE 13.1
Common match formulations. The major chemicals are shown. The match sticks are dipped in various salt solutions like ammonium sulfate, sodium phosphate, or ammonium phosphate. This dip provides a controlled fireproofing which prevents smoldering and afterglow. These salts also keep the match from turning to dusty ashes when it burns.

BAKING POWDER AND BAKING SODA

Yeast cells provide carbon dioxide which leavens bread or makes it light. In cakes the leavening agent may be air. This is trapped through use of beaten egg whites or by "creaming"—stirring sugar and butter together. More often the major part of gas in cakes is made by chemical reaction in the batter. Here baking powder is used.

Baking powders are made by mixing starch, sodium bicarbonate ($NaHCO_3$), and a solid which is an acid or can form acid in water. The

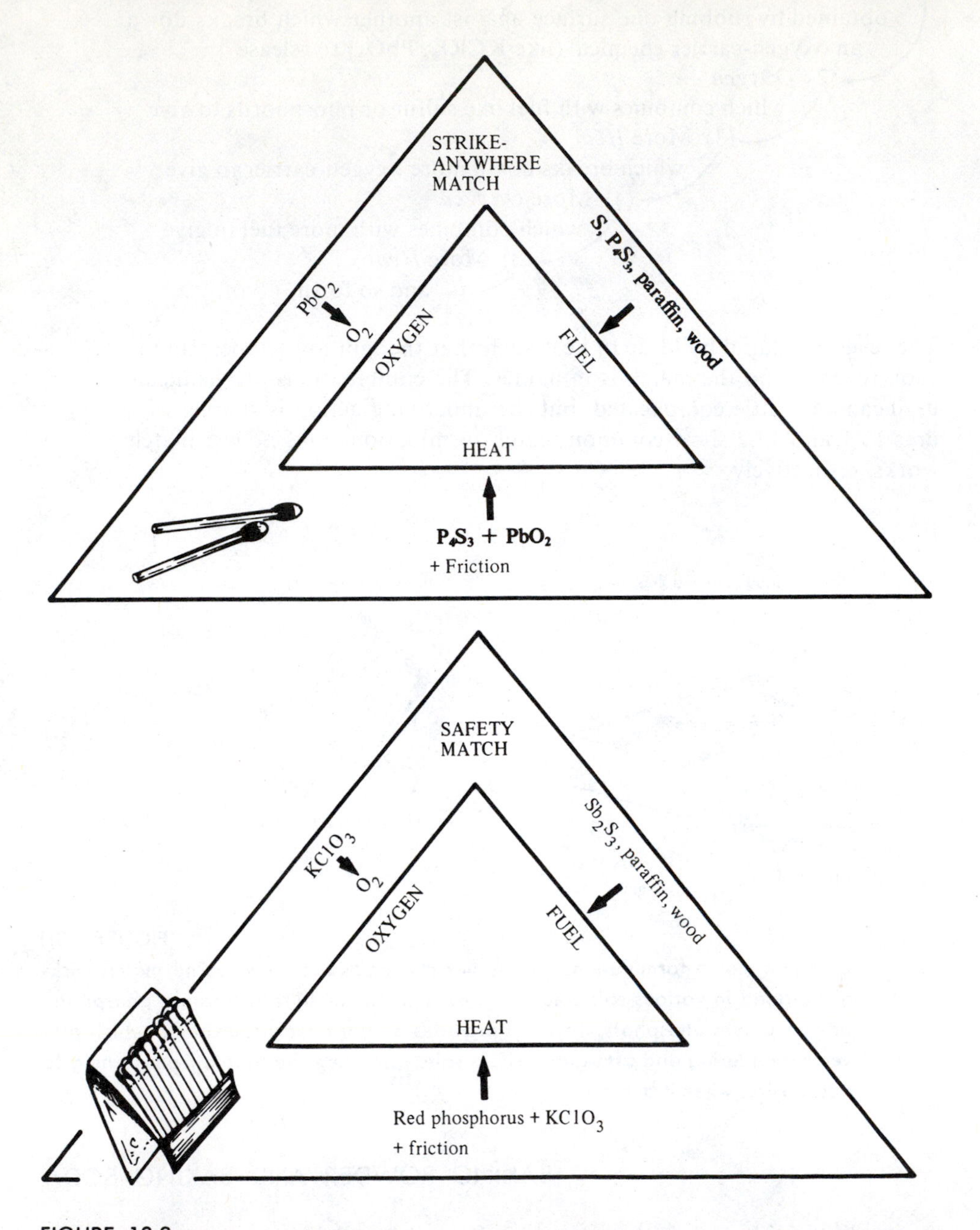

FIGURE 13.2
How a match works. In the center is the fire triangle. The outer triangles show how the two types of matches provide the needs for heat, oxygen, and fuel. The detailed chemical reactions can be quite complex, but the main concept is to provide input for the fire triangle.

starch, which provides the major bulk, helps to keep the mixture dry and prevents caking.

There are various sources of acid in baking powders. A good example is a phosphate baking powder which may use calcium dihydrogen phosphate, $Ca(H_2PO_4)_2$. A tartrate baking powder may contain $KHC_4H_4O_6$, cream of tartar, along with tartaric acid, $H_2C_4H_4O_6$. The major need is for some source of hydrogen ion in water solution (the definition of an acid). Carbon dioxide gas is obtained by reaction of hydrogen ion with the bicarbonate.

$$NaHCO_3 + H^+ \rightarrow Na^+ + H_2O + CO_2\uparrow$$

The beauty of baking powder formulations is that they all use solid forms of acids. These do not give hydrogen ions until the dry powder is mixed with water in the batter.

Baking soda is a name for sodium bicarbonate. Certain recipes use it in combination with sour milk, sour cream, or buttermilk, all of which supply the acid hydrogen ions necessary for reaction with the baking soda. Baking soda is also used as indigestion relief and as a soothing dressing on insect bites and superficial burns. It has a natural alkaline reaction which neutralizes excess acid of any kind. In the stomach this reaction occurs: $HCl + NaHCO_3 \rightarrow H_2O + CO_2\uparrow + NaCl$. This is the origin of "burping" following an alkalizer.

SOAP

The exact date when soap was first made is not known. One legend places the time at about 1000 B.C. and the place a hill near Rome called Sapo Hill. Here people offered burnt animals as sacrifices to the Roman gods. Fats from the animals dripped down through the wood ashes and some observant person noticed that the Tiber River, polluted from this source, had remarkable cleaning power. Actually soap was formed by reaction of the animal fats with alkali in the ashes. The word *soap* is derived from Sapo Hill where conditions were just right for this accidental discovery.

The production of soap in quantity developed slowly. History records a kind of rough soap in use in France about 100 A.D. By about 700 A.D. soap-making was an important craft in Italy and Spain. England did not start soap-making until after the year 1000.

From your knowledge of the various materials of the environment, you can imagine that dirt is a mixed-up mess of "worn-down" materials—mostly long chains. It varies considerably in composition from place to place. You will be glad to know that there is no formula for dirt. But there is for soap.

What Is Soap, Chemically?

Soap is a reaction product of a fat with an alkali. A fat is a rather complex ester of long-chain acids with the alcohol glycerol (chapter 10). Soap-making can be summarized by the following equation.

$$
\begin{array}{l}
CH_2-O-\overset{\overset{\large O}{\|}}{C}-C_{17}H_{35} \quad + \quad Na{:}OH \\
| \\
CH-O-\overset{\overset{\large O}{\|}}{C}-C_{17}H_{35} \quad + \quad Na{:}OH \quad \longrightarrow \quad 3C_{17}H_{35}\overset{\overset{\large O}{\|}}{C}-O-Na \quad + \quad \begin{array}{c} CH_2OH \\ | \\ CHOH \\ | \\ CH_2OH \end{array} \\
| \\
CH_2-O-\overset{\overset{\large O}{\|}}{C}-C_{17}H_{35} \quad + \quad Na{:}OH
\end{array}
$$

A solid fat (glyceryl tristearate) + Alkali (sodium hydroxide) → A soap (sodium stearate) + Glycerol (by-product)

A typical soap is better visualized as

$$CH_3CH_2CH_2CH_2CH_2CH_2CH_2CH_2CH_2CH_2CH_2CH_2CH_2CH_2CH_2CH_2\overset{\overset{\large O}{\|}}{C}-O^-Na^+$$

This is the $C_{17}H_{35}COONa$ molecule shown above. But here you can see better the long, nonpolar chain of carbon atoms which is a necessary part of the soap action. Also on the end is a necessary polar grouping of the sodium salt

$$-\overset{\overset{\large O}{\|}}{C}-O^-Na^+$$

What soap does, then, is to entangle very large numbers of long chains or "tails" in the dirt or grease to be washed out of clothes, off dishes, and so on. Then the polar ends, or "heads," stick out of the tiny globs or droplets which are thereby easily carried off in water. The soap forms a sort of bridge between water and oil. This is another example of emulsification, mentioned in chapter 10 on foods.

An *emulsion* is a mixture of liquids wherein one of the liquids contains tiny droplets of the other which are evenly distributed throughout. An emulsion is not a true solution like sugar in water, or salt in the ocean. In other words, the liquids do not dissolve in each other as individual molecules. One is dispensed in the other as tiny droplets, each droplet containing many molecules.

The action of soap involves breaking up dirt into small globs, each of which is "pin-cushioned" by millions of the tail ends of molecules. The hydrocarbon portion, or tail, dissolves in the greasy glob. The $-COO^-Na^+$ ion migrates slightly into the water, leaving the droplet with a negative charge from the negatively charged heads which are at the surface. Each droplet tends to repel other droplets and this is the conceptual basis for the emulsification which occurs. Then the dirt can be flushed away in water. (See Figure 13.3.)

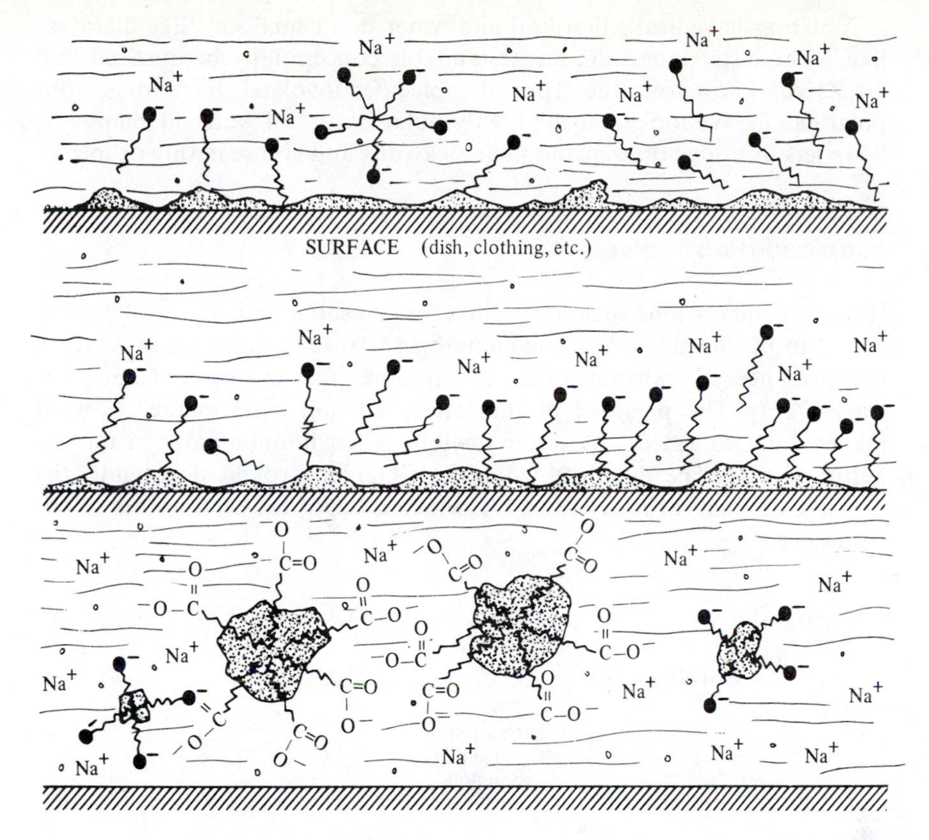

FIGURE 13.3
How soap works. Food residues involve oil and greasy deposits. The soil on clothes consists of various dusty debris which is imbedded in natural oils rubbed off your skin. These are all broken up into tiny, wettable droplets by the action of billions of soap molecules.

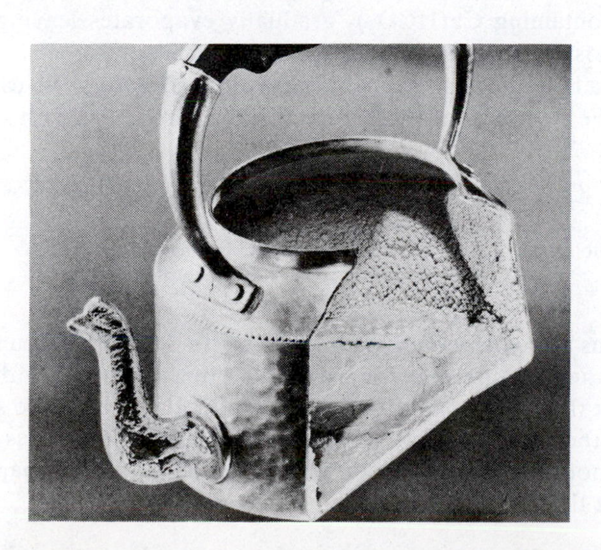

Hard water scale can best be observed in limed-up tea kettles. Here is an extreme example of long-time accumulation of calcium carbonate scale. (Courtesy Culligan International Company.)

You may have heard that "oil and water don't mix" or "like dissolves like." These statements are understandable conceptually because oil and fat, as you know from the types of molecules involved, have long, nonpolar chains. We looked at the highly polar nature of water in chapter 7. Soap lets us bridge the gap and float away dirt and grease in tiny droplets.

Soap and Hard Water

If water contains ions such as calcium, magnesium, or iron in solution it is said to be "hard" water. The commonest cause of hardness in water is the presence of calcium salts in solution, for example, $CaSO_4$ or $Ca(HCO_3)_2$. The presence of the latter, calcium bicarbonate, is what makes the hardness of the water obvious when boiling water in the tea kettle or pots. The dissolved salt $Ca(HCO_3)_2$ is broken down into the rocklike calcium carbonate, $CaCO_3$, which is deposited on the sides of the container.

$$Ca(HCO_3)_2 \xrightarrow{\Delta} \underset{\text{Crust or fur in tea kettles and pots}}{CaCO_3} + \overset{H_2O \ + \ CO_2 \ \uparrow}{H_2CO_3}$$

This is the exact opposite of the way the water became hard in the first place, that is, by water dissolving limestone rock as it trickled through the earth or ran down in streams (Figure 13.4).

The action in your tea kettle is similar to that in underground caves where stalactites form on the ceiling and stalagmites on the ground. Here the water containing $Ca(HCO_3)_2$ gradually evaporates leaving the limestone redeposited as beautiful long spikes and spires.

Making a lather with soap is difficult with hard water. The calcium ions (or Mg^{++}, Fe^{++}, and so on) form an insoluble compound with soap:

$$\underset{\text{A soluble soap}}{2C_{17}H_{35}COONa} + Ca^{++} \rightarrow \underset{\text{Insoluble calcium soap curd (calcium stearate)}}{(C_{17}H_{35}COO)_2Ca} + 2Na^{+}$$

This explains why you get a ring on the bathtub or a sort of scum on water in a basin when you use soap for washing. When you wash with soap you first convert the calcium salts to curdy compounds. Then more soap molecules are lathered up to go to work on the cleaning job. Thus with hard water you not only get a scum or ring but you also waste soap. How do you prevent this?

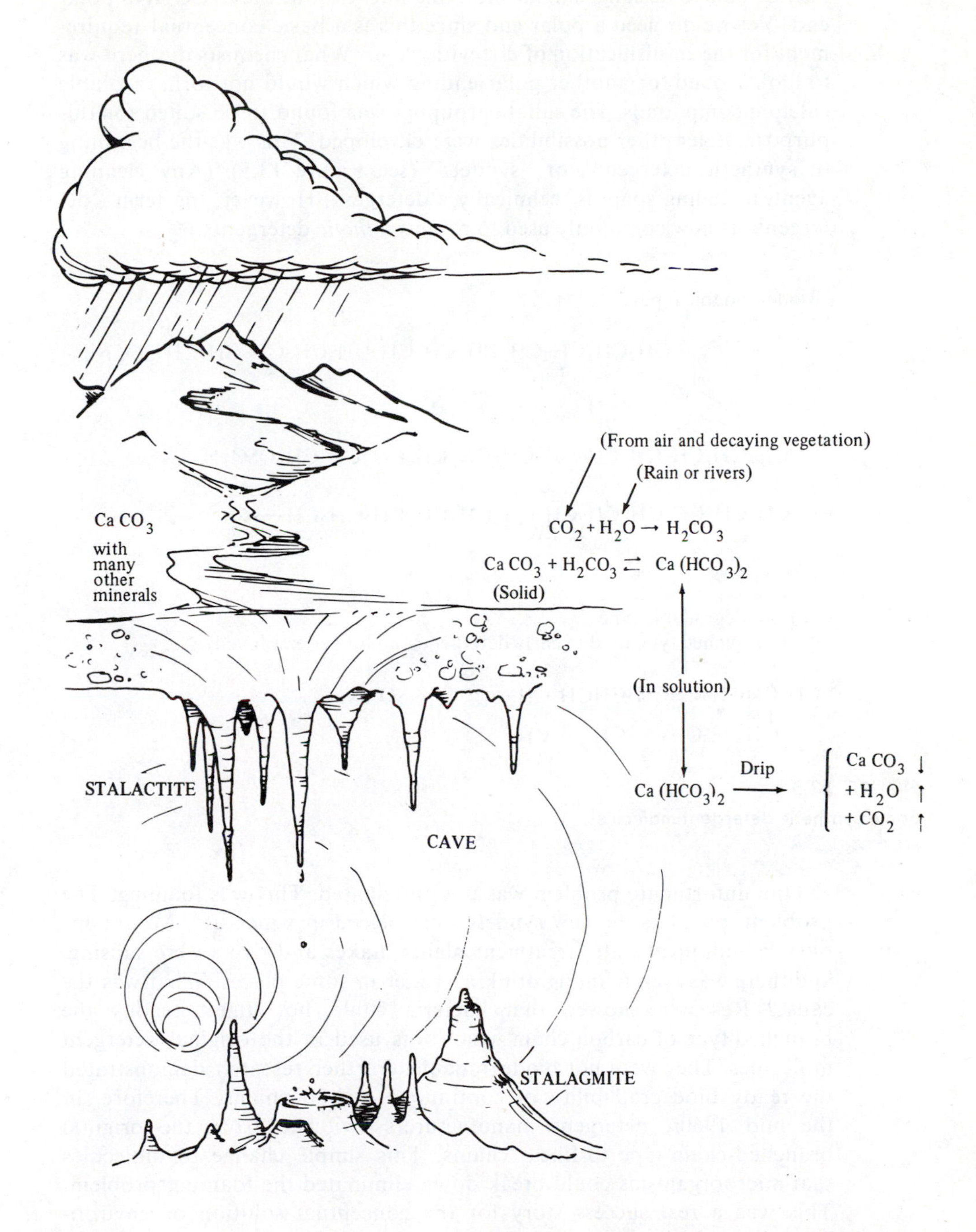

FIGURE 13.4
Water becomes hard by dissolving limestone rock.

Solving the Hard Water Problem: Synthetic Detergents

Since we know conceptually the mechanism by which soap works, we may be able to design a similar molecule but without the —COONa polar end. Yet we do need a polar end since this is a basic conceptual requirement for the emulsification of dirt with water. What chemists did here was to look around for another polar ending which would not form insoluble calcium compounds. The sulfate grouping was found to be suited for this purpose. Later other possibilities were developed. This was the beginning of synthetic detergents, or "syndets" (see Figure 13.5). (Any cleaning agent, including soap, is technically a detergent. However, the term "detergent" is now commonly used to mean *synthetic* detergents.)

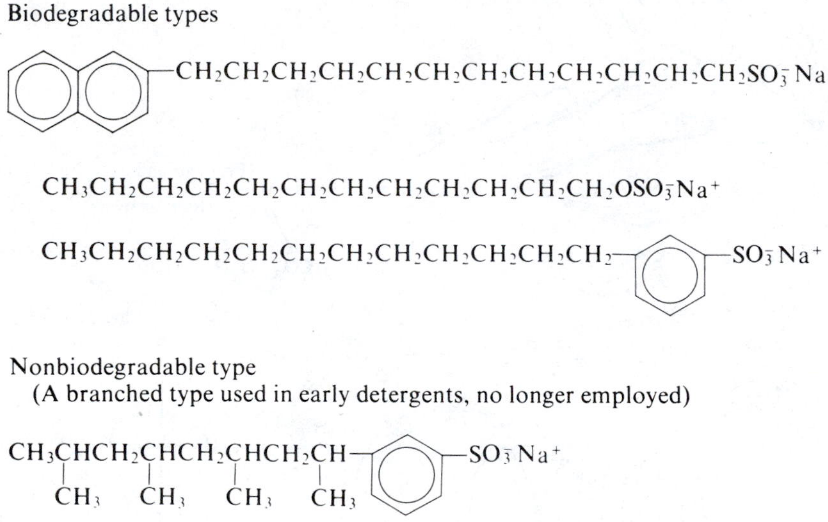

Biodegradable types

$$C_{10}H_7-CH_2CH_2CH_2CH_2CH_2CH_2CH_2CH_2CH_2CH_2CH_2CH_2SO_3^-Na^+$$

$$CH_3CH_2CH_2CH_2CH_2CH_2CH_2CH_2CH_2CH_2CH_2CH_2OSO_3^-Na^+$$

$$CH_3CH_2CH_2CH_2CH_2CH_2CH_2CH_2CH_2CH_2CH_2CH_2-C_6H_4-SO_3^-Na^+$$

Nonbiodegradable type
(A branched type used in early detergents, no longer employed)

$$CH_3CH(CH_3)CH_2CH(CH_3)CH_2CH(CH_3)CH_2CH(CH_3)-C_6H_4-SO_3^-Na^+$$

FIGURE 13.5
Some synthetic detergent molecules.

One unfortunate problem was not anticipated. This was foaming. The problem spread as the new syndets were placed in wider use. Mountains of suds built up at water treatment plants. Lakes and rivers were sudsing, and there was even foaming drinking water in some places. What was the cause? Research showed that bacteria could not digest readily the branched type of carbon chain which was used in the original detergent molecules. They were not biodegradable. Further research demonstrated the ready biodegradability of continuous carbon chains. Therefore, in the mid 1960s, detergent manufacturers switched from the original branched-chain type to linear chains. This simple change to molecules that microorganisms could break down eliminated the foaming problem. This was a real success story for the conceptual solution of environmental problems. Soaps have linear chains so that they are readily decomposed by bacteria.

Toilet Soaps

The syndet story is a dramatic example of the fact that structural arrangements of carbon atoms in molecules have great effect on their operating properties. Another example involves toilet soaps (also called bar soaps). Here we recognize that the fully saturated long chain soap molecules provide a harder, more insoluble soap than the unsaturated ones. Also the longer chains are harder and more insoluble than the shorter type. These findings are related to the oleomargarine problem where we saw that unsaturated chains are not as easily aligned side by side and are therefore more likely to be liquids than solids (chapter 10).

Olive oil has a very high percentage of unsaturated oleic acid and is used in making fine toilet soap such as Castile soap. Palm oil also contains rather high oleic acid content, and it finds wide application in toilet soaps. Soap made from plain beef fat or tallow is very hard because of the high content of long chain saturated molecules. Soap chemists blend coconut oil, which has a high percentage of shorter chain molecules, with tallow as starting material for a softer soap. Cottonseed oil which contains a high percentage of linoleic acid (with two double bonds per molecule) is also used as a cheap soap ingredient contributing to a softer end product.

How about floating soap? Here warm soap is beaten to trap billions of very tiny air pockets before it is extruded and cut into bars.

Shampoos, like soaps, come in a great many varieties. They are complex mixtures of various soaps, synthetic detergents, water softening agents, mineral oil, lanolin, dyes, and perfumes.

Choosing Soaps at the Supermarket

Acids are normally present in soiled garments and these react with soap to reduce its effectiveness by converting the sodium salt molecules to the poorly ionized fatty acid. For example:

$$C_{17}H_{35}COO^-Na^+ \; + \; H^+ \; \longrightarrow \; C_{17}H_{35}COOH \; + \; Na^+$$

Ionic soap molecule — From acid in soiled garment — Non-ionic fatty acid molecule

In order to provide greater alkalinity to neutralize the acid, soap manufacturers have "built" their soap products with alkaline products. These also help to counteract the hardness of water. At the supermarket you have: (1) unbuilt or light-duty soap (like Ivory flakes, Ivory snow, Lux flakes) and (2) built or heavy-duty soap (like Fels Naptha).

The unbuilt soap products are approximately 95 percent soap. They also often contain a brightener. This is a fluorescent organic dye which becomes fixed on fabrics during washing. Then in sunlight the dye molecules absorb ultraviolet light and give off visible light. This provides a brighter visual effect. The dyes can be selected to favor the release of light in the blue range to counteract the usual yellowing of clothing.

The built or heavy duty soaps contain from 55 to 80 percent soap along with varying amounts of alkaline chemicals. The builders commonly used include washing soda ($Na_2CO_3 \cdot 10H_2O$), borax ($Na_2B_4O_7 \cdot 10H_2O$), water glass (sodium silicate, Na_2SiO_3), and trisodium phosphate (Na_3PO_4). These materials increase sudsing, improve cleaning action, and help to prevent hard water formation of soap scum. The built soaps also usually contain a brightener.

SYNDETS

Synthetic detergents (called simply "detergents") are also available in the two general types—unbuilt for light-duty applications and built for the more grimy laundering jobs. The unbuilt syndets, which also contain a brightener, are nonalkaline and are therefore safe for wool and silk and dyed fabrics. The built syndets are similar to built soaps containing various alkaline materials. They also contain brightener and often other additives such as carboxymethyl cellulose which helps keep the dirt in suspension so that it rinses out without redepositing on the clothing.

Detergents may be good suds-formers like soap. Others are low-suds products designed for automatic washers where high sudsing interferes with the mechanical action. There are hundreds of syndets on the market and the variety is further multiplied by the fact that they come in liquid, powdered, or tablet form.

Enzymes used in some detergents are obtained chiefly from a spore-forming bacteria (Bacillus subtilis) commonly found in the air. Protein and fat molecules are often a part of dirt on clothing. The enzymes break up the chains or digest these organic molecules in a similar way to enzyme action in the body. The enzymes require more closely controlled washing conditions, especially temperature in the range of 50 to 70°C (122 to 158°F). In addition, they are destroyed by bleaching agents. They have to be used cautiously also because some people are sensitive to them. Breathing the fine powder should of course be avoided.

The Phosphate Controversy

You sometimes hear people talking about phosphates as if they were poisons. Actually the opposite is the case. They are foods. Phosphates serve us in many important ways: as building material with calcium in bones and teeth, in brain and nerve tissue, in blood as part of control processes and metabolic reactions, in certain types of proteins, and in special kinds of molecules involved in control of protein structure and the transfer of genetic information. It is because of their nutrient properties that phosphates have been implicated in excess growth of algae in lakes (chapter 15).

The following experience shows how complicated a problem can be when it involves chemicals and environmental processes. Environmental concern reached a peak in 1970 and some organizations were boycotting sale of phosphate detergents. In March 1970 the Secretary of Commerce and manufacturers met to discuss reducing phosphate levels in detergents. The manufacturers recommended against hasty action in adopting a suggested new phosphate substitute, NTA (sodium nitrilotriacetate). However, government pressure was mounting for a phosphate substitute. One big manufacturer announced in full-page newspaper ads at the end of March that it would start using NTA in some products. On May 1, 1970 the Department of the Interior's water quality agency publicly demanded a cut in phosphate use by all manufacturers. New plants had been built to produce NTA to meet the rapidly growing demand. Then in December, 1970 the picture abruptly changed. The Department of Health, Education, and Welfare called an urgent meeting in Washington, D.C. The Surgeon General announced that studies with rats indicated that NTA under some conditions may cause birth defects. The government people urged detergent manufacturers to "voluntarily" cease using NTA. By 1971 almost 100 brands of phosphate-free detergents were being sold in the U.S. Some localities banned the sale of phosphates in detergents. In September, 1971 the U.S. Surgeon General announced that some of the newer nonphosphate detergents were causing injuries because of the high content of caustic alkali materials. The Surgeon General issued a joint statement with the Environmental Protection Agency and the Food and Drug Administration recommending the use of phosphate detergents as a less harmful alternative.

The problem of phosphates in detergents is still not solved, but the experience should be useful. This may be a classic example of the need for extensive research on environmental problems.

Detergent Characteristics

The magazine *Changing Times* had an article at the height of the phosphate controversy with the title, "Soaps, Detergents and Why Nobody Can Say What's Best."

From an analysis of all the pros and cons, plus a little knowledge of the chemistry involved, we could come up with a summary of practical hints:

Soap, even though it is preferred by environmentalists, does not clean as well as detergents (synthetic) and can produce curdy deposit in hard water areas.

Nonphosphate detergents should be used with care. In general they substitute washing soda (Na_2CO_3) for phosphates and this has high alkalinity. You will not necessarily get a rash if a little gets on your skin. But alkalies are dangerous if swallowed, breathed as dust, or if they get in your eyes. This is especially important if small children are around. The FDA estimates that around 20,000 toddlers swallowed phosphate detergents every year when they were the major home laundry detergent. This caused few problems. But highly alkaline nonphosphates can cause considerable damage. If they get into the eyes the recommendation is to wash freely in running water and see a doctor.

There are optimum ways to use nonphosphates. In hard water they precipitate compounds of Ca^{++}, Fe^{++}, and Mg^{++} ions which cause clothes to look dingy. This can be minimized by adding the detergent to water before adding clothes (preventing the precipitate from forming in the clothing fibers), by use of hot water (to cut down on amount precipitated and make it less adherent), and by making sure clothes are well rinsed (to remove deposit on the surface of the cloth). In hard water areas phosphate-free detergents produce a build-up of scale (precipitated hard water compounds) which can interfere with the functioning and life of pumps, agitators, and filters in washing machines as well as cause excessive abrasive wear on clothes. Recommendations here are inspection to check for build-up of scale and removal by occasional use of phosphate detergents or other methods.

SOFTENING HARD WATER

Water softening consists of removing the "hardness" ions such as Ca^{++}, Mg^{++}, and Fe^{++}. The most modern and most common method is now ion-exchange. Water is simply passed through a column containing materials which exchange the ions for others. In the ordinary home water softener, the system merely replaces Ca^{++}, Mg^{++}, and so forth with Na^{+} ions. Industrial needs for very pure water are met by removal of all ions.

The materials first used in ion-exchanger tanks were the zeolites which are naturally occurring silicate minerals (chapter 12). Later artificial

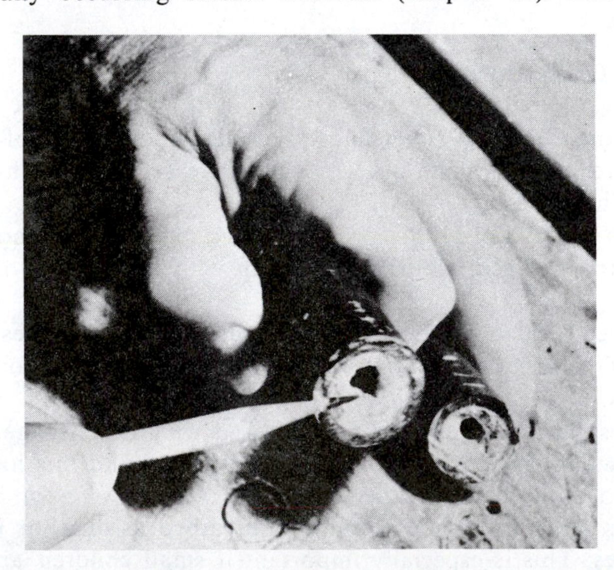

Hard water scale deposits in hot water pipes, valves, dishwasher jets, and shower heads as solid rock, where it chokes off water flow. (Courtesy Culligan International Company.)

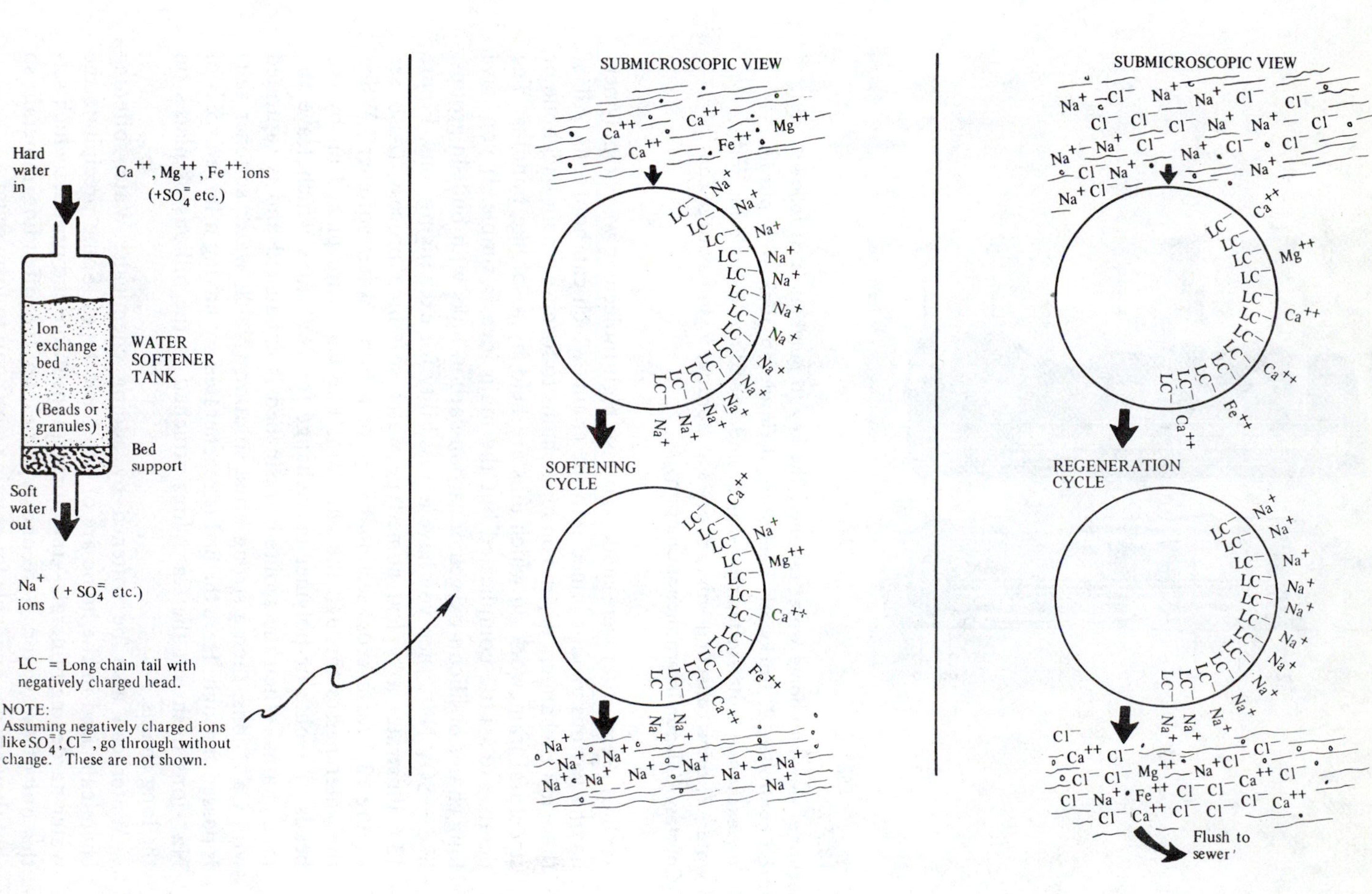

FIGURE 13.6
A home water-softening process.

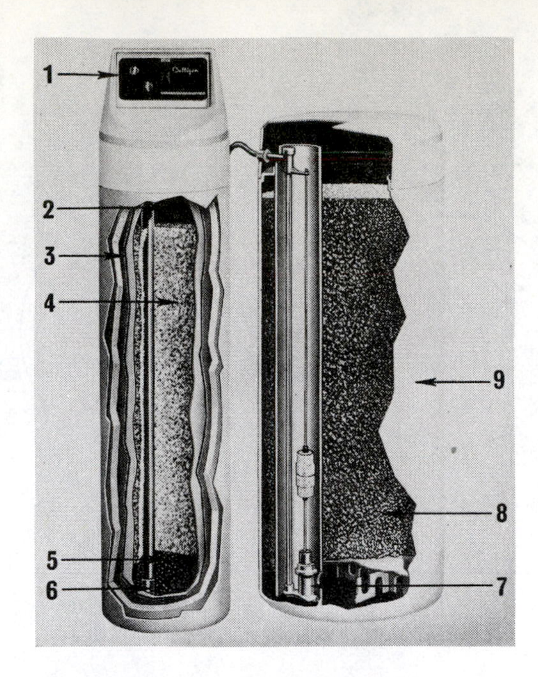

FIGURE 13.7
Cross-sectional view of a home water softener. The key to numbers is as follows: 1. recharge controller; 2. backwash freeboard; 3. insulated tank with plastic liner; 4. ion exchange water-softening resin; 5. underbedding; 6. collector for treated water; 7. brine collecting valve; 8. dry salt storage; 9. salt storage container. (Courtesy Culligan International Company.)

zeolites (Permutite) were made and also high-molecular weight polymers having various groups joined into the chains which can hold ions. All of these ion exchangers depend on long chains to form the main backbone or structural framework on which ions are held by a saltlike bonding. The details are a little complicated but the main idea is simple. If you have long chains of silicon-oxygen or carbon-carbon links with built-in groups, like $—SO_3^+Na^-$, then you have a possibility for exchanging ions. Figure 13.6 illustrates a typical home-type water softening process. (Also see Figure 13.7 for a cross-sectional view of a home water softener.) When the water passes through the unit, ions like Ca^{++} are picked up by the beads of zeolite or polymer in exchange for Na^+ ions which leave the long chains. When the zeolite or resin bed is exhausted and is saturated with Ca^{++} ions, then a strong brine solution (NaCl, water softener salt) is passed through. Here the bed is regenerated by having a large excess of Na^+ ions flush out the Ca^{++} ions attached in the billions of places on the long chains.

Water can also be softened by adding chemicals. Water-softening chemicals used in the home are of two basic types. Some precipitate the water-hardening minerals—they form an insoluble end product which settles out of the water. A second type ties up the offending minerals so that they are not available for reacting with soap to form scum.

An example of the first type is washing soda, sold under the Arm & Hammer trademark:

$$Ca^{++} + 2Cl^{-} + 2Na^{+} + CO_3^{=} \rightarrow CaCO_3\downarrow + 2Na^{+} + Cl^{-}$$

Hard water ion — Obtained by dissolving washing soda ($Na_2CO_3 \cdot 10H_2O$) — Precipitate of insoluble solid (forms cloudiness) — Ions left in water (now "soft")

An example of a softening agent which ties up the offending hard water ions is the complex phosphate, sodium hexametaphosphate $Na_6P_6O_{18}$. (Calgon is a trademarked product of this type.)

The following shows the reaction in water solution as ions:

$$2Ca^{+2} + 2SO_4^{-2} + 6Na^{+} + P_6O_{18}^{-6} \rightarrow Ca_2(P_6O_{18})^{-2} + 6Na^{+} + 2SO_4^{-2}$$

Calcium free to form soap curd — Calcium tied up in a complex ion so that it cannot form soap curd

MISCELLANEOUS HOUSEHOLD CLEANERS

There are many other household cleaning jobs besides that of washing clothing. Here special products have been designed for particular applications. You are already aware of the fact that light petroleum fractions could be used to dissolve oil stains. Several household products now use organic solvents in various formulations for this purpose. In addition, special combinations of materials have been designed to accomplish a particular type of cleaning job. Here concepts are tied in closely with operational requirements.

Powdered Cleansers and Bleaches

The common kitchen and bathroom cleaners are mixtures of materials selected for a particular application. To create the abrasive action needed to remove crusted stains on sinks, pots, or kitchen utensils, finely powdered sand or pumice (a light spongy stone from volcanoes) is added along with the usual soap, syndets, builders, and perfumes. In addition many of these powdered cleansers contain bleaching agents.

Bleaches are used to remove the dingy yellow coloration which develops in aged cellulose-type materials. In the early days light-colored clothing was bleached by hanging it out on the clothesline where the sun and air cause some whitening and germicidal action. However, several chemicals have been developed which achieve a high degree of bleaching action rapidly. They are essentially oxidizing agents. The most common bleach in home use today is sodium hypochlorite (NaOCl) which is sold in a 5 percent solution of water, such as Clorox. It is interesting that this fairly dilute solution is still very powerful. New brides have often found out to their distress that liquid bleach should not be dumped into dry laundry in the washing machine. It must be diluted with water before adding to clothing. Directions on bleach bottles point this out. But many people read the directions only after they have a few holes burned in their laundry.

Another common bleaching agent is called bleaching powder. The formula is usually written $CaOCl_2$ or $CaCl(OCl)$. The active oxidizing part is the OCl^- ion which is also present in liquid bleach. Bleaching powder is also used to kill germs and algae in swimming pools. A special "high test" bleaching powder has been made which is largely $Ca(OCl)_2$.

Toilet Bowl Cleaners

Since toilet bowls are continually full of water with a large surface for evaporation, there is a potential buildup of crusted salts like $CaCO_3$. The surface deposit can also contain magnesium and iron salts from the hard water along with various organic coloring agents.

Once the problem is recognized, the solution conceptually is rather simple: use a strong solution of acid which dissolves the crusted $CaCO_3$. This is exactly the procedure chemists have followed. The chief cleaning agent for toilet bowls has been a solid acid salt, sodium hydrogen sulfate, $NaHSO_4$. This is still used quite extensively. Recently a more convenient product in liquid form has become available. This is a solution of hydrochloric acid, HCl, of which the usual concentration is about 9 percent. Commercial cleaners generally also contain a disinfectant and fragrance chemical.

Bowl cleaners provide hydrogen ions from strong acid in the cleaner which react with the crusted salts to give ions in solution which can then be flushed away.

$$\underset{\text{Crusted bowl deposit}}{CaCO_3} \xrightarrow[\text{From HCl or } NaHSO_4 \text{ in solution}]{H^+} \overset{H_2O \,+\, CO_2 \\ \uparrow}{H_2CO_3} + \underset{\text{Salt in solution (as chloride, or sulfate)}}{Ca^{++}}$$

Bowl cleaners are very effective in preventing build-up of hard water deposits. They are powerful chemicals, however, and can cause skin irritation and eye injuries as well as deterioration of clothing. If any gets in the eyes, the required first-aid is immediate flushing with water.

Drain Cleaners

Clogged drains in home sinks result from a combination of matter including fat and grease from foods, hair, coffee grounds, lint, and smaller bits of food residues. The chief culprit is greasy, fatty material which may be liquid when washed down the drain and then solidifies on cooling. Consequently, the conceptual approach to cleaning drains is different from that of the bowl cleaners.

You can of course appreciably cut down on the need for drain cleaning by simply not building up excessive fatty deposits in the first place. In other words, pour off food fats and grease into separate cans where they can be solidified and even sold for a few cents a pound to certain butcher shops. They are recycled to manufacture plastics or soap, and you end up with fewer drain problems.

In many of these cases of drain blockage a simple procedure can open up the drain. A first try at opening a drain can be pouring in boiling water—not just a cup or two but a couple of gallons. This may work if the drain is not completely clogged. The build-up of heat may melt the greasy deposit. Sometimes a plumber's "snake" or coiled wire is useful in breaking through a blockage. Then hot water may complete the clean-up.

In the event chemical drain cleaners are necessary, they must be applied with care. The usual drain cleaner is based on lye or NaOH, which acts on the fat to form soap. Lye is a strong alkali which can damage the skin and cause severe eye injury.

Many of the solid cleaners also contain chips of aluminum. These serve to release hydrogen gas to keep the drain open and prevent solidification of the cleaner itself. Producing bubbles is a common reaction of alkali on aluminum metal and can be represented as:

$$2Al + 2NaOH + 6H_2O \rightarrow 2NaAl(OH)_4 + 3H_2$$

The release of hydrogen gas introduces an additional explosion hazard. Care must be exercised to keep flames, sparks, or cigarettes away.

A more likely hazard with drain cleaners is skin or eye injury. Most injuries occur because people use a plumber's plunger or snake with, or just following, the application of alkaline drain cleaner. The possibility of splashing the chemical into the eyes is so great that most chemists would recommend never using any mechanical device with the chemical cleaners.

Liquid drain cleaners are also dangerous because they are usually concentrated solutions of sodium hydroxide in water. Newer enzyme cleaners based on petroleum solvents which help to dissolve grease do cut down on possibility of skin injury. But they add the usual hazards of petroleum products—fire and the dangerous possibility of children swallowing the material.

Never Mix Cleaners

Some people try to make a toilet bowl cleaner more effective by adding bleach, or sometimes ammonia solution is mixed with chlorine bleach. Such mixtures can be dangerous. No cleaning product should be mixed with another. You can get reactions releasing a lot of heat with splattering or even poison or irritant gases. Chlorine is the poison gas released when bowl cleaner is mixed with bleach. This is highly irritating to the mouth, eyes, and lungs, and pneumonia may result.

Chemical Oven Cleaners

Since a major component of oven spatter is fatty materials from roasts, alkalies are useful in removal. As with the alkaline drain cleaner, soap forms right on the oven wall from residual fat by reaction with sodium hydroxide.

The original alkali oven cleaners contained jellied or paste sodium hydroxide solutions in order to provide a thick coating for longer contact with the oven wall. Now the trend is toward aerosol products. These use sodium hydroxide as a cleaning agent along with thickeners and propellant.

The high alkalinity necessary to remove oven stains (pH greater than 13) emphasizes that special care is required. These alkalies can cause severe skin burns and serious eye damage. The aerosol type is especially dangerous because of fire hazard and explosion of the can if carelessly left on a hot stove or in the oven. Such accidents have not been uncommon.

ADHESIVES

Adhesive is a broad term which includes glue, mucilage, paste, and organic cements. The word *glue* originally referred to adhesives prepared from animal proteins such as hides, hoofs, and cartilage. Other natural sources for adhesives used in the early days were starch, gum, and dextrins. The adhesive property of dextrins is discussed in chapter 10 on foods.

A natural requirement for adhesives is to provide contact between surfaces. Most solid surfaces appear smooth but they are actually very rough when viewed through a microscope. Liquids are often the medium for sticking substances together because they provide extensive molecular contact between surfaces by filling in spaces to form tight bonding.

There are two major ways of providing bonding. One uses a binder along with a solvent to prepare a paste or liquid. The bond is made when the solvent evaporates or is absorbed by the material being bonded. The second type requires mixing of two substances which react chemically to form a polymer which bonds the surfaces. The epoxy adhesives are of this type.

Selection of the best adhesive product depends upon the type of materials to be bonded. In bonding paper, wood, or other porous materials, the solvent-type adhesive is often used. For bonding two pieces of metal, an adhesive which cures by chemical action is preferred. However, for bonding a porous to a nonporous surface, like fabric to metal, the solvent-type may still be used since the solvent molecules can escape. In many wood and paper adhesives, the solvent used is water. When the adhesive binder is a synthetic polymer, rubber, or natural gum, organic solvents are used.

Besides natural gums, wheat flour, corn starch, and similar plant materials, adhesive chemists now use long chain, synthetic molecules like those outlined in chapter 9.

AEROSOL SPRAYS

One of the most common containers around the house is the aerosol spray can. Originally this type of spray device was used chiefly for insecticides. Now pressurized can dispensers are used for practically everything from hair sprays to whipped cream and even toothpaste. Some of the applications may be a bit far-fetched as retailers attempt to capitalize on the apparent irresistability of pressurized cans to shoppers.

The term aerosol basically refers to a dispersion of very fine (colloidal) particles of a liquid or solid in a gas. The particles are suspended in gas as tiny droplets or solid specks. Common synonyms are fog or smoke. Because these dispersions have in recent years been achieved chiefly by pressurized cans, the term aerosol has come to be applied in everyday speech to the container itself.

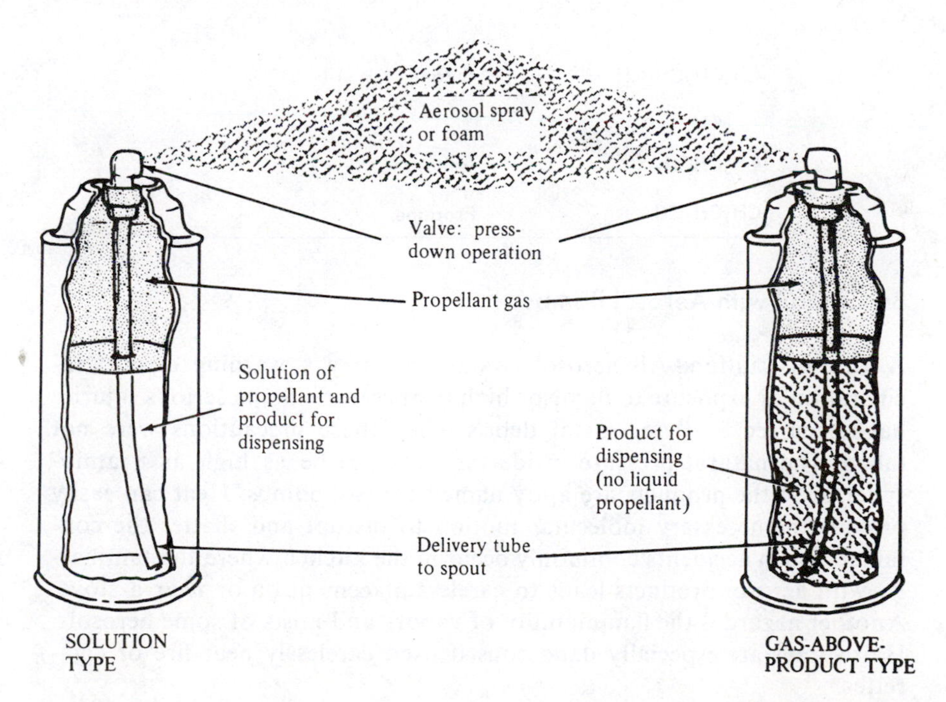

FIGURE 13.8
Two types of aerosol spray.

Two Types of Aerosol Spray

The two basic methods of producing an aerosol spray are: (1) forming a solution of the material to be dispensed in a liquified propellant, or (2) using a separate propellant gas which lies above the liquid which in turn is pushed out by the gas when a valve is pressed. (See Figure 13.8.)

The propellant gas must be suitable for the product to be dispensed. For example, food materials like whipped cream, cheese spreads, and pancakes have gas propellants which give no flavor or smell, like nitrogen, carbon dioxide, and dinitrogen oxide (also called nitrous oxide, N_2O), or mixtures of these. Products like paints, hair lacquers, and insecticides can use a variety of propellants. Examples are shown in Table 13.1.

TABLE 13.1
Some Propellants for Aerosol Sprays

CCl_3F (Cl—C(Cl)(Cl)—F)	Trichloromonofluoromethane
CCl_2F_2 (Cl—C(F)(F)—Cl)	Dichlorodifluoromethane
$CH_3CH_2CH_2CH_3$	Butane
$CH_3CH(CH_3)CH_3$	Isobutane
$CH_3CH_2CH_3$	Propane

Be Careful with Aerosol Bombs

A note of caution: All aerosol containers carry a warning against incineration or exposure to flame or high temperature. Some serious injuries have occurred by flying metal debris when these precautions were not taken. The natural pressure inside the cans can be as high as 8 atmospheres. So the products are aptly named aerosol bombs. Heat can easily provide the necessary molecular motion to disrupt and shatter the containers. Such accidents commonly occur in the kitchen where the familiarity with aerosol products leads to careless placement on or near a stove. Another hazard is the flammability of vapors and mists of some aerosols. Hair sprays are especially dangerous if used carelessly near fire or cigarettes.

There is also a growing concern about possible overexposure to propellant products. The fluorinated hydrocarbons can have toxic effects which apparently vary from person to person. Excess breathing of some propellant gases can cause behavior problems, mental confusion, reduced comprehension, and irritation of the breathing passages, skin, and eyes. These effects can be delayed and may occur quite a while after exposure. This is another case where operational judgments need to be backed up by conceptual knowledge of what is involved. The propellant gas is released freely in the neighborhood of application, but it is invisible and therefore often neglected operationally.

PAINTS

Paints were used as long ago as 50,000 years to decorate the walls of caves. Museums contain many relics of the past painted with natural coloring materials. People noticed very early that plant and clay materials could be ground with water to make primitive paints.

Paint is a thin film applied to a surface for protection and decoration. The most important purpose is protection of wood and metal surfaces against weather, oxygen of the air, sun, and chemical and mechanical damage. Today in the U.S. about three-quarters of a billion gallons of paint are used each year. These paints are complex mixtures which contain three major components:

1. *Pigment.* The average person thinks of pigment in terms of adding color to enhance beauty. More importantly, the pigment provides protection both to the surface and also the paint film, especially from ultraviolet radiation. Most clear films will not stand sunlight exposure over extended periods.
2. *Binder or nonvolatile vehicle.* This is the major body or nonvolatile binding ingredient. It gives adhesion to the surface, holds the pigment, and provides a protective barrier. Many possible binders are available, including vegetable oils such as linseed oil (flax plant), tung oil (from seeds of a tree grown in China and Southern United States), soybean oil, castor oil, fish oil, and various synthetic polymers or resins, or combinations of these with natural oils.
3. *Thinner, solvent, or volatile vehicle.* This is the volatile liquid which evaporates as the paint dries. The common vehicles are now usually hydrocarbon solvents or water. Evaporation of the organic thinner or the water leaves the pigment and binder in a continuous film.

Varnish is a transparent coating achieved by not using pigment. Varnishes can be made using natural resins from trees or synthetic polymers such as phenolic resins. Shellac is a special transparent coating material made from secretions of scale insects on trees in Southern Asia. The secretion is dissolved in alcohol which evaporates on application, leaving a glossy surface coating. Lacquers are fast-drying materials based chiefly on cellulose nitrate. Lacquers also contain added chemicals called plasticizers to make the finished coating tough and less brittle (chapter 9). Clear lacquers are often used as a final finish on furniture. Pigmented lacquers are commonly used for metal coatings. Nitrocellulose lacquers were once used extensively for automobiles. Today coatings based on synthetic plastics have been found to dry harder and tougher and require fewer coats.

Water-Thinned Paints

Research during World War II on rubber and plastics led eventually to a whole new series of paint ingredients based on synthetic polymers (chapter 9). The most startling development here is the "second generation" paints which use water as thinner. Most of these latex paints today have acrylic or polyvinyl acetate polymers suspended as very tiny droplets in water. They are therefore also called emulsion paints.

In emulsion paints (sometimes also called water-based paints) the binder and pigment are suspended as very tiny droplets in water. On application the water evaporates to allow formation of a tough surface film which interlinks and cannot again dissolve in water. These paints have been highly successful in the home because of the obvious advantages of easy clean-up, fast drying, and elimination of organic solvents. Here is a good example of how concept helps the paint chemist provide a better product. The long chain molecules used in the coatings would be very viscous if used alone and practically impossible to apply. By dispersing the molecules as tiny droplets in water, they are easily applied. Then the coalescing and linking together provides a tough durable film as the smaller molecules of water evaporate into the air.

A Few Practical Points on Paints

If you have done any painting, you realize that quality of paint can vary widely. Prior to 1972 many paint cans merely said "one coat covers." This does not mean much since one paint might require a much thicker coating than another. You might need one-and-a-half gallons to paint a room which one gallon of a better paint might also cover. The cheaper paint could have less hiding power because of a lower percentage of expensive pigment, like TiO_2. Now manufacturers are required to indicate how many square feet the paint will cover per gallon.

Home paints should no longer contain toxic lead compounds which have resulted in illness and even death to children who gnawed the paint off window sills or ate peelings off walls. The old standard white lead has largely been superseded by zinc oxide (ZnO) and titanium dioxide (TiO_2), even in industrial usage. Some lead paints are still used and, therefore, some danger from lead paints remains, especially in older houses or in cases of careless application of the wrong kind of paint.

COSMETICS

Cosmetics are products other than soap which are applied to the body for cleansing or beautifying. The word is derived from the Greek root *kosmos* which refers to order. Cosmetics include a tremendous number of products.

The line between cosmetics and drugs is a little hazy even though everyone knows what cosmetics are and what drugs are. If the chemical is for curing a skin condition, removing hair, or changing a body function like some deodorants and antiperspirants, then it is a drug. Drugs are considered in chapter 14.

The use of cosmetics probably goes back to prehistory when someone found that certain clays could be used to smooth and color the skin. Up to 1938 when the original food and drug act of 1906 was revised to include cosmetics, these materials were subject only to local controls. Now federal controls regulate cosmetics in interstate commerce, and most states have laws modeled on the federal regulations. The laws control such things as adulteration, labeling, composition, directions for use, and misbranding.

Creams

All creams are very fine mixtures of liquid chemicals wherein the individual particles are larger than ordinary molecular size. They are thus technically emulsions.

The behavior of creams is dependent on whether tiny droplets of an oil are dispensed in water or tiny droplets of water dispensed in an oil. Most common are the oil-in-water emulsions, since they feel much smoother and less oily and are very easily removed from the hands. However both types are available in large numbers.

The old designation *cold cream* derived from the cold sensation on the skin when the cream was spread to clean and lubricate the tissues. Cold creams originally contained a few oily substances in rose water. Now cold creams contain several ingredients. Here is a typical formulation: distilled water, almond oil, white beeswax, lanolin (purified wool fat), mineral oil, spermaceti (a wax from the head of the sperm whale), borax ($Na_2B_4O_7$), and perfume.

Many additional materials are added to cold creams, including zinc oxide, castor oil, hormones, vitamins, olive oil, avocado oil, peanut oil, and emulsifying agents like lecithin. Thus creams can serve for carrying special purpose chemicals and then they receive names according to the application being emphasized, such as deodorant creams and hormone creams.

Vanishing cream is a special type of emulsion of stearic acid [$CH_3(CH_2)_{16}COOH$] and water. Usually the formulation involves mixing potassium or sodium hydroxide with the stearic acid to form a small amount of soap (potassium or sodium stearate). Other additives may include mineral oil, glycerol, hydrogenated cottonseed oil, and perfume.

The "vanishing" nature of the cream is not because it disappears completely by absorption into the skin. What it does is hide natural blemishes or rough areas by forming a thin film on the skin. Then face powder pro-

An automatic filling machine for creams and lotions. (Courtesy Max Factor.)

vides the finished smooth surface. Vanishing creams usually have a pearly appearance.

Some cleansing creams have been developed based on mixtures of solid and liquid hydrocarbons only, without water. They are used mostly for cleaning very oily skin and can cause excessive drying in normal skin since petroleum hydrocarbons have this property. These are sometimes called quick liquifying creams.

Lotions

Lotions are probably even more familiar to most people than creams. The word is from the Latin (*lotio*) for a washing. A lotion is a fluid cosmetic product based on water or water-alcohol mixtures. Lotions are usually applied to the skin, but the term is also used to describe solutions for use in setting and waving hair. Many lotions are emulsions; some are solutions. Solutions sometimes contain additives to make them cloudy since there is apparently a preference for the rich appearance of milky or opaque cosmetic products.

Lotions come in endless types. Look in your neighborhood drugstore and you will find hand and face lotions, suntan lotions, deodorant and antiperspirant lotions, preshave and aftershave lotions, body rubs, baby lotions, medicated lotions, foot lotions, and on and on.

A typical hand or face lotion may contain combinations of the following ingredients in a predominantly water system: lanolin, paraffin, glycerine, alcohol, borax, mineral oil, coloring agent, perfume agent, and emulsifying agents (like stearic acid salts).

Depending on the purpose and manufacturer, various other additives can be used, such as antiseptics, almond oil, cocoa butter, geranium oil, camphor, spermaceti, and beeswax.

Powders

Powders are used in a variety of cosmetic applications like face powder, baby powder, body powder, talcum powder.

The first powder for cosmetic purposes was probably made from a starch from rice to which some natural coloring material was added. This was found effective in smoothing over rough spots in the skin and cutting down on the "shine" which results from oil and moisture natural to skin surfaces. The original purpose remains, but modern cosmetic chemists have investigated many vehicles more suitable than food starch. The result is a rather complex mixture of chemicals selected to provide opacity or hiding power for blemishes and shine and at the same time easy application, good adherence to skin, and considerable durability. Here is a partial listing of basic ingredients: starch, talc, titanium oxide, kaolin, magnesium oxide, zinc oxide, zinc stearate, magnesium carbonate, and magnesium stearate.

A modern development is the so-called pressed powder, which is much more convenient than the loose product. These caked powders contain a small quantity of binding chemicals like lanolin and mineral oil. This is another example of how concepts can lead to success in providing proper operational properties. Here long chain molecules were incorporated which would bind slightly but still allow smooth application.

Rouge formulations are very similar to those for face powders. The amount and kind of dye is varied to suit the end color requirements. Paste rouge is obtained by incorporation of petrolatum (long chain petroleum fraction) or various waxes.

Lipsticks

Lipsticks are also not a modern invention. However, the formulations available today are considerably improved over the primitive materials of the first "face painters." The cosmetic chemists now have available a vast conceptual background on the behavior of certain types of fatty or oily materials which they use in selecting materials for blending into the complex mixture represented by lipstick. Here are a few of the operational properties desired: soft and easy application, adherence to lip surface, avoidance of smearing, resistance to mild wearing action of eating and drinking, nontoxicity, nonirritability, and ability to retain color dyes and perfume.

Some of the most common chemicals used to achieve these operational advantages are themselves mixtures which the cosmetic chemist

Lipstick and eye make-up products undergo intricate processes of compounding, blending, and milling to ensure uniformity and quality. (Courtesy Max Factor.)

uses without any attempt to isolate and incorporate molecular substances. This would make it extremely expensive and would not add any advantage. Like all cosmetics, you can never give the formula for lipstick. The cosmetic chemist more likely refers to a formulation for a particular lipstick. This is really a recipe for mixing many ingredients like the following: castor oil, cocoa butter, lanolin, perfume, beeswax, carnauba wax, paraffin, butyl stearate, isopropyl palmitate, polymerized ethylene glycol, stearic acid, and coloring agents.

The newer liquid lipsticks are also quite variable in composition. A common formulation is based on a solution of cellulose derivative along with dye chemicals. The cellulose derivative can be cellulose acetate or cellulose nitrate obtained by reacting the side OH groups of cellulose with chemicals to alter solubility and properties. Other polymers can also be used, and the usual formulation involves a blend, including additives like glycerol, dibutyl phthalate, or dioctyl adipate to soften or elasticize the film which is deposited on the lip.

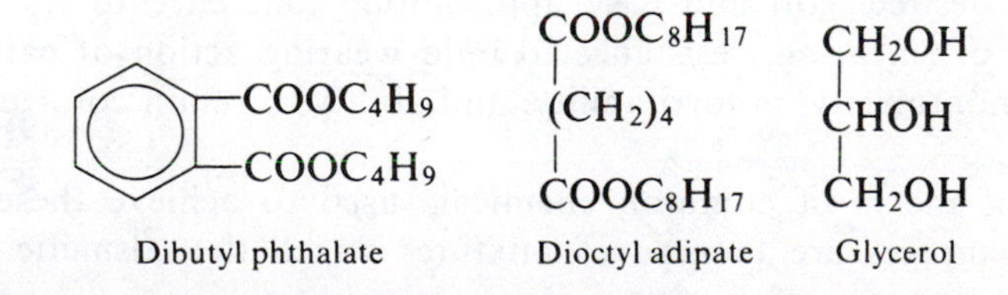

Deodorants and Antiperspirants

Perspiration is not necessarily unpleasant in odor when first secreted by the sweat glands. Bacteria cause decomposition which gives "body odor." Sweating is of course a vital function for control of temperature, elimination, and lubrication of the skin. The odor of perspiration is changed slightly by the food a person eats: onions, garlic, spices, asparagus, and certain types of cheese give a special odor.

Antiperspirants are used to cut down on perspiration. The initial products contained salts of aluminum such as aluminum sulfate or aluminum chloride. It was found that these compounds caused rotting or deterioration of clothing. They were also irritating to many people because they are acidic. Formulations were changed by addition of neutralizing agents like urea. A common ingredient today in antiperspirants is aluminum chlorhydroxide complex—$Al_2(OH)_5Cl$, which has some basic (—OH) character incorporated.

The original deodorant (after soap and water) was essentially a perfume or odor-masking agent. *Eau de cologne* was used for many years. Modern deodorants go to the root of the problem and incorporate chemicals to check growth of bacteria (bacteriostats). The most common chemical in use up to 1972 was hexachlorophene. It is now banned except by prescription. (See the section below on antiseptics.)

Structural formulas for several of the chemicals used in deodorants are shown below.

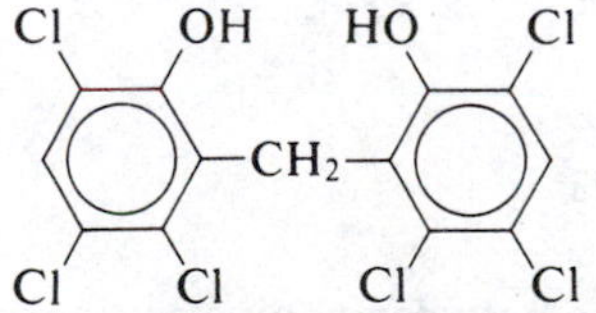

Cl OH HO Cl
(ring)—CH_2—(ring)
Cl Cl Cl Cl

Hexachlorophene
(now banned in U.S.)

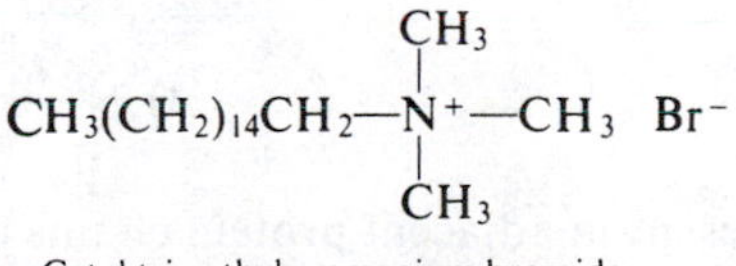

$$CH_3(CH_2)_{14}CH_2—N^+(CH_3)_2—CH_3 \quad Br^-$$

Cetyl trimethyl ammonium bromide
(a quaternary ammonium salt)

ZnO_2

Zinc peroxide

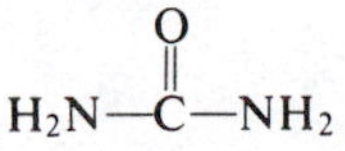

$$H_2N—\overset{O}{\overset{\|}{C}}—NH_2$$

Urea

$$(CH_3)_3CCH_2C(CH_3)_2—C_6H_4—O(CH_2)_2OCH_2N^+(CH_3)_2CH_2—C_6H_5 \quad Cl^-$$

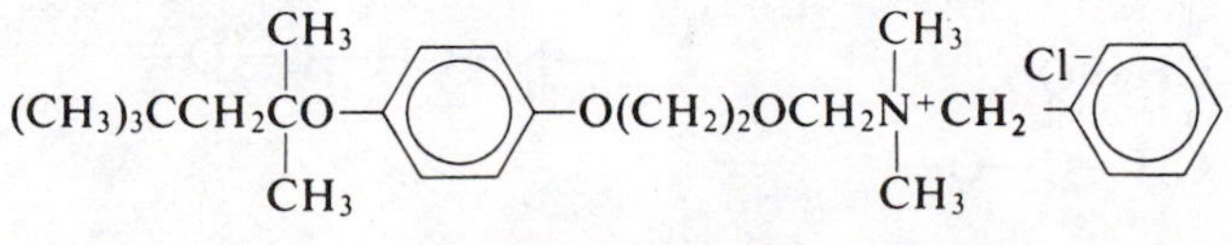

Benzethonium chloride

Deodorants and antiperspirants can be incorporated into a variety of forms like creams, sticks, powders, solutions, emulsions, and aerosols. These are cosmetic products according to the Food, Drug, and Cosmetic Act if they just deodorize. They are drugs if the claim is made that they cut down perspiration. Then the active antiperspirant chemical must be listed on the label.

HAIR PRODUCTS

Hair is a rather complex protein. Various amino acids react to form long chains in the build-up of complex protein molecules. A photograph and drawing indicating the nature of the complex hair structure are shown in chapter 10. Concepts about the nature of the protein molecules have been helpful in developing products for hair treatment.

Hair Permanent Waving

The protein chains in hair arrange themselves in an orderly fashion with various types of bonds formed between them. Some of these bonds between chains are hydrogen bonds. Others consist of links between two sulfur atoms in adjacent chains. Thus if the amino acid cysteine

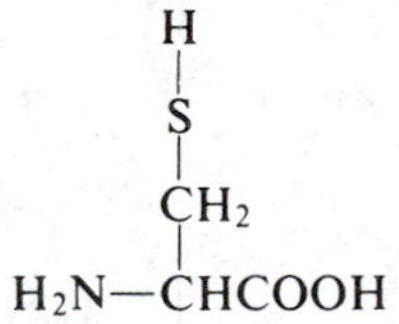

```
        H
        |
        S
        |
       CH2
        |
  H2N—CHCOOH
```

is present in adjacent protein chains a linkage between chains can occur:

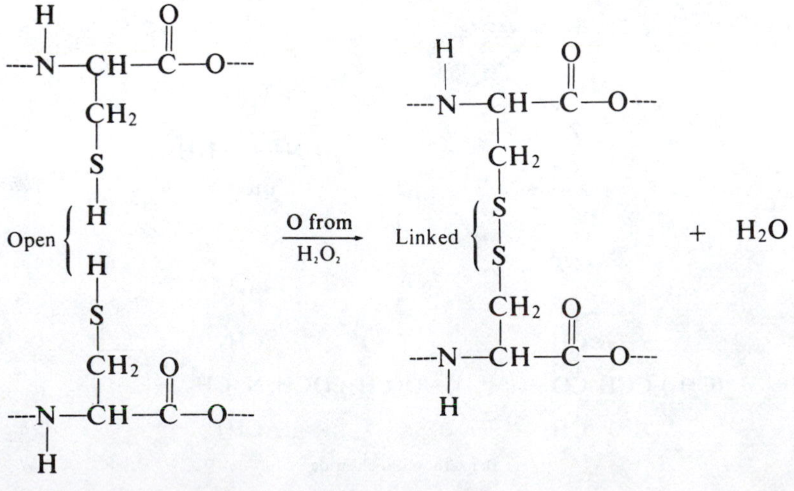

```
     H       O
     |       ||
 ---N—CH—C—O---                                H       O
        |                                      |       ||
       CH2                                 ---N—CH—C—O---
        |                                         |
        S                                        CH2
        |                                         |
      { H                                       { S
 Open {              O from      Linked {    |            +  H2O
      { H            ------->                  { S
        |              H2O2                       |
        S                                        CH2  O
        |                                         |   ||
       CH2  O                              ---N—CH—C—O---
        |   ||                                 |
 ---N—CH—C—O---                                H
     |
     H
```

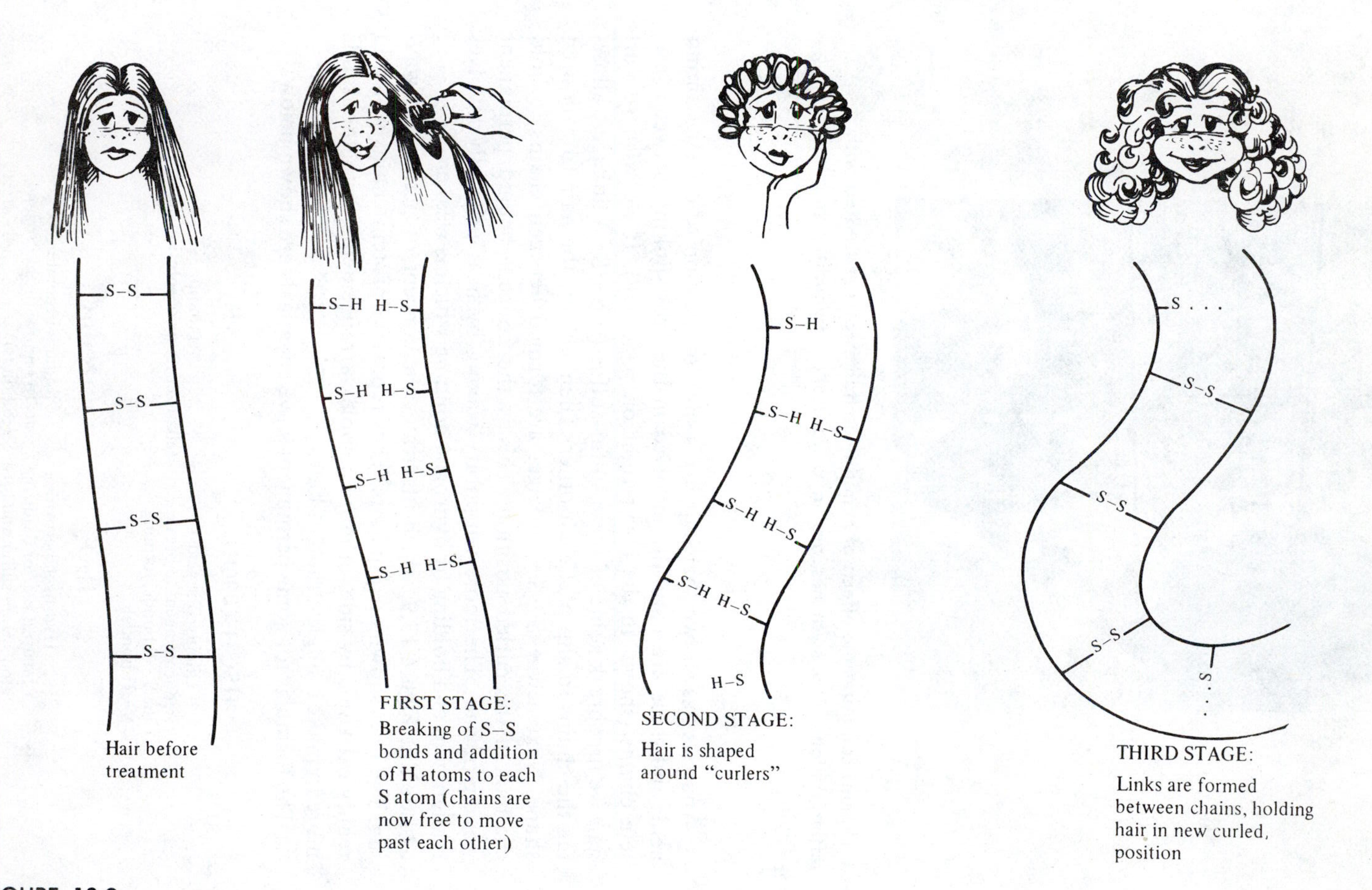
S–S
S–S
S–S
S–S
Hair before treatment
S–H H–S
S–H H–S
S–H H–S
S–H H–S
FIRST STAGE:
Breaking of S–S bonds and addition of H atoms to each S atom (chains are now free to move past each other)
S–H
S–H H–S
S–H H–S
S–H H–S
H–S
SECOND STAGE:
Hair is shaped around "curlers"
S . . .
S–S
S–S
S–S
S . . .
THIRD STAGE:
Links are formed between chains, holding hair in new curled, position

FIGURE 13.9
The concept behind permanent waving.

A physical chemist compares the effects of humidity on hair swatches treated with various setting lotions. The swatches are in a controlled humidity chamber. (Courtesy Clairol.)

Analysis has shown that hair has a high proportion of cysteine amino acids and these are naturally involved in the cross-linking between protein chains (shown in the lower right on page 378). If we wish to curl hair we first break some of these sulfur-sulfur (—S—S—) linkages, allowing the chains to slip past each other. Then we curl the hair to a desired shape. Next new —S—S— linkages are formed between chains to hold the hair in the curled position. This is the concept behind permanent waving of hair. The more temporary waving by heat primarily involves realignment and holding by hydrogen bonding which is weaker.

Look at Figure 13.9, which illustrates the concept of permanent waving. The lines shown are not hairs. They represent protein chains which in reality exist side by side, in more complex arrangements, with thousands more to make a single strand of hair.

The formulas for some permanent wave chemicals are shown below.

$HSCH_2COOH$

Thioglycolic acid
(reducing agent to break the S—S bonds, forming S—H bonds)

NH_3

Ammonia
(alkalizer to make solution basic)

H_2O_2

Hydrogen peroxide

$NaBrO_3$

Sodium bromate

("neutralizer" or oxidizing agents to remove the hydrogen atoms from sulfur and form S—S bonds)

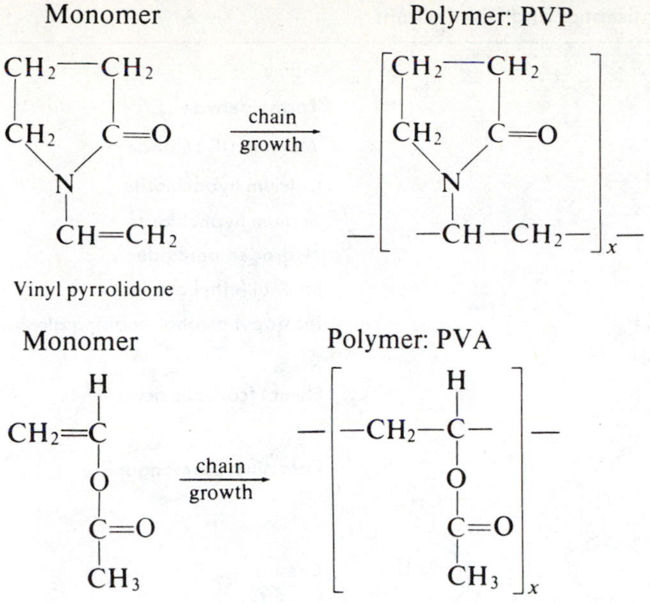

Monomer — Polymer: PVP

$$\underset{\text{Vinyl pyrrolidone}}{\begin{array}{c} CH_2-CH_2 \\ | \qquad\quad | \\ CH_2 \quad C{=}O \\ \diagdown \; \diagup \\ N \\ | \\ CH{=}CH_2 \end{array}} \xrightarrow[\text{growth}]{\text{chain}} -\left[\begin{array}{c} CH_2-CH_2 \\ | \qquad\quad | \\ CH_2 \quad C{=}O \\ \diagdown \; \diagup \\ N \\ | \\ -CH-CH_2- \end{array}\right]_x-$$

Vinyl pyrrolidone

Monomer — Polymer: PVA

$$\begin{array}{c} H \\ | \\ CH_2{=}C \\ | \\ O \\ | \\ C{=}O \\ | \\ CH_3 \end{array} \xrightarrow[\text{growth}]{\text{chain}} -\left[\begin{array}{c} H \\ | \\ -CH_2-C- \\ | \\ O \\ | \\ C{=}O \\ | \\ CH_3 \end{array}\right]_x-$$

Vinyl acetate

Note: x may be in the hundreds or thousands

FIGURE 13.10
Long chain molecules for hair sprays.

Hair Sprays

The earliest hair sprays were essentially lacquers which produced a firm, somewhat thick, film. This type is still available for the person who wants an overall rigidity and stiff, stand-up hair effects.

The more common types provide a softer film and use softer polymer coatings like PVA (polyvinyl acetate) and PVP (polyvinyl pyrrolidone) (see Figure 13.10). These wash out readily. Commercial products contain many special additives like damp-proofing ingredients, silicones, lanolin, dyes, sunscreens to absorb ultraviolet radiation, or even shellac as a coating aid.

ANTISEPTICS AND DISINFECTANTS

Antiseptics are substances that hinder or prevent the growth of microorganisms in or on the human body. Disinfectants are chemicals which kill bacteria and other disease organisms. Disinfectants are used for instruments, utensils, clothes, and living areas, whereas antiseptics are applied on living tissues. Table 13.2 gives the formulas of some antiseptics and disinfectants.

TABLE 13.2
Formulas of Some Antiseptics and Disinfectants

I_2	Iodine
$HCHO$	Formaldehyde
$HgCl_2$	Mercury (II) chloride
$Ca(OCl)_2$	Calcium hypochlorite
$NaOCl$	Sodium hypochlorite
H_2O_2	Hydrogen peroxide
C_2H_5OH	Ethanol (ethyl alcohol)
$CH_3CHOHCH_3$	Isopropyl alcohol (rubbing alcohol)
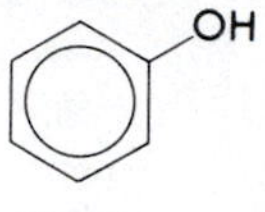 OH	Phenol (carbolic acid)
$KMnO_4$	Potassium permanganate
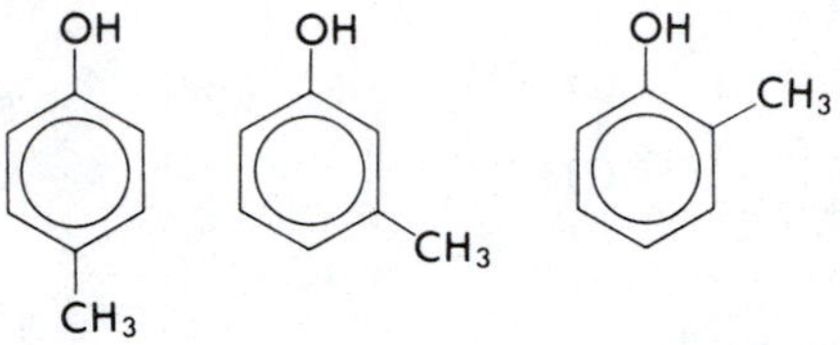 OH, CH_3; OH, CH_3; OH, CH_3	Cresols
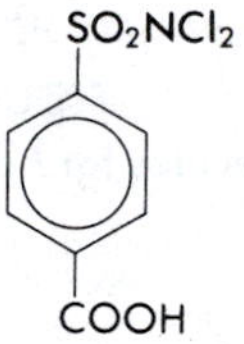 SO_2NCl_2, COOH	Halazone
$C_6H_5CH_2—N^+(CH_3)_2—C_{12}H_{25}\ Cl^-$	Quaternary ammonium compound (one example only; varied chains are possible)
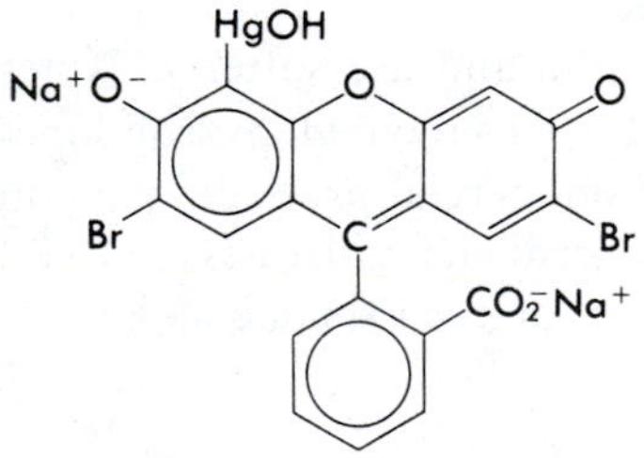 HgOH, Na^+O^-, O, O, Br, Br, C, $CO_2^- Na^+$	Mercurochrome
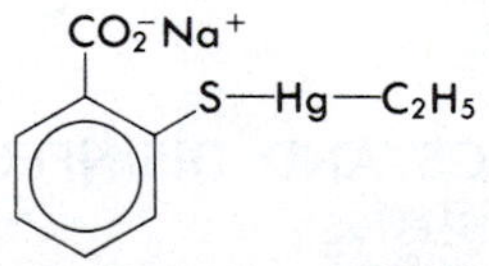 $CO_2^- Na^+$, S—Hg—C_2H_5	Merthiolate
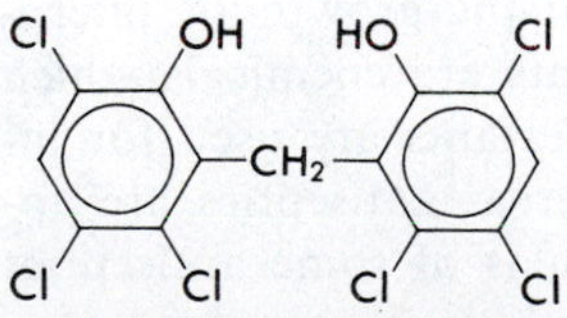 Cl, OH, HO, Cl, CH_2, Cl, Cl, Cl, Cl	Hexachlorophene

The attempt to control disease has a long history. Early peoples used smoke to preserve foods from decay. During the Middle Ages sulfur and special kinds of wood were burned in houses affected by the bubonic plague. Although people did not have a clear concept of the connection between contagious disease and microorganisms, they worked by experience and laid the foundation for our modern approach to disease-carrying bacteria. Even before Pasteur published his findings on bacteria as agents for disease (beginning in 1860), hypochlorite solutions were used in the treatment of wounds and for purification of drinking water. Their direct descendants are modern laundry bleaches, which are also used as disinfectants.

In the late nineteenth century Joseph Lister (1827–1912), the founder of antiseptic surgery, introduced coal tar chemicals like phenol for cleaning wounds in surgery and for decontaminating air in surgery areas. The same type of product is still used in household products like Lysol, which is a soap solution of cresols. Modern surgery is aseptic rather than simply antiseptic. This means freedom from all forms of bacteria, which is achieved by sterilizing instruments with heat and chemicals.

Iodine solutions were used on skin and wounds beginning around 1840 and in the late nineteenth century were used extensively prior to surgery. Tincture of iodine is a solution of iodine in alcohol. The usual strength is 2 percent iodine. A major problem was the sting caused by the alcohol. To eliminate this problem organic compounds were developed which release free iodine. These are called iodophors. An example is the complex formed by iodine with polyvinyl pyrrolidone.

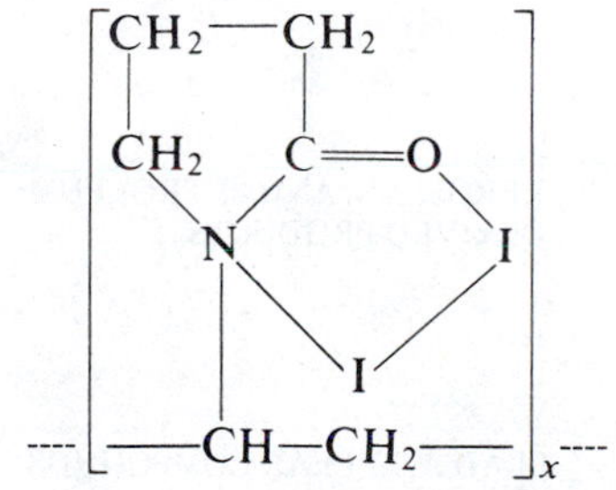

```
 [CH2——CH2          ]
 | |        |         |
 | CH2      C==O      |
 |    \    /     \    |
 |      N          I  |
 |      | \       /   |
 |      |   \    /    |
 |      |     I       |
----[——CH—CH2——]x----
```

Here iodine molecules are available by breaking away from the PVP. The sting is avoided.

Hexachlorophene is another of the organic compounds. It was used extensively as an antibacterial agent in many soaps, shampoos, cosmetics, and deodorants for about thirty years before the Food and Drug Administration in late 1972 banned it from virtually all nonprescription drug and cosmetic products. Hexachlorophene was patented in 1941 by a Swiss company. In 1971, tests at FDA laboratories indicated that hexachlorophene could damage nervous tissue, and consequently the FDA warned against routine bathing of infants with a 3 percent solution of the compound. Then in 1972 the deaths of forty infants in France resulted in a virtual ban. The babies had been treated with a talcum powder which contained an excess of hexachlorophene, 6 percent.

You can see that care is required for proper use of common chemicals found in the home. Just because the material is in the detergent carton or in a bottle or can for bowl and drain cleaning does not mean it is safe. There is the danger of the overly familiar product taking on an aura of harmlessness. All chemicals are capable of doing damage in the wrong place or if used improperly. If you learn nothing else in this chapter other than respect for proper application of chemical products, it will have been a valuable study.

All chemical products are designed for a particular use. Someone has an idea and decides to try a particular combination of chemicals to carry it out. This is simply an attempt to achieve certain operational properties based on conceptual analysis of the problem. The manufacturer will in most cases put proper warnings on labels along with recommendations for use. Unfortunately, some people do not read directions. In other cases

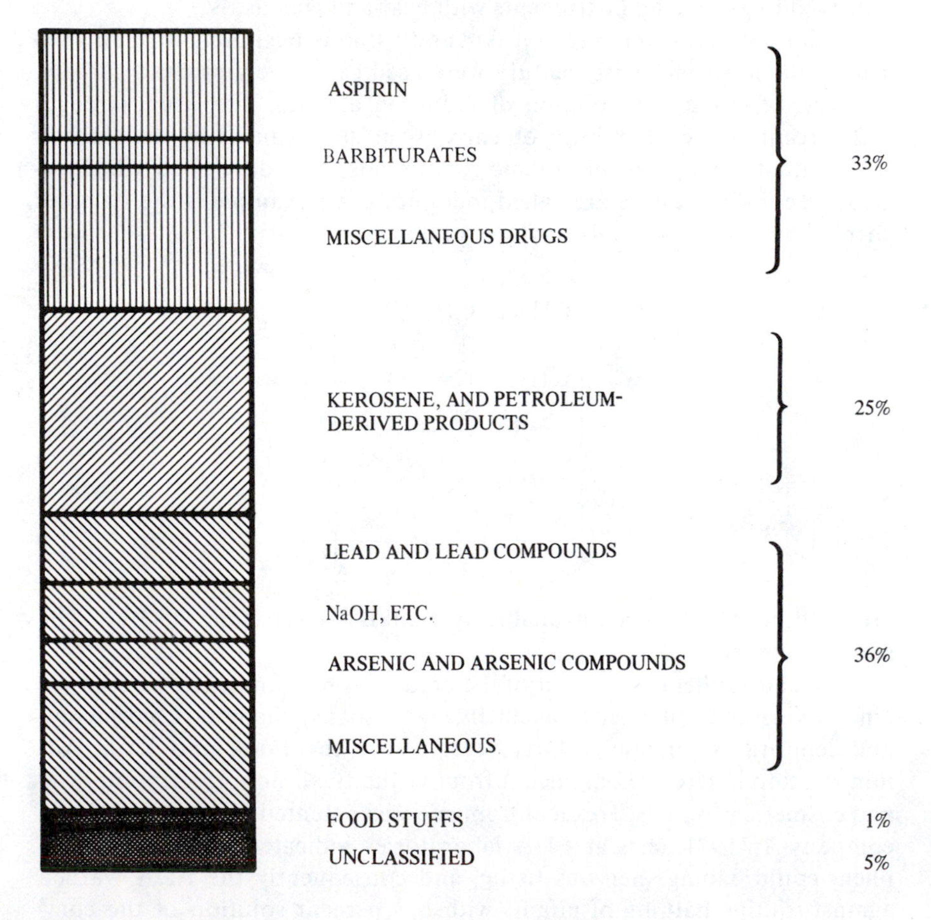

FIGURE 13.11
Substances responsible for death from accidental poisoning of children under five years of age (approximate distribution in developed countries).

human error can occur. The ultimate protection depends on knowledge of the potential problems and respect for keeping chemicals in their proper place.

Many children are injured or killed by chemicals. The four- or five-year-old can have no knowledge of chemical hazards. If his big brother leaves kerosene or gasoline in a coke bottle after cleaning auto parts, the toddler will presume it is something to drink. This happens too often and the child dies. Or someone tells the child that a medicine is candy to get him to eat it. Then later the aspirin bottle is left within reach and he helps himself. This hazard increased with the production of fruit-flavored aspirins for children. These are now sold only in small-size bottles. In many cases of even aspirin overdoses the child may die. Figure 13.11 shows the distribution of child deaths from accidental poisoning by particular chemical substances. What is the answer? Knowledge and care on the part of those who handle chemicals around the home.

PROJECTS AND EXERCISES

Experiments

1. This experiment will show the operation of soap molecules. Get two jars, preferably with tight-fitting covers or caps. (Or you could use a tall glass which you could cover with your hand.) Fill one jar half full of warm water. Then add some liquid oil like cooking oil or mineral oil to a depth of about one-quarter of an inch. Cover the jar and shake vigorously for a minute. Observe how the oil behaves. What happens? Allow the jar to stand until the globules of oil collect again on the surface. Next fill the second jar about half full of warm water and add about the same amount of oil. Then add about half a cup of liquid soap or a strong soap solution obtained by dissolving powdered soap or synthetic detergent in warm water. Shake the mixture vigorously as you did before. Compare the behavior with the first jar. What conceptual explanation can you offer for the behavior?
2. There are many ways of investigating the behavior of soap. Take a large clean plate or deep pie tin and rinse it with water until it is very clean. Fill it with cold water to within about a half inch of the top so that you have a large surface of water. Let it stand for awhile on a table or the floor until all movement of water ripples completely settle down. Then sprinkle some talcum powder or fine bath powder over the surface of the water. You want just a light coat of powder over the whole surface. Now wet a bar of toilet soap and touch it gently to the water near one edge. What is the effect? Knowing what you do about the soap molecule, can you come up with an explanation? There are several ways to describe the effect. Concepts are best for explaining the operation you observe.

Exercises

3. A bright twelve-year-old hears that you are studying chemistry and asks you how matches start fires. Explain the process conceptually without going into all the details of chemistry equations.

4. It is interesting that the earliest people who started fires did it by rubbing stones or sticks together, while today we also start fires by rubbing matches on some surface. Using chemistry concepts, indicate what happened when people first rubbed sticks together and then contrast the mechanism with that of our modern matches. Use your general knowledge of the environmental materials involved to explain the first methods of making fire.
5. When you spill pancake syrup on clothes or on your hands it can be washed away easily by using water alone. On the other hand, if you spill a cooking oil or get butter on your fingers, it does not wash away easily unless you use soap. Briefly explain the difference, using concepts that would make more sense than an operational description.
6. The production of lightness in cakes is attributed mechanically to the incorporation of gas in the batter. Most cakes use a combination of leavening agents, for example: (a) creaming of sugar and butter, (b) use of beaten egg white, (c) addition of baking powder, and (d) formation of steam during baking. Consider individually the four sources of gas above and write down the formulas for gas involved in each case. Do you think each tiny gas pocket in the cake is a mixture or a substance? Which of the four items above would you expect to provide the major amount of gas? As a final item, if you know a good baker you might ask him or her why the sugar and butter are creamed. Do you have any additional insights now on this common household operation?
7. Someone says that we should only use natural cleaning agents like soap because synthetic or man-made detergents are not easily degraded by bacteria. How could you answer this objection intelligently?
8. An intelligent adult who has not had the advantage of studying chemistry asks you what hard water is, the major problems it causes, and how we eliminate the problems. Briefly answer these questions.
9. Briefly outline the reason for using phosphates in soap and detergent products. How would you answer someone who says he does not want these "poisons" in his laundry soap?
10. Look over the possibilities for softening hard water. Choose the one method you would prefer, describing briefly the methods you evaluated and the advantages or disadvantages of each.
11. Outline the dangers of household cleaners along with precautions to avoid injury.
12. A bright twelve-year-old asks you how glue sticks things together. Give a brief explanation, describing types of adhesives usable for certain applications.
13. Give a conceptual description of a paint.
14. Why is it especially necessary today to read composition labels or directions on paint cans before thinning with added solvent?
15. Examine the listing of possible ingredients for cosmetic powder products. Pick out three in the list for which you could give an operational reason why it would be included.
16. "Cold creams" make the skin feel cold, even though the cream itself is not cold. How can you explain the sensation theoretically? (Some change occurs which we considered in detail in chapter 7 on water.) What is the mechanism for the cold feeling?
17. Pick out any cosmetic product you like and then describe how you would for-

mulate it. Include a list of ingredients, reasons for using them, and the way you would prepare the materials into a finished product.

18. What is the conceptual basis of permanent waving?
19. Look around at home or in a supermarket for a few disinfectant or antiseptic products. Describe briefly what ingredients are used and what precautions taken to avoid injury.
20. What is your reaction to the wide-scale use of hexachlorophene before the 1972 ban?
21. Choose one of the quotations opening this chapter and write a short paragraph indicating what it means to you now, especially pointing out how different meanings may be implied.

SUGGESTED READING

1. Information on home chemicals is difficult to find in easily available reference sources. Some general information is available in encyclopedias.
2. Bennett, H., *The Chemical Formulary,* D. Van Nostrand Company, Inc., New York, and Chemical Publishing Co., Inc., New York. This is an extensive work running through many volumes beginning with Volume 1 in 1933. New volumes come out as the publisher accumulates more information on newer products. It is said to be "a collection of valuable, timely, practical commercial formulas and recipes for making thousands of products in many fields of industry." It is a classic in this field. If you have any curiosity as to what goes into various products or how the manufacturer produces them, you would do well to consult the *Chemical Formulary*.
3. Good sources of current information on household products are the periodicals *Consumer Reports* and *Consumer Bulletin.* These provide evaluations of competitive products and also precautions on dangerous products revealed by independent testing laboratories.

chapter 14

Drug Use and Abuse

We forget ourselves and our destinies in health, and the chief use of temporary sickness is to remind us of these concerns.

—R. W. EMERSON

A disease known is half cured.

—PROVERB

Joy, Temperance and Repose,
Slam the door on the doctor's nose.

—HENRY WADSWORTH LONGFELLOW

The word drug today is often associated with *abuse, addiction, habit, overdose, control,* and the like. But in themselves drugs are neither good nor bad. They are simply chemicals, and you have seen many examples of the same chemical having good effects in some cases and bad effects in others. Excess sodium chloride in the blood can cause death. Too much sugar can be dangerous to a person with diabetes. Even an excess of water can cause convulsions. Consequently, judgments on drug usage are difficult.

Our knowledge of mechanisms of drug action is only in its infancy. Yet we have come a long way chemically, from observing the effects of certain plant extracts, for example, to isolating and identifying their active components. Knowing the molecular architecture, or the exact arrangement of atoms in the molecule, has permitted us to make synthetic drugs in larger quantity, in purer form free of side effects, and even to make improved medicines by planned alteration of molecules.

Drugs are commonly considered to be substances used in treating disease. Pharmacology is the science of drugs, their preparation, uses, and especially their effects. In pharmacology a drug is defined as any material which by its chemical nature alters the structure or functioning of a living organism. In its broadest sense, then, the word drug includes many varieties of chemicals from vitamins and alcohol to sex hormones and LSD.

More simply, a drug is a chemical other than food that, when taken into the body, produces a change in it. If the change helps the body, the drug is a medicine. This was the common early meaning of the word drug. However, if a drug causes a change which harms the body the drug is a poison. The same chemical can be a medicine and a poison depending on conditions of use and the person using it.

Drugs are used extensively by doctors for control of disease, the restoration of health, or the easing of pain. Estimates place the number of drugs currently in use at something more than 5000. Most of these have been introduced within the last twenty or thirty years. Every year several hundred new drugs appear while a similarly large number of older drugs become obsolete.

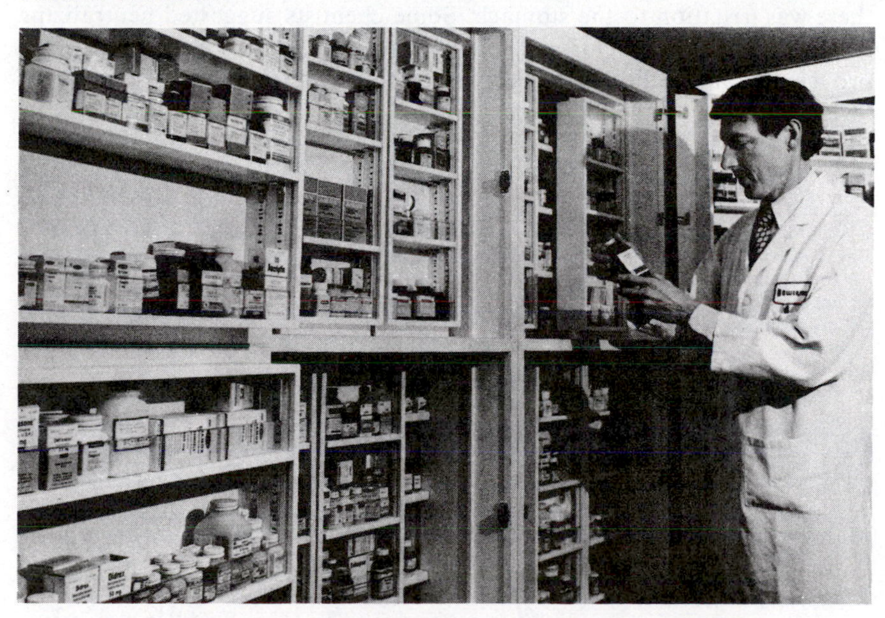

As a result of growing conceptual background on molecular structure, today's pharmacists have available a great variety of helpful drugs. (Photograph courtesy of Syntex Corporation.)

ASPIRIN: THE MOST COMMON PAIN-RELIEVER

The history of aspirin usage shows clearly how gradual refinement of a concept can result in better drug products in large enough quantities to satisfy growing demands.

Aspirin's History

A London doctor, Thomas Sydenham (1624–1689), is generally given credit for the initial investigation that led to aspirin. He became impressed with the reports of Jesuit missionaries on the medicinal properties of the bark of the cinchona tree, extracts of which were used successfully by certain Indians in South America in treatment of malaria. Sydenham then investigated bark of various trees and found that extracts of the barks of the willow and poplar were effective in reducing fever.

Scientists began the slow process of determining why. In 1826 the active principle of willow bark was isolated and called *salicyn* (Latin: willow tree). Around 1835 a chemical called salicylic acid was obtained from willow extracts. In 1852 salicylic acid was made without using any natural willow materials. Then in 1874 a process suited for large-scale production was developed. In the meanwhile the willow extracts and synthetic material had been found increasingly helpful in medicine, especially in treatment of rheumatic fever.

There were, of course, side-effect problems with salicylic acid as there usually are with all drugs, especially when they are first introduced. One of these was irritation to the stomach. Some chemists suggested neutralizing the acid with sodium hydroxide. This was found to be a considerable improvement, and sodium salicylate became the major drug for rheumatic fever.

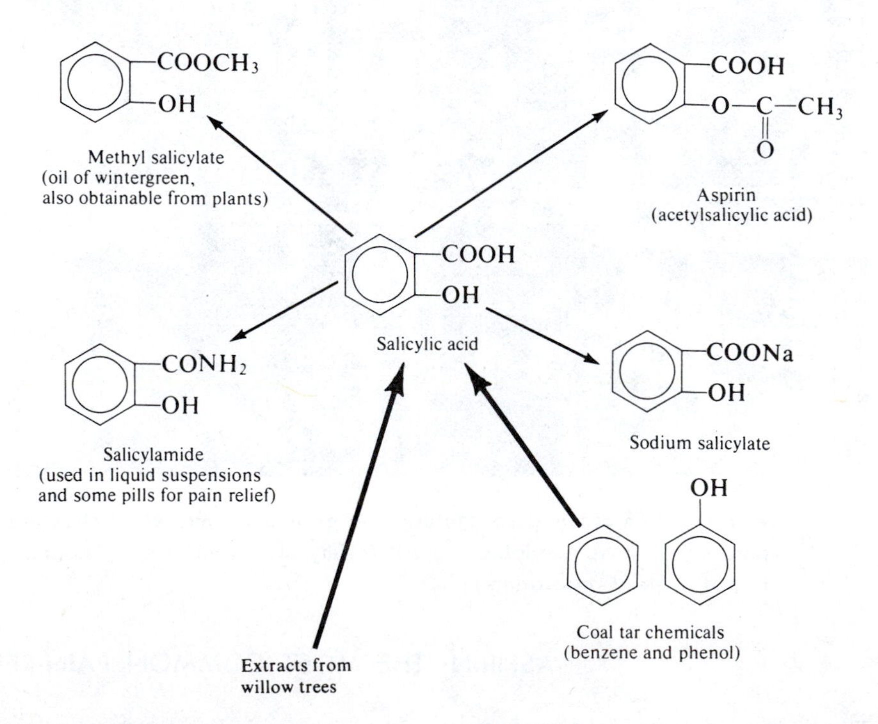

FIGURE 14.1
Some of the compounds related to aspirin.

However, sodium salicylate had a rather unpleasant taste and still caused stomach irritation in some cases. Attempts were made to alter the structure of the molecule while still retaining the medicinal value and without contributing further side effects. Several variations in the molecular structure were investigated. Toward the end of the nineteenth century, the German chemist, Felix Hofmann, investigated various related compounds including a derivative called acetylsalicylic acid that had been prepared in 1853. He tried this compound on his father, who suffered from arthritis and could not tolerate the stomach irritation caused by sodium salicylate. It was very effective and produced much less gastric irritation. In 1899, the new drug was produced commercially by the Bayer chemical plant where Hofmann worked and was given the trade name Aspirin, which indicates it is "from the Spirea willow," the source of the original salicylic acid. The term aspirin is now a generic name in the United States. Figure 14.1 illustrates the formulas of some of the compounds derived from salicylic acid.

Aspirin's Effects

Aspirin, or acetylsalicylic acid, is now one of the most widely produced drugs. In the United States alone, aspirin is manufactured in an amount sufficient to make over 25 billion standard 5-grain tablets per year. (A grain is about 0.065 grams or 0.0023 ounces.) It is the most widely used analgesic (a chemical that relieves or stops pain), and is much preferred over morphine because it does not involve physiological dependence. It is inexpensive and fast-acting and hence is used freely for everyday aches and pains. Aspirin is also very effective as an antipyretic, or fever-lowering agent and also as an anti-inflammation agent. It thus finds wide application from rheumatism and arthritis to hangover and the common cold.

It is interesting that one simple chemical can act so effectively on so many facets of the body's defensive system such as inflammation, fever, and pain. In spite of extensive research to clarify the mechanism of the action of aspirin, much of the mystery of its operation still remains.

Which Aspirin Should You Buy?

There is no difference between substances that have the same chemical structure, providing they are pure. And the drug laws require conformity here. Thus aspirin is aspirin, regardless of the source.

There are, of course, other analgesics. And partly in an attempt to get a more distinctive product to advertise, as well as to provide enhanced properties, manufacturers have often combined other ingredients with aspirin. (See Figure 14.2.)

One of the most promoted aspirin combinations is "buffered" aspirin, sold under various trade names, which uses a small amount of antacid or alkalizer in an attempt to neutralize some of the acidity of the stomach. The acidity comes from hydrochloric acid, HCl, which is natural to stom-

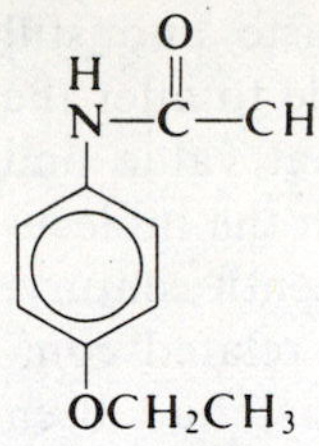

Phenacetin

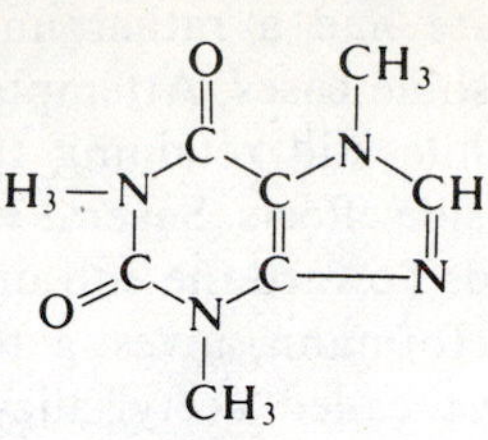

Caffeine
(Anacin is an aspirin-caffeine mixture)

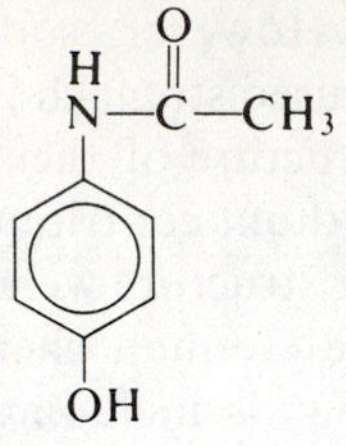

Acetaminophen
(In Excedrin, along with aspirin, caffeine, and salicylamide. Tylenol, Tempra and Lyteca contain *only* this compound.)

FIGURE 14.2
Some chemicals used in analgesics along with or in place of aspirin.

ach action and also from the aspirin itself. Note the acid group—COOH in the aspirin molecule (page 390).

Phenacetin and caffeine were until recently very common additives found in three-component pills along with aspirin. These could be purchased as APC tablets or under various trade names. Caffeine is a stimulant and also a diuretic, but there is apparently no definitive proof that it enhances the effect of aspirin. The original purpose of adding caffeine to aspirin tablets may have been to stimulate the circulatory system in dispersing the chemicals in the bloodstream and to aid the kidney in excretion. But the situation of combination products for analgesics has changed recently because of some reservations about phenacetin. Phenacetin is about equal to aspirin in relief of pain and in reducing fever although not as effective as aspirin in reducing inflammation from arthritis. Recent studies have suggested, however, that long continued use might be involved in kidney damage or blood abnormalities. Some of the major manufacturers of APC pills have dropped phenacetin, some using aspirin and caffeine alone in their product while others have replaced phenacetin with a chemical called acetaminophen. This relatively new drug is a much more expensive analgesic than aspirin. Its action is similar to phenacetin in pain and fever relief, but it is not as effective as aspirin for reducing arthritis inflammation. However, it is useful for people who cannot take aspirin because of irritation or, more seriously, allergic hypersensitivity. No long-term studies have been completed on its effect on the kidneys.

The best way to avoid stomach irritation is to take aspirin when food is in the stomach or at least by drinking a full glass of water. Many people gulp the aspirin tablets down with just a swallow of water, often between meals. This could contribute to stomach upset. (See the experiment on aspirin at the end of this chapter.)

ANTACIDS: THE ALKALIZERS

A class of drugs whose mechanism is much easier to understand than that of analgesics is the antacids. As the name implies, they go against, or neutralize, acid. The stomach normally operates with a low pH caused by hydrochloric acid but in cases of excess acid secretion (called hyperacidity), a method is needed to remove or neutralize the extra acid. Neutralization uses a basic or alkaline chemical which provides OH^- ions which combine with the H^+ ions to form neutral water. A common example is milk of magnesia or $Mg(OH)_2$. This is an alkalizer. If milk of magnesia is taken in excess, it acts as a laxative.

The alkalizing action of $Mg(OH)_2$ occurs by reaction with the hydrogen ions.

$$2H^+ + 2Cl^- + Mg(OH)_2 \rightarrow 2H_2O + Mg^{++} + 2Cl^-$$

This can be summarized as follows:

$$2HCl + Mg(OH)_2 \rightarrow 2H_2O + MgCl_2$$

Another common ingredient for control of acidity in the stomach is precipitated chalk or calcium carbonate. Its action can be summarized in this equation:

$$2HCl + CaCO_3 \rightarrow H_2O + CO_2 + CaCl_2$$

Here the hydrogen ions from the stomach acid are combined to form H_2CO_3, carbonic acid, which decomposes to give the final products water and carbon dioxide. The acid is thus neutralized. In some cases calcium carbonate causes constipation, but this can be offset by blending with other antacids such as magnesium carbonate.

The old home remedy for excess acidity is sodium bicarbonate, $NaHCO_3$. Its action is simply removal of hydrogen ions, viz., $HCl + NaHCO_3 \rightarrow H_2O + CO_2 + NaCl$. There are hundreds of highly advertised products in this field. Some of these contain simple antacids. Others, for instance Alka-seltzer, contain aspirin along with a source of carbon dioxide for "fizz" ($NaHCO_3$ plus a solid acid).

THE USE OF CHEMICALS AGAINST DISEASE

Our conceptual understanding of the actions of sulfa drugs and some of the antibiotics is more advanced than with most drugs. The work of the German bacteriologist, Paul Ehrlich (1854–1915) at the beginning of the twentieth century laid the groundwork for chemotherapy, the treatment of disease with chemicals. While working with microscope slides of cellular tissues, Ehrlich noticed that various types of cells picked up different dyes and that some tissues were especially selective in taking a stain from specific dyes. He guessed that it might be possible to take advantage of this natural biological selectivity to fight organisms which cause disease. In other words, he hypothesized that a specific chemical might be picked up by an invading organism while remaining essentially harmless to the

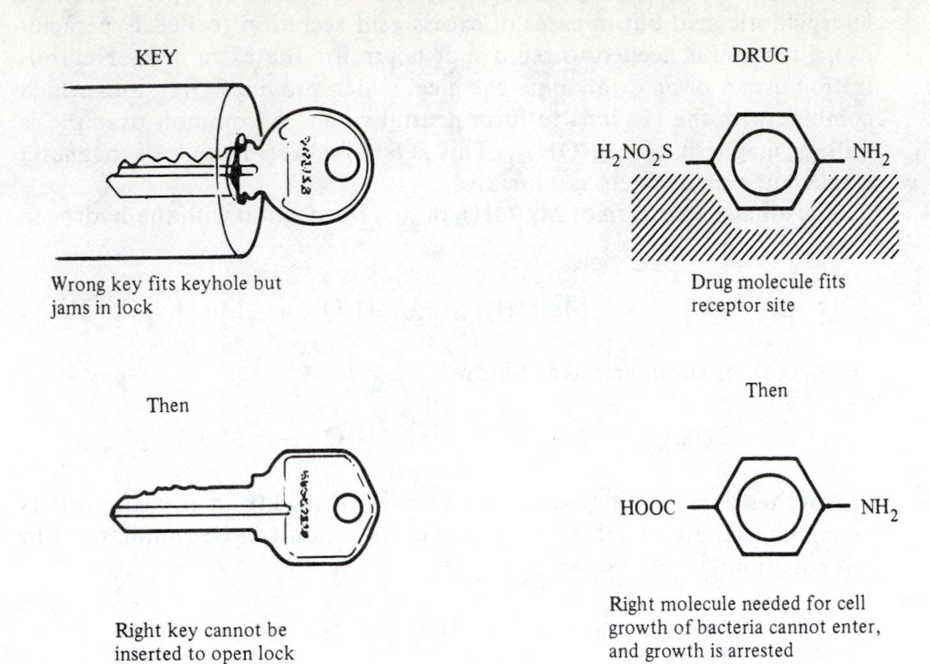

FIGURE 14.3
Simplified analogy for one type of drug action. The mechanism is more complex, involving shapes of three-dimensional molecules.

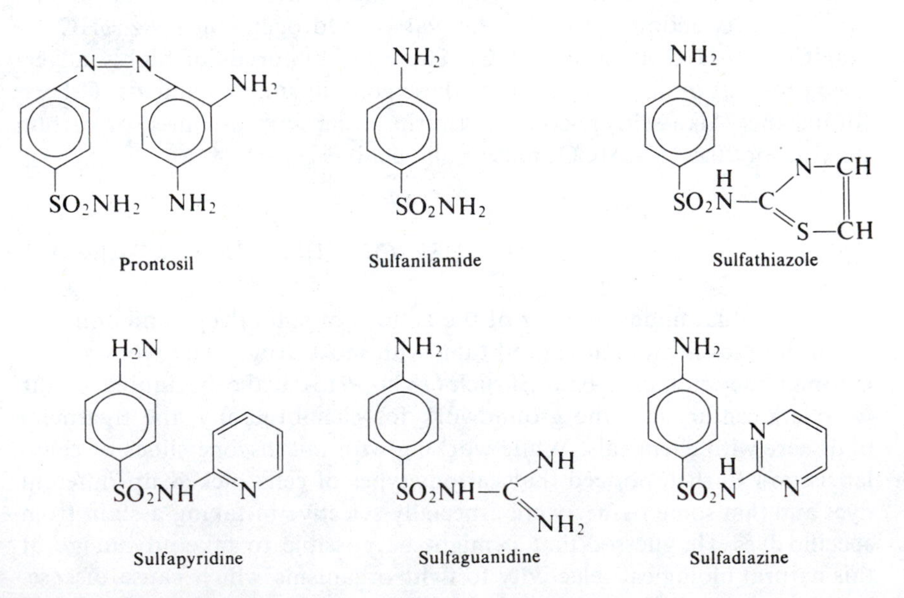

FIGURE 14.4
A few examples of sulfa drugs.

host. The best-known drug he developed was "salvarsan 606" which was the first effective agent against syphilis. The 606 label indicates that he patiently tried 605 other chemicals before the successful result with "606." Ehrlich thus gave backup to the idea that there are certain receptors or sites of chemical reaction where specific chemicals can be attached and interfere in some way with the action of living organisms. The receptors appear to be proteins of the living tissue, and much research has been done to clarify behavior of specific chemotherapeutic agents.

Ehrlich and his co-workers advanced considerably the concept of three-dimensional receptors in cell materials requiring specific substances in order to function. This is the *lock-and-key* concept. You know that only one cut of a key can open a lock. However, a lot of other keys may fit and jam the lock. Ehrlich proposed that a chemical which is the same shape as a natural chemical but which contains other kinds of atoms may be absorbed at a receptor site. This could then "jam" the vital functioning of the particular organism, such as bacteria which cause disease. (See Figure 14.3.)

THE SULFA DRUGS

This lock-and-key concept of chemicals was further fortified by the discovery of the sulfa drugs in the 1930s. Researchers in a German dye laboratory had invented a new dye which was routinely tested for drug action and found to control streptococcus organisms. The dye was named Prontosil and was the first of the sulfa drugs. The formula is shown in Figure 14.4.

Once the formula of a new chemical is known, research on similar structures can proceed more logically. This is exactly what happened. Experiments indicated that Prontosil was modified in the human body and that the active principle was a chemical called sulfanilamide. Clinical tests showed that sulfanilamide had several disadvantages when used directly. It was only slightly soluble in water and tended to crystallize from solutions. Even more important, when taken orally it eventually reaches the kidneys. When the dose is large or continued for some time, as is often necessary with sulfa drugs, the kidneys may be damaged by accumulation of sulfanilamide. Other toxic side effects made it obvious that sulfanilamide could not be given directly by mouth. Workers in various laboratories began to search for related compounds and the development of a whole series of sulfa drugs resulted. These were essentially related to sulfanilamide but were much freer of side effects, that is, they were less toxic to the host. The sulfa drugs were the first of the "miracle drugs"; they served effectively as agents against infection until the introduction of the antibiotics in the 1940s.

Sulfa drug research also added new insights to the lock-and-key concept of drug action. Investigations on drug mechanism indicated that sulfanilamide was antagonistic to a naturally occurring chemical, para-aminobenzoic acid (PABA). Compare the two formulas which follow.

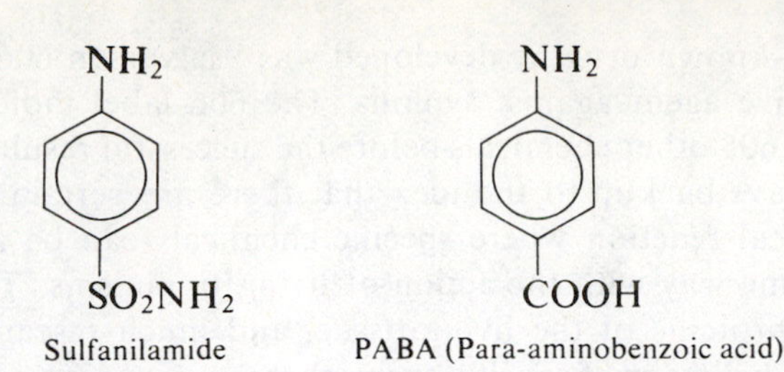

Sulfanilamide PABA (Para-aminobenzoic acid)

Sulfanilamide stopped growth of bacteria, but if PABA was added they started growing again. The hypothesis was that PABA was essential for proper action of a bacterial enzyme. The close similarity of sulfanilamide to PABA meant that the sulfa could attach itself into the enzyme system. If sufficient sulfa drug molecules were present they could preempt most of the crucial spots on the enzyme. The enzymes then could not function properly and bacterial growth could be stopped.

ANTIBIOTICS

Antibiotics like penicillin and streptomycin are probably more familiar to you than the sulfa drugs. Antibiotics are chemicals which are made by microorganisms and which can retard the growth or even kill other microorganisms. The possible existence of antibiotics was recognized by the founders of bacteriology, especially the great French scientist Louis Pasteur (1822–1895). In 1877 Pasteur noticed that anthrax bacilli growing in urine were killed when the specimen was contaminated by other organisms from the air. He and his co-workers also noticed that animals who were infected with anthrax disease often did not die from it if they were also infected by other germs at the same time. Pasteur's insight is indicated by his prophetic remark that "The time may come when we may utilize harmless microbes for combating harmful ones."

The Long Road to Penicillin

Other bacteriologists gradually developed this idea but the antibiotic era of medicine really started with the discovery of penicillin by the British bacteriologist Alexander Fleming (1881–1955). In 1929 Fleming reported his observations on the adverse effect of a mold on the growth of staphylococci bacteria. He carried out further experiments with a broth obtained from the mold, *Penicillium notatum*. He found this broth could cure some infections in mice and he called the broth penicillin. Later *penicillin* was applied only to the active chemical found in the broth.

There was a long and tedious period of more than ten years from the time that Fleming made his original observation to the achievement of a practical drug product. It was in 1940–41 when techniques were developed to produce a purified and safe product for clinical use. Then a team of research people working at Oxford under the bacteriologist Howard

Among the scientific tools used in the pharmaceutical industry to develop new drugs is paper chromatography, being used here by a researcher for identification of antibiotics. (Courtesy Abbott Laboratories.)

Florey (1898–1968) and the biochemist Ernst Chain (1906–   ) finally developed the process for isolation of the chemical penicillin.

Only very small supplies of the active agent became available since it was very difficult to extract penicillin from the mold. The Oxford group was growing the mold in small flasks and about six hundred flasks were necessary to produce enough product to treat a patient for one day. Because of the war, any great expansion of the mold-growing project was impossible. Then the Oxford research people turned to the United States and the U.S. Department of Agriculture laboratory at Peoria, Illinois, where entirely new methods were gradually developed for growing the mold. Still it took about one and one-half years of work to produce enough penicillin for two hundred patients.

Finally in the summer of 1943 two breaks occurred in the project. A new strain of *Penicillium* was found on a moldy cantaloupe in a Peoria supermarket which gave enormously increased yields of penicillin. Also three American pharmaceutical companies teamed up to develop an entirely new technology for growing the molds and for purification and isolation of the chemical penicillin. The method uses vats holding up to 15,000 gallons, improved culture fluids, and growth of the mold in "deep cultures" which are stirred constantly by streams of air. The urgent need for penicillin to treat infected battle and bombing wounds in World War

Left photo: In the early 1940s, penicillin was produced in flasks like these. The flasks contained a nutrient into which the penicillin mold was injected and subjected to conditions that would lead to optimum production of the crude antibiotic. Right photo: Multi-story fermentation tanks used in present-day penicillin production dwarf a workman, left foreground. Antibiotics are produced in the U.S. alone in excess of 10 million pounds per year. (Photos courtesy of Merck & Co., Inc., Rahway, N.J.)

II hastened the ultimate victory. The final result was enough penicillin for every soldier who needed it and enough for civilians as well.

All during the long struggle to develop penicillin, scientists were searching for methods of isolating a particular molecular species. Determination of the molecular structure of penicillin was a necessary part of the search. Now scientists can produce varieties of penicillin by altering various groupings on the penicillin molecule (see Figure 14.5).

The success of penicillin spurred research people to look for similar powerful agents against infectious disease. Many other antibiotics were subsequently found in the 1940s and 1950s. Interestingly enough, most of these came not from molds but from microorganisms in the soil. The formula of streptomycin, the first antituberculosis drug, and a few of the other important antibiotics, is shown in Figure 14.6.

```
     O                 S
     ‖               /   \
R—C—NHCH—CH      C(CH3)2
           |       |        |
       O=C——N———CHCOOH
```

A generalized structure for penicillin

```
                          O                 S
                          ‖               /   \
(C6H5)—CH2—CNHCH—CH      C(CH3)2
                                |       |        |
                            O=C——N———CHCOOH
```

Penicillin G

FIGURE 14.5

The elusive penicillin molecule. R can be varied to make different types of penicillins. For example, Penicillin G, which is widely used, has R as C_6H_5—CH_2—

Streptomycin

Chloromycetin

Generalized formula for tetracyclines

R	*R′*	*Name*
H	H	Tetracycline
Cl	H	Aureomycin or Chlorotetracycline
H	OH	Terramycin or Oxytetracycline

FIGURE 14.6
Formulas of several important antibiotics.

Limitations and Cautions With Antibiotics

Thousands of antibiotics have been investigated, but only a very few have been found suitable. The major problem is toxicity. Many cautions have to be observed with those antibiotics in general use, for example:

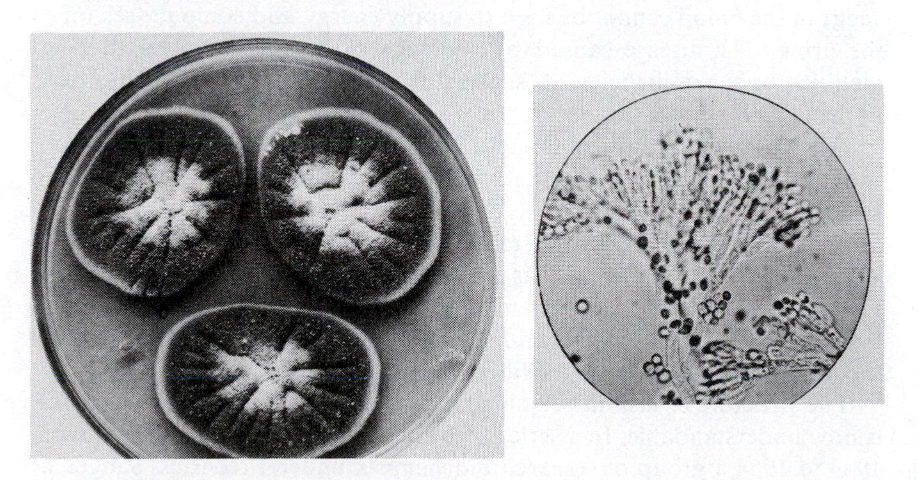

Photo on left shows penicillin mold growing in a culture dish. On the right is a photomicrograph of penicillin crystals. (USDA photos.)

1. Some people develop an allergy to a particular antibiotic. Subsequent doses can cause strong reactions and even death.
2. Disease-causing bacteria may in some cases develop a resistance to antibiotics.
3. "Broad-spectrum" antibiotics (like tetracycline) may upset the natural balance in the body by killing off beneficial bacteria.
4. If full dosage of an antibiotic is not taken, there may be a resurgence of the bacteria after a slight dormancy.
5. Antibiotics should not be taken for colds, which involve virus particles; antibiotics are ineffective in combatting viruses. (At present there is no drug that can combat viruses.)

HORMONES: VITAL CHEMICALS FOR CONTROL

Hormones are chemicals which are secreted directly into the blood stream by the endocrine, or ductless, glands. Hormones (Greek: setting in motion) are important chemical messengers which influence the activity of organs, cells, and tissues in various parts of the body. It is only within the twentieth century that investigations have begun to reveal the function and structure of some of these important chemical regulators.

Insulin

Some hormones are proteins which are complex long-chain molecules containing nitrogen (chapter 10). Insulin is one of the important protein hormones. The pancreas, a gland near the stomach, produces both insulin and pancreatic juice which aids digestion. When insulin is insufficient the sugar in the blood cannot be used to supply energy and some passes off in the urine. The disease caused by shortage of insulin is called diabetes mellitus, or sugar diabetes. A successful treatment consists of injections of insulin.

Originally insulin was obtained from the pancreas of dogs, but now it is obtained from pancreatic glands of sheep, cattle, pigs, and even certain kinds of fish.

It was known in 1889 that removal of the pancreas would result in sugar diabetes. However, it was not until 1922 that a Canadian doctor, Frederick Banting, developed a method of extracting the hormone insulin from animal pancreas glands and showed that it could be used to control the blood sugar in diabetes mellitus.

The structure of insulin remained a puzzle for many years. The reason is now understandable. In a series of brilliant investigations running from 1944 to 1954 a group of research biochemists under Frederick Sanger of Cambridge University finally determined the structure. Insulin consists of a chain of fifty-one amino acids (see Figure 14.7).

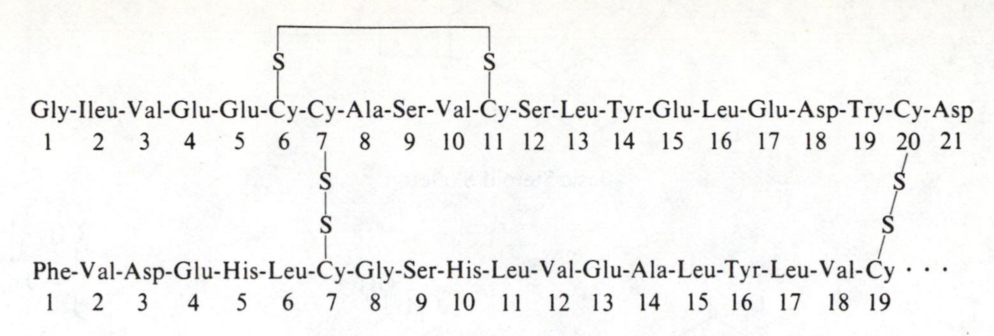

FIGURE 14.7

Structure of beef insulin. The protein hormone consists of two chains of amino acids, cross-linked through sulfur bridges. This is a rather simple protein (see chapter 10). Insulin from other species differ slightly in amino acids at a few positions. Amino acid code is as follows:
Gly = glycine; Ileu = isoleucine; Val = valine; Glu = glutamine; Cy = cytosine; Ala = alanine; Ser = serine; Leu = leucine; Tyr = tyrosine; Asp = asparagine; Try = tryptophan; Phe = phenylalanine; His = histidine; Arg = arginine; Lys = lysine; Thr = threonine.

Adrenalin: The Hormone for Action

Not all the hormones are as complex as insulin. Adrenalin, secreted by the adrenal glands which are located at the upper part of the kidneys, has a rather simple molecular structure.

HO, HO (on benzene ring) — $CHOHCH_2NHCH_3$

Adrenalin is secreted under stress of anger or fear, preparing the body for defensive action—"fight or flight"—by a series of changes including: faster heart action; contraction of blood vessels; increase in rate of blood coagulation; and release of sugar from glycogen storage.

Suggestions have been made that some illness, both mental and physical, may be related to the fact that in most stress situations in our culture we do not go into the action which adrenalin prepares us for. We can neither fight nor flee, but rather we tend to bottle up our reaction and suppress the normal effects of adrenalin.

Sex Hormones

Many important and interesting hormones belong to a chemical group called steroids. Steroids have a four-ring system (Figure 14.8). All the

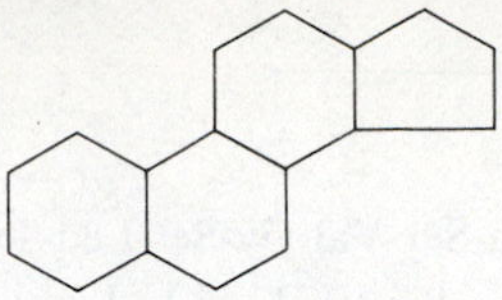

Basic Steroid Skeleton

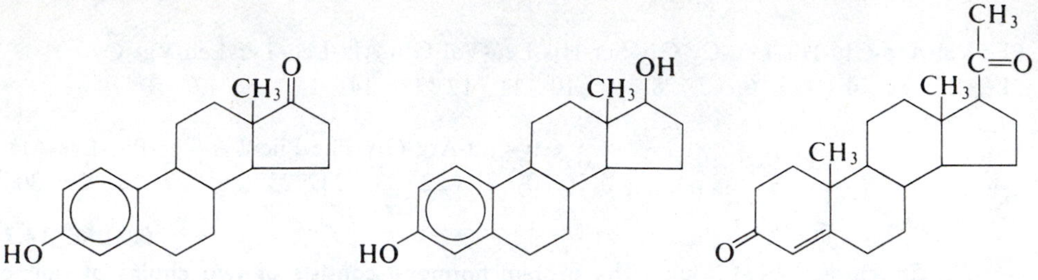

Estrone

Estradiol

Progesterone

Female Sex Hormones

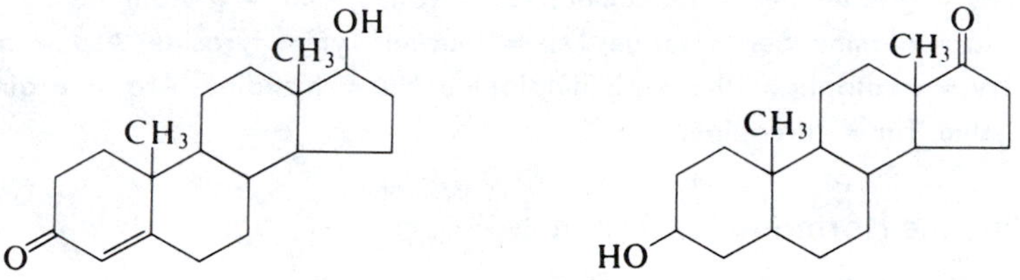

Testosterone

Androsterone

Male Sex Hormones

FIGURE 14.8
Some sex hormones.

carbon rings are joined together into the steroid skeleton which then can have varied atoms or groups on the four rings. This makes for many possibilities. In addition, any particular molecule can have various spatial arrangements of the atoms.

Important steroids are the male and female sex hormones. Three types of sex hormones are now recognized.

1. Estrogens: female sex hormones produced mainly in the ovaries, which regulate female functions such as the menstrual cycle and the development of secondary female sex characteristics such as breast changes at puberty.

2. Androgens: male sex hormones produced mainly in the testes, which regulate secondary male sex characteristics such as deepening of voice, growth of beard at puberty, and functioning of the prostate gland.

3. Progestrogens: female sex hormones formed chiefly in the ovaries and concerned with regulation of the various stages of pregnancy.

One very interesting aspect of the formulas is that there is very little difference between male and female hormones. The major differences they produce are not at all obvious from the slight differences in molecular structure. The exact way they are involved in producing such important physiological differences is largely unknown.

Sex hormones are nevertheless used considerably in medicine as replacements in cases of deficiency or as inhibitors. Estrogens are used to prevent atrophy of the uterus and breasts after surgical removal of the ovaries. Androgens are used in cases of testicular deficiency.

World War II Pushes Steroid Research

The outbreak of World War II accelerated the progress of steroid research. There were rumors that Germany was administering extracts of adrenal glands to their aviators so that they could operate better under stress. The research efforts led to clarification of the structures of many steroids and preparation of many new compounds by alteration of groupings on the steroid skeleton.

In the early 1940s a researcher at Pennsylvania State University, Russell Marker, made an interesting discovery. He found that a certain Mexican yam contained a large proportion of a basic steroid. The structure of the molecule was determined and the Mexican yam became a source for a variety of related steroid molecules.

Discovery of "cortisone" as a very effective drug for relief of rheumatoid arthritis was announced in 1949. Cortisone was obtained from the outer covering or "cortex" of the adrenal glands. Cortisone therapy gave such dramatic relief to sufferers from rheumatoid arthritis that it spurred new research efforts. About 30 other hormones have now been isolated from the adrenal cortex. (See Figure 14.9.) These hormones are some of the most powerful drugs ever developed. Unfortunately, however, they do not offer cures but only very marked relief of symptoms. They also may produce severe side effects. Cortisone is used today only when less dangerous therapies are found to provide no adequate relief.

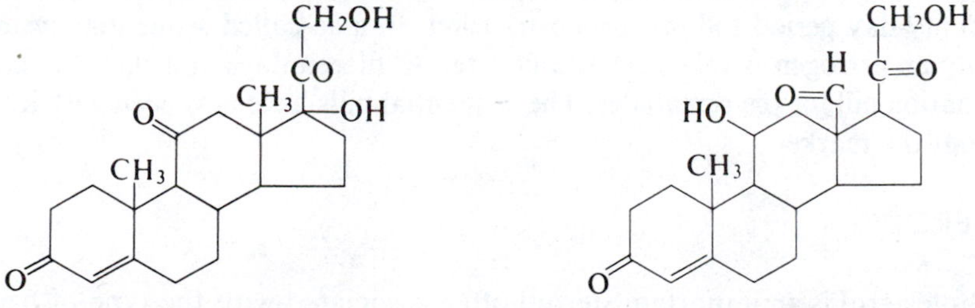

CH_2OH CO CH_3 OH O CH_3 O

(a) Cortisone

CH_2OH H C=O O=C HO CH_3 O

(b) Aldosterone

FIGURE 14.9

Cortex hormones. (a) Cortisone has been found effective for relief of rheumatoid arthritis; (b) Aldosterone is an important hormone from the cortex which controls absorption and excretion of inorganic ions and water in the kidneys.

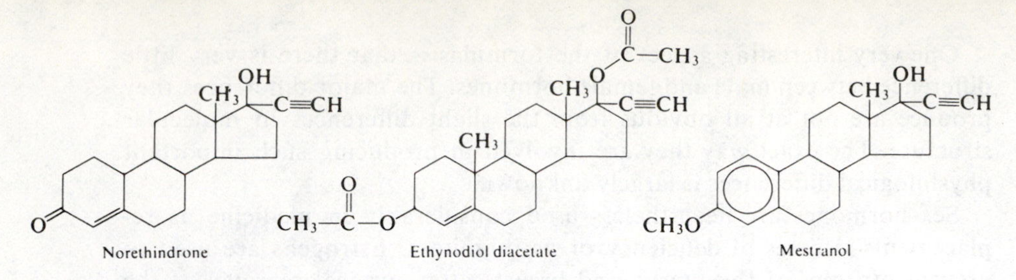

FIGURE 14.10
Examples of steroids used in birth control pills.

Development of "The Pill"

The conceptual knowledge of molecular structures contributed directly to development of the pill for birth control. In the 1930s researchers showed that ovulation or release of an egg from the ovary could be blocked in rabbits by the injection of progesterone, the hormone produced in large quantities during pregnancy. Progesterone caused painful side effects when injected and was not effective in ordinary dosages by mouth.

The success with cortisone-related hormones during the war years stimulated research into potential use of hormones to prevent conception. Beginning in the early 1950s, research workers started intensive investigations into steroid chemicals related to progesterone. The groundwork had already been laid by Marker's work on altering natural steroids which were much more plentiful than progesterone. Many synthetic steroids were made. Tests showed a few with good oral acceptance as pills and effective suppression of ovulation. Figure 14.10 provides the formulas of several synthetic birth control molecules. You can see the close similarity to the natural sex hormones shown in Figure 14.8.

The first oral contraceptive went on sale in 1960. It was quickly joined by many others as different pharmaceutical firms brought out their own versions. The impact of oral contraceptives was enormous and the full implications, both medical and social, have yet to be evaluated. The usual product contains an estrogen along with the progesterone-like steroid (progestrogen). In some applications the combination estrogen-progestrogen pill is taken daily in the twenty-day period following menstruation. In a so-called sequential system, only an estrogen is taken for the first ten to fifteen days and then the combination pill for the remainder. The sequential pills have now been withdrawn from the market.

CHOLESTEROL

Cholesterol is an important steroid often associated with the type of heart disease called hardening of the arteries.

Cholesterol was discovered early in the nineteenth century but its chemical structure was not determined until 1930. It is probably the most common of all steroids and is found in varying quantities in almost all living organisms. In humans the central nervous system contains a high content of cholesterol. The brain tissue consists of about 10 percent cholesterol on a dry weight basis. Normal blood plasma contains a small percentage of cholesterol. It occurs in fatty excretions such as dandruff and ear wax. Gallstones are predominantly cholesterol.

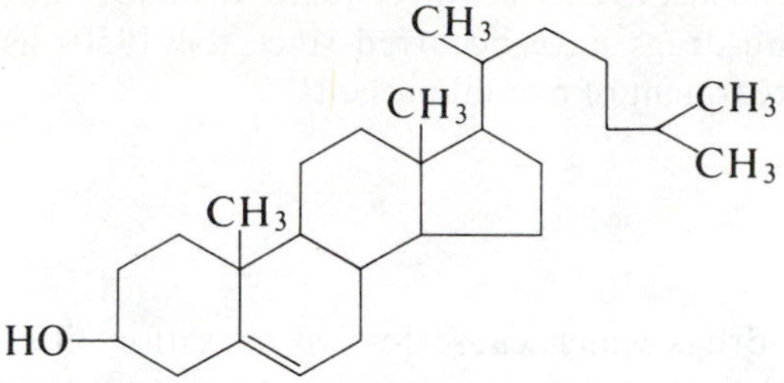

Cholesterol

Obviously then we cannot function without this important steroid. Most organisms, including humans, have the ability to make cholesterol from simpler molecules obtained from food. It is believed to be a precursor for most other steroids like sex hormones and adrenal cortex hormones. However, science still has much to learn regarding the functioning of cholesterol in the body.

Studies have attempted to correlate cholesterol level in blood serum with atherosclerosis, a condition in which fatty deposits are built up in the arteries, causing them to narrow, and contributing to blockage of blood supply. It was early recognized that people with high blood cholesterol levels have a greater tendency toward atherosclerosis, and that the fatty deposits in arteries are rich in cholesterol. Consequently, a connection between high blood cholesterol and narrowing of the arteries was surmised. This assumed relationship is the reason for diets prescribed to lower cholesterol levels in the blood. However, the actual mechanisms operating here are apparently complex. The body can make cholesterol from any diet—fats, carbohydrates, or proteins. Also it has been found that cholesterol is made chiefly in the liver and that diets containing a low amount of cholesterol may trigger the liver to produce more.

All kinds of guesses have been made to account for hardening of the arteries. High fat, high carbohydrate, and high protein diets have all been suggested as contributing to this disease; some people believe stress or lack of exercise are implicated. So far our conceptual background is not sufficient to give us definite answers.

Eating foods containing polyunsaturated fats has been found to lower cholesterol levels. Saturated fats raise these levels. Research is continuing to determine exactly why this occurs.

DRUGS AFFECTING THE CENTRAL NERVOUS SYSTEM

The chemicals which affect the central nervous system, causing a change in an individual's behavior, are difficult to classify. We will limit our discussion to those drugs called "psychoactive"—chemicals that have some effect on the mind: (1) anesthetics, (2) depressants, (3) tranquilizers, (4) stimulants, (5) narcotics, and (6) hallucinogens.

Some "mind-drugs" go back into prehistory. Others have been developed within the last twenty or thirty years. Probably the most dramatic advances in mind-drugs have occurred since the 1950s as a response to the wide-scale treatment of mental patients.

Anesthetics

Anesthetics are drugs which cause loss of sensation like the feelings of pain, touch, cold, and so on. Anesthesia is the condition of loss of sensa-

TABLE 14.1
Examples of Chemicals Used in Anesthesia

$CHCl_3$	Chloroform
N_2O	Nitrous oxide (laughing gas)
$CH_3CH_2—O—CH_2CH_3$	Ether (really diethyl ether, the original anesthetic, an ether in general being R—O—R′)
$CH_3—O—CH_2CH_2CH_3$	Methyl propyl ether
Three-membered ring: CH_2 bonded to CH_2—CH_2	Cyclopropane
Ring structure: HN—C(=O)—C(C_2H_5)($CH(CH_3)(CH_2)_2CH_3$)—C(=O)—N=C(NaS)—(back to HN)	Thiopental sodium (pentothal sodium), a barbiturate
$CF_3CHBrCl$ or F—C(F)(F)—C(H)(Cl)—Br	Halothane
$CHCl_2CF_2—O—CH_3$	Methoxyflurane
$[NH_2—C_6H_4—C(=O)—O—CH_2CH_2—\overset{+}{N}H(C_2H_5)_2]\ Cl^-$	Procaine hydrochloride (novocain), one of a variety of local anesthetics where the side chains are varied from one to five carbon atoms
$C_2H_5O—C(=O)—C_6H_4—NH_2$	Benzocaine, another local anesthetic

tion, especially to feeling of pain. Anesthetics have made possible tremendous advances in surgery. The history of anesthetics goes back to the mid-1800s when several people began testing various chemicals to achieve anesthetic effects. An American dentist, William Morton, is often given credit for the development of the first anesthetic, even though historians indicate others were also involved. In 1846 he made the first public demonstration of the use of ether for a tooth extraction. Subsequently, ether was used successfully in an operation at Massachusetts General Hospital. The method spread rapidly to England, France, and other countries.

Once the concept of a general anesthetic was established it was only natural that scientists would investigate other compounds. The original ether compound is easy to administer by inhalation and gives good muscle relaxation along with very little effect on blood pressure, pulse rate, or respiration. Disadvantages are after-operation nausea and irritation of respiratory passages. Now many other compounds find use in anesthesia. Often combinations are used. Some examples are given in Table 14.1 and in the section on barbiturates.

Depressants: Alcohol

Ethyl alcohol is the most widely used depressant drug. It is generally called simply alcohol since it is the most common alcohol in use (chapter 9). It is a depressant because it retards the brain's control mechanisms.

Alcohol is sometimes said to be an anesthetic, which it is if taken in excess. This points up the difficulty in classifying drugs by operational or behavior effect on the body. Even a single chemical may fit into several categories depending on the dosage. For example, a small dosage of a depressant drug can produce sedation or calming. The drug could be called a sedative. A little larger dosage can induce sleep and then the drug could be called a hypnotic. An even higher dosage could bring on deeper anesthesia. The general trend of depressants may be indicated as follows:

Increasing dosage ⟶

Sedation → Hypnosis → Anesthesia → Coma
↓
Death

Alcohol is chiefly absorbed in the small intestine. Only about one-fifth of the alcohol absorbed is picked up by the blood vessels in the stomach. Consequently, the presence of food in the stomach delays the transfer of alcohol into the intestines and makes for a lower absorption rate. Particularly proteins and fats, which are slowly digested, have the effect of delaying transfer of the food-alcohol mixture to the small intestine.

It is interesting that carbon dioxide has the effect of relaxing the pyloric valve which serves as a gate between the stomach and the small intestine. This is the reason why champagne, sparkling wines, and whiskey mixed with carbonated beverages have a faster effect. Carbon dioxide in an alcoholic drink speeds passage into the intestine where most absorption of alcohol occurs. Of course the greater the concentration of alcohol the greater the absorption: an ounce of whiskey will provide a much higher alcohol content in the blood than an ounce of beer.

Alcohol can have an irritating effect on the stomach. It dilates the capillaries and causes the stomach lining to be engorged with blood, and also causes an increased secretion of hydrochloric acid by the stomach. This is the reason it is not recommended for people with ulcers. The practice of taking aspirin with alcohol is poor since the aspirin itself can irritate the stomach lining and the combination could cause bleeding.

Misconceptions. There are probably more misconceptions about this drug than any other. One is that drinking alcohol makes you warm. Actually alcohol causes body temperature to fall, not rise. Alcohol dilates the capillaries in the skin so that they carry more blood, and the skin flushing with warm blood may give the drinker the illusion that he is warmer. Actually the process causes cooling. More evaporation from the warm skin occurs. This requires energy and hence the overall effect of alcohol is to lower body temperature. Dilation of the capillaries of the skin can become permanent in the case of people who drink in excess.

Another misconception is that vigorous exercise, hot showers, or stimulants like coffee can make a drunk sober because these methods speed the elimination of alcohol, unchanged, through the skin or kidneys. But tests show that only a small fraction of alcohol is eliminated unchanged while the major portion must be oxidized. Oxidation occurs chiefly in the liver, in a couple of stages, eventually giving acetic acid.

$$\underset{\text{Ethyl alcohol}}{CH_3CH_2OH} \xrightarrow{[O]} \underset{\text{Acetaldehyde}}{CH_3CHO} \xrightarrow{[O]} \underset{\text{Acetic acid}}{CH_3COOH}$$

Even though the rate of removal of alcohol through oxidation may vary slightly due to size of the liver and disease, the average rate is equivalent to about two-thirds to three-fourths of an ounce of whiskey, or about a glass of beer, per hour. The acetic acid can then be used by other body processes and can be converted to carbon dioxide and water:

$$CH_3COOH \xrightarrow{[O]} CO_2 + H_2O$$

The liver is then the bottleneck in the whole system of alcohol elimination. No way has yet been found to speed up the liver's job. When more alcohol is ingested than the liver oxidizes, then the blood level of alcohol can build up to a point of intoxication.

Is Alcohol a Food? Although some people say alcohol is a food, this can apply only insofar as alcohol can be used as a fuel, just as other molecules containing carbon and hydrogen, like sugar and starch. Alcohol then can serve as a source of energy. But it is a grave misconception to consider alcohol as a substitute for food. It contains no vitamins, protein, fat, or minerals, all of which are necessary in a balanced diet. One of the results of heavy drinking over a continued period of time is cirrhosis of the liver, a chronic disease in which liver cells become shriveled and hardened due to an increase in fibrous tissue. Cirrhosis is not considered to be only a direct result of alcohol but an indirect one caused by malnutrition. The excess intake of alcohol as an energy source becomes a substitute for necessary food nutrients and this results in the disease.

The Hangover: A Headache to Explain. There are many "hangover" theories among people who drink. Symptoms of hangovers vary considerably from person to person, but the usual ones include headache, depression, fatigue, and sometimes nausea. Studies have not thoroughly pinpointed the mechanisms in hangover. Some of the effects may be due to alcohol, but it is believed that many other factors are involved, all related to drinking. Fatigue is a natural result of excess activity carried on well past the time the drinker would ordinarily give in to sleep. The excessive fatigue may result in headache by buildup of fatigue chemicals in his system. In addition, various other chemicals may be incorporated in the alcoholic beverage during its manufacture from malt, hops, grapes, yeast, grain, and so forth. These other chemicals in liquors are commonly called congeners. Many have been identified. Liquors with a high content of congeners like bourbon are considered to have a higher potential for hangover than those like gin and vodka which have a low congener content.

Are There Alcohol Addicts? The problem of excessive alcohol intake often brings up the question of addiction. Here there is a considerable variety of opinions among alcohol experts. The term addiction apparently means many different things to many people, and to avoid the difficulties of a common definition more recent practice is to use the term alcohol dependence. There is no doubt that certain people have shown considerable psychological and physiological dependence on alcohol. There is also no medical question of the severity of withdrawal symptoms which an alcohol-dependent person experiences when his supply of liquor is suddenly cut off. The combined physical and psychological effects are summed up in the name delirium tremens. This is an extremely violent nervous and mental disorder accompanied by involuntary shaking of the body, or convulsion, and terrifying hallucinations. The abrupt withdrawal after prolonged physical dependence has even caused death in many cases.

Barbiturates

A major class of depressants are the barbiturates, commonly called sleeping pills. They can be used for calming or for sleep; that is, they are sedatives or hypnotics depending on the dosage.

Today there are in use more than twenty-five barbiturates, all of which are derived from a chemical called barbituric acid. This was first made in 1864, but it was not until much later that the sedative or hypnotic power of derivatives of the acid, or barbiturates, was recognized.

$$O{=}C(N{-}H)_2 \;+\; (C_2H_5O{-}C{=}O)_2CH_2 \;\longrightarrow\; O{=}C(N(H){-}C{=}O)_2CH_2 \;+\; 2C_2H_5OH$$

Urea | Diethyl ester of malonic acid | Barbituric acid | Ethyl alcohol

$$O{=}C(N(H){-}C{=}O)_2C(R)(R')$$

General formula for Barbiturate

FIGURE 14.11
Some barbiturate chemistry.

The first barbiturate was offered for sale in 1903. This is the chemical barbital and it was given the trade name Veronal. Figure 14.11 and Table 14.2 show the general method for making barbiturates. By looking at the simple process of making barbituric acid shown you can see how variations in starting materials could lead to different barbiturate structures. This could never have been accomplished without the conceptual knowledge represented in the formulas.

High-speed, multiple-die tablet machines are used today to turn out pills in enormous quantities. (Courtesy Parke, Davis & Company.)

TABLE 14.2

Variations Obtained in Barbiturates by Using Slightly Varied Molecular Structure

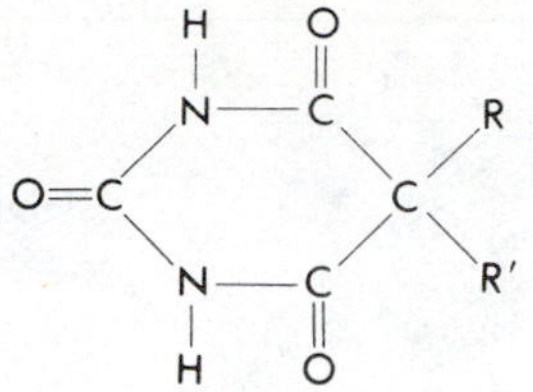

```
         H    O                              H    O
         |    ||                             |    ||
         N —— C     R                        N —— C     R
       /        \  /                       /        \  /
  O = C           C                   S = C           C
       \        /  \                       \        /  \
         N —— C     R'                       N —— C     R'
         |    ||                             |    ||
         H    O                              H    O
```

Barbiturate

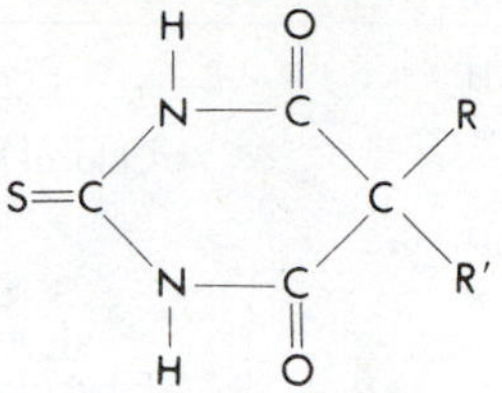

Thiobarbiturate (sulfur in place of oxygen on left carbon atom)

Atomic Chain Groups on Right Carbon Atom		*Name (Trade Name)*	*Use*
R	R′		
CH_3CH_2-	CH_3CH_2-	Barbital (Veronal)	Long-acting barbiturates: treatment of convulsive disorders, like epilepsy
CH_3CH_2-	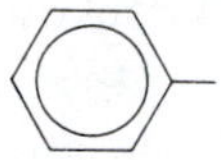 C_6H_5-	Phenobarbital (Luminal)	
CH_3CH_2-	$CH_3\overset{CH_3}{\overset{\mid}{C}}HCH_2CH_2-$	Amobarbital (Amytal)	Intermediate-acting barbiturates: treatment of patients who awaken during the night
CH_3CH_2-	$CH_3CH_2\overset{CH_3}{\overset{\mid}{C}}H-$	Butabarbital (Butisol)	
CH_3CH_2-	$CH_3CH_2CH_2\overset{CH_3}{\overset{\mid}{C}}H-$	Pentobarbital (Nembutal)	Short-acting barbiturates: small doses for induction of general anesthesia before operations; large doses for minor surgery or emergency military surgery; small doses also for patients who have difficulty falling asleep
$CH_2{=}CHCH_2-$	$CH_3CH_2CH_2\overset{CH_3}{\overset{\mid}{C}}H-$	Secobarbital (Seconal)	
$CH_2{=}CHCH_2$	$CH_3CH_2CH_2\overset{CH_3}{\overset{\mid}{C}}H-$	Thioamylal (Surital)	Ultrashort-acting thiobarbiturates: intravenous injections to induce unconsciousness smoothly and rapidly before operations, followed by use of other anesthetics
CH_3CH_2-	$CH_3CH_2CH_2\overset{CH_3}{\overset{\mid}{C}}H$	Thiopental (Pentothal)	

Physicians now have a wide range of barbiturate chemicals to choose from according to the particular problem they are dealing with. Table 14.2 shows the variety available for use as sedatives, hypnotics, or as adjuncts in anesthesia.

Knockout Drops: A Nonbarbiturate Depressant. The barbiturates are the most common depressants used as sedatives and hypnotics, but nonbarbiturate drugs are also available for these purposes. One of the oldest of these is

TABLE 14.3
Some Non-barbiturate Depressant Drugs

$Cl{-}C(Cl)_2{-}C(OH)_2{-}H$	Chloral hydrate	Ring structure labelled C_2H_5, O, N, H, O, with phenyl ring	Glutethimide (Doriden)
CH_3CH_2OH	Ethyl alcohol		
$CH_3CH_2{-}C(C{\equiv}CH)(OH){-}CH{=}C(Cl){-}H$	Ethchlorvynol (Placidyl)	Ring structure labelled O, CH_3, H, C_2H_5, C_2H_5, N, H, O	Methyprylon (Noludar)

chloral hydrate. It is the effective ingredient in "knockout drops" or, together with alcohol, in a "Mickey Finn."

One of the problems with nonbarbiturate sleeping pills is that people tend to feel they are safer than the barbiturates. However they can be equally dangerous and lead to the same abuses as barbiturates. For example, Placidyl causes habituation, and tolerance develops with extensive use. It also can cause exaggerated depressant action if taken with alcohol. Table 14.3 gives the formulas of some nonbarbiturate depressant drugs.

Abuse of Barbiturates and Other Depressants. The barbiturates are theoretically available only through a doctor's prescription. Nevertheless, they are among the most commonly abused drugs. "Drugs of abuse" is a special term referring to those obtained and used in illicit traffic without a doctor's recommendation. Some drugs of abuse are produced outside the normal pharmaceutical channels. These are called bootleg products. In the case of the barbiturates, most materials are merely diverted from legitimate channels. The abusers of barbiturates are looking for immediate relaxation, mild euphoria, and a feeling of well being.

Popular brand-name depressants usually are coded by identifying color to indicate dosage. Also many carry trademarks or other identifying symbols. Drug abusers usually refer to these in a special jargon. The red capsules are very common and are called "redbirds." Yellow phenobarbital capsules are called "yellow jackets," blue capsules are "blue heavens" and red and blue capsules are "Christmas trees." Depressants in general are called "downers," an obvious nickname.

Depressants can be very dangerous drugs whether obtained by doctor's prescription or not. Of the chemicals that cause death in the United States, sleeping pill formulations, mainly barbiturates, come second after alcohol, followed by carbon monoxide and salicylates. There are over 1,000 deaths each year in the United States from acute barbiturate poisoning. This is in addition to about 3,000 suicides. Estimates place the number of barbiturate addicts at about 200,000.

One of the major drug problems resulting in accidental death is the use of barbiturates with or after alcoholic beverages. When both of these depressants are present in the body together a greatly enhanced depressant effect is obtained. This synergistic effect is responsible for many of the deaths reported in the newspapers as due to an overdose of sleeping pills.

Barbiturate abusers can cause much anguish to themselves and others. They frequently exhibit such symptoms as difficulty in thinking, impaired reaction time, poor memory, and slurring of speech. Their moods can change from sluggishness to happiness to suicidal depression. Some people even become paranoid and commit violence against others. The brutal murder of eight student nurses in Chicago in 1966 was committed by a man long addicted to barbiturates and alcohol.

Withdrawal symptoms of a long-time barbiturate user can be even more devastating than in a heroin addict. Withdrawal has to be made gradually over a period of weeks in order to minimize the extreme strain on the body. If withdrawal is abrupt, there may be cycles of improvement followed by severe reaction, including extreme anxiety and restlessness, cramps, convulsions, and nausea. Delirium and hallucinations may follow. Death can occur in the convulsion and delirium stage.

Tranquilizers

The middle of the twentieth century saw the introduction of two psychoactive drugs which have had as much importance in the treatment of the mind as did the introduction of anesthetics for surgery in the mid-nineteenth century. These drugs are now often called tranquilizers even though tranquilizing was not the purpose when the first two (reserpine and chlorpromazine) were introduced. Tranquilizers have radically changed the procedures of mental hospitals, and have been instrumental in cutting down the patient population in mental wards.

Reserpine was derived from the roots of an Indian plant, *Rauwolfia serpintina.* It had been used for centuries in Hindu medicine for all sorts of ailments, but reports of these uses were largely ignored. Then a modern Indian doctor reported success in using reserpine to reduce high blood pressure or hypertension. It was tried in the United States for the same purpose. Doctors noticed that one of the effects of reserpine was that patients became calm and much less anxious about their condition. It then was tried with some success on schizophrenic mental patients. Subsequently, reserpine found more general use as a tranquilizer than as a treatment for hypertension.

The second major tranquilizer was originally made as an antihistamine. A short digression on histamine will show the relationship between the two and how antihistamine research led to tranquilizers.

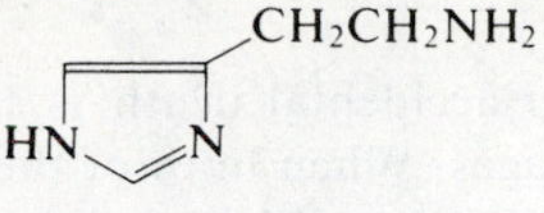

Histamine

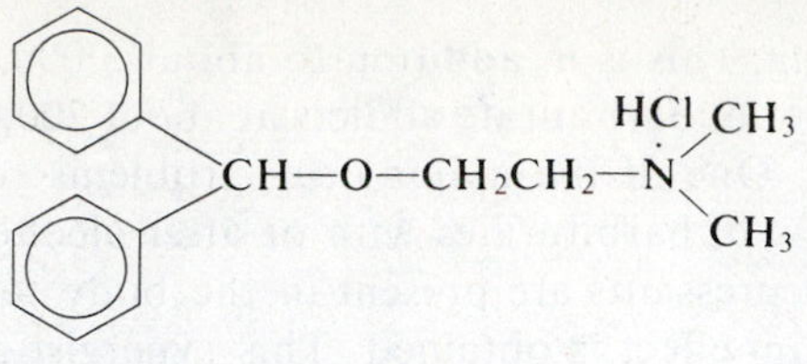

Benadryl, an antihistamine

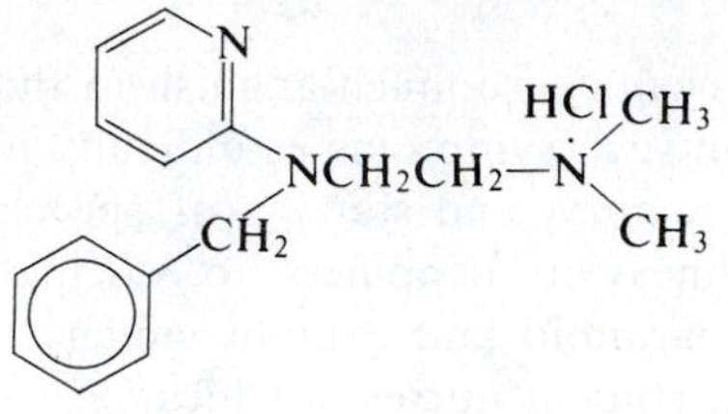

Pyribenzamine, an antihistamine

FIGURE 14.12
Some allergy chemicals.

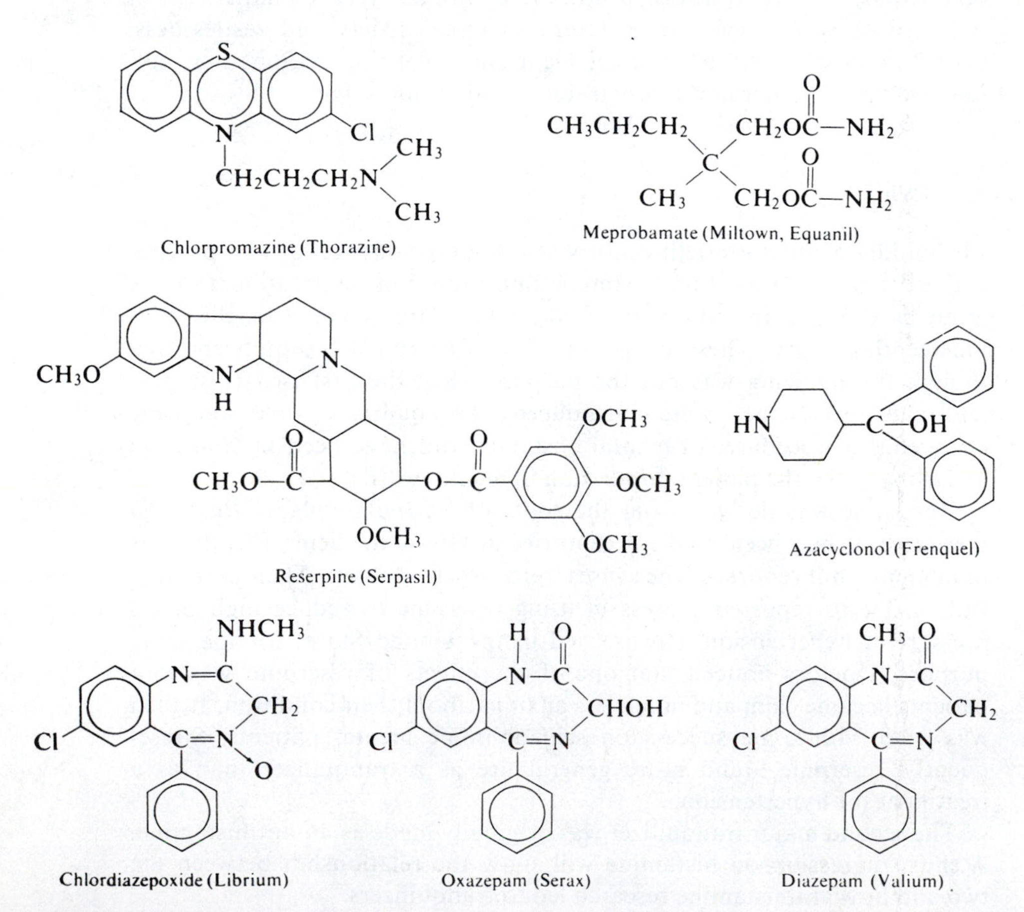

FIGURE 14.13
Examples of tranquilizers.

Histamine: The Allergy Chemical. Histamine is a chemical which occurs naturally in small amounts in plant and animal tissues. It has varied effects on the body, including dilation of blood vessels, swelling, increased secretion of fluids from eye and nose tissues, and increased production of stomach acid. Histamine is involved in symptoms of bronchial asthma, hay fever, and other allergies.

Chemists working on a histamine-allergy hypothesis tried to make other chemicals which would have structures similar to histamine. The guess was that these chemicals might occupy sites in the cells and block the attachment of histamine. Many effective antihistamines were developed with this concept in mind. The key was found to be branches of carbon chains on nitrogen. Figure 14.12 gives the formulas of histamine and two antihistamines. (It should be noted here that a common side effect with many antihistamines is their tendency to produce drowsiness and sedation.)

In the early days of antihistamine research, a French pharmaceutical company made and tested the compound called chlorpromazine (now sold under the trade name Thorazine). The French researchers noticed that patients were unusually calm after receiving the chemical, even after surgery. They then tried it on very disturbed schizophrenic and manic patients. It was found extremely effective in calming the patients. This success, following that with reserpine, resulted in the widespread use of chlorpromazine as a tranquilizer and also the synthesis of many other similar compounds. Examples are shown in Figure 14.13.

Use of Tranquilizers. Tranquilizers are now in widespread use to relieve tension and anxiety and sometimes relax muscles. They act by depressing the central nervous system but are in general considered milder than the barbiturates. Although not basically sleeping pills, doctors are using them more frequently to calm people down and avoid use of barbiturates.

Physicians now have several different products available as tranquilizers for two classifications of use: major tranquilizers, or antipsychotic drugs, and minor tranquilizers, or antianxiety drugs.

Reserpine and chlorpromazine are examples of major tranquilizers. These and others are used in treating extremely hyperactive, hyperexcited people and in treatment of acute and chronic psychotic conditions. They are thus widely used in mental hospitals. Even large doses have been found not to cause major addiction problems. There are side effects, however, which may include jaundice, slurred speech, muscle tremors, drowsiness, and blood changes.

Meprobamate (also sold under trade names Miltown and Equanil) is an example of a minor tranquilizer. It also was not first introduced as a tranquilizer but as a muscle relaxant. Then doctors noticed it had value as a minor tranquilizer for moderate cases of anxiety and tension. Now there are several chemicals available for this purpose. Some trade names are Librium, Valium, and Serax. The tranquilizers are not generally considered drug abuse chemicals. However, they may in some cases be associated with habituation and psychic dependence.

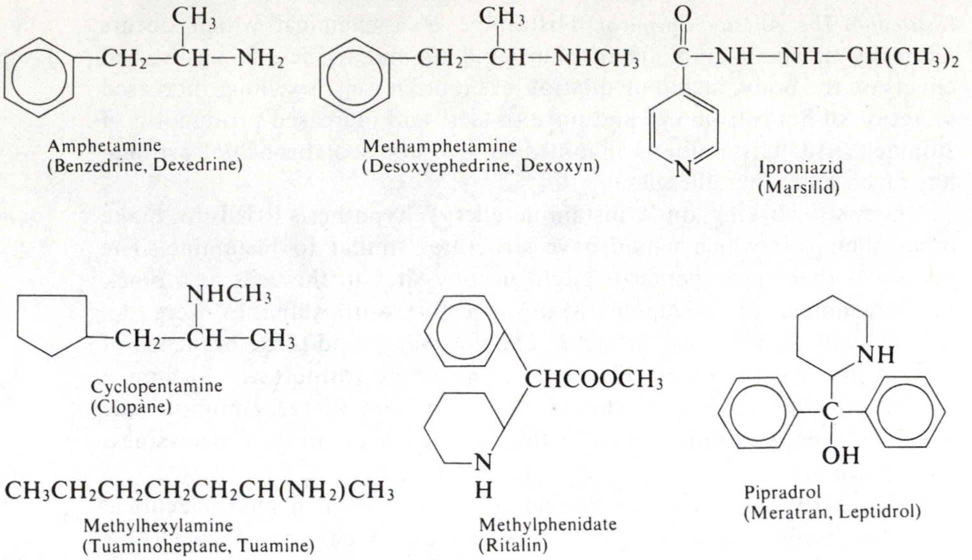

Amphetamine
(Benzedrine, Dexedrine)

Methamphetamine
(Desoxyephedrine, Dexoxyn)

Iproniazid
(Marsilid)

Cyclopentamine
(Clopane)

Methylphenidate
(Ritalin)

Pipradrol
(Meratran, Leptidrol)

$CH_3CH_2CH_2CH_2CH_2CH(NH_2)CH_3$

Methylhexylamine
(Tuaminoheptane, Tuamine)

FIGURE 14.14
Stimulants.

Stimulants

Chemicals which have the opposite effect on the central nervous system from that of depressants are called stimulants. You have already met a mild stimulant, caffeine. This is a commonly used chemical found in coffee and tea and in small amounts in many cola-flavored soft drinks.

Stimulants affect the central nervous system by producing alertness, excitation, extended wakefulness, and alleviation of fatigue. As with other classes of drugs discussed above, there are many chemicals which have these effects, and many were originally developed for different purposes than stimulation of the central nervous system. Iproniazid (trade name, Marsalid) is a common antidepressant drug. It was originally developed for treatment of tuberculosis. Then people noticed that it had the property of relieving mental depression. Amphetamine, now a common stimulant, was originally introduced for treatment of asthma.

Amphetamines. Amphetamines are now probably the most widely used (and abused) stimulants. Amphetamine was originally made in the nineteenth century but it was not until 1932 that it began to see wide use. It was then introduced under the name benzedrine in an inhaler to reduce nasal congestion. It was later used during World War II in order to keep exhausted troops alert. Subsequently other derivatives of amphetamine were investigated. Now there are several stimulants based on the amphetamine structure as well as others of the nonamphetamine type (see Figure 14.14). These drugs produce mood elevation with increased wakefulness, alertness, concentration, and confidence. The ability to perform physically is also enhanced.

Abuse of Amphetamines. The amphetamine-type stimulants commonly abused are benzedrine, dexedrine, and methedrine, a shortened version of methamphetamine. The latter has also been called "speed," "splash," "crystal," or "meth." It is available in both tablet form and as a liquid in ampules. When injected into a vein it produces an immediate high, an extreme form of stimulation characterized by hyperactivity and excitation. In some individuals this stimulation can lead to paranoid delusions and violence. An overdose can cause death. The drug fiend stereotype, often in the past applied to the opium user, may be more appropriately applied to some of the individuals who abuse stimulants like speed.

Because amphetamines are commonly prescribed drugs, the potential for abuse is considerable. You have probably seen newspaper stories on the use of amphetamines by celebrated artists, politicians, and jet-setters. They are apparently used by many types of people, from truck drivers (who nickname the pills "co-pilots") to students using them to stay awake before exams. People take them when dieting because of their effect of curbing appetite. Some amphetamines have nicknames among drug abusers. Dexedrine capsules with a brown top are called "brownies." Available as pale orange, three-sided tablets, they are called "orange hearts." Stimulants generally are called "uppers" or "pep pills." Other descriptive terms are "peaches," "roses," "bennies," "dexies" and "coast-to-coast" for the long-acting capsules.

Large quantities are made by regular drug manufacturers, but because amphetamines are generally considered easy to make a considerable amount of bootleg material is also available to drug abusers.

It is obvious from the large-scale prescription of amphetamines that some people are helped by the stimulants. However, tolerance of the drug can occur and this means the body needs more and more in order to achieve the same effect, increasing the risk of an overdose. Overdoses can cause extreme excitability, anxiety, and even hallucinations. Strong psychological dependence is another possible effect. Withdrawal can result in extreme fatigue, depression, and lassitude. There is apparently a great temptation to use barbiturates to get some sleep and relief from the depression caused by stimulant-withdrawal.

There are even several combination stimulant-depressant drugs which attempt to get an operational balance between stimulation and depression of appetite, which amphetamine provides, and control of overstimulation, which barbiturates provide. These are used as aids to people who are dieting and also for mild depression problems. Dependence can develop and use over a long period may result in tolerance. The balance between stimulant and depressant action is very delicate. Overdoses can result in excessive depression or stimulation.

Cocaine: Anesthetic and Stimulant. Cocaine was originally developed as a local anesthetic in the mid-1800s. It is derived from the leaves of the coca plant which grows in the uplands of Bolivia, Peru, and Chile. It has been used for centuries by Indians living in the Andes mountains for its stimulant properties. They apparently rely on the drug, which they obtain by simply sucking or chewing coca leaves, to sustain themselves in a life of toil and deprivation in the rugged mountains.

It is interesting that Sigmund Freud (1856–1939), who is known for developing psychoanalysis, was greatly interested in the use of cocaine. He promoted it as a local anesthetic and did the first study of the effects of the drug. In fact he tried cocaine as a cure for morphine addiction in a doctor patient. He did succeed in curing the morphine addiction but at the same time the doctor became addicted to cocaine.

Cocaine was eventually replaced as an anesthetic by better products like procaine (Novocain). Cocaine was a drug of abuse at the turn of the century, and it has recently reappeared on the United States drug scene.

The favorite methods of taking cocaine are injection and "snorting," or sniffing through the nose. The effect of cocaine is much shorter than that produced by the amphetamines but is similar otherwise. The body does not appear to develop tolerance but psychological dependence does occur. Cocaine addicts are known to "shoot up" or inject cocaine many times a day to sustain euphoria and avoid the severe depression which occurs when the effect of the drug begins to wane.

Chronic use results in digestive disorders, nausea, loss of appetite and weight, insomnia, skin abscesses, and occasional convulsions. Prolonged sniffing of the white crystal powder—or "snow" in the drug jargon—perforates and wears away the partition between the nostrils. Paranoid delusions with hallucinations occur. The mental feelings of criticism often trigger compulsive and violent acts against others. Overdoses can also cause death from heart or respiratory arrest.

Narcotics

The term narcotic is used in various ways. It is derived from a Greek word meaning to make numb. Thus in pharmacology narcotics are drugs which will induce sleep or stupor and relieve pain. In legal usage *narcotic* describes a drug which is habit-forming or addictive. Opium is the drug which often comes to mind when the word narcotics is used. The narcotics, sometimes called opiates, do include opium and its active component, morphine. But they also include heroin, which is morphine chemically altered to make it nearly ten times stronger. Narcotics also include a series of synthetic chemicals which have a morphinelike action.

Morphine, methadone, and meperidine are used medically and are not often seen on the black market. Heroin is the chief narcotic of abuse. It is not used in medicine, and all heroin in the U.S. is smuggled into the country.

If a narcotic is given in large enough dosage to cause sleep it is said to be a hypnotic. If it merely relieves pain by numbing the nerves it may be called an analgesic or anodyne. You can see again that descriptive labels and the demarcation between drug categories may be quite imprecise. Cocaine has been legally classified as a narcotic even though it is a stimulant which does not create tolerance or physical dependence. Marijuana is not a narcotic under federal law but is considered a narcotic under some state laws. In pharmacology marijuana is a mild hallucinogen and it will be considered below in that category.

Opium. Opium is made from the juice of poppies native to Greece and the Orient (and illegal in the United States). A white milky juice is obtained from the unripe capsules or seed pods. The juice is refined to opium, a white powder with a sharp, bitter taste.

Morphine was isolated from opium in 1803, followed by other chemical components. It is now recognized that opium owes its analgesic and narcotic properties to over twenty-five nitrogen-containing chemicals present in a complex mixture. However, the chief effects of opium are due to morphine.

Opium is rarely used in medicine today. Doctors instead use compounds made from opium or synthetic materials with similar effects. The three most common derivatives of opium are morphine, codeine, and papaverine, shown in Figure 14.15. You can see how codeine can be obtained from morphine by adding a methyl group. These narcotic chemicals are used to relieve pain, ease convulsions, bring on sleep, and control diarrhea. They have also been used in cough syrup.

Narcotics Abuse. Heroin is the most common narcotic drug of abuse because it is so potent in producing intense, long-lasting euphoria. Drug traffickers dilute or cut heroin so that it normally ranges between 3 and 10 percent when sold to the user. A common adulterant used in cutting is lactose or milk sugar. A typical addict mixes the cut material into a liquid solution and injects it into a vein. This is called mainlining. It can also be injected just under the skin (skin popping) or sniffed through the nose. The addict's mainlining equipment, referred to as the "works," includes a needle and syringe or bulb for injection, a cord to use as tourniquet for location of veins, and a teaspoon for preparing the solution for injection. Addicts often contract hepatitis, tetanus, tissue infections, and abscesses

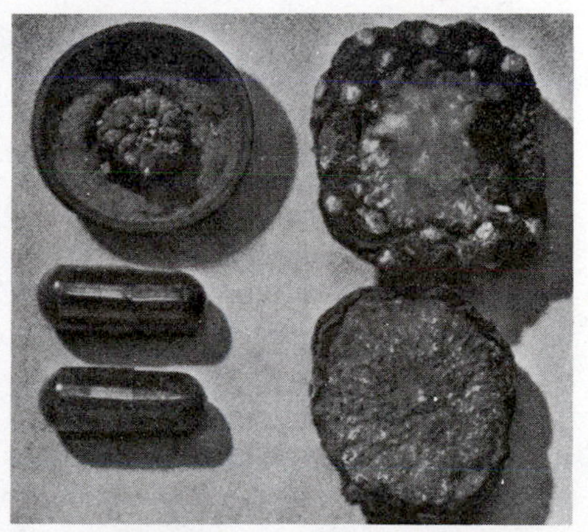

At left, opium poppy and derivatives: crude and smoking opium, codeine, heroin, and morphine. At right, peyote cactus, buttons, and ground buttons. (Photos courtesy of the Drug Enforcement Administration.)

Morphine Morphine Morphine

Codeine Heroin

Papaverine Narcotine

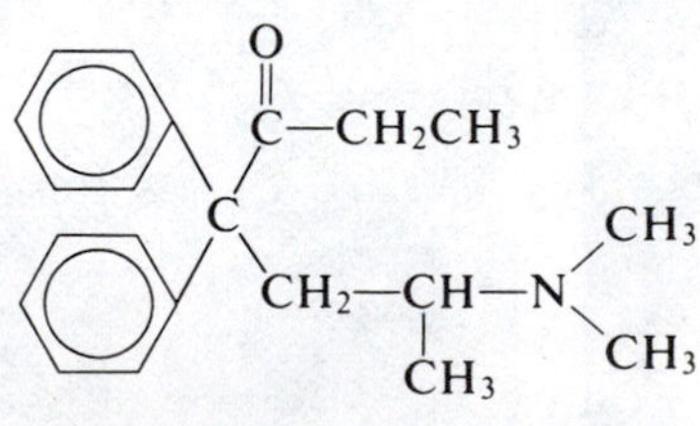

Methadone, a narcotic used to treat heroin addicts

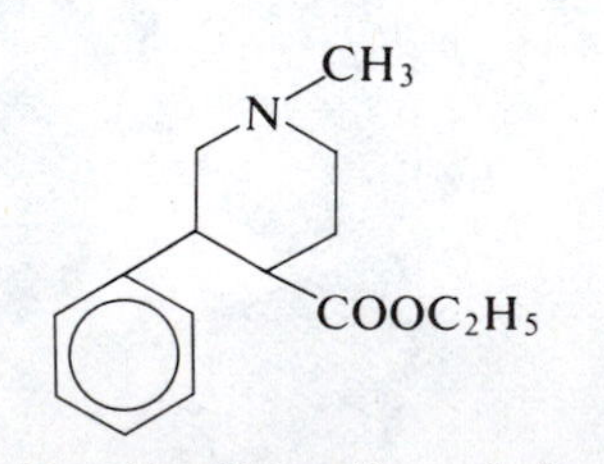

Meperidine (Demerol), a synthetic narcotic not as potent or as habit-forming as morphine

FIGURE 14.15
Examples of narcotics. Note the varied ways of showing the molecules.

of the skin because they share contaminated equipment. The contaminated equipment, overdoses, or other body abuse all contribute to a large number of deaths for addicts. In New York City alone the death toll is in the neighborhood of 1000 per year.

The end effect of abuse is narcotic addiction. The sad sequence of events is the same as with amphetamine addiction. The drug user is hooked, which means his body requires larger doses to get the same effect. The body thus develops a natural defense or tolerance to the drug. If heroin use is stopped withdrawal sickness occurs, varying in intensity. Its symptoms include shaking, sweating, chills, diarrhea, nausea, and abdominal and leg cramps. Psychological dependence may be as great as the physical. The user develops a craving for the drug for emotional reasons. He feels he has to have it and will go to any ends to "get a fix" in order to be able to escape facing up to whatever he finds unpleasant in reality, and also to avoid the pain of withdrawal symptoms. The comic John Belushi's death in 1982 was attributed to drug overdose by a mixed injection of cocaine and heroin ("speedballing").

Hallucinogens

The term hallucinogen is derived from the Latin and Greek words suggesting "a wandering of the mind." A hallucination is seeing or hearing things which have no basis in reality. The experience occurs only in a person's mind.

When one hears the word hallucinogen he commonly thinks of LSD, although the class includes other chemicals. Marijuana is usually included as a mild hallucinogen because it can cause hallucinations in very high doses. The word *psychedelic* has recently become descriptive of drugs like LSD. It is derived from Greek words meaning soul and visible, and is tied in with the concept of mind expansion which some drug users apply to their experiences. Another term often used by medical researchers is *psychotomimetic,* which means producing, mimicking, or imitating a psychosis (a severe form of mental disturbance or disease).

Hallucinogens from Mushrooms and Cactus. Mescaline is a hallucinogen made from the peyote cactus, specifically peyote buttons which are the heads of a small gray-brown cactus plant which grows in Mexico and the southwestern United States. The buttons have been used for their hallucinogenic properties by Indians in religious rituals. Psilocybin is the chemical which is the active principle in a small Mexican mushroom (Psilocybe Mexicana). Its chemical structure has similarities to that of LSD. DMT is a chemical which occurs in certain South American plants. Indians in some of the South American countries and some people in Haiti have used snuff made from these plants in their religious rituals. DMT is now produced synthetically. It is often used mixed with parsley or tobacco and smoked. It can also be injected. The high can begin within minutes of taking DMT and can last from one to three hours depending on the dose. Figure 14.16 gives formulas for a few of the hallucinogens.

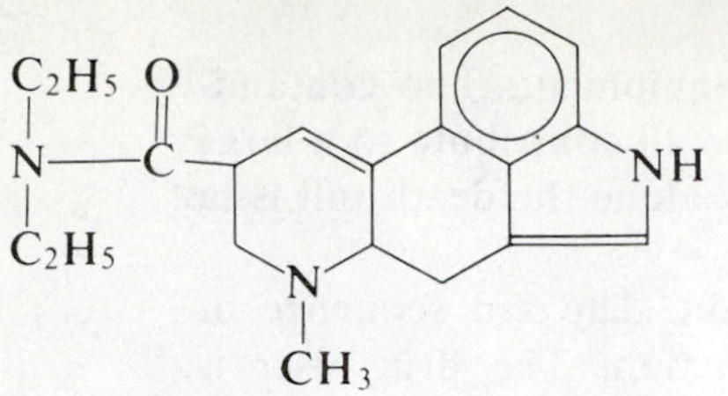

LSD (lysergic acid diethylamide)

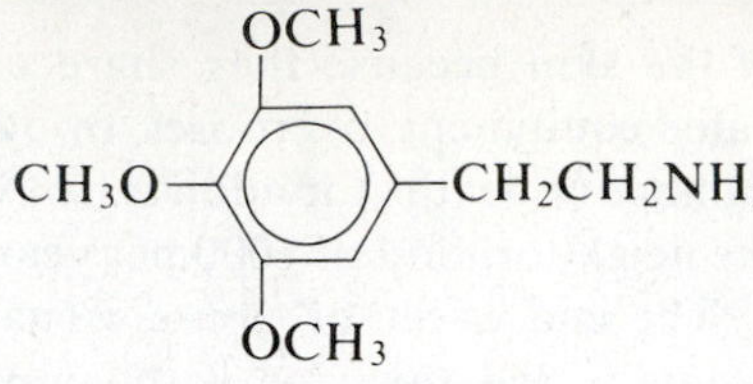

Mescaline

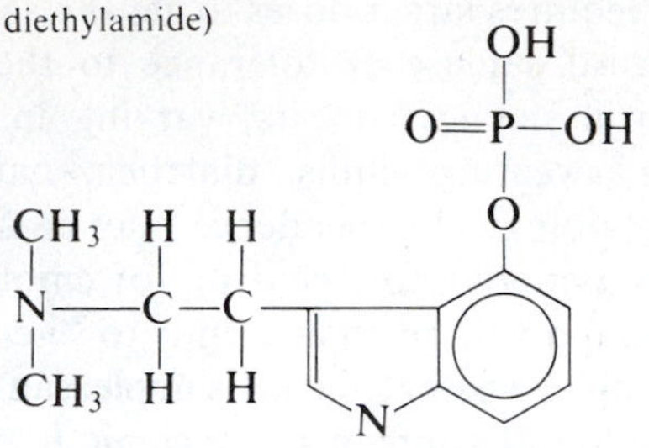

Psilocybin

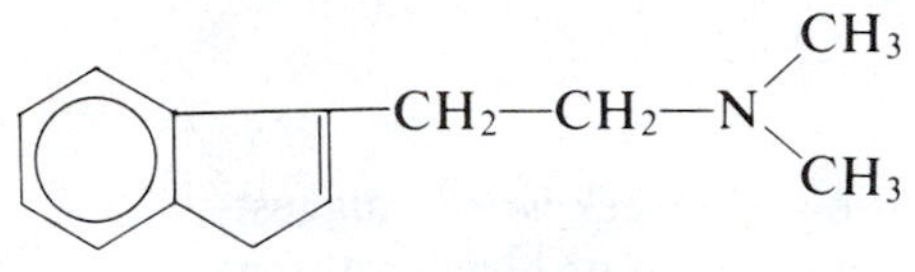

DMT (dimethyltryptamine)

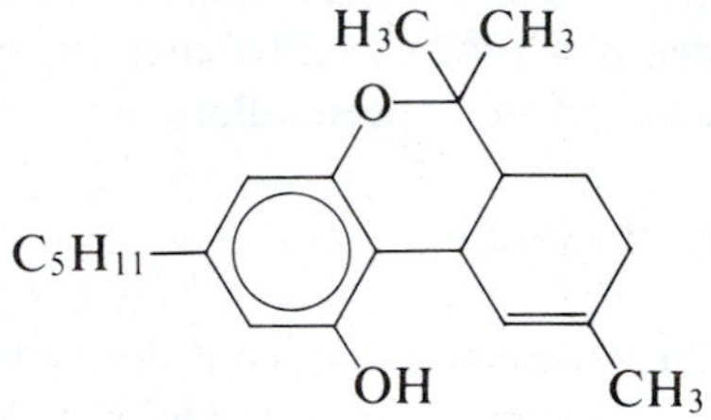

THC (tetrahydrocannabinol)

FIGURE 14.16
Formulas for a few of the hallucinogens.

A Trip with LSD. LSD is a man-made chemical called lysergic acid diethylamide. It was first made in 1938 by the Swiss chemist Albert Hoffman from a fungus, ergot, which grows on rye grain. Five years later, he accidentally ingested some LSD and experienced the first LSD "trip." Here is his description of the experience:

> In the afternoon of 16 April, 1943, when I was working on this problem, I was seized by a peculiar sensation of vertigo and restlessness. Objects, as well as the shape of my associates in the laboratory, appeared to undergo optical changes. . . . I noted with alarm that my environment was undergoing progressive change. Everything seemed strange and I had the greatest difficulty in expressing myself. My visual fields wavered and everything appeared deformed as in a defective mirror. I was overwhelmed by a fear that I was going crazy. The worst part was that I was clearly conscious of my condition. . . . I was unable to concentrate on my work. In a dreamlike state I left for home where an irresistible urge to lie down overcame me. I drew the curtains and immediately fell into a peculiar state similar to a drunkenness, characterized by an exaggerated imagination. With my eyes closed, fantastic pictures of extraordinary plasticity and intensive color seemed to surge towards me. . . .

Hallucinogens like LSD are now made chiefly in clandestine laboratories. The original manufacturer in Switzerland has taken LSD off the market and the only legitimate material is the small quantity being used in controlled research projects. There are no proven therapeutic uses for the drug. Research is under way in areas such as treatment of alcoholics and narcotics addicts, studies of genetic effects and birth defects, effects on terminal cancer patients, and studies to trace the path of the chemical in the body. Unfortunately our conceptual knowledge of how the drug operates is still not very well developed.

The effects of LSD vary greatly according to amount used, personality of the user, and the conditions under which the drug is used. It is usually taken in a pill or capsule form. However, the material is so powerful and is a colorless, odorless, and tasteless chemical, so that it can be placed on almost anything. It has been hidden on sugar cubes, chewing gum, cookies, blotting paper, and even the back of postage stamps.

An LSD trip lasts usually from eight to twelve hours. Its physical effects are enlarged pupils, flushing of the face, chilliness, and a rise in heartbeat and blood pressure. The psychological effects are more profound. Sensation is basically altered, vision being especially affected. The person may see unusual patterns. Depth perception is thrown off and the meaning of the object perceived may be transformed. One sense experience may be translated or merged into another. Thus sensations may "cross over": music and sounds may be seen, while colors may be heard, and odors felt. This is called synesthesia. Illusions and hallucinations can occur. Reverie is common and delusions are possible. The sense of time and personal identity are strangely altered. Users have reported a sensation of losing the normal feelings of boundaries between one's body and the space surrounding.

The emotional variations occurring on a trip may be extreme. A person may run the gamut from bliss to horror all within a single experience. Having two strong and opposite feelings at the same time is a common experience. Users report feeling happy and sad, depressed and elated, relaxed and tense at the same time. Violent crimes and self destruction have been tied to LSD usage.

The users consider a good trip one in which pleasant sensations predominate. A bad trip, or "bummer" in the drug jargon, is the opposite. Dread and horror predominate and the images are terrifying. The terror is increased because the sense of time is changed and a few minutes may seem like hours. There is apparently no way of predicting who will have a bad trip. Even users who have a history of good trips may find the next one horrifying.

Overall evaluation indicates that the chemical LSD can be extremely dangerous. The user may experience feelings of panic and occasional "flashbacks" of the trip days or months after the last dose. Depression and suicide as well as accidental death have resulted because the user developed paranoid feelings that he is super powerful. People have been known to walk in front of moving cars or attempt to fly from high windows. While there is some question as to whether LSD can directly cause mental illness in a stable individual, there is ample evidence that it can bring about acute and sometimes long-lasting mental illness in susceptible persons.

At left, marijuana plant. At right, manicured marijuana, cigarettes, and seeds. (Photos courtesy of the Drug Enforcement Administration.)

Marijuana. Marijuana is a dried plant material from the Indian hemp (Cannabis sativa). The plant grows wild in many parts of the world. The tough stringy stalks have been used since very early times for making rope and clothing. It was not until the early years of the twentieth century that the use of the hemp plant as marijuana was introduced in the U.S., and it is only within the last ten years that it has undergone a sharp increase in use. The numbers using marijuana are not known, but estimates indicate that a considerable number of people use marijuana as a way of life, like alcoholics who use a chemical to deal with their problems and as a crutch to face reality. Habitual users are called potheads. But there are many occasional users, too.

When used as a drug, the leaves and flowering tops of the plant are dried and crushed. They are then rolled into thin homemade cigarettes, called joints, which are crimped or twisted on the ends to keep the fine marijuana from falling out. Pipes with small bowls are also used.

Traffickers in marijuana frequently include all parts of the plant in the final product, including seeds and stalks. They may even include grass, tea, and oregano as adulterants. Hashish or "hash" is the potent dark brown resinous material which is collected from the flowering tops of the plant. It is at least five times stronger than crude marijuana.

Marijuana in its natural state varies widely in strength, depending on where it is grown, the climate, and whether it is wild or cultivated for smoking or eating. The proportion of flowering tops in the final mixture also has a major effect on the strength. The conceptual reason for this was not known until the 1960s. Then researchers isolated the principal active molecular substance in marijuana. It is called delta-9-tetrahydrocannabinol or THC for short. (See Figure 14.16.) The flowering tops have the highest THC content, the leaves have smaller amounts, while stalks and seeds have practically none. THC can produce a trip lasting four to six hours. It is probable that synthetic THC may soon become a more common drug of abuse since the molecule can be synthesized fairly readily.

Effects of Marijuana. There are many slang terms for marijuana, the most common being hemp, tea, grass, pot, and weed. When smoked, marijuana introduces chemicals into the bloodstream through the tiny alveoli in the lungs. The users hold the smoke in the lungs to increase absorption. The user's mood and thinking is affected within minutes. The immediate effect is a reddening of the whites of the eyes, increased heart beat, and some coughing due to irritating effect of the smoke. Dryness of the mouth, sleepiness, and increased hunger are also common. The long-term physical effects are not yet known.

Marijuana is not considered addictive, that is, it does not lead to physical dependence. Tolerance for the drug does not occur and withdrawal symptoms like those seen with heroin do not develop.

The effects of the drug on emotions and feelings vary considerably, depending on the person, the expectations, the circumstances, and, of course, the strength and quantity of the marijuana. It can act either as a stimulant or depressant even though the active principle (THC) is a hallucinogen. The senses of hearing, taste, and touch are generally enhanced. Time is usually distorted and five minutes may seem like an hour. Space also may be enlarged. The mind tends to wander and thought may become dreamlike. The notion that one is thinking more effectively is often reported later. (However, studies do not support this feeling.) Illusions or misinterpretations of sensations occur. True hallucinations involving experience of nonexistent sensations are apparently rare as are delusions or false beliefs. Frequently the individual becomes withdrawn. Occasionally uncontrolled crying or laughing occurs. The experienced smoker may have some control over the high feeling or the passive euphoria. Some users find the high pleasant; some find it frightening or unpleasant; others develop unfounded suspicions accompanied by fear, anxiety, or even panic. These negative effects are apparently more likely to occur with the young user whose personality is not yet well established.

Logical thinking is more difficult for a person under the influence of marijuana even though he may think his judgment is improved. Performing a complex operation requiring good reflexes and clear thinking may be impaired. Driving a car is particularly dangerous.

The way marijuana works is still largely unknown even though extensive research is going on in this area. Some users compare it to alcohol and tobacco and suggest it is equally "desirable." Alcohol is of course far from harmless. It is responsible for over 25,000 deaths on U.S. highways every year and immeasurable anguish and suffering related to these deaths. The social and economic costs of alcohol dependency must also be considered. And tobacco is no health food. It is a strong factor in lung cancer, heart disease, emphysema and related diseases. It contributes generally to a shortened life span.

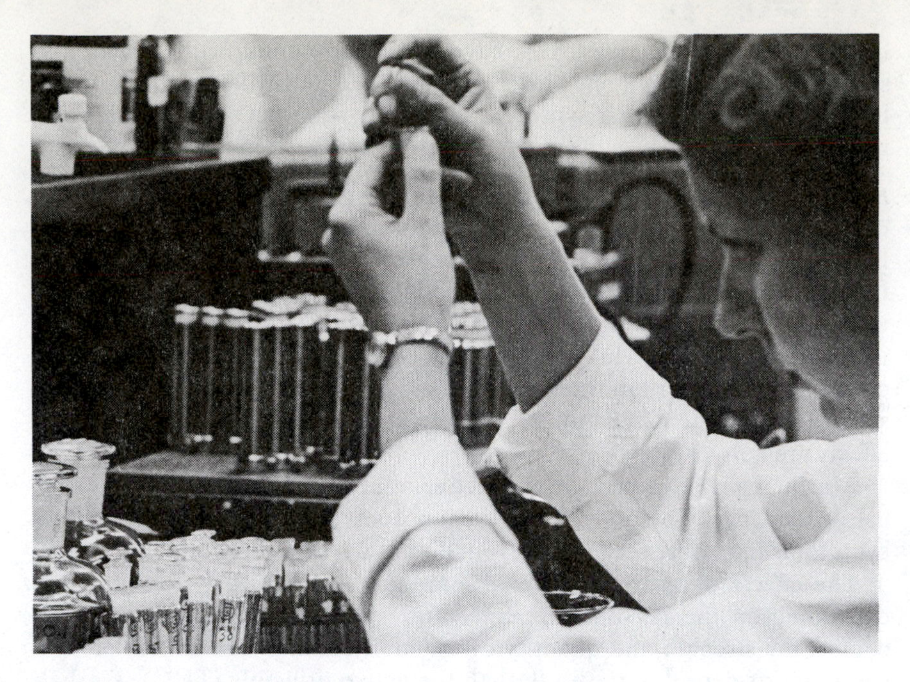

To learn more about how a drug is metabolized in the body, research scientists study the effects of an experimental drug on liver enzymes in animals. (Courtesy Abbott Laboratories.)

These are all powerful chemicals. What is needed is increased conceptual understanding of all drugs.

The relief of pain has been an aim of people from the earliest times. Today we have the advantage of being able to identify and alter the molecular structure of specific drug molecules. Continuing progress into conceptual understanding of drug action should contribute to control of disease and the betterment of people's lives, even though there is always, in the background, the paradox of drug abuse.

PROJECTS AND EXERCISES

Experiments

1. This experiment is a simple one using aspirin tablets. Get two cups. In one place about one-half cup of cold water. In the other pour about one-half cup of boiling water. When the swirling settles down carefully drop an aspirin into each cup. Let them stand and observe what happens. Does the aspirin dissolve? After about fifteen minutes swirl or stir the water in both cups and then let it settle. Do you get a solution now? How do you suppose the drug gets into a person's system when swallowed? Next add another one-half cup of water to each cup and stir again. What does the behavior of the mixture indicate now? Include any additional ideas you think are pertinent here. On the basis of your observations, can you explain why "gulping" aspirin with

just a swallow of water might give more stomach irritation than when a full glass of water is used?

2. This is a thought experiment. Assume that there are chemicals between nerve endings of adjacent cells continually being produced by normal healthy nerve cells. One type of chemical is involved in transmission of impulses from cell to cell. This allows you to "know" that you see or touch something. Now assume that another type of chemical between nerve cells blocks undesired signals from the subconscious part of the brain so that the brain can attend to whatever present project the person is involved with. Also assume we have a drug molecule which reacts with and ties up the natural blocking chemical between cells. What do you think might happen? If the bonding between the drug and the blocking chemical eventually eliminates the drug molecules and the initial dose of the drug is used up so that the body produces more blocking chemical, what might follow? What might be the result if the person, because of an extreme overdose of the drug or permanent damage to the cellular mechanism involved in making the blocking chemical, does not produce any more of the blocking material? Now assume another situation where the person's body defense mechanism happens to temporarily tie up the drug near the blocking site and later releases the drug. What might this produce? The guesses you make in this thought experiment are involved in one hypothesis among many being investigated as a possible explanation of an LSD "trip."

Exercises

3. Ethical drugs is a name given to those which can be sold only with a doctor's prescription (they are also called prescription drugs). Drugs which are sold without a prescription are often called proprietary drugs. The old name for these was patent medicines. Name three ethical drugs and three proprietary ones. Which type is most heavily advertised to the general public?
4. What are the differences between sulfa drugs and antibiotics?
5. Explain briefly, using examples, how the concept of lock and key can be helpful in developing more effective drugs.
6. The production of penicillin was said to be a final triumph of the concept of Pasteur. Explain why. Then give a brief description of the problems of developing the drug and indicate how Pasteur's concept was finally put into practical operation.
7. Both insecticides and antibiotics can be abused by applying them too broadly. Describe how this can occur. How could the dangers of overapplication be minimized?
8. Consider the widespread use of the pill and some of the medical reservations as to its use. Suggest some ways to approach the possible problems.
9. Show how concepts were necessarily involved in the development of chemicals used in the pill.
10. Look up some background information on cholesterol. You will find this chemical often mentioned in health books, newspapers, and magazines. Then summarize recommendations made in regard to diet. Do you notice a lack of exact knowledge here? Does your background in chemistry help you decipher the status of knowledge on the cholesterol problem?
11. Doctors have to be very observant of the effects of drugs on patients. One of the reasons is because of possible side effects. However, observation may also

determine that a particular drug has an entirely unexpected desirable effect. New applications of some drugs have been made because of the work of a very observant doctor. Give two examples, indicating what symptoms were originally being treated and the symptoms these drugs are now used for.

12. Abusers of certain drugs are often called "speedfreaks" or "methheads." Review the names of the drugs mentioned in this chapter and then suggest which chemical type is probably involved. Would you consider the terms appropriate for the truck driver who takes a "co-pilot"? Why or why not?
13. There are many classifications of drugs. One classification according to use or purpose distinguishes drugs (a) to fight disease, (b) to prevent disease, (c) to aid body functions, and (d) to ease pain. Give two examples from each class. Do you think a drug could fit in more than one of these classes? Would you consider this classification more operational-oriented or concept-oriented? Why?
14. The classification of drugs is not always clear-cut so that a drug is often placed in a certain group even though it may have some properties which would allow its inclusion in a different group. Give two examples of drugs which fit this description. Do you think they are classified correctly? Why or why not?
15. Could codeine become a problem in drug abuse? Why do most confirmed addicts use heroin rather than codeine?
16. What do you think is the answer to the problems of drug abuse?
17. There have been considerable difficulties in evaluating clinically drugs like LSD and marijuana. Consider the problems and list as many reasons as you can why it has been difficult to get meaningful evaluations of the effects of these drugs on people using them.

SUGGESTED READING

1. There are many general topics in this chapter that you could profitably investigate further by reference to encyclopedia articles. Suggested topics include the following: drug, drug addiction, alcohol, anesthesia, stimulant, depressant, narcotic, and so forth. In addition there are often articles on specific drugs like caffeine, insulin, opium, morphine, and so on.
2. Barron, F., Jarvik, M.E., Bunnell, S. Jr., "The Hallucinogenic Drugs," *Scientific American,* 210, No. 4, pp. 29–37 (April 1964).
3. Gates M., "Analgesic Drugs," *Scientific American,* 215, No. 5, pp. 131–136 (November 1966).
4. Consumer Reports Editors, *The Medicine Show, Some Plain Truths About Popular Remedies for Common Ailments,* Consumers Union, Mount Vernon, N. Y., 1971 (Revised Edition).
5. Cohen, Sidney, *The Beyond Within, The LSD Story,* Atheneum, New York, 1964. This is a very comprehensive coverage of the background of LSD by a medical doctor who is himself an expert in this field.
6. Carroll, C.R., *Alcohol, Use, Nonuse and Abuse,* Wm. C. Brown Company Publishers, Dubuque, Iowa, 1970.
7. Di Cyan, E., and Hessman, L., *Without Prescription,* Simon and Schuster, New York, 1972.

8. Cohen, S., *The Drug Dilemma,* McGraw-Hill Book Co., New York, 1969. This is a short book covering many aspects of the drug scene by one of the researchers in psychopharmacology. The effects, side effects, treatment and other aspects of all abused drugs are covered.
9. White, K. L., "Life and Death and Medicine," *Scientific American,* 229, No. 3, pp. 23–33 (September 1973). This whole issue is devoted to the role of medicine in the life of modern man.
10. Mellinkoff, S. M., "Chemical Intervention," *Scientific American,* 229, No. 3, pp. 103–112 (September 1973).
11. Axelrod, J., "Neurotransmitters," *Scientific American,* 230, No. 6, pp. 58–71 (June 1974). Concerns chemical messengers by means of which nerve cells communicate.

chapter 15

Concepts for Cleaner Water

Clearly the problem of man and nature is not one of providing a decorative background for human play, or even ameliorating the grim city. It is the necessity of sustaining nature as a source of life, . . . and most of all of rediscovering nature's corollary of the unknown in the self, the source of meaning.

—*IAN L. McHARG*

Cleanliness and order are not matters of instinct; they are matters of education, and like most great things you must cultivate a taste for them.

—*DISRAELI*

The River Rhine, it is well known,
Doth wash the city of Cologne;
But tell me, Nymphs, what power divine
Shall henceforth wash the River Rhine?

—*SAMUEL TAYLOR COLERIDGE*

Pollution comes from the Latin root verb meaning to soil or defile. It implies dirt or contamination. For our purpose, we can think of pollution as chemicals in the wrong place or in the wrong amounts.

Water consists of polar molecules (chapter 7) and can therefore easily dissolve various salts. Thus in its natural state, trickling over rocks and over inorganic residues, it can pick up a large amount of minerals. In addition, it can dissolve considerable quantities of organic chemicals, like sugar and alcohol, whose molecules contain compatible polar groups. Water has indeed been called the "universal solvent" even though it does not dissolve large quantities of oils like those from petroleum. Yet even here it can be heavily contaminated by floating surface oil when petroleum leaks occur.

THE WATER CYCLE

Water covers more than 70 percent of the earth's surface. This makes it a very common substance, which may partly account for the fact that it is often abused. Most of the earth's surface water (97 percent), however, is in the oceans. Only about 3 percent is considered "fresh," or unsalty. Most of this is unavailable because it is frozen as ice in the earth's polar regions. Less than one percent of the surface water then is freely available for "use." This small percentage still represents a considerable amount of water. And it is available to people by tapping into nature's water cycle (see Figure 15.1).

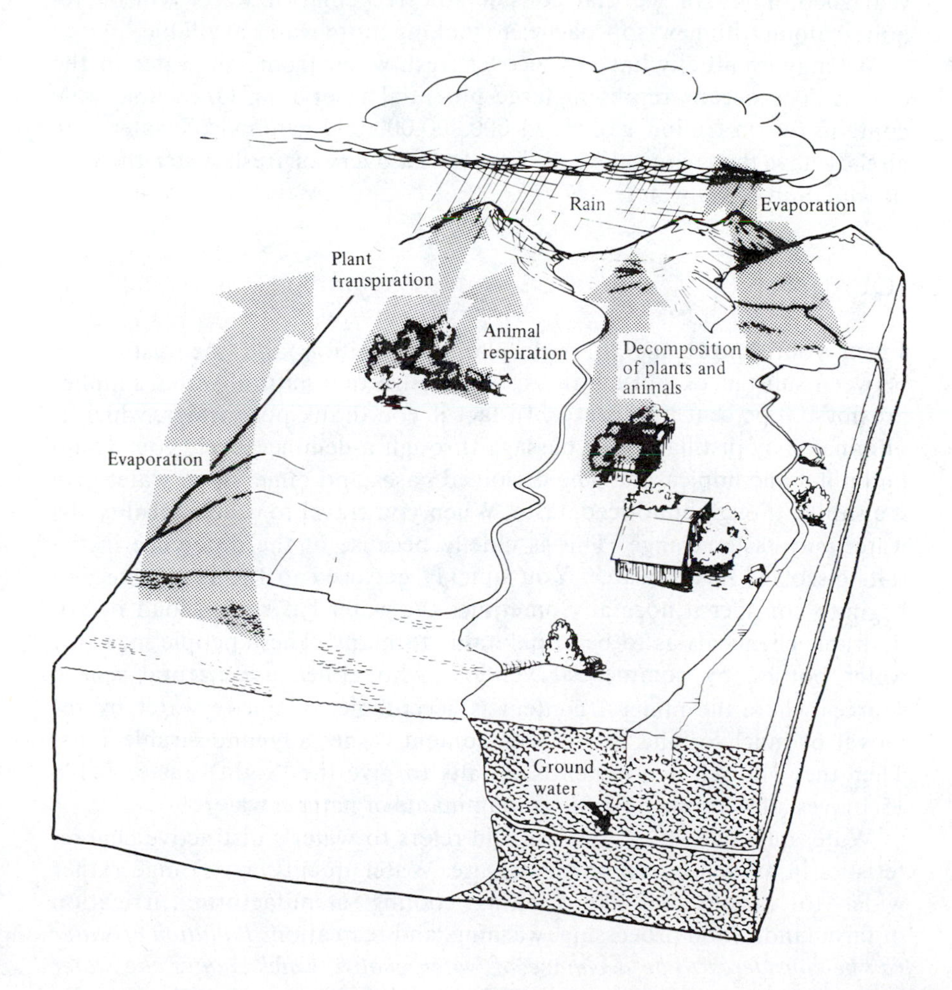

FIGURE 15.1
The water cycle.

By studies of quantities of rain and snow which fall during long periods of time, estimates have been made of the total amount of precipitation we get per day. For the United States the figure is in the neighborhood of $4\frac{1}{2}$ trillion gallons (4,500,000,000,000 gallons) per day. This can be quite misleading since it represents an overall average.

Also, the total water precipitation is not all "available" water. A large quantity, approximately 70 percent, is returned to the air by evaporation, or is used by plants. Another large portion, estimated at about 23 percent, is involved in runoff from the land. This is eventually returned to the oceans. Only about 7 percent of the total U.S. precipitation is now used by people as fresh water. This is the water available for industrial, agricultural, and municipal uses. The total of these applications now runs well over 300 billion gallons per day. Studies of increasing growth and more broadened use of water indicate that demands will probably triple by the year 2000. You can see that considerable recycling of water will be required along with new approaches to making more water available.

We may eventually have to recover fresh water from salt water in the oceans. The oceans represent large potential reservoirs. One cubic mile contains over a trillion gallons (1,000,000,000,000 gallons) of water. We already have the technical know-how for recovery of fresh water through desalination of sea water.

WATER QUALITY

From your knowledge of molecular composition and the distinction between substances and mixtures, you realize that natural water supplies are not 100 percent pure water. In fact if you drank pure water, which is obtainable by distillation or passage through a demineralizer, you would find it flat and unpleasant. The dissolved gases and minerals in water give it a subtle, though unnoticed, taste. When you travel to a new locality the water may taste strange. This is chiefly because of the difference in the salts dissolved in the water. You quickly get used to the new taste and begin to consider it normal. Sometimes the water has such a load of undesirable chemicals as to be unpalatable to many. These people may buy water bottled by commercial vendors who either use natural spring sources where the mineral content is acceptable, or purify water by removal of much of the excess salt content, especially undesirable ions. Then they add selected minerals as salts to give the "right" taste. Table 15.1 gives a few of the possible contaminants of natural water.

Water quality is, then, relative and refers to water's distinctive characteristics in reference to a particular use. Water quality may range rather widely for various uses like drinking, cooling, manufacturing, irrigation of farm lands, food processing, washing, and recreation. *Pollution of water can be considered to be a change of water quality which makes the water unsuited for a particular end use.* The impairment to water quality usually involves the addition of some material, but it can also be caused by a

TABLE 15.1

Examples of Foreign Materials in Natural Waters

Suspended dusts	Dissolved salts
Inorganic	Positive ions
Organic	H^+, Na^+, K^+, Ca^{+2}, Mg^{+2}, Fe^{+2}, Fe^{+3}, NH_4^+
Dissolved gases	Negative ions
CO_2, N_2, O_2, H_2S, CH_4	Cl^-, F^-, SO_4^{-2}, CO_3^{-2}, NO_3^-, HCO_3^-, PO_4^{-3}, $H_2PO_4^-$, HPO_4^{-2}
Living organisms	(Sources of ions are compounds such as $Ca(HCO_3)_2$, $MgSO_4$,
Algae, bacteria, viruses	$CaCl_2$, $MgCl_2$, $Mg(HCO_3)_2$, Na_2SO_4, NaCl, etc.)

change in temperature such as that caused by a steam plant using water for cooling purposes (thermal pollution).

Water can be polluted naturally by storms producing excess sediment, oceans backing up into rivers, oil seepage, or decay of dead animal life. The major pollution problems come from the activities of people (see Figure 15.2).

Types of Water Pollution

The U.S. Public Health Service places water pollutants in eight classes:

Water Pollutant	*Examples*
1. Wastes which demand oxygen	Dead plants and animals, animal waste products (urine and fecal matter), industrial wastes from food-processing plants, slaughterhouses, meat-packing plants.
2. Agents which cause disease	Bacteria, viruses.
3. Plant nutrients	Water running off irrigated lands containing nitrates, phosphates and potassium compounds.
4. Synthetic organic chemicals	Chlorinated pesticides like DDT, detergents, dyes.
5. Various inorganic chemicals	Acid wastes like HCl, H_2SO_4 and a great variety of industrial salts, and heavy metals like mercury and cadmium.
6. Sediments	Clay, silt, sand, gravel.
7. Heat	Warm water from cooling uses.
8. Radioactive minerals	Effluent from mining and processing of radioactive minerals, or from fallout in tests of nuclear bombs.

Our approach to pollution problems is to consider the molecules of the polluting chemicals. What kind of molecules are they? How do they get mixed up with the molecules of water? And how can we best separate them again?

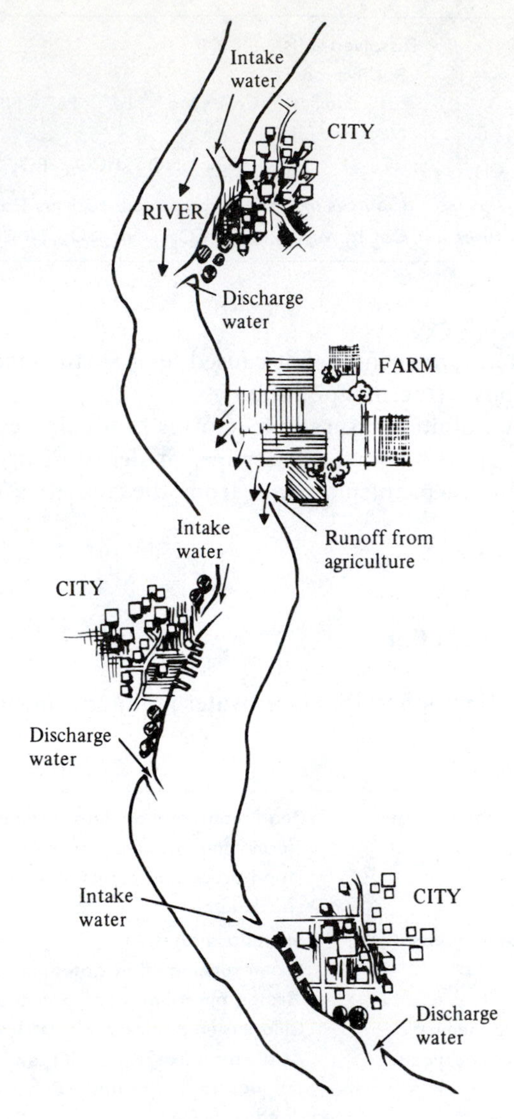

FIGURE 15.2
The problem of water reuse. Discharged water of poor quality upstream can be a hazard to users below.

Oxygen-Demanding Wastes

Some organic waste products are decomposed by bacteria which use dissolved oxygen in the process. Examples are municipal and farm sewage.

The chemicals in these organic effluents are fundamentally compounds of carbon. The molecules also contain hydrogen along with nitrogen, oxygen, sulfur, phosphorus, and a few other elements. What they all have in common is a requirement or demand for oxygen in the process of breakdown to simple end products. You have seen this process in the burning of a carbon-containing compound like gasoline.

Microorganisms such as bacteria use the organic waste material as part of their own requirements. Bacteria which decompose the molecules of organic wastes using the oxygen of dissolved air are *aerobic.* These bacteria obtain energy by "burning" the organic molecules. This can produce the simple end products carbon dioxide, CO_2, and water, H_2O. The mechanism is similar to your own digestion and use of food ingredients. The larger the quantities of organic wastes, the more oxygen will be demanded by the bacteria to accomplish the decomposition.

This is the origin of the term biochemical oxygen demand or BOD. The BOD is simply the amount of oxygen required during the breakdown of the wastes by bacteria. The BOD is thus a measure of the amount of such wastes in a particular water sample.

The oxygen depletion by aerobic bacterial action can destroy fish life. In addition, if the wastes result in essential depletion of the dissolved oxygen then the aerobic bacteria die. These contribute more waste, to raise the BOD even more. Other anaerobic (without air) bacteria take over. These do not require dissolved oxygen. They use oxygen in oxygen-containing compounds instead. But the end products they produce are not simply H_2O and CO_2 as in the case of the aerobic bacteria. Instead the end products of anaerobic decay include noxious products like the following:

H_2S Hydrogen sulfide, the poison gas that smells like rotten eggs. (The same H_2S gas comes from the sulfur in the protein of rotten eggs.)

NH_3 Ammonia or ammonia derivatives, which smell like rotten fish and meat. A derivative of ammonia is one where a hydrogen atom is replaced by some type of carbon chain. Two molecules with very disagreeable odors found in decaying flesh are

$$H_2NCH_2CH_2CH_2CH_2NH_2 \qquad H_2NCH_2CH_2CH_2CH_2CH_2NH_2$$

Putrescine Cadaverine.

CH_4 Methane or marsh gas which is flammable and explosive.

Water Can Carry Disease

The bacteria that cause breakdown and degradation of organic material in water are beneficial, as are many microbes. However, water supplies may also carry disease-causing microorganisms which often enter water by contamination with raw sewage. Diseases such as typhoid fever, dysentery, cholera, hepatitis, and polio can come from contaminated water. Because of knowledge of the nature of transmission of these diseases, they have been practically eliminated in most developed countries through careful checks on water quality, installation of water treatment facilities, and vaccinations.

Tests for microbe contamination of water use a harmless type of bacteria as an "indicator" organism. These are bacteria which live as natural inhabitants of the large intestine (or colon) of humans and other vertebrates. They are the *Escherichia coli,* sometimes called *E. coli* or coliforms. They are benign in that they cause no illness. However, they are always present in feces. In addition, they do not multiply once discharged into natural waters. Consequently, their presence in water supplies is a good indication that sewage wastes have recently entered the stream. If this is the case, then it is possible that pathogenic microbes may also be present along with the harmless coliforms.

Pollution From Fertilizers, Salts, Eutrophication

Another source of water pollution which is not usually visually detectable is plant nutrients, chiefly the primary plant nutrients such as nitrates, phosphates, and potassium compounds. (Plant nutrients are covered in more detail in chapter 11.) Although essential for plant growth, they can be major contaminants if they get into water supplies. Contamination occurs when both rain and irrigation water drains from fertilized agricultural land. Both fertilizers and pesticides can be problems here.

High concentration of salts of any kind are detrimental to domestic and industrial usage of water. Hard water is caused chiefly by calcium salts (chapter 13). In the case of salts that are plant nutrients, we can get a very odd type of pollution which has been given the name *eutrophication.* The word is from the Greek meaning well nourished.

Eutrophication is the process by which a body of water such as a lake becomes enriched with nutrients. This results in excessive growth of algae, which are water plants (see Table 15.2). Like land plants, algae depend on nutrients, carbon dioxide, water, and sunshine to grow. And if the quantities of nutrients are supplied in lavish amounts, they grow profusely. This is often called algae bloom. The condition can result in large masses of slimy algae near the surface. Further deterioration can then occur. Algae and aquatic weeds eventually die and begin to decompose through normal bacterial action—which consumes oxygen. The oxygen content of the lake is then reduced so that fish die. The increase of dead organic matter blocks penetration of sunlight needed for plant growth, and more plants die.

Eventually the lack of oxygen can lead to increased growth of anaerobic bacteria which produce an obnoxious state of putrid odors and tastes.

TABLE 15.2
Approximate Composition of One Type of Algae

*Element**	*% by Weight*
Carbon	46.5
Nitrogen†	8.1
Phosphorus†	0.7
Potassium†	0.8
Calcium	0.5
Sulfur	0.3
Iron	0.3
Magnesium	0.2
Sodium	0.04
Manganese	0.03
Zinc	0.005
Copper	0.004
Boron	0.0004

*These elements are present in compounds. The major remaining composition is made up of hydrogen and oxygen which are available from the water in which the algae grow.
†Primary plant nutrients.

If the process goes this far you have no longer a lake but a swamp. The whole process is often called natural aging of lakes.

Lakes have turned into swamps long before people came along. And a swamp is often eventually converted into a meadow in nature's recovery processes, sometimes spanning hundreds of thousands of years. However, people have been contributing to deterioration of lakes by hastening the eutrophication process. Now that the mechanism of eutrophication is recognized, we know that the solution consists in preventing the excessive addition of nutrient chemicals to lakes.

BIODEGRADABILITY

A material is said to be biodegradable when it can be readily broken down into simpler end products by bacteria. The problem of biodegradability was emphasized in the case of detergents (chapter 13) chiefly because of the obvious foaming problems. Many other organic chemicals eventually get into the environment. These are either burned directly to simpler end products or are subject to eventual breakdown by microorganisms. Sometimes this natural degradation takes place very slowly, or sometimes the quantity of organic molecules is too large for bacteria to handle. Then disagreeable odors and tastes may be detected in water supplies contaminated by organic chemicals; or unpleasant tastes and odors may appear in fish, clams, or oysters taken from contaminated waters. The solution is control of the discharge.

DDT: A Good Example of a Bad Problem

In many other cases where no obvious effects of chemical use are noticed, there is need for much more extensive research. We have too little knowledge of the pathways, biodegradability, survival time in the environment, and potential hazard of many of the organic chemicals which are such an important part of our modern society.

The problem of DDT buildup in the environment is a good example. The persistence of DDT was at first thought to be a distinct advantage when the insecticide was introduced in the 1940s (chapter 11). The DDT stayed around the areas treated for long periods of time, increasing its effectiveness and requiring fewer applications. It was not until the 1960s that the problems of DDT buildup in the environment were recognized as serious.

The basic difficulty is the durability of the DDT molecule. There are two factors here. First of all, it is not readily degraded by bacteria. Secondly, it is not broken down readily in animal tissue. In fact, the biological half-life of DDT is estimated at about seven or eight years, the period of time required for an animal to break down half of the DDT it takes into its tissues. The DDT molecule has no —OH grouping so that it can be eliminated in the normal animal excretion processes. Therefore it builds up in fatty tissues, causing disruption of the reproductive cycle in fish and fish kills. In addition, certain levels of DDT in commercial fish have been labeled hazardous for human consumption.

DDT Can Move up the Food Chains

Oceans and lakes do not usually have a high concentration of DDT. The problem lies in the ability of plants and animals to concentrate a chemical like DDT in fatty tissue and to pass it along in complex food chains, shown in Figure 15.3. Food chains are the cyclic processes whereby simple organisms serve as food for more complex ones, and these in turn for others, up to the higher forms of life.

Food chains start with plants which convert CO_2 and H_2O to plant structures in photosynthesis. Plants also absorb other nutrients or chemicals in their environment. Animals eat the plants. Then other animals eat these, thus continuing in a chain up to higher forms. In all these complex interchanges the DDT molecules tend to concentrate in fatty tissues of animals. As the food molecules move up the chain, DDT molecules move with them and build up to higher and higher concentrations.

Research has also shown that DDT can be concentrated in tissues and eggs of many kinds of birds, causing thinner egg shells, which break before incubation is completed. The insecticide is believed to affect an enzyme which controls calcium metabolism. If the enzyme molecules are deactivated by the DDT, they are not available to provide the proper calcium carbonate layering in the egg shell. DDT buildup in food chains is thus implicated in drastic drops in populations of certain carnivorous birds. The danger from DDT is receding since its ban. (See Chapter 11.)

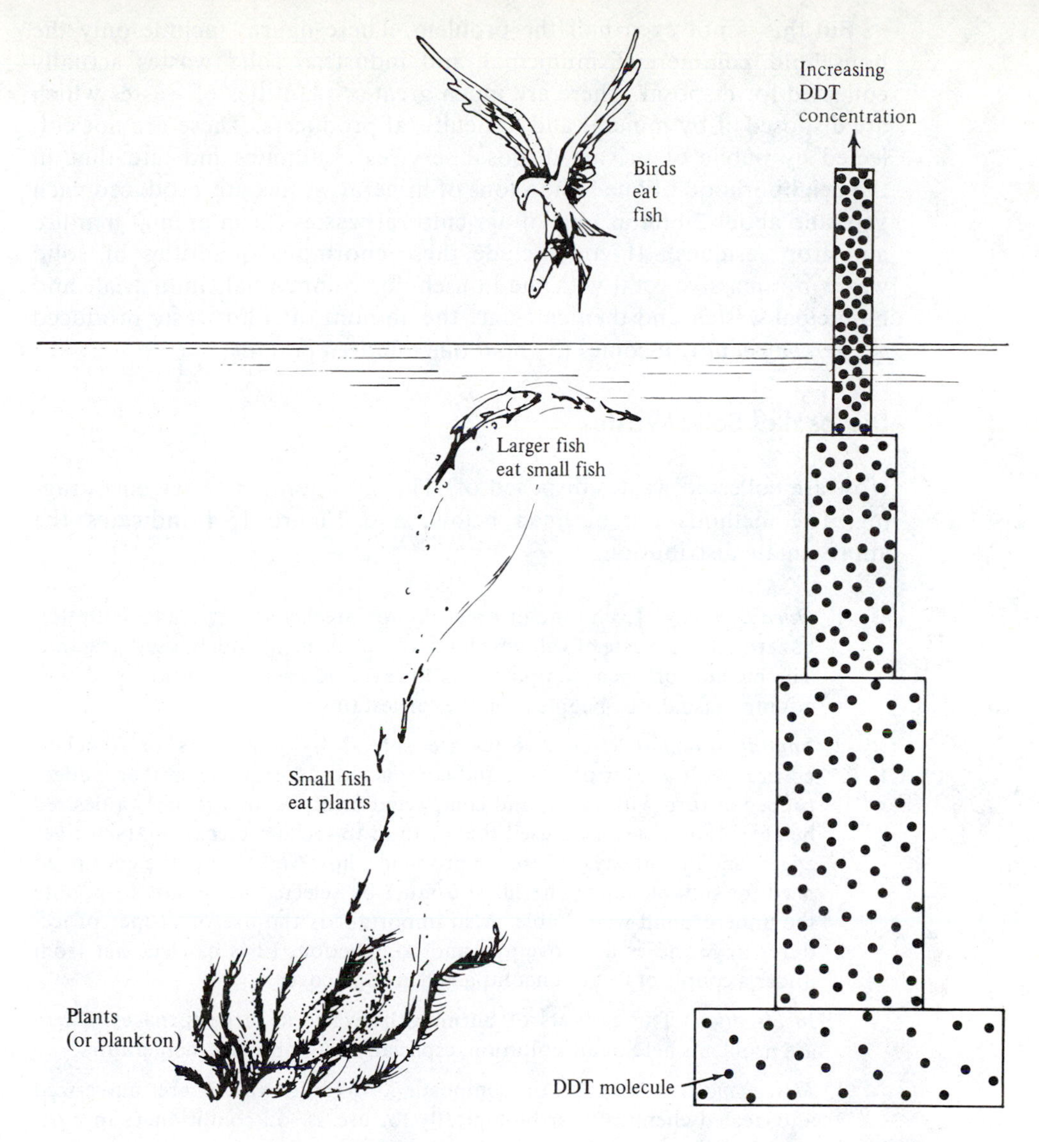

FIGURE 15.3

The concept of a food chain. Here DDT is biologically concentrated as it moves up to higher life forms.

SOLID WASTES

The exact quantities of solid waste being produced in the U.S. are impossible to determine. However, reliable estimates have been made in national surveys and the figures are astounding. The average solid waste collected now runs around six pounds *for each person* in the country *each day*. Estimates indicate it will be around eight pounds per person per day by 1980.

But this is not even half the problem. These figures include only the household, commercial, municipal, and industrial solid wastes actually collected for disposal. There are much greater quantities of wastes which are disposed of by mineral and agricultural producers. These are not collected by public or private disposal services. Estimates indicate that in the neighborhood of one billion tons of mineral wastes are produced each year and about 2 billion tons of agricultural wastes (farm animal manure and crop residues). If you include these enormous quantities of solid waste in a massive total with the household, commercial, industrial, and municipal wastes, and then calculate the amount of solid waste produced *per person per day,* it comes to about one hundred pounds!

Disposal of Solid Wastes

How are collected wastes disposed of? The situation is not yet encouraging. The methods are outlined below, and Figure 15.4 indicates the approximate distribution.

1. *Open Dumping.* Even though open dumps are health, fire, and pollution hazards, and a waste of valuable land, this is the most widely used practice. The method of open dumping was always the easiest solution and was adopted casually by people from the earliest times.
2. *Sanitary Landfills.* Solid wastes are spread by bulldozers in trenches, covered with a layer of earth, and compacted. More trash is then added, topped in turn with earth, and compacted until the area reaches a desired height. Many cities have used this method to reclaim creeks, marshy river edges, and inland bays. There are problems, however. One is the continued need for suitable land. The fill site must be selected so as not to pollute the underground water table. Also important is the use of proper procedures to get necessary coverage and compaction. Flies can get out from under a couple of feet of uncompacted ground cover.
3. *Incineration.* This consists of burning the waste in large furnaces. There are problems here of air pollution, especially with obsolete incinerators.
4. *Miscellaneous Disposal.* In composting, organic wastes are pulverized and treated chemically or biologically for use as soil conditioners in agriculture. Other small outlets are hog feeding, ocean dumping, and recovery for reuse.

Reuse of Solid Wastes

Theoretically all materials are recyclable. The basic concept behind all reuse or recycling is the Law of Conservation of Mass. And behind this there is an even simpler concept. Atoms never wear out.

There are many practical problems with recycling, but there is a great potential as well. Some of the ideas being tried are considered below.

1. *Sorting trash items before refuse pickup.* Many communities now have recycling centers but their operations are usually limited. Most success

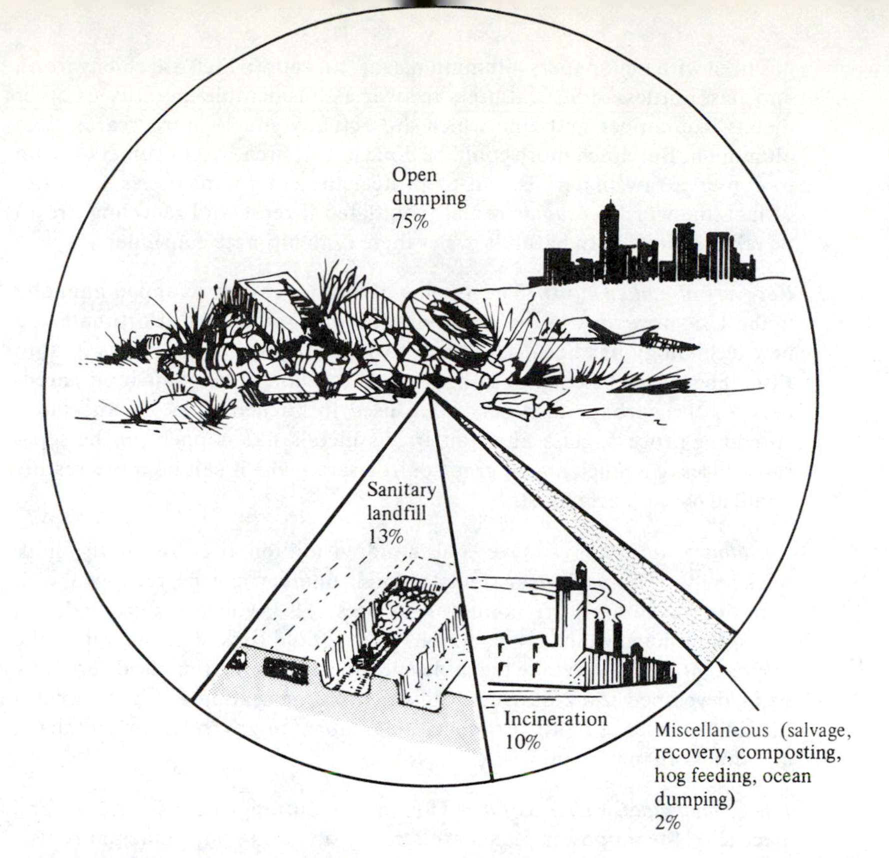

FIGURE 15.4
Solid waste disposal.

Open dumps present many pollution problems including breeding of flies and rats, pollution of underground water table, and pollution of the air when burning is used. (Courtesy of EPA, E. St. Louis, Illinois Metro-East Journal.)

has been with newspapers, aluminum cans, tin cans (which are chiefly iron), and glass bottles. Industrial users recover a considerable quantity of other metals like copper and zinc which are actually much more scarce than aluminum. But much more could be done in this area by consumers—both to conserve raw materials and to protect the environment. As a matter of fact, the word *consumer* is really outdated if reuse and recycling are to be taken seriously. In nature's cycles there is no ultimate consumer.

2. *Recovery of junked autos.* The number of motor vehicles scrapped annually in the U.S. now runs in the neighborhood of eight million. Fortunately, a new technology has been developed to reclaim junked cars. This is auto shredding. It uses tremendous high speed machines like giant food shredders, or the garbage disposers often used in kitchen sinks. In the auto-shredding process, glass and nonferrous metals like copper can be separated to leave a much higher grade of iron scrap which can be more readily handled by the steel industry.

3. *Reclaiming junked tires.* Over one hundred million tires reach the junk piles in the U.S. each year. Only a small number can be recapped, and even these eventually are nonusable as tires. Also, you can only make so many door mats, which is a common outlet for old tires. The answer to the disposal problem is to take the molecules apart. Practical methods are now being developed using destructive distillation, or heating at high temperature in the absence of air. Products obtainable are gas, oils, carbon black, and other organic chemicals.

4. *Use of municipal wastes as fuel.* This means burning wastes for heat and electricity. Steam power plants are already operating on municipal refuse, especially in Europe. In 1970 Chicago built a $20 million incinerator which also develops power and became the first American city to incinerate all of its refuse.

5. *Complete refuse recovery.* A considerable investment and many engineering innovations are necessary for complete refuse recovery. The idea is to bring large quantities of collected refuse to a central processing plant which sorts, grades, and treats the material, turning out salable products such as metals, glass, electrical power, agricultural mulch, cinders, concrete additives, and building materials.

   Operationally the problem looks impossible when you think of the complex mixture which trash is. Conceptually it is a little simpler. For example, trash may be separated into organic and inorganic molecules since the organics may be burned, giving chiefly CO_2 and H_2O. Recovery of metals from the inorganic refuse represents problems not theoretically different from those already encountered with low-grade ores.

Open dumps are by far the cheapest method of handling solid wastes ($0.60 per ton). Next in costs come sanitary landfills ($2.80 per ton) and incineration ($9.90 per ton). The complete recovery plants, at least initially, would cost even more. Eventually, considering the usable products obtained, they may represent a worthwhile investment. One of the products would be a clean environment and this may very well prove to be worth whatever it costs.

The Plastics Scrap Problem

Plastics products have been unfairly criticized as a major and almost insurmountable problem in solid wastes because of their durability and the fact that they are not biodegradable. This is undoubtedly a problem but not sufficient to condemn plastics out of hand considering their tremendous contribution to convenience and health in our modern age. All plastic products do eventually end up as solid waste. Is it all that bad? It can be if people just dump the plastics on the land or in the rivers. But knowing that plastics are long-chain carbon compounds suggests a solution. They are burnable. Plastics like polyethylene can be incinerated readily and actually contribute fuel and bulking value which is desirable in modern incinerating plants. While plastics like PVC (polyvinyl chloride) also contribute hydrochloric acid gas (HCl) as an additional end product (along with CO_2 and H_2O), modern incinerator technology can minimize corrosion and pollution problems. Polystyrene, which is so common in egg cartons, meat trays, and disposable coffee cups, can also produce black smoke and soot when burned. This is simply carbon from the styrene chains, not fully oxidized to CO_2. While the soot is unpleasant, it is not any more difficult to handle than the end products from many other organic materials. Modern incinerator practice traps the soot in the furnace so that it can be fully oxidized to CO_2. The black smoke is not necessarily a signal saying, "Don't use polystyrene." Rather it says, "Don't use open dumps."

The plastics problem illustrates that judgments made on strictly operational lines may be very superficial, and that conceptual understanding of molecules can point the way around apparently impossible situations. This does not mean the practical answers will be easy or cheap. But answers are to be found if human desire and effort can produce the necessary investment.

INORGANIC CHEMICALS

Inorganic chemicals is a broad category which includes mineral acids like HCl, salts of various kinds, and finely divided metals. These inorganic materials can enter streams and lakes from city and industrial waste waters, mine drainage, agricultural sources, and natural processes.

Acids and Salts

A major source of acid pollutants is water drainage from operating and abandoned mines. The U.S. Environmental Protection Agency estimates that about four million tons of sulfuric acid drains from mines every year. This is formed by chemical reaction between water, air, and sulfide compounds like FeS_2. When the coal seams are disturbed, in both surface and underground mines, water and air can enter to react with the sulfide compounds from the earth. Certain types of microbes are believed to be involved in the complex reaction to form sulfuric acid.

The control problem has not been solved. Among the methods being used are the sealing of abandoned mines to keep out water and air and the chemical treatment of the water flowing from mines. But a great deal of research is still needed in both the isolation of the many possible water pathways and the removal of chemicals for reuse in other industrial processes.

Acids are not the only inorganic water pollutants. You may not think of table salt (NaCl) as being a "bad chemical," but it can be very detrimental in fresh water supplies. Natural waters contain many kinds of salts besides NaCl, including compounds of calcium, magnesium, and iron. None of these are toxic in the usual meaning of that term, but they can have damaging effects on fish and plant life. This is the problem involved in buildup of salts in irrigation water.

Salt content of water is naturally increased during the irrigation of agricultural crops as plants selectively remove water molecules and some minerals, leaving higher concentration of salt. Also a large amount of water is evaporated to the air in irrigation, causing salts to build up in the soil and run-off water. Other salts can enter the water through fertilizers and by dissolving natural minerals in the ground.

Mercury: A Dangerous Element

Some salts of heavy metals such as arsenic, mercury, lead, and cadmium have direct toxic properties. Mercury, originally called quicksilver or liquid silver, was known by the ancients. It occurs naturally in the environment in small quantities in rocks, soil, water, and living organisms. In ores mercury compounds occur in more concentrated form with small quantities of free mercury.

The fact that mercury is the only metal which is a liquid at room temperatures made it attractive to the early alchemists. It has also found many uses in our modern age including the following: thermometers and barometers; liquid electrodes for manufacture of chlorine and sodium hydroxide from sodium chloride; electrical apparatus like mercury vapor street lamps, fluorescent lights, silent electrical switches, and batteries; high vacuum pumps; scientific instruments; fungicide in paint, paper, and agricultural applications; and as an alloy ingredient in mixture with other metals (called amalgams).

The application of mercury is thus fairly widespread. Total consumption in the U.S. runs close to 10 million pounds per year. World production is double this figure. Because most mercury compounds are extremely insoluble in water it was not considered to be an environmental hazard in water pollution until very recently. Now the picture is drastically changed.

The Mercury Scare

The first major episode involving mercury contamination of water occurred in Minimata Bay in southwest Kyushu, Japan, between the years 1953 to 1960. Here a strange disease surfaced with symptoms including numbness and tingling of hands, feet, or lip, unsteady movements, tremor, fits and blindness. By the time the cause of the disease was determined over one hundred cases of "Minamata disease" had been reported. Over forty deaths occurred. And the cause? Its discovery required a long and difficult scientific detective job. Eventually a mercury compound was found to be the agent. All of the victims had eaten fish and shellfish caught in Minimata Bay. The fish in turn contained small quantities of methyl mercuric chloride, CH_3HgCl. The mercury was traced to a factory which discharged waste sludge containing mercuric compounds.

Other incidents of mercury poisoning were reported in Iraq (1956 and 1961), Pakistan (1961), Guatemala (1965), and Alamogordo, New Mexico (1970). The cases in Guatemala involved farmers and their families who ate bread made from seed which was treated with mercury compounds and which was supposed to be used for planting only. The New Mexico case involved a farmer who fed mercury-treated grain to his pigs, and later the family ate the butchered pigs.

The mercury scare became more widespread in the 1970s beginning with research by Norvald Fimreite, a Norwegian student at the University of Western Ontario, Canada. He found that fish from Lake St. Clair, which lies between the state of Michigan and the province of Ontario, contained dangerous levels of mercury. This started other investigations. Later mercury was found in tuna and swordfish. The FDA removed over a million cans of tuna from the market and the swordfish practically disappeared from the U.S. market. Tuna canners are now able to control the mercury content by not catching the larger tuna and by avoiding areas where fish are likely to have a higher mercury content than the maximum allowable by the FDA (0.5 ppm). Swordfish, being at the top of the food chain, tend to have mercury contents above the maximum and are therefore also subject to special selection.

What We Know about Mercury

Most mercury compounds are toxic when present in high enough concentration. But the immediate danger is greatest with compounds like methyl mercuric chloride, CH_3HgCl, and dimethyl mercury, $(CH_3)_2Hg$. Mercury compounds can damage all tissues. The methyl mercury type is especially damaging to tissues of the central nervous system in the brain. "Hatter's Disease," which was a peculiar occupational hazard of people who worked on felt hats, is now recognized as due to mercury poisoning. Hatters used mercury in the treatment of felt for hats and the long, continuous exposure caused brain damage, the symptoms of which were tremors, slowing of speech, and personality changes. The term "mad as a hatter" has become a classical simile from Lewis Carroll's description of the mad hatter in *The Adventures of Alice in Wonderland.*

Until the 1960s it was not realized that inorganic mercury (free metal or compounds) could be changed to the more toxic methyl type of compound in the environment. In fact some very insoluble mercury compounds were once used in medicine, for example, as antiseptics and for the treatment of syphilis. These compounds are not toxic and are eliminated apparently without damage to the patient. In addition, silver-mercury amalgams have been used for filling teeth with no apparent difficulty. In other words, until comparatively recently, mercury, while recognized as potentially poisonous in certain circumstances, was not considered a threat to people or the environment.

In 1968, Dr. John M. Wood, of the University of Illinois, showed that the very toxic methyl mercury compounds could be formed from inorganic mercury by microbes in natural systems. Dr. Wood tried without success to alert people to the dangers implied by his research. It took the 1970 tuna and swordfish scare to bring out the full impact of Wood's findings. The chlorine and alkali industry was able, in a very short time, to eliminate what had apparently been a long-continued loss of large quantities of inorganic mercury to the waters of rivers and lakes. Other mercury users have also tightened controls.

The FDA is continuing surveillance and control of mercury levels in fish. Other federal recommendations made in the early 1970s included the discontinuance of methyl mercury compounds for seed dressing and the reduction of industrial mercury discharges to levels no higher than those found naturally in the environment. In addition, many states have set up regulations of mercury discharge and prohibited mercury compounds in paints where they had been added to prevent mildew.

In spite of tighter controls, considerable mercury contamination is continuing. There are small concentrations of mercury in fossil fuels like coal and oil which, because of the large quantities used, contribute significant amounts of mercury to the environment. The mercury first enters the atmosphere and eventually finds its way into water bodies. In addition, solid wastes may contribute mercury from such sources as discarded lighting fixtures, mercury lamps, switches, drugs, batteries, and broken thermometers.

The mercury problem is very complex. Even if all mercury contamination were eliminated (which is practically impossible), there would still be the problem of naturally occuring mercury and also of microbes working on lake and river bottom muds and converting the inorganic mercury there into the highly toxic methyl mercury form.

Interesting research was carried out by Professor Edwin Wilmsen at the University of Michigan. In fossil fish discovered at a 750-year-old archaeological site in Peru, mercury levels were as high as the highest ever found in marine fish, indicating that for thousands of years fish have picked up mercury in their tissues in concentrations much higher than their surroundings. Although this shows that mercury is a part of the natural environment, we must be aware that mercury levels may be thrown out of balance by manipulation.

Is Cadmium a Potential Hazard?

Some researchers feel that cadmium could be another element whose behavior we are neglecting at our peril—just as we did with mercury. In fact, Dr. Henry Schroeder of the Dartmouth Medical School, a recognized authority on trace metals in organisms, has expressed the idea that metals like cadmium may be more insidious and serious pollutants than the organic pollutants and gross contaminants which have had most attention from environmentalists.

An interesting picture develops when we consider cadmium. You can check the Periodic Table and see that it is in the same family as zinc, placed right below it in the chart. And mercury is right below cadmium. In other words, all three are in the same family. Cadmium is believed to be a nonessential trace element, but it is always closely related to zinc in the environment and invariably is a contaminant in zinc ores and zinc products. Because of their close chemical resemblance, cadmium may replace zinc in the body.

Dr. Schroeder believes that the cadmium replacement may be a factor in a common and very important American ailment—hypertension, or high blood pressure. Zinc is required in the breakdown of fats, and when it is replaced by cadmium the system loses its ability to properly handle these materials, which then build up in the arteries. The result may be hypertension and heart disease. Dr. Schroeder thus believes that cadmium may be a major factor in hypertension.

A major source of cadmium in the body is believed to be water. When soft water flows over zinc-coated surfaces like galvanized pipes it may pick up traces of cadmium as well as zinc. Another source is food. Zinc is so common in the environment that it appears in many foods, and foods that contain zinc also contain cadmium. We have apparently been living with cadmium for a long time.

We need much more research on how metals in the environment such as cadmium, lead, nickel, beryllium, and antimony are related to public health. Most organic molecules are degradable eventually but no metal is degradable. Once a metal is dug from the earth, it may stay with us for a long time.

Sediments

Sediments are particles of soil and worn-away materials which are suspended in and carried by water. The term sediment implies that the particles will settle to the bottom—eventually. Silt refers to very fine particles of earth, sand, clay, and so forth carried by moving water.

Erosion of land surfaces, producing irregular gullies and crevasses, is often thought of chiefly as a physical problem. However, definite chemical problems are involved as well. Estimates place the total sediment delivered to waterways in the U.S. at close to 4 billion tons per year. About three-quarters of this comes from agricultural land. Calculations of nutrient losses from this land are staggering: 5 billion pounds of nitrogen, 3 billion pounds of phosphorus, and 65 billion pounds of potassium!

When the sediment reaches the lakes, rivers, and oceans other problems arise. Here are a few of the known possibilities:

1. Aquatic life is killed, for example, by sediment covering spawning beds and clogging fish gills.
2. Photosynthesis is reduced by sediment blocking penetration of sunlight and thus preventing production of oxygen by aquatic plants.
3. Water is clouded by sediment which requires removal before domestic and industrial use.
4. Recreational values of lakes, streams, and oceans are adversely affected.
5. Reservoirs are filled and the life of farm ponds shortened.
6. Undesirable chemicals are carried by adsorption on silt particles, such as phosphates which contribute to eutrophication, and insecticides which kill fish and other desirable aquatic life.
7. Pumps, turbines, and irrigation equipment are eroded.

A few suggested control measures are:

1. Use of good ground-covering plants.
2. Shaping of land on slopes by contouring and terrace-building to prevent rapid rain runoff.
3. Cutting down on overgrazing by cattle.
4. Growing different crops in alternate strips.
5. Use of protective tree growth to lessen the effect of wind.
6. Leaving considerable crop residue on land after harvesting.
7. More careful control of industrial operations.
8. Better planning of cities to provide green belts and balanced vegetation.
9. Control and regulation of strip-mining operations.

HEAT POLLUTION

Industrial cooling and electrical power plants produce tremendous quantities of heat. A simple way to get rid of the heat is by circulating water which is then returned to rivers, lakes, or oceans at higher temperatures.

Heat pollution is often referred to as thermal pollution, the term derived from the same root as *thermometer*—Greek for *heat*. The discharge of heated water can have serious consequences for aquatic life. Here are some of the difficulties.

1. The amount of dissolved oxygen in the water is decreased. Higher temperature means more kinetic energy of molecules and consequently, the molecules of warm water are moving faster than those of cold water. This is relative, but the overall effect is that the amount of oxygen which can stay dissolved in warm water is less.
2. The elevated temperature can seriously affect the reproductive cycles of fish. For example, spawning may occur at the wrong time or eggs may hatch out prematurely. Then the environment may not provide the proper food for the small fish and large numbers may die.
3. The temperature increase may reach lethal limits for particular species of fish. If high enough, of course, all fish are killed. Fish are especially sensitive to temperature changes because they are cold-blooded. They do not have the very effective regulation system for temperature of warm-blooded mammals, which works by the evaporation of water molecules into the air. The evaporation process uses a lot of heat and helps to keep you cool. Fish obviously cannot do this.
4. Higher temperatures can contribute to excessive growth of algae.
5. Disease organisms may thrive at the higher temperatures.

Controlling the heat pollution problem will be difficult. The demand for power is increasing enormously and hence the demand for cooling water is also growing. There is theoretically no way to convert heat to energy without loss.

WASTE WATER TREATMENT

From this brief review of only a few of the potential water pollutants, you can see that recovery of "pure" water presents complex problems. However, knowing the source and nature of the pollutant and understanding the recycling processes which nature uses for handling waste can help technology devise better treatment processes for waste water.

Domestic sewage is a much less complicated problem than industrial wastes, which will not be treated in detail here. In fact, sewage has a higher proportion of one substance than Ivory Soap, which is advertised as 99.44 percent pure soap. Sewage averages higher than 99.5 percent water. It is that tiny fraction of less than one-half of one percent which gives all the trouble.

There are several levels of approach for handling sewage. For many centuries open dumping or discharge of raw sewage into rivers was a common practice. As population centers grew this presented aesthetic and health problems. The early history of cities involved many epidemics of waterborne diseases such as cholera, dysentery, diarrhea, and typhoid. Slowly systems were developed for more efficient removal of sewage wastes.

Cesspools and Septic Tanks

A cesspool is simply a pit, hole, or pool in the ground. It is a collection device which retains solid matter and lets liquid escape into the surrounding soil.

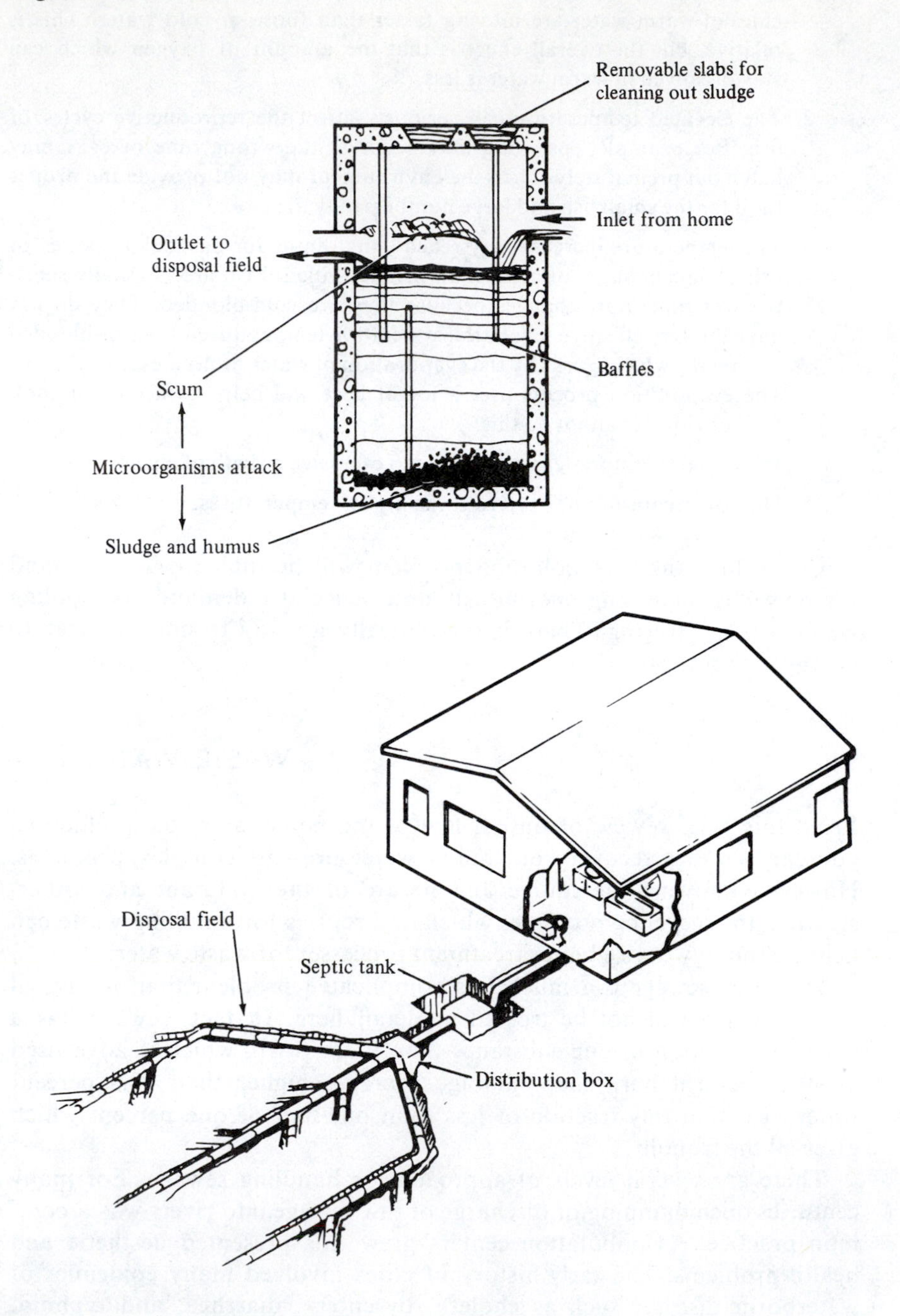

FIGURE 15.5
Septic tank used in rural areas.

The septic tank represents a refinement of the simple cesspool (see Figure 15.5). It contains baffles which permit only relatively clear liquid to flow out into the ground for distribution through a field of loose pipes. The sludge and scum in the tank are attacked by naturally occurring microorganisms. This forms gas, which escapes, and humus, which collects on the bottom. A properly functioning tank has to be cleaned every few years to remove the humus. This is often used as fertilizer.

Septic tanks are still widely used in the U.S. Both septic tanks and cesspools can contaminate nearby underground water supplies.

Primary Treatment: Physical Separation

In primary treatment, shown in Figure 15.6, large solid bodies are first removed by passage through a screen. Modern plants combine a grinding step along with screening. In a "grit chamber," sand, grit, cinders, and small stones are allowed to settle to the bottom. This is chiefly to prevent damage to pumping equipment.

After grit removal the sewage still contains dissolved inorganic and organic chemicals along with suspended solids. The next settling step takes place in what is often called a sedimentation tank. The sewage may take from one-half to three hours to pass through. Suspended solids sink to the bottom where they are pumped off as raw sludge. The liquid goes out an overflow at the top of the tank. If only primary treatment is being used, this liquid is then treated with chlorine before being discharged into a stream, river, or ocean.

Chlorine reduces the number of disease-carrying bacteria. However, all microorganisms are not killed. Primary treatment is considered a very inadequate sewage process even though it is the only treatment used by about 30 percent of U.S. cities.

Two views of primary treatment equipment. On the left the unit is shown in operation. On the right, the unit is empty for repairs, showing the rotating vanes on the bottom which sweep the solids out for collection. (Courtesy City of San Buenaventura Sanitation Division.)

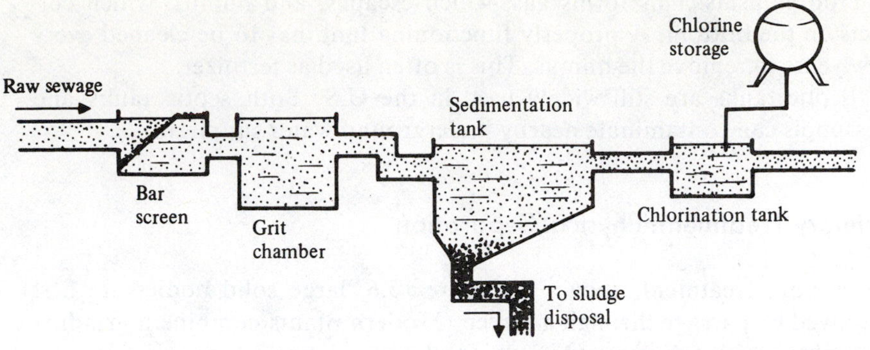

PRIMARY TREATMENT ONLY

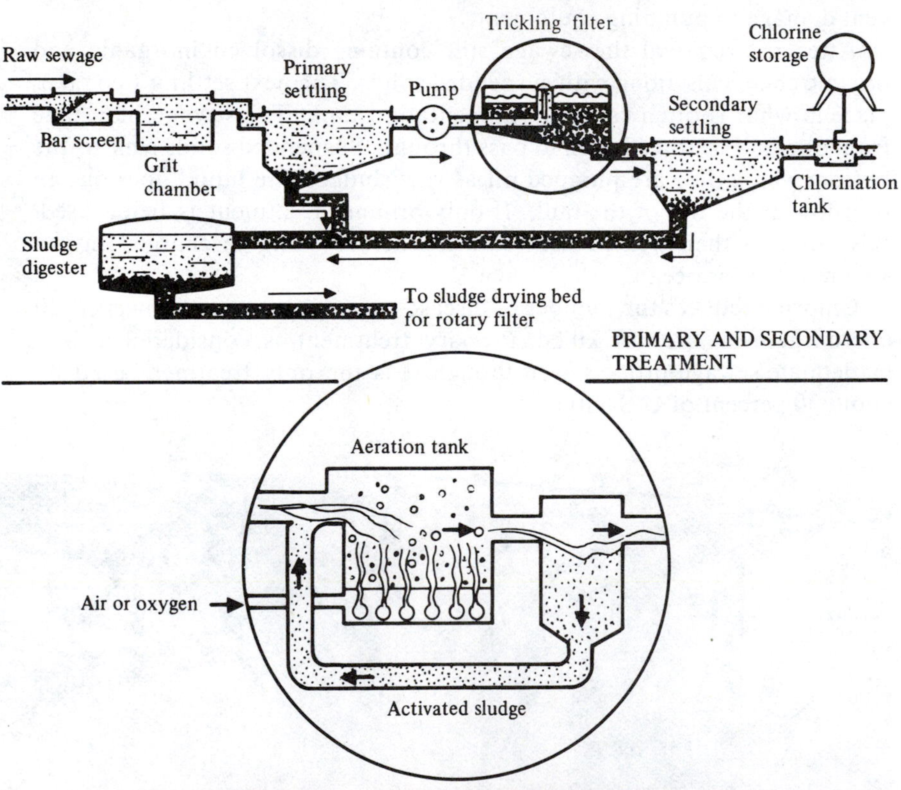

ACTIVATED SLUDGE METHOD USED IN PLACE
OF TRICKLING FILTER IN SECONDARY TREATMENT

FIGURE 15.6
Sewage treatment.

Aerial view showing construction of a modern waste water treatment facility. (Courtesy City of San Buenaventura Sanitation Division.)

Secondary Treatment: Where Bacteria Do Some of the Work

In secondary treatment, shown in Figure 15.6, the liquid overflow from the primary sedimentation tank is not chlorinated immediately but led to a bacterial treatment system. The two chief types are trickling filters and the activated sludge system.

A trickling filter is not really a filter but a bed of stones from 3 to 10 feet deep. The liquid from the primary sedimentation tank is sprayed over the bed. Bacteria collect on the stones and consume most of the organic matter in the sewage.

Modern plants tend to favor the activated sludge process in place of the trickling filter. This process provides for an intimate mixing of air, sewage, and sludge activated or loaded with bacteria. The bacteria can get oxygen more efficiently from air in the swirling aeration tank. A recent innovation which is even more efficient uses oxygen in place of air.

The treated sewage water, after settling and sludge removal, then goes to a chlorination tank before discharge to a stream or lake.

The disposal of sludge represents up to half of the operating costs of sewage treatment plants. Many different approaches are used including the following: (1) incineration; (2) burial in deep holes, abandoned mines, or in sanitary landfills; (3) use as fertilizer, humus, or soil conditioner; (4) pumping to strip mines and marginal farmlands to provide a topsoil capable of returning the land to vegetation for parks or agriculture; and (5) dumping in the ocean.

At left, a trickling filter showing rotating piping which delivers the sewage water over stones covered with bacteria. The process is also called biofiltration. The photo at right shows the swirling foam generated on top of activated sludge tanks. (Photos courtesy of San Buenaventura Sanitation Division.)

Advanced Treatment: Removal of Phosphates

Secondary treatment still leaves a considerable variety of contaminants in the water, among them dissolved inorganic salts and bacteria-resistant organic molecules. Methods are now being developed for improving the water from secondary treatment plants. This is called advanced (or sometimes tertiary) sewage treatment.

A major problem here is phosphates and nitrates which contribute to overnourishment of lakes. A method for removing phosphates will show how chemical concepts are intimately involved. The advanced treatment plant at Lake Tahoe, California, removes phosphates by addition of lime (CaO). This first makes the solution alkaline:

$$CaO + H_2O \rightarrow Ca(OH)_2$$

Then the calcium ion forms an insoluble compound (precipitate or "floc") with the phosphate ion:

$$3\,Ca^{+2} + 2PO_4^{-3} \rightarrow Ca_3(PO_4)_2\downarrow$$

The equation above emphasizes the fact that calcium ions combine with phosphate ions to give an insoluble compound, calcium phosphate. (More complex calcium compounds may be involved.) There are positive ions associated with the phosphate ions and negative ions with the calcium ions. A total equation would include all the ions to balance the system, even though the calcium and phosphate ions are the main ones because

As shown in left photo, modern waste water treatment plants use rectangular tanks for settling. Here an operator is shown cleaning surface equipment. In photo at right, solids formed in tertiary treatment are removed by pumping through mixed media filtering tanks (sand and anthracite coal). (Photos courtesy of San Buenaventura Sanitation Division.)

they form the floc of $Ca_3(PO_4)_2$. An example of a total equation is the following:

$$3\,Ca(OH)_2 \quad + \quad 2\,Na_3PO_4 \quad \rightarrow \quad Ca_3(PO_4)_2\downarrow \quad + \quad 6\,NaOH$$

The lime can be recovered and reused again and the phosphorus can be made available for use in fertilizer.

Here are a few other advanced treatment methods:

1. *Purification with activated carbon.* Many dissolved organic molecules which resist bacterial action may be removed by use of very adsorbent granular carbon (called activated carbon). The carbon can be regenerated after use by special heat treatment.
2. *Oxidation with chemicals.* This process removes dissolved organic molecules by oxidizing them—chiefly to carbon dioxide and water. The chemical oxidants can work faster than bacteria used in the secondary treatment. Also they may oxidize molecules left over from the bacterial oxidation. Examples of chemicals which can be used are ozone (O_3) and hydrogen peroxide (H_2O_2).
3. *Air blowing to strip out gases.* This is one method of removing nitrogen compounds. Most of the nitrogen in municipal waste water is in the form of ammonium compounds. If the water is made alkaline, as by the addition of lime, then the ammonium compounds can be converted to free ammonia.

$$NH_4^+ \quad + \quad OH^- \quad \rightarrow \quad NH_4OH \quad \xrightarrow[\text{blowing}]{\text{Air}} \quad \underset{\text{blown out as gas}}{NH_3\uparrow} \quad + \quad H_2O$$

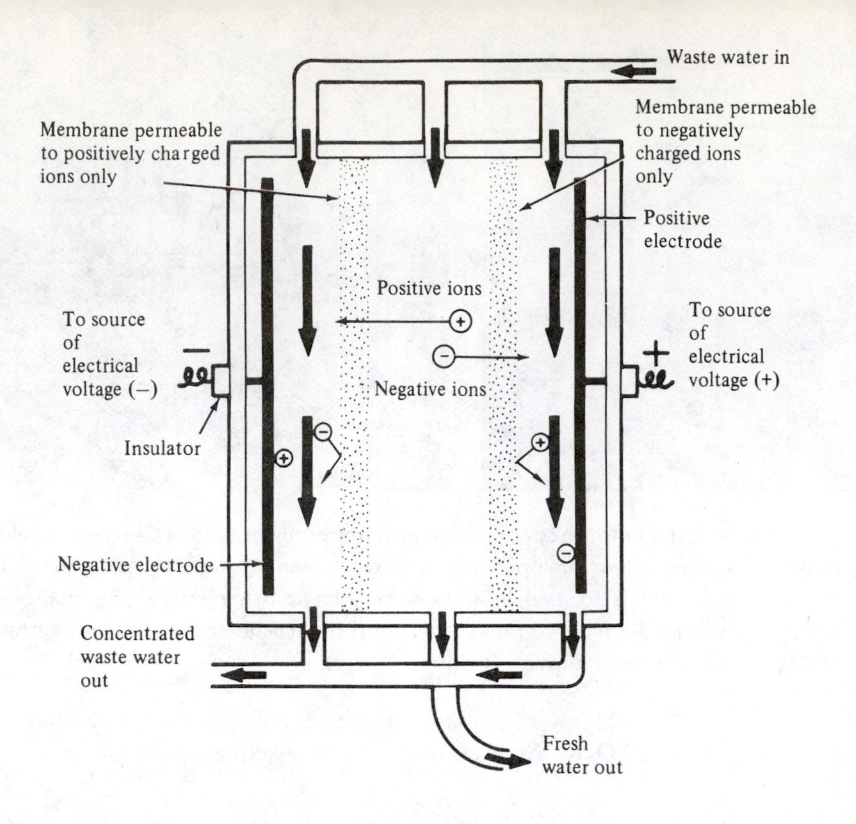

FIGURE 15.7
Electrodialysis.

Modern approach to water pollution problems involves individual waste water treatment facilities for each industry. This is a view of a waste water treatment facility for a large petrochemicals plant in Puerto Rico. This facility could handle the entire sewage treatment needs of a city the size of Albany, New York. (Courtesy Union Carbide Corporation.)

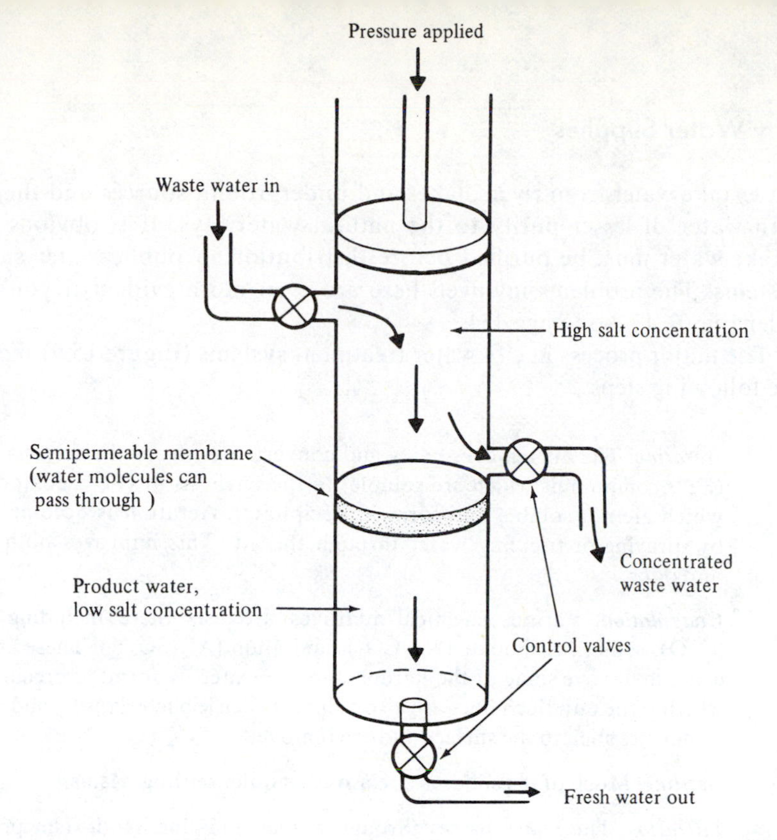

FIGURE 15.8
Reverse osmosis.

4. *Removal of inorganic ions.*
   (a) *Electrodialysis.* Electrodialysis has been used successfully in demineralizing brackish ground water (see Figure 15.7). The concept is simple: apply a voltage across a tank containing semipermeable membranes so that one electrode is positive and the other negative. Then the negative ions in solution will migrate through the membrane toward the positive electrode while the positive ions will be attracted toward the negative electrode. The action is similar to electroysis (chapter 12).
   (b) *Reverse osmosis.* This is another process using membranes (see Figure 15.8). It is being developed for application to waste water after having been proved successful for demineralizing water supplies in homes and industrial plants. Osmosis is the passage of water molecules through a semipermeable membrane. The membrane in this case must be permeable to water molecules but not ions. The normal situation is for water to go through the membrane from the side of low concentration to that which has the higher concentration of ions. This is reversed by applying pressure to the more concentrated solution. Manufactured membranes are usually of cellulose acetate. The exact mechanism by which they work is not fully understood.
   (c) *Ion exchange.* This is the technique which has already been highly developed to remove hard water ions in home water softeners and also for treatment of industrial wastes (chapter 13).

City Water Supplies

Cities take water from rivers, lakes and underground sources and then return water of lesser purity to the natural waterways. It is obvious that intake water must be purified before distribution in public water supply systems. The problems involved here are even more evident if you consider Figure 15.2 on page 434.

The major process in city water treatment systems (Figure 15.9) include the following steps:

1. *Aeration.* The air removes gases and converts iron from the ferrous form (Fe^{+2} compounds which are soluble) to the ferric form (Fe^{+3} compounds which are not soluble and form a precipitate). Aeration is accomplished by spraying or trickling water through the air. This improves both taste and odor.
2. *Coagulation.* Various chemical additives are used here including lime (CaO), sodium carbonate (Na_2CO_3), and alum ($Al_2(SO_4)_3$). These chemicals can remove some of the hardness in the water by forming precipitates which settle out. Such flocs are also helpful because bacteria, silt, and other impurities stick to the surface and are removed.
3. *Settling.* Much of the solids are removed in quiet settling basins.
4. *Filtration.* The water passes through gravel and fine sand. The process removes very small suspended particles which would give the water a cloudy appearance.
5. *Disinfection.* Most water treatment systems now use chlorine to kill disease-causing bacteria. Chlorine may be added before coagulation or after filtration. Some cities add it in both places.
6. *Optional fluoride addition.* In recent years, many city water systems have added fluoride (NaF) to the water to help reduce tooth decay. Fluoridation was originally prompted by the discovery that certain areas of the country had a very low incidence of tooth decay. Investigation showed the presence of natural fluoride in the drinking water. Then tests showed that addition of fluoride was beneficial to children whose teeth were being formed. The fluoride ions get into the complex tooth structure which is predominantly calcium phosphate, $Ca_3(PO_4)_2$. Fluoridated toothpastes and topical fluoride treatment attempt to provide the same benefit.

   There has been some community opposition to adding fluoride to the water system. One concern is the fact that sodium fluoride is a rat poison. However, the amount added to water supplies is carefully controlled at one part per million of water. A person would have to drink over 1000 gallons of treated water containing one ppm of fluoride in order to receive a lethal dose. And he would have to drink that quantity in one day since fluoride is readily eliminated by the kidneys. Another objection is that fluoride forms brown spots, or mottling, on teeth. This occurs from some natural waters containing an excess of fluoride—in the range of 5 ppm. However, treated water at 1 ppm does not cause this trouble.

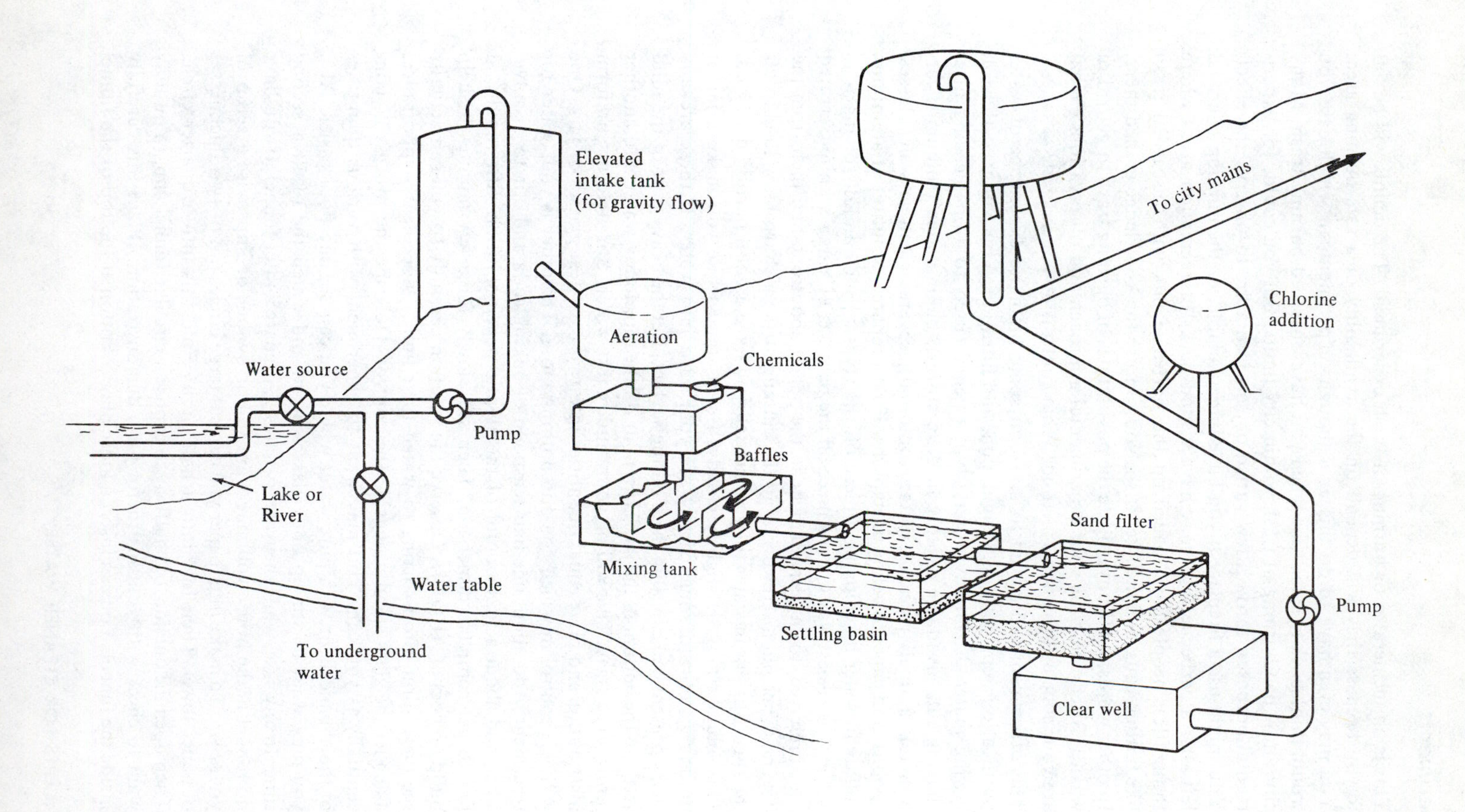

FIGURE 15.9
Water treatment for domestic use.

PROJECTS AND EXERCISES

Experiments

1. This is an incinerator or burning-of-trash experiment. The details will be left to you because they will depend on the availability of a safe burning area. You may even have to do this as a thought experiment, in which case you would think back to some time when you have observed burning trash. Many cities now ban burning of trash within city limits. But you could, for experimental purposes, throw a few varied objects into a fire to observe the effect. You might use a fireplace if you have one available, or the burning charcoal left after a barbecue. You may have visited parks or recreation areas where thoughtless people have loaded up the barbecue pits with discarded trash like bottles, cans, and the like. If so, you can use the knowledge gained there. In other words, *safely* investigate the possibilities of burning trash to get rid of wastes. Briefly describe your experimental approach and observations. What are your conclusions on this method? What controls do you suggest?
2. This experiment is designed to give you some firsthand experience of water purification steps: settling, coagulation, and filtration. You will need the following materials: muddy water; styptic pencil (used to stop bleeding after shaving cuts, available cheaply in drug stores); baking soda (sodium bicarbonate); a small piece of cotton (optional); several clean jars or glasses; spoons; and paper toweling or napkins. First prepare a solution of aluminum sulfate in water by chipping a piece off the styptic pencil and dissolving it in water. The easiest way to cut a piece off the pencil is by using a knife (serrated edge preferred) to cut a small niche in the pencil about one-half inch from the flat end. Then put a screwdriver in the cut and strike a sharp blow to chip off the slice. Stir this in water till you can tell you have a good portion dissolved. This solution may be slightly cloudy. Let it stand while you make a jar of muddy water. Stir some earth from your yard into water. If there are any twigs present you can skim them off the surface. Let the heavy part of the mud settle to the bottom of the jar. Then pour off the top cloudy water into another jar or large glass. This will be your muddy test water. Stir it to get uniform distribution and pour equal quantities into two clear test glasses or jars. One will be a control in which you add nothing more. The other is your water for treatment. Add about one-half teaspoon of baking soda and stir to dissolve. Then add about a teaspoonful of the styptic pencil solution. Stir rapidly at first to get complete dispersion. Then stir slowly for a few minutes to help build up flocs of $Al(OH)_3$. Observe the solution carefully so you can actually see these form. Stop stirring to watch their light, "flocculent" appearance. Then let the glass stand alongside the control. Describe the process as conceptually as you can. You may want to experiment with various quantities of the chemicals or with pure water where you can see the flocs easier. Also you may want to stir up both glasses again and watch the result a second time. In any case, while the water is settling, prepare a filter funnel. If you have no funnel in the house you may want to borrow one or improvise a setup of your own. To make filter paper you can cut large circles out of two thicknesses of paper towel. Then fold into a half circle. Fold this into a quarter circle. Then open the outside fold halfway to expose a circular funnel cup. (You may want to check with your instructor for a demonstration.) Place this carefully inside the funnel. You can insert a loose plug of cotton at the top of the funnel

stem to help support the filter cone and prevent holes or tearing. Next, carefully pour the top, partially clear water into the filter paper cone, the funnel being supported in a clean glass. You may have to make a few trial adjustments here. When finished, you should have very clear water. However, please DO NOT drink it, since the proportions of chemicals used were not necessarily adjusted carefully and other purification steps might be required. Compare your experiment with the diagram and descriptions of water purification found in this chapter.

3. This is a thought experiment. You may remember the damage caused when the oil tanker *Torrey Canyon* went aground off the English coast. For this experiment try to think of one of the larger tankers springing a leak and sinking off the coast of the United States. Presume this might happen, say, five years from today. That gives you time to think about the situation without the pressure for immediate action as was the case with the *Torrey Canyon*. What ideas can you come up with to control the flow of oil? You can use some concepts about the nature of oil and water. In the *Torrey Canyon* emergency, some people thought that synthetic detergents would be good for emulsifying the oil. A large amount of detergent was used but the sad experience was that this caused more damage to marine life than oil spills where no detergent was used.

Exercises

4. Review the data given in this chapter on the eventual disposal of the many gallons of water that fall each day in the U.S. Then, recognizing that we are increasing our needs for fresh water yearly, outline methods you believe would be feasible to meet a future requirement for 1000 billion gallons per day.
5. Describe what is meant by BOD, how it is measured, and what it indicates about water quality.
6. Following are the formulas for two of the amino acids present in proteins:

$$NH_2(CH_2)_4\underset{}{C}H(NH_2)COOH \qquad CH_3S(CH_2)_2CH(NH_2)COOH$$

Lysine (left; NH_2 attached to the CH carbon) — Methionine (right; NH_2 attached to the CH carbon)

Refer to the section in this chapter dealing with anaerobic decay and suggest which of the chemical end products listed there might be expected from these amino acids. How could the properties of these end products be considered a signal to avoid contact with decaying organic matter?

7. Give two examples of how conceptual analysis through the water cycle could help us solve water problems.
8. Explain what is meant by the term biodegradable. Then give three examples of materials that are not biodegradable and three that are.
9. Briefly describe the concept of food chains and give an example of a potential chemical hazard involving a food chain.
10. Someone has proposed a new way to dispose of solid wastes—to compact them under high pressure for dumping into trenches known to exist deep in the ocean. Compare this method with three other possibilities and briefly suggest advantages and disadvantages of each.

11. Give an operational and conceptual description of recycling. Then explain the scientific principle behind the practice.
12. Describe briefly the hazard of mercury in the environment, how the mercury gets there, and what can be done about it.
13. Write a paragraph describing what you believe should be the approach by the government on the mercury-in-fish problem.
14. Considering the Kinetic-Molecular Theory, how would you account for the fact that fish cannot easily regulate their temperature when the environment warms up?
15. Briefly compare the primary, secondary, and advanced methods for waste water treatment. Then suggest why the advanced treatment has not yet become as common as secondary.
16. Someone remarked that the major shift in emphasis from the primary to secondary to advanced treatment for waste water is from an operational to a conceptual approach. Briefly justify this statement, giving specific examples to back up your argument.
17. Pick out two methods of advanced waste treatment and explain their chemistry.
18. In considering the growing water pollution problem it has been said that we are going to have to make big improvements in the treatment of water we drink or in the treatment of water we discard. Consider the possibilities here and briefly outline what you think would be the best practice to follow, giving reasons.
19. Someone asks you why tests are made for coliform bacteria in water supplies when these particular bacteria are not the harmful ones. How would you answer?

SUGGESTED READING

1. There are many topics in this chapter which you might want to investigate by reference to encyclopedia articles. In addition, you will want to keep your eyes open for current articles on pollution in magazines and newspapers.
2. Peixoto, J. P., and Kettani, M. A., "The Control of the Water Cycle," *Scientific American,* 228, No. 4, pp. 46–61 (April 1973).
3. Pennman, H. L., "The Water Cycle," *Scientific American,* 233, No. 3, pp. 98–108 (September 1970). This is one of the articles in a whole issue devoted to the "biosphere," that part of the earth in which life exists. Some of the others are also referenced below.
4. Woodwell, G. M., "The Energy Cycle of the Biosphere," *Scientific American,* 223, No. 3, pp. 64–74 (September 1970).
5. Deevey, E. S. Jr., "Mineral Cycles," *Scientific American,* 223, No. 3, pp. 148–158 (September 1970).
6. Woodwell G. M., "Toxic Substances and Ecological Cycles," *Scientific American,* 216, No. 3, pp. 24–31 (March 1967).
7. Benarde, M. A., *Our Precarious Habitat,* W. W. Norton Co., New York, 2nd edition, 1970.

8. American Chemical Society, *Cleaning Our Environment, The Chemical Basis for Action,* American Chemical Society, Washington, D.C., 1969. Also *A Supplement to Cleaning Our Environment,* 1971.
9. Ehrlich, P. R., Holdren, J. P., and Holm, R. W., eds, *Man and the Ecosphere,* W. H. Freeman and Co., San Francisco, 1971. *Readings from Scientific American.*
10. Johnson, H. D., ed., *No Deposit—No Return,* Addison-Wesley Publishing Co., Reading, Mass., 1970.
11. Goldwater, L. J., "Mercury in the Environment," *Scientific American,* 224, No. 5, pp. 15–21 (May 1971).
12. Siever, R., "The Steady State of the Earth's Crust, Atmosphere and Oceans," *Scientific American,* 230, No. 6, pp. 72–79 (June 1974).
13. MacIntyre, F., "The Top Millimeter of the Ocean," *Scientific American,* 230, No. 5, pp. 62–77 (May 1974).
14. Bascom, W., "The Disposal of Waste in the Ocean," *Scientific American,* 231, No. 2, pp. 16–25 (August 1974).

chapter 16

Concepts for Cleaner Air

I never saw a man who looked
With such a wistful eye
Upon that little tent of blue
Which prisoners call the sky,
And at every drifting cloud that went
With sails of silver by.

—*OSCAR WILDE*

Fresh air keeps the doctor poor.

—*DANISH PROVERB*

Nature is a mutable cloud which is always and never the same.

—*EMERSON*

Air pollution was not of much concern in the years preceding World War II. The atmosphere was considered a giant waste basket capable of absorbing whatever we put into it. Much of the pollutant material was in gas form and this was considered negligible simply because it is invisible. This "waste basket mentality" is undergoing a radical change as people come to realize the problems of polluted air through the obnoxious impact of pollution on the sense organs and several air pollution disasters.

AIR POLLUTION DISASTERS

A disaster is usually thought of as an event which happens suddenly or unexpectedly like an earthquake or a tornado. But the classic cases of air pollution disasters were not quite of the sudden variety. Although they were precipitated by a particular set of circumstances, a long-term buildup of air pollution factors was the primary cause.

Donora, Pennsylvania. October, 1948. Donora is a city located on the inside of a horseshoe-shaped valley of the Monongahela River about 30 miles from Pittsburgh. A heavy fog descended on the city and the air was calm. People could smell the sulfur dioxide in the air a little more than in the usual Donora fogs. The valley contained a steel mill, sulfuric acid plant, a zinc smelter, and other factories. Sulfur dioxide was implicated along with mixtures of other pollutants. Twenty people died.

London, England. December, 1952. There was extensive fog. London was most severely affected. It is situated in the broad valley of the Thames River and has a complex of potential pollutant sources. In a five-day period there were between 3500 and 4000 deaths in excess over the normal. Autopsies showed no special mode of death other than irritation of the respiratory tract. Measurements were made of air samples during the episode. Suspended smoke and sulfur dioxide were found to be high.

What have we learned from these air pollution disasters? At least an awareness that the air is an essential resource for the health of living things. Another lesson is that there is no one, easily eliminated toxic substance involved. We have to consider the total air environment and preserve its quality.

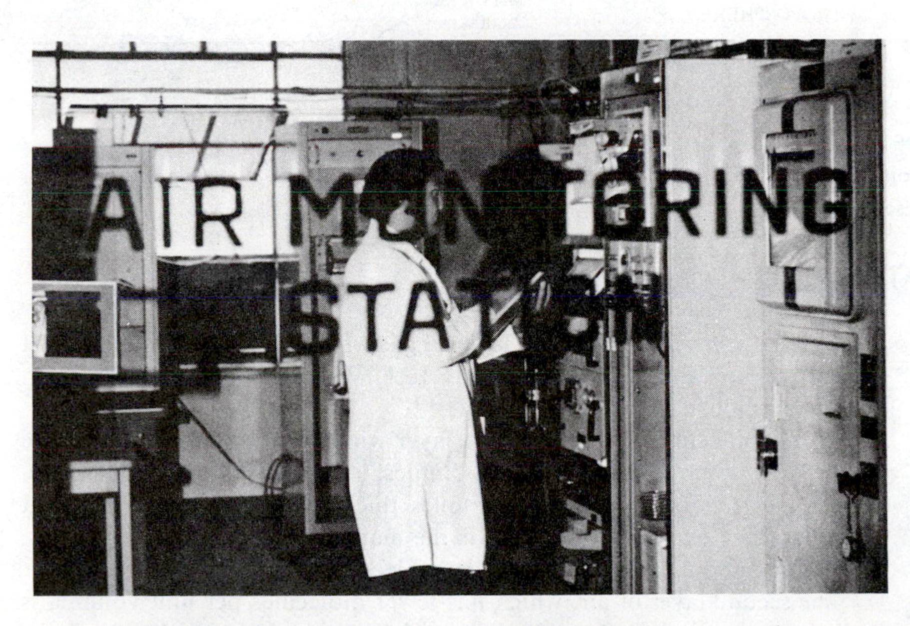

One of the lessons we have learned from air pollution episodes is an awareness that the air is an essential resource. Most large cities now have elaborate air monitoring stations and thus avoid air pollution disasters by taking necessary action when the monitors indicate a danger. (Courtesy Los Angeles County Air Pollution Control District.)

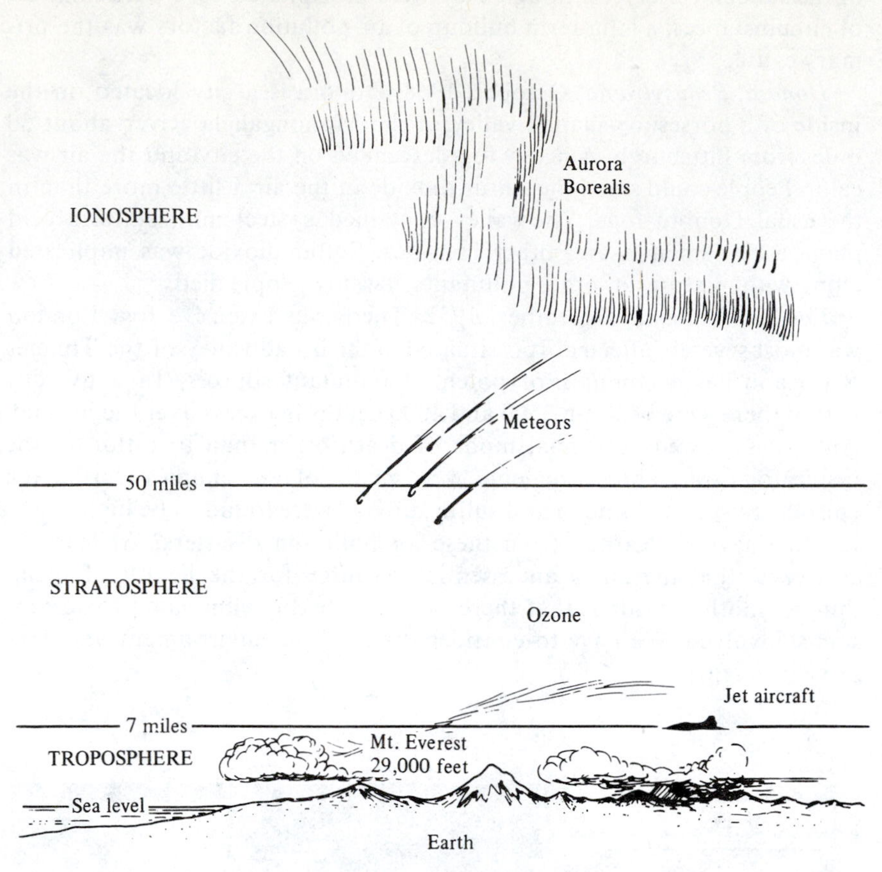

FIGURE 16.1
The main layers of the atmosphere, with distances indicated approximately and not to scale.

WHAT IS AIR?

Air, or the *atmosphere,* is a thin layer of gas which surrounds the earth just above the land and the oceans (Figure 16.1). The layer of air on the earth is very much like the skin on an apple.

Most of the air molecules are in a layer which covers the surface of the globe to an average depth of about 7 miles. This layer of the atmosphere, called the *troposphere,* is about 5 miles thick over the poles and about 10 miles thick over the equator. It is the major reservoir of air molecules, containing about 90 percent of all the molecules in the atmosphere.

The second layer of air, which has fewer molecules per unit volume, is called the *stratosphere.* It extends from 10 to about 50 miles. It contains a layer of ozone (O_3) from about 12 to 21 miles up. Above the stratosphere is a layer containing ions which shield the earth from high energy rays originating in outer space. The ions also serve as a layer to bounce radio waves back to earth. This layer, called the *ionosphere,* extends from 50 to

TABLE 16.1
Composition of Clean, Dry Air Near Sea Level

Name of Gas	*Formula*	*% by Volume*	*ppm*
Nitrogen	N_2	78.09	780,900
Oxygen	O_2	20.94	209,400
Argon	Ar	0.93	9,300
Carbon dioxide	CO_2	0.0318	318
Neon	Ne	0.0018	18
Helium	He	0.00052	5.2
Methane	CH_4	0.00015	1.5
Krypton	Kr	0.0001	1
Hydrogen	H_2	0.00005	0.5
Dinitrogen oxide	N_2O	0.000025	0.25
Carbon monoxide	CO	0.00001	0.1
Xenon	Xe	0.000008	0.08
Ozone	O_3	0.000002	0.02
Ammonia	NH_3	0.000001	0.01
Nitrogen dioxide	NO_2	0.0000001	0.001
Sulfur dioxide	SO_2	0.00000002	0.0002

300 miles above the earth. It eventually blends away to outer space where molecules are very rare.

The major air mass, then, is the layer of molecules extending 5 to 10 miles above the earth. Even this quantity of air is not all available for diluting pollutants, however. Tests have shown that there is no great diffusion of surface air beyond 2 miles (12,000 feet), and many pollutants never go much higher than one-third of a mile (2000 feet). (The diameter of the earth is 8000 miles so the available air layer is a pretty thin skin on the apple.)

What is the nature of the troposphere? Table 16.1 gives its composition. You can see that four gases actually comprise 99.99 percent of dry air.

WHAT IS AIR POLLUTION?

What is air pollution? It is the addition to air of substances which alter its normal composition such as to produce a measurable effect on people, animals, plants, or other materials. As in the case of water pollution it consists of chemicals in the wrong place or in the wrong concentrations.

Air pollution is relative. When you breathe you are changing the composition of normal air. You breathe about 20 times per minute, taking in approximately eight-tenths of a quart of air each time. In one day you take in over 50 pounds of air, which is about 10 times your total intake of food and water combined. You can live without food for five weeks and for five days without water. But if you are without air for five minutes you are in serious danger of death. Your breathing contributes very slightly to a change in air by producing a higher content of carbon dioxide from your burning of food for energy and heat.

Indeed, the major source of air pollution is burning fuel—for heat or energy or to get rid of wastes—on a much larger scale than in your body.

Air Pollutants and Their Sources

The major air pollutants are the following:

General Description	*Alternate Designation*
Carbon monoxide	CO
Sulfur oxides	SO_x
Hydrocarbons	HC
Nitrogen oxides	NO_x
Particles	Particulates

Sulfur oxides is a general class of considerable importance and includes both sulfur dioxide, SO_2, and sulfur trioxide, SO_3. Sometimes the symbol SO_x is used to indicate the inclusion of all oxides of sulfur just as NO_x is used as an all-inclusive symbol for nitrogen oxides.

TABLE 16.2
Sources of Most Common Pollutants (in Millions of Tons Per Year United States)

	CO	*SO_x*	*HC*	*NO_x*	*Particles*	*Totals*
Autos, trucks, etc.	66	1	12	6	1	86
Factories	2	9	4	2	6	23
Electric power plants	1	12	1	3	3	20
Space heating	2	3	1	1	1	8
Refuse disposal	1	1	1	1	1	5
TOTALS	72	26	19	13	12	142

Table 16.2 lists the sources and amounts of major pollutants emitted in an average year in the 1960s. The percentage breakdown for these major sources of pollutants is:

Autos, trucks, etc.	60%
Factories	17%
Electric power plants	14%
Space heating	6%
Refuse disposal	3%

You can see the reason for the heavy emphasis on cleaning up auto exhaust. Figure 16.2 shows this is still a major problem. However, individual chemicals differ in their molecular makeup, and therefore rankings in order of weight produced do not necessarily run parallel with rankings in

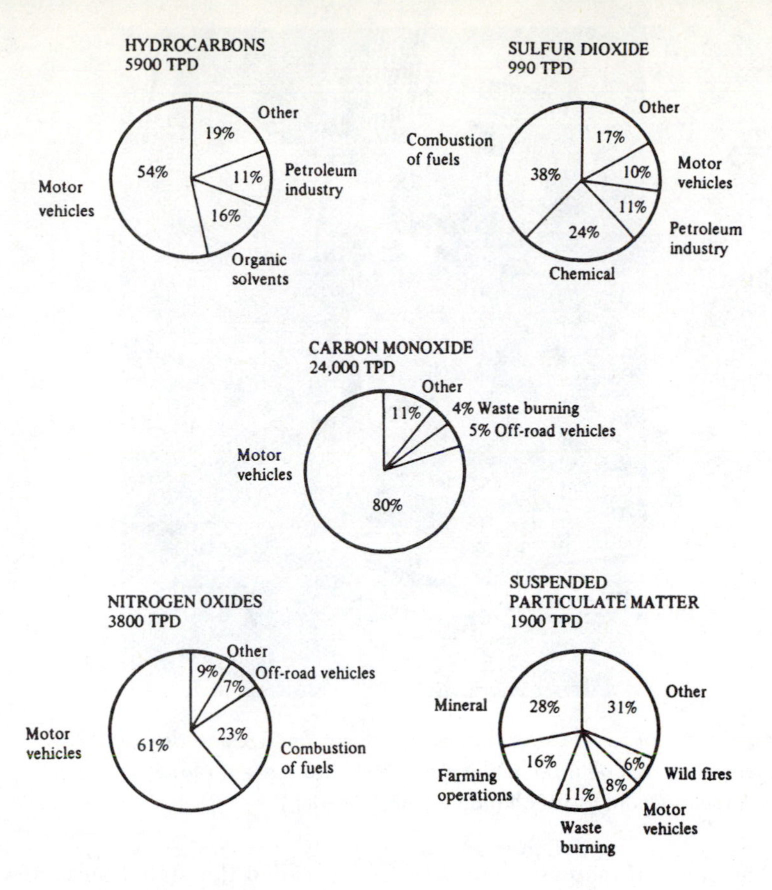

FIGURE 16.2
Latest data available from the California Air Resources Board give a general picture of the sources of major air pollutants in California. (TPD = tons per day.)

the order of damage to people and the environment. Some very toxic chemicals might be released in only very small quantities. They would not even appear in the above table. Yet these could cause a severe pollution problem under particular circumstances.

The Special Case of Carbon Dioxide

Carbon dioxide is a major product of combustion of organic materials. The tonnage emitted far exceeds those of the five major pollutants listed. However, it is taken up so quickly and efficiently by nature's carbon cycle that it has not caused any acute pollution problem.

But there may be a more long-range pollution problem here. Because of the enormous quantities of fossil fuels being taken from the earth and burned (coal, oil, natural gas), the global concentration of carbon dioxide is increasing. Calculations show an increase of carbon dioxide from 296 ppm in 1900 to 318 ppm today.

Photo showing the major source of air pollution. Even on a clear day, there is considerable emission of pollutant molecules which are gaseous and invisible. (Courtesy Los Angeles County Air Pollution Control District.)

The effect of increased levels of CO_2 is called the greenhouse effect. A greenhouse is a building with glass roof and sides in which green plants can be grown even in a cool climate because of retention of the sun's heat. Carbon dioxide in the air acts similarly to the glass. It allows the passage of the short wavelength radiation from the sun. This warms the earth which in turn emits longer wavelength (infrared) radiation (heat rays). These can be absorbed by the CO_2 molecules, preventing the loss of the energy. The CO_2 can then radiate the energy to the earth or transfer it to other molecules near the earth's surface. Thus the temperature increases. Both O_2 and N_2 have only two atoms per molecule and do not absorb the infrared energy which the three-atom carbon dioxide molecule does.

The picture is not a simple one and there is considerable controversy as to the eventual outcome. Some have predicted a gradual increase of earth temperature that will melt the polar ice caps, raise the level of the oceans, and flood coastal cities. Calculations of temperatures recorded over a span of time indicate a slight global rise of 0.4°C between 1880 and 1940. Then there was a drop of 0.2°C between 1940 and 1967. One explanation for the drop is the increase in particle pollution in the air. This would deflect some sun energy and keep the earth cooler. Other factors yet unknown may also be responsible. Some scientists suggest that one form of pollution may be balancing off another type, at least as far as changing the average global temperature is concerned.

Carbon Monoxide (CO)

Normal burning of carbon with plenty of air gives carbon dioxide, as in the barbecue reaction (chapter 7).

$$C + O_2 \rightarrow CO_2 + \text{energy}$$

If a compound of carbon is burned completely (with plenty of air) it will also give CO_2 as the end product of the carbon atoms in the molecule. Thus methane, CH_4, which is the chief component of natural gas, burns as follows:

$$CH_4 + 2O_2 \rightarrow CO_2 + 2H_2O + \text{energy}$$

Gasoline is a complex mixture of hydrocarbons. One example of the many types of molecules present is octane, C_8H_{18}. When this is burned with an adequate supply of oxygen the overall reaction is:

$$2C_8H_{18} + 25O_2 \rightarrow 16CO_2 + 18H_2O + \text{energy}$$

The problem of carbon monoxide pollution is simply that the internal combustion engine cannot provide that extremely large amount of oxygen for complete burning of the gasoline molecules. You can see from the equation above that for every 2 molecules of octane you would need 25 molecules of oxygen. And air is chiefly nitrogen, so the very nature of the internal combustion engine gives considerable quantities of carbon monoxide. You actually get a mixture of CO and CO_2 along with the other exhaust products.

We know a great deal about how carbon monoxide acts as a poison but very little about what happens to it when it enters the atmosphere. The toxic property is caused by the formation of a fairly stable compound between CO and the hemoglobin of the blood.

Where does all the carbon monoxide go? There is still considerable uncertainty on this point. We put enough CO into the air each year to double the concentration in about five years, yet measurements indicate a somewhat steady level of CO in the air of about 0.1 ppm. The large quantities of CO we are dumping into the air must be going somewhere. You might at first assume it is being converted directly into carbon dioxide by reaction with O_2:

$$2CO + O_2 \rightarrow 2CO_2$$

However, this reaction has been carefully studied and found to be insignificant under the low temperature conditions in the atmosphere. Some recent research has shown that microorganisms in the soil may be responsible for removing CO from air, but it is not yet certain whether this is the main removal method. Research is continuing.

Sulfur Oxides (SO_x)

Sulfur is a plant nutrient because some of the molecular building blocks which are necessary for proteins contain sulfur atoms. Fossil fuels come from organisms which lived millions of years ago, and these organisms needed sulfur atoms in their proteins just as you do today. In fact you are very likely using some of the same sulfur atoms which some prehistoric plant or animal used many, many years ago. A large part of the sulfur compounds present in the atmosphere come from the biological decay of organisms or the burning of fossil fuels like coal and oil. This is a good example of the statement: Pollutants are resources in the wrong place.

Fairly reliable estimates have been made on the sulfur pollution of the atmosphere. About two-thirds is from natural sources, chiefly decay of organisms containing sulfur. A small natural source is exhalations from volcanoes. The activities of people contribute only one-third of the sulfur compounds in the air. Most of this is emitted as SO_2.

The fact that man-made sulfur pollution is only one-third of the total can be misleading. The problem here is one of distribution and concentration. The natural decay processes are fairly widely spread so that emis-

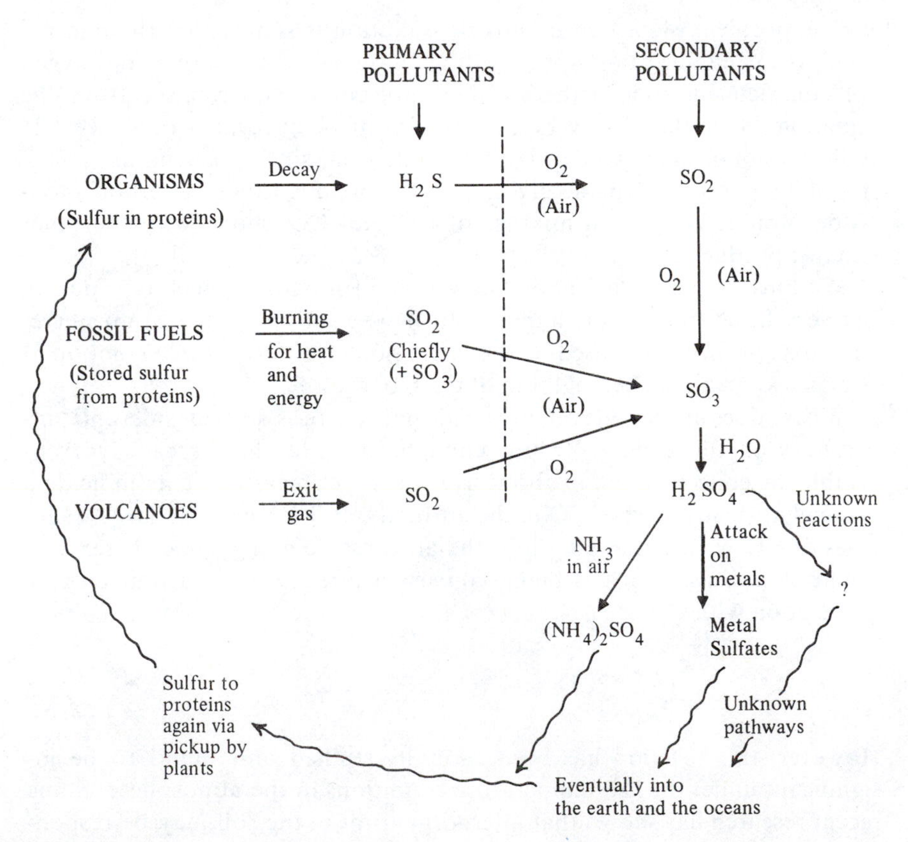

FIGURE 16.3
Sulfur cycling in the air.

sions are more easily absorbed in nature's own cyclic systems, whereas the pollution by people is apt to be concentrated where damage can be considerable before the molecules are dispersed and absorbed into the sulfur cycle (see Figure 16.3).

Where Does the Odor of Rotten Eggs Go? The large proportion of sulfur reaching the air from natural biological decay goes through the intermediate compound hydrogen sulfide, H_2S. This is a foul smelling gas—the odor of rotten eggs. Eggs are high in proteins containing sulfur atoms. Where do all the foul odors of decaying organic materials, or even one rotten egg, go? Nature has ways of neutralizing these odors as she prepares to circulate the atoms for use again. Hydrogen sulfide is easily oxidized to SO_2:

$$2H_2S + 3O_2 \rightarrow 2H_2O + 2SO_2$$

This SO_2, along with smaller amounts from burning of coal and oil, is then further oxidized in the air to SO_3.

Sulfur dioxide is a colorless gas while sulfur trioxide is a liquid at ordinary temperatures. In the burning of fossil fuels only a small fraction of SO_3 is formed, and the conversion rate of SO_2 to SO_3 is very slow at normal temperatures. We can convert SO_2 to SO_3 by heat and use of catalysts. Then the SO_3 forms dense white clouds as condensation occurs. It forms tiny droplets of sulfuric acid with the moisture in air.

$$2SO_2 + O_2 \rightarrow \overbrace{2SO_3 \xrightarrow[\text{from air}]{H_2O} H_2SO_4}^{\text{Clouds of mist}}$$

Damage by SO_x Pollution. There are three main categories of SO_x pollution damage.

1. *Damage to plants.* Plants are more sensitive to sulfur pollution than animals. Damage can vary from leaf injury, yellowing, and slowing of photosynthesis, to death of the plant.
2. *Materials damage.* You have seen above how the corrosive acid, H_2SO_4, can be formed from all kinds of sulfur pollutants. This is believed to be the major factor in damage to various materials (see Figure 16.4). Many materials are affected, including metals which corrode; paints which lose their durability; building materials like concrete, marble, limestone, and mortar which are leached and eroded; textile fibers, fabrics, paper, and leather which become brittle or lose their strength. A startling example of the effect of polluted air on stone is Cleopatra's Needle, an obelisk which was moved to New York City in the late nineteenth century. In less than one hundred years it has shown more deterioration than in the more than two thousand years it stood in Egypt.
3. *Damage to human and animal health.* Sulfur oxides irritate the respiratory system, causing both temporary and permanent injury. When small par-

METAL CORROSION

$$Zn + H_2SO_4 \longrightarrow ZnSO_4 + H_2$$

$$Fe + H_2SO_4 \longrightarrow FeSO_4 + H_2$$

BREAKING OF POLYMER CHAINS

$$\sim N(H)-(CH_2)_6-N(H)-C(=O)-(CH_2)_4-C(=O)-N(H)\sim \;\; + \; H_2O \xrightarrow{H_2SO_4 \text{ Catalyst}} \sim N(H)-(CH_2)_6-NH_2 \;\; HO-C(=O)-(CH_2)_4-C(=O)-OH \;\; HN(H)\sim$$

[Nylon polyamide]

Broken chain

ATTACK ON STRUCTURAL MATERIALS

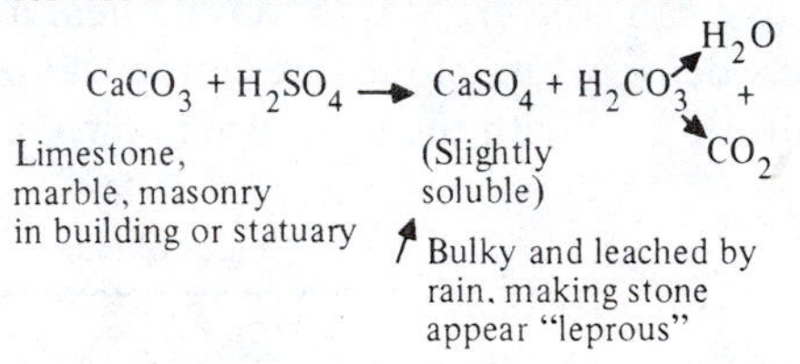

FIGURE 16.4
Examples of mechanisms for damage of materials by SO_x.

ticles (often found in polluted areas) are breathed in along with sulfur oxides, the irritation and injury may increase significantly. This was undoubtedly the case in the pollution disasters described earlier.

Nitrogen Oxides (NO_x)

Although many different oxides of nitrogen are known, only three are important constituents of the atmosphere (N_2O, NO, NO_2). And only two of these are emitted by the activities of people so as to be of concern in pollution studies. These are nitric oxide, NO, and nitrogen dioxide, NO_2. Since both of these occur together in air pollution, the symbol NO_x is often used to include both NO and NO_2.

You can probably guess where the NO_x pollution originates. You know that air contains almost 80 percent nitrogen, and can assume that the large numbers of nitrogen molecules in the air can react during burning. The high temperatures provide enough energy to activate nitrogen mole-

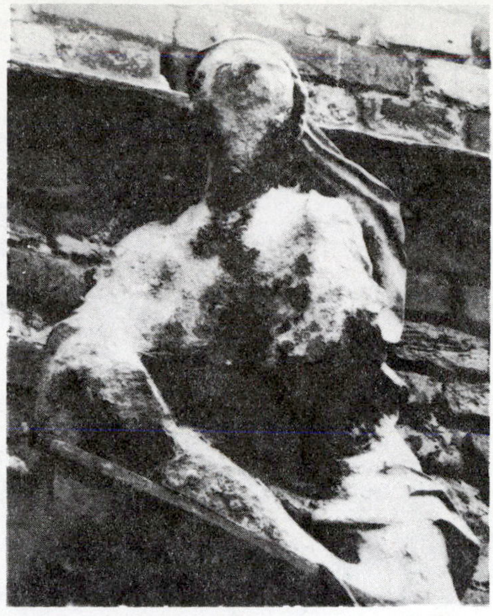

Damage to sandstone statuary caused by air pollution. Stone sculpture is located outside a German castle built in 1702. Left photo was taken in 1908 and shows only minor damage. Right photo, taken in 1969, shows almost complete destruction. (From E. M. Winkler, *Stone: Properties, Durability in Man's Environment,* Springer-Verlag, Vienna-New York, 1973, 230 pages; also, see *Science,* Vol. 181, 31 August 1973.)

cules which are somewhat inert at ordinary temperature. The following reaction occurs:

$$N_2 + O_2 \rightarrow 2NO$$

The nitric oxide may be considered a byproduct of the desired production of heat and energy by burning a fuel in air. The largest proportion of man-made nitric oxide comes from autos and electric generating plants.

When nitric oxide, NO, which is colorless, comes in contact with air, it reacts fairly readily with O_2 to form NO_2, which is a reddish brown gas with a pungent, choking odor.

$$2NO + O_2 \rightarrow 2NO_2$$

The quantities of NO_x emitted by man-made burning are much less than those produced from natural sources, which are bacterial action and lightning storms. The problem is that people produce large quantities of NO_x in concentrated areas, while nature distributes her NO_x more evenly around the globe. (See chapter 11 for a discussion of the nitrogen cycle.)

Nitric oxide or NO, at the concentrations that occur even in polluted urban air, is not an irritant and is not considered a health hazard. The hazard is its ability to undergo oxidation and produce the more toxic

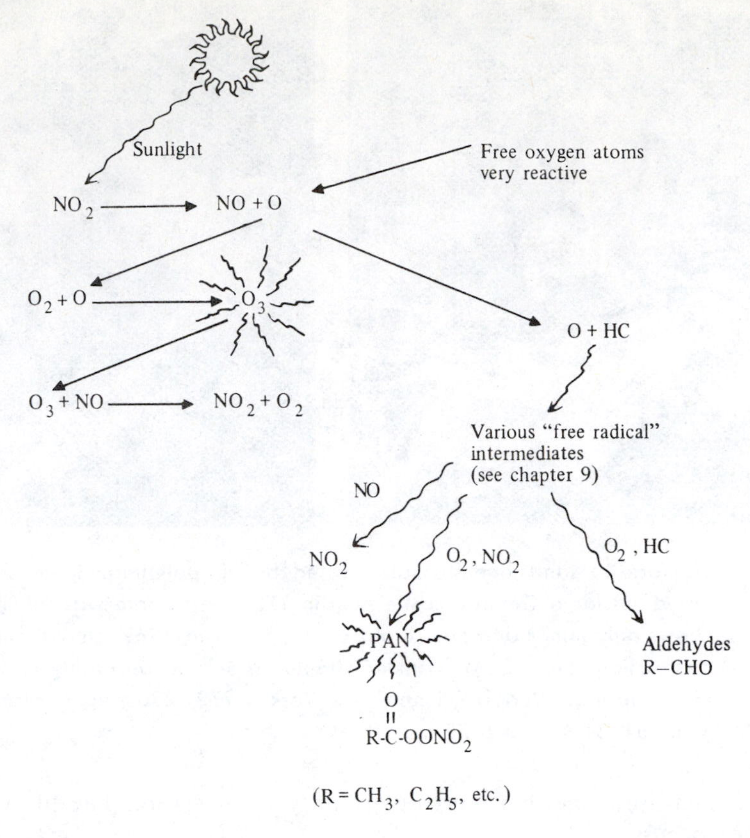

FIGURE 16.5
Sunlight, NO_2, and hydrocarbons: some of the complex chemistry involved.

NO_2. The direct toxicity of NO_2 apparently mainly acts to damage lung tissues.

Nitrogen dioxide not only directly affects the health, but it is also a prime agent in the production of other secondary pollutants, namely nitric acid, ozone, and hydrocarbon reaction products which are strong irritants. There has been much research on the complex interactions involved here. The chemical reactions are very complicated, involving NO_2, sunlight, and hydrocarbons, along with many other intermediate secondary pollutants.

Research so far has uncovered a few of the pathways and some of the secondary pollutants. The most important end-product pollutants which come from the complex interaction of NO_2 and ultraviolet light are ozone and a group of compounds called collectively PAN (for peroxyacylnitrates). These products are often called photochemical oxidants because they are caused by the action of light on chemicals and have a strong oxidizing power. Some of the complex reactions leading to these eye irritants are illustrated in Figure 16.5.

Hydrocarbons

Most of the hydrocarbons that enter the atmosphere come from natural sources. In fact the hydrocarbons put into the air by people are estimated at only 15 percent of the total. Again, nature is selective—limiting the types of hydrocarbons, and usually distributing them more broadly. The haze which gives the Great Smoky Mountains their name is related to long-chain hydrocarbons, called terpenes, released by certain evergreen trees. You know that methane, CH_4 (sometimes called marsh gas) is the common gas released from bacterial decay. Methane does not enter into the complex reactions that lead to PAN.

Gasoline hydrocarbons are lost to the air in many ways, for example, by loading tank trucks, filling service station storage tanks, filling auto gas tanks, spills, and emissions from the car in use. Some of these are now controlled but much loss still occurs. Large quantities of hydrocarbons and related organic solvents also enter the atmosphere in application of paints, varnishes, lacquers, roofing tars, and asphalt-type road coatings.

The major pollutant damage from hydrocarbons in the air is not caused directly by the hydrocarbon molecules themselves. Organic solvents can, of course, be quite toxic if breathed in high enough concentration, for example, by glue sniffing, where the volatile solvent is inhaled. This can be fatal. However, the volatile hydrocarbons distributed in the air as gaseous pollutants are quite dilute. There is no evidence that these cause direct harm to people. The polluting power of hydrocarbons comes from the fact that they enter into the complex interaction system with sunlight and NO_2 to contribute significant quantities of photochemical oxidants like ozone and PAN.

The ozone molecule, O_3, does not contain any of the atoms from the original hydrocarbon molecule as PAN does. However, research on the complex chemistry involved shows that hydrocarbon molecules contribute significantly to the increase in ozone concentrations, because the hydrocarbon molecules form intermediates which in turn react with NO. This eliminates NO molecules which would react with ozone:

$$NO + O_3 \rightarrow NO_2 + O_2$$

Thus the hydrocarbon interferes with a normal mechanism for neutralization of O_3.

The Importance of Ozone in Polluted Air

Ozone is the most concentrated component in polluted air of the photochemical type. Ozone is the pungent-smelling gas often noticed around electrical equipment like motors and generators. It is also sometimes detected after lightning storms. It is produced from oxygen by ultraviolet light or electrical discharge: energy + $3\,O_2 \rightarrow 2\,O_3$.

Ozone injury on upper surface of spinach leaf. (Courtesy University of California, Riverside.)

The concentration of ozone is often used as an indicator of the relative severity of pollution of the photochemical type. There is, however, a fairly wide variation in sensitivity of different people.

Ozone is also known to have detrimental effects on vegetation and materials like rubber, paint, and textiles. Damage to leaves of plants and cracking of rubber occur at much lower concentrations than those known to effect humans. This is perhaps fortunate, since we can get signals from the damage caused to plants.

Smog: The Popular (?) Name for Air Pollution

Distinctions are made between two types of "smog": (1) "Classical" smog of the "London type" is produced chiefly from burning of coal and oil. The chief offenders here are SO_x, smoke, and soot, with often absorbed chemicals. This was the first type of air pollution given the name smog (from smoke + fog) in the early part of this century. The intensity of episodes is diminishing as less coal is burned in cities and more controls are installed. (2) "Photochemical" smog of the "Los Angeles" type results from burning hydrocarbon fuels, chiefly gasoline in autos. The major ingredients here are NO_x, hydrocarbons, sunlight and the photochemical oxidants discussed above. The term smog was first applied incorrectly to this type of pollution. Not only are smoke and fog not involved but, ironically, if they are present in quantity, the special Los Angeles variety of pollution cannot occur. You recognize the reason: sunlight is essential to photochemical "smog" and it could not operate efficiently through smoke and fog. However, the generalized use of the word smog is now well established to refer to air pollution of whatever kind.

Los Angeles is still the chief city that comes to mind when one men-

The two faces of a great metropolis—New York City. The "clear" view on the left shows relatively low visible pollution, while the right photo shows considerable smog during an inversion. (Photos courtesy EPA, New York Daily News photo.)

tions photochemical smog, but most major cities have various degrees of the same type of pollution.

Normally air is warmer at the ground and colder above. Energy from the sun is the chief source of heat for the earth's surface. The molecules of air near the surface are then warmer than those above. This means they move faster and expand so that the air is lighter in weight. This warm air rises carrying pollutants to the upper atmosphere where they are spread out and eventually absorbed in various cycles. In a *thermal inversion,* a layer of warm air moves in at higher altitude and traps a colder layer near the ground. The situation is inverse, that is, inverted or turned upside down. This is thermal inversion. The result is that pollutants are trapped in the cooler lower layer. Currents develop in the cool layer but they cannot move up when they reach the lid of warmer air. They therefore build up in the lower layer causing dangerous pollution conditions.

The inversion persists until the weather changes to permit the trapped air to circulate more freely. Los Angeles has a peculiar geographical situation in being located in a basin where the surrounding hills prevent easy lateral movement of air.

Particulate Matter

Small solid particles and liquid droplets, often called particulates, are not gases like the other air pollutants considered previously. Nevertheless, these are present in the air in enormous numbers and, in many cases, are definitely known to be involved in human disease, especially that related to the lungs. The terms particles and particulates are used here interchangeably.

Sizes of particles vary widely from the relatively large chunks which get in your eye occasionally to the tiny invisible particles—much more numerous—which can be even more harmful.

The solid and liquid particles come from a wide range of sources which partly accounts for their complex chemical composition. However, even particles from a single source may contain a great variety of different molecules. In fact, particles are mostly complex mixtures of materials on much larger than molecular size. Table 16.3 gives the composition of coal "fly ash" which is the light ash that flies in the air from burning fuels. The main mechanism by which particles are removed from the air is by settling out under the influence of gravity.

TABLE 16.3
Approximate Composition of Fly Ash from Burning Coal*

Component	%	*Component*	%
Carbon	0.4–36	Carbonate (as CO_3^{-2})	0–3
Iron (as Fe_2O_3 or Fe_3O_4)	2–27	Silicon (as SiO_2)	17–64
Magnesium (as MgO)	0.06–5	Phosphorus (as P_2O_5)	0.07–47
Calcium (as CaO)	0.1–15	Potassium (as K_2O)	2.8–3
Aluminum (as Al_2O_3)	10–58	Sodium (as Na_2O)	0.2–1
Sulfur (as SO_2)	0.1–24	Undetermined	0.08–19
Titanium (as TiO_2)	0–3		

*Results are ranges from a wide variety of coal furnaces. The exact composition is not determined and for simplicity components are expressed as indicated even though the actual molecular composition is much more complex.

Here are a few sources which contribute to the complex particulate pollution problem: erosion, grinding, spraying, industrial smoke, auto exhausts, agricultural tillage, surface mining, refuse burning, fossil-fueled power plants, and forest fires.

What are the effects of particulates? Below is a brief rundown of known effects.

1. *Effects on climate and weather.* Particles in the atmosphere can scatter and absorb sunlight. For this reason, cities receive up to 20 percent less solar radiation than rural areas. In particular cases the reduction of sunlight can be as high as one-third in summer and two-thirds in winter. Reduced visibility is an obvious problem in transportation involving autos and aircraft. Particles can also influence the formation of clouds, rain, and snow by acting as nuclei upon which the polar molecules of water can condense.
2. *Effects on materials.* The damage here can be much more than the unsightly films on your shelves or the extra costs of cleaning and laundering. Particulates can also cause corrosion of metals, erosion and darkening

of buildings and statues, soiling of painted surfaces, and damage to electronic equipment. Many particulates are actively corrosive either by nature or by absorption of acid-forming molecules from the air like SO_2.

3. *Effects on vegetation.* It is difficult to separate the damage to plants caused by particulates from the overall pollution damage. However, there is little doubt of the fact that dusts interfere with photosynthesis and cut down on quality of air which is most suited to plant growth.
4. *Effects on animal and human health.* The damage here is chiefly to the respiratory system.

Polluted atmospheres have been found to contain many different types of toxic particles, including metals like beryllium, nickel, and lead; asbestos; and soot containing absorbed toxic agents. Carbon in the form of soot is heavily implicated in the London type of smog. Carbon has long been recognized for its ability to absorb other chemicals. It is often used in gas masks as a filtering medium. However the very property of providing a large surface area for absorption can be a detriment when soot particles are breathed.

One of the most insidious forms of damage involves what are called carcinogens—chemicals which can cause cancer. These can be absorbed on the surface of soot which then carries them into the respiratory system. Many specific chemicals isolated from polluted air have been found to cause cancer in laboratory animals. One of the most common of these is a condensed benzene ring system called 3,4-benzpyrene (BP):

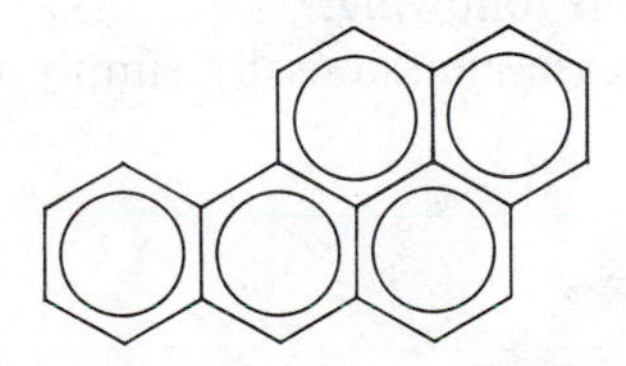

This chemical has been found in the air of many cities. Carcinogens in city air are believed to contribute to the higher cancer rates that occur in urban areas compared with rural areas. These are found to be significantly higher even after making corrections in the calculations to take account of the greater amount of smoking that occurs in cities. The chemical BP is also found in cigarette smoke. Auto exhaust, refuse burning, and coal smoke contribute significant quantities to the air. Many people believe England's high lung cancer rate may be due to the emission of carcinogens like BP in enormous quantities from coal combustion. The BP reaches the lungs in particulate form.

THE BROAD PICTURE OF AUTO POLLUTION CONTROL

Up to the present time there have been five general approaches to the problem of auto pollution control: (1) modification of the internal combustion engine to reduce pollutants; (2) addition of reactors to complete

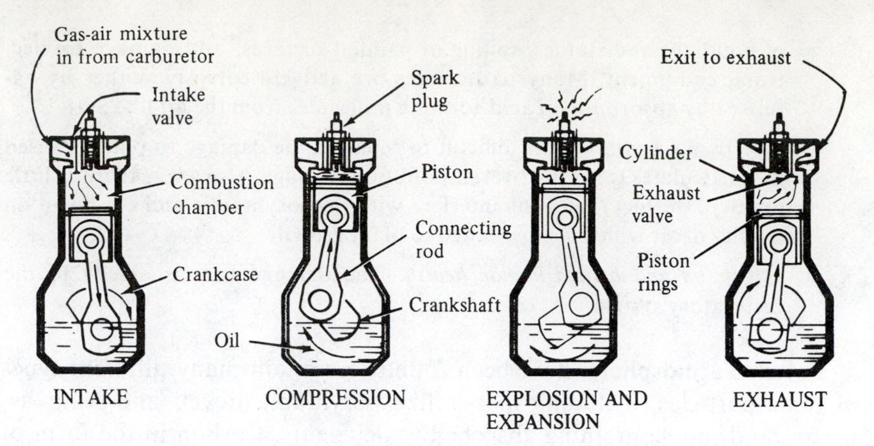

FIGURE 16.6
The basic internal combustion engine, using a four-sequence cycle.

the combustion of unburned exhaust molecules like CO and hydrocarbons; (3) modification of gasoline mixtures; (4) development of new fuels like natural gas, alcohol, or liquefied petroleum gas (LPG); and (5) complete replacement of the internal combustion engine. (The abbreviation ICE will be used below in reference to the internal combustion engine.) Figures 16.6 and 16.7 show the nature of the ICE and the distribution of emissions with no controls.

(1) Modification of the ICE. There have already been sweeping changes in this category including the following:

(a) Control of crankcase emissions by simply using a pipe to recycle

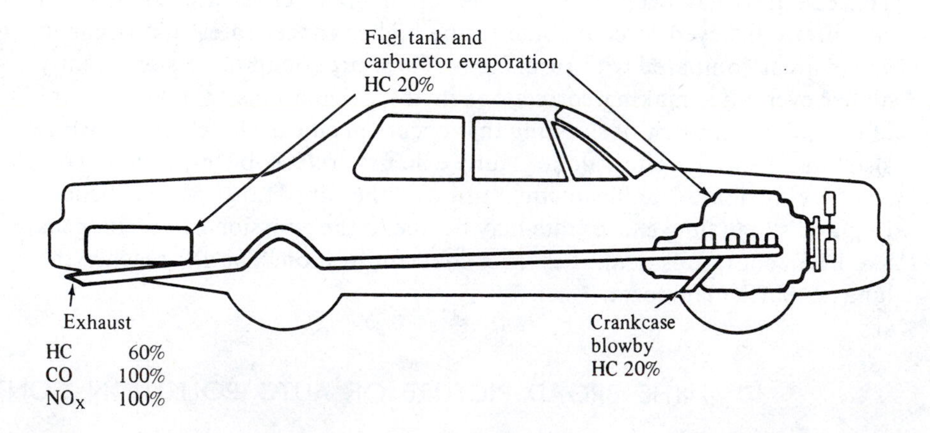

FIGURE 16.7
Approximate distribution of emissions for a car not equipped with any controls.

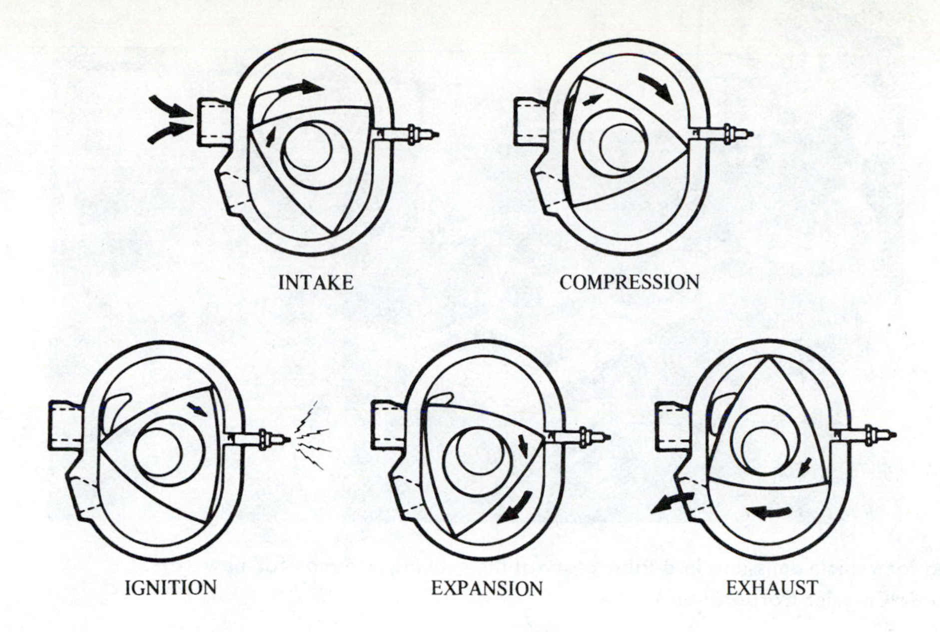

FIGURE 16.8
Sequence of Wankel rotary engine cycle events.

the gases (mostly hydrocarbons) to the fuel feed system for return to the combustion chamber. Crankcase emissions are sometimes called blowby and are the gases which escape around the piston rings and enter the space above the oil in the crankcase. These were formerly vented to the atmosphere.

California pioneered in legislation for control of auto pollution. Crankcase controls on all new cars sold in California were required beginning with 1963 models. Federal legislation later required them on all new cars sold nationwide beginning with 1968 models. Under growing pressure for auto regulation, the manufacturers had installed crankcase controls on all cars sold in California beginning with 1961 models and on all cars sold nationwide beginning with 1963 models.

(b) Modification of combustion chamber design. Examples are development of combustion chambers with turbulence in gas-air mixture to get more complete combustion; and use of a rotary combustion chamber.

The Wankel is the most prominent example of the rotary combustion chamber engine. It uses a sort of three-lobed rotating piston in a standard four-sequence cycle (see Figure 16.8). This engine has had some success in Europe and Japan. It is not theoretically any better on exhaust emissions than the standard ICE, but the manufacturers have had good success with emission control devices.

(c) Control of evaporation of hydrocarbons from the carburetor and fuel tank. Evaporative controls were required on new cars beginning with 1970 models in California and nationally in 1971. There are a variety of

Testing for vehicle emissions in a laboratory at the proving grounds for new cars. (Courtesy Chrysler Corporation.)

devices used. They collect vapors from the fuel tank and carburetor and deliver them either to the crankcase or to an activated-carbon cannister. In either case the vapors are subsequently recycled to the fuel induction system and burned in the engine.

(d) New carburetor designs to provide for a leaner gas-air mixture which gives lowered CO and HC emissions. Rich gas-air mixtures are high in gas, low in air; lean mixtures are low in gas, high in air, and consequently give more complete combustion to CO_2.

(e) Design of devices to reduce NO_x emissions. Oxides of nitrogen are the most difficult to control. Examples of the approach here are the two types designed for installation on older cars produced before manufacturers were required to control air pollution. The exhaust gas recirculation (EGR) method reduces NO_x by diluting the fuel/air mixture, thereby reducing peak combustion temperatures and pressures conducive to the formation of NO_x. The vacuum spark advance disconnect (VSAD) device achieves the same result by modification of the distributor spark advance. Generally, the spark is retarded either by disconnecting the vacuum spark advance mechanism or by electronic means with an initial retard in the basic setting.

(2) Addition of Reactors to Get More Complete Burning. Two major types of exhaust reactor are being tried at the present time.

(a) The thermal exhaust reactor is really a special kind of exhaust manifold and takes the place of the usual manifold. It allows exhaust gases enough time at high temperatures to react more completely. Extra air is injected in order to achieve more complete burning. The problem with thermal reactors is that they decrease fuel economy and require expensive alloy metals capable of withstanding the high temperatures.

(b) The catalytic exhaust reactor is usually located in the muffler position. Catalytic reactors, often called catalytic converters, operate at lower temperatures by use of a catalyst bed which speeds up oxidation of hydrocarbons and carbon monoxide. Various catalysts have been investigated like platinum and chromium oxide. The great problem is that lead compounds coat the active catalyst surface and reduce its effectiveness.

(3) Modification of Gasoline Mixtures. There are as yet no emission standards for lead as there are for HC, CO, and NO_x. Also no technology exists for removing lead compounds from auto exhaust. Leaded gasoline averages about 2.5 grams of lead per gallon of gasoline.

Certain types of hydrocarbon molecules provide high octane ratings. The gasoline mixture obtained from conventional refining of petroleum does not contain enough of these. Lead compounds have been added as the cheapest way to increase octane rating. Another way around the situation, given the high compression engines we have in use today, is to redesign the gasoline mixture. This can be done—but it requires extra processing. In other words, petroleum-refining companies have the technical ability to change the molecular structures and increase the proportion of high octane molecules in the overall gasoline mixture (chapter 8). In the early 1970s companies started marketing low-lead and no-lead gasolines. They could not do this by selling the usual gasoline mixtures without lead added. This would have caused increased knocking in the common high compression engine. Consequently the gasoline mixtures were altered.

(4) New Fuels. Some success has been achieved with liquefied petroleum gas (LPG) which is a mixture chiefly of propane (C_3H_8) and butane (C_4H_{10}). The exhaust hydrocarbons from this mixture are much less troublesome than those from gasoline. In addition, there is no lead requirement so that catalytic reactors can be used. LPG fuels have been used for public and fleet-owned vehicles. There is a greater fire and explosion hazard with LPG and also the supply is limited.

Some experimental use has also been made of both compressed and liquefied natural gas, which is mostly methane, CH_4. This provides an attractive, clean-burning fuel but the supply situation is even worse than with LPG.

(5) Complete Replacement of the Internal Combustion Engine. This does not necessarily mean going back to horses which, incidentally, contributed to pollution problems also. But the development of a different type of combustion system does appear to offer considerable promise. There are some people who believe that no amount of added-on corrective devices can be the long-term answer to the inherent problems of the internal combustion engine.

There are over 100 million cars on the road in the U.S. and 10 million new ones are turned out each year. All of these use a system of enclosing the fuel and air in a cylinder and then exploding it to hammer down the pistons that turn the driveshaft. At highway speeds the average car uses about 20,000 explosions per minute. Explosions are inherently inefficient ways to burn fuel. A flame front set off by the spark plug moves rapidly to the relatively cool cylinder walls where it is quenched, leaving the fuel incompletely burned. Then the exhaust valves open to let the incompletely oxidized gas mixture out into the atmosphere. Or the mixture can be passed through exhaust reactors where some additional burning occurs—wasting heat and energy.

When engines are designed or adjusted for extra power, the fuel-air mixture is rich in gasoline—contributing to less complete combustion and more CO and HC emission. When the mixture is made less rich (more air is used), then the NO_x emissions increase. This is the battle that engineers have been fighting unsuccessfully for so many years.

Are there alternatives to the internal combustion engine? There are good possibilities here, three major ones being considered below:

(a) Gas turbines. These are continuous combustion engines where the fuel is sprayed into a combustion chamber which is also supplied with high pressure air. The gaseous products of combustion are then allowed to expand through the vanes of a turbine into the atmosphere. The concentration of pollutants is low because of the large volume of air used, but total pollution per mile is similar to that of the standard ICE. Costs are high.

(b) Battery-powered engines. These engines use batteries similar to those you are familiar with for starting the ICE. Many other types of battery are now available besides the standard, lead-acid storage cell. Battery power has been used successfully in fleet cars and in some public bus systems. Problems are limited range, need for recharging facilities, and expense for special battery types. The electrical energy has to be provided from some source, such as a coal-burning power plant. Electric trolleys and buses were at one time quite common in cities. Unfortunately, they were dropped in most places in favor of the ICE, powered either by diesel oil or gasoline. The diesel engine, now so common in trucks and buses, emits less CO than the gasoline ICE but comparable quantities of NO_x and HC.

(c) Steam engines. Everyone has heard of the Stanley Steamer which can be seen in antique collections of old cars. Steam engines have probably been the most popular power plants in history except for animal muscles.

The steam engine is an external combustion engine. This is a major advantage over the ICE and accounts for the fact that it produces only about one percent as much pollution. The fuel can be burned under better conditions—at low pressures and relatively low temperatures. Most important, burning can take place with an abundant supply of oxygen. The basic system is simple. Fuel is burned in a furnace surrounded by coils and provided with an adequate air supply. The heat turns water (or other working fluid) in the coils into a gas under pressure. The gas is fed to a cylinder where it moves a piston which in turn moves the crankshaft connected to the wheels. There have been many modifications of the early engines of the 1700s, including steam turbines which are used in modern steamships.

Many people like the idea of a return of the steam car, and not just from nostalgia for the good old days. It has other advantages:

Steam engines can burn any kind of fuel from kerosene to high test gasoline.

Fuel is not burned by exploding it, which means the engine is quiet and does not need a muffler.

The steam engine can provide very fast acceleration. In fact, a clutch and reduction gear are used to reduce the instant getaway capability.

There are also disadvantages. Otherwise we would have more Stanley Steamers around today. Early steam cars had problems such as:

Slow start-up, requiring about a half hour warm up of boilers.

Visibility restrictions because steam was vented to open air.

High water requirements (the Stanleys got between 1 and 2 miles on a gallon of water!).

Water freezing in cold weather.

There has been considerable research by steam car developers. Some of the problems have been overcome. Modern steam engines warm up in less than one minute and use recycling to prevent loss of working fluid. Research has not yet come up with a suitable substitute or modification for water as a working fluid to eliminate completely the danger of freezing. The antifreezes like alcohol used in the ICE are not satisfactory in the steam engine where vaporization is required to drive pistons.

THE GENERAL PICTURE OF INDUSTRIAL EMISSIONS

It is difficult to generalize about the pollutants which can be generated by industrial processes. Some examples are given in Table 16.4. Each kind

TABLE 16.4
Examples of Emissions Associated with Specific Industries

Steel mills	Particles, smoke, CO, fluorides
Nonferrous smelters	SO_x, various metals, particles
Petroleum refineries	SO_x, HC, smoke, particles, odors
Portland cement plants	Particles, SO_x
Sulfuric acid plants	SO_x, H_2SO_4 mist
Iron and steel foundries	Particles, smoke, odors
Kraft pulp mills (for paper)	SO_x, particles, odors
Hydrochloric acid plants	HCl mist and gas
Nitric acid plants	NO_x
Soap and detergent plants	Particles, odors
Alkali and chlorine plants	Cl_2 gas
Fertilizer plants	Fluorides, particles, ammonia
Aluminum plants	Fluorides, particles
Phosphoric acid plants	Acid mist, fluorides
Food processing plants	Odors, particles

of factory can "specialize" in certain pollutants. However, most of them are involved to some degree in the five most common pollutants, CO, SO_x, particles, NO_x, and HC.

In recent years there has been a growing trend toward the principle that each industry should be responsible for its own cleanup problems. Modern plants have pollution controls built into the overall process. The older and smaller operations are the ones that give a lot of the difficulty.

Control of Industrial Particulates

One of the common methods of controlling any kind of pollution from both factories and electric power plants is a tall stack. The tall stacks do not reduce emission of pollutants but dilute the pollutants in the air and achieve lower ground-level concentrations. Smokestacks were at one time considered a good sign of prosperity. The only really acceptable stack exhaust today is a white plume of steam—not smoke which is made up of solid particles. Many stacks are now being converted to achieve this goal. Some of the methods are shown in Figures 16.9 through 16.12.

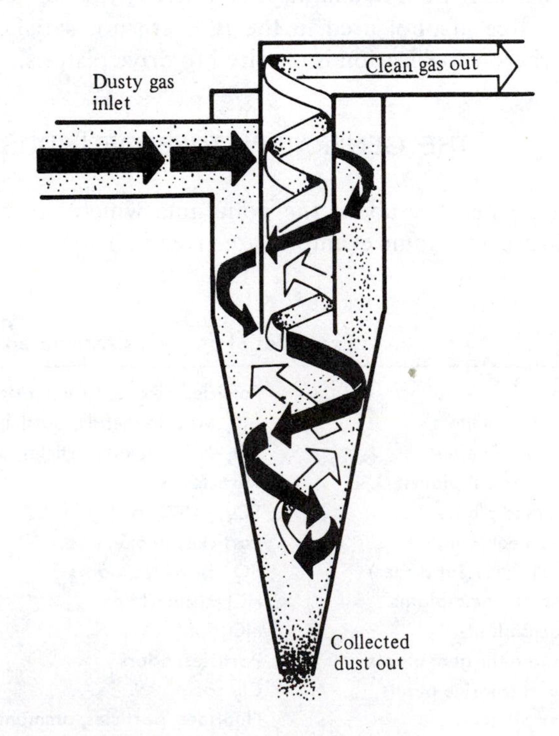

FIGURE 16.9
Cyclone collector. The whirlwind effect throws out particles against the sides, where they fall to the bottom to be collected.

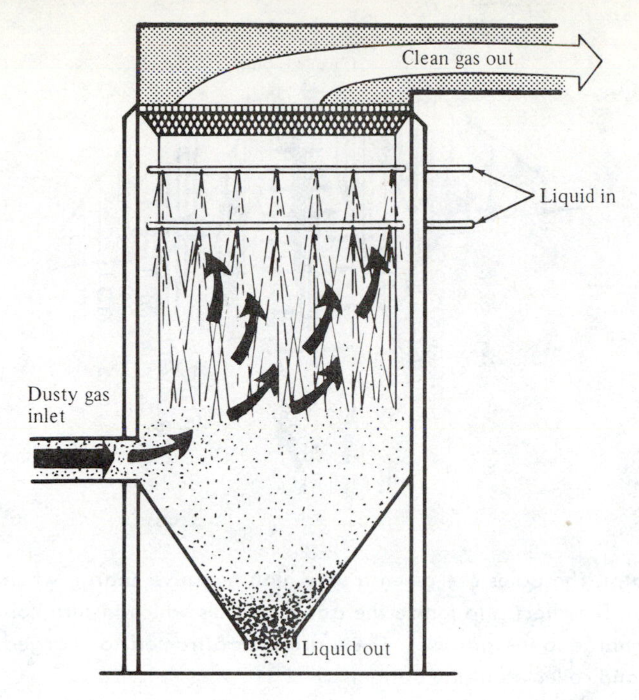

FIGURE 16.10
Liquid scrubber. The water is sprayed like a shower, removing the impurities.

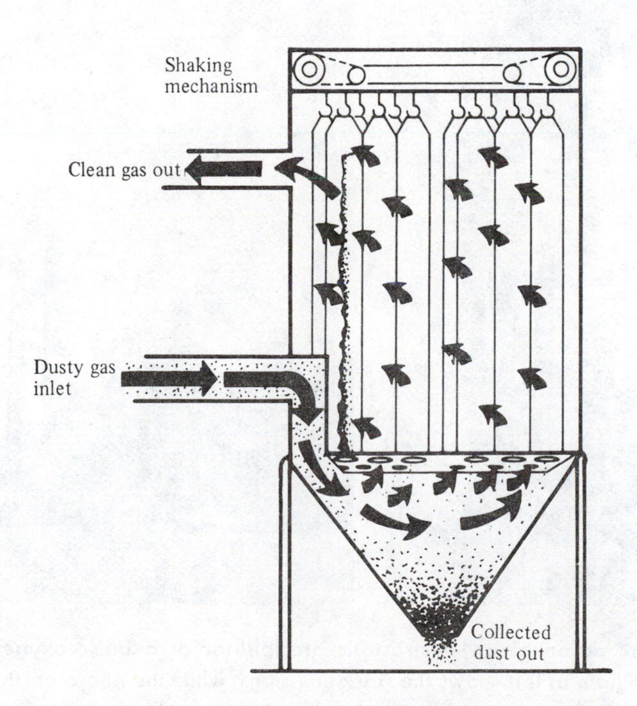

FIGURE 16.11
Bag house filter. Filters are merely devices for removing particles from a fluid. Like the filters used in autos, vacuum cleaners, and coffee makers, this type of filter was designed for a specific need.

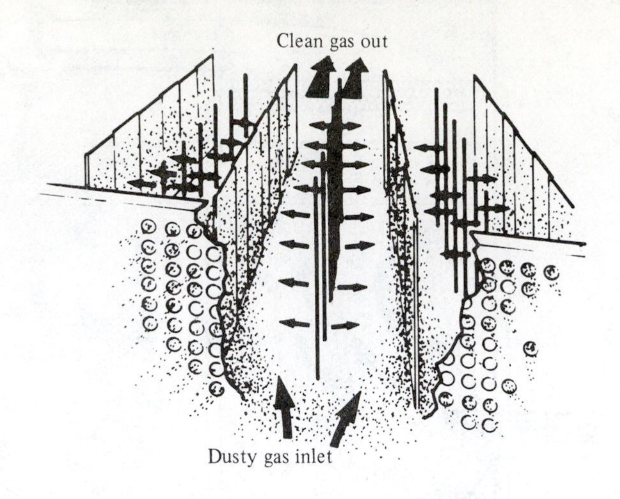

FIGURE 16.12
Electrostatic precipitator. The poles are given a very high negative charge while the plates are positive. The effect is to ionize the gas molecules which in turn collide with and give a charge to the particles. These are then attracted to charged collecting electrodes and collected at the bottom.

Demonstration of the action of an electrostatic precipitator at a basic oxygen steel-making plant. Photo at left shows the stack emissions, while the photo on the right shows the precipitator doing its job of removing solid particles from furnace fumes. (Courtesy American Iron & Steel Institute.)

Control of Industrial SO_x Emissions

Several methods of controlling industrial SO_x emissions are the following:

1. *Use of low sulfur fuel.* The ideal situation would be to choose a fuel low in sulfur. There are such fuels, for example, natural gas, but this is not available in the quantities needed. Some petroleum fractions are low in sulfur but, again, are limited in quantities needed. There are various grades of coal available, some low in sulfur (less than 1 percent). The average sulfur content of coal mined in the U.S. runs about 3 percent. The drawbacks to low-sulfur coal are that it is located far from major metropolitan markets, and is generally lower in heating value. Even if it were used, however, it would contribute some SO_x pollution.
2. *Removal of sulfur before burning.* The easiest fuel to clean up is petroleum. The refining industry has had experience in removal of sulfur from gasoline fractions since sulfur compounds can produce foul odors and corrosion problems. The technical know-how is available for doing the same thing with fuel oils but this will require special refining equipment and the purified oil will cost more. Can coal be cleaned up? The answer is yes and no. Theoretically it is quite possible to remove the sulfur atoms and leave chiefly carbon and ash, but so far the practical achievement on an economically sound basis has not been possible.
3. *Removal of SO_x from flue gases.* This is the approach which is expected to alleviate the sulfur pollution problem in the near future. There are several methods now under study which involve absorption of SO_2 from flue gases.

Modern approach to meeting increasing demands for electrical power. Photo shows the Turkey Point Power Plant located about 25 miles south of Miami on the western shore of Biscayne Bay. The two units on the left are the fossil fuel type, while the two units at the right are nuclear powered. (Courtesy Florida Power & Light Company.)

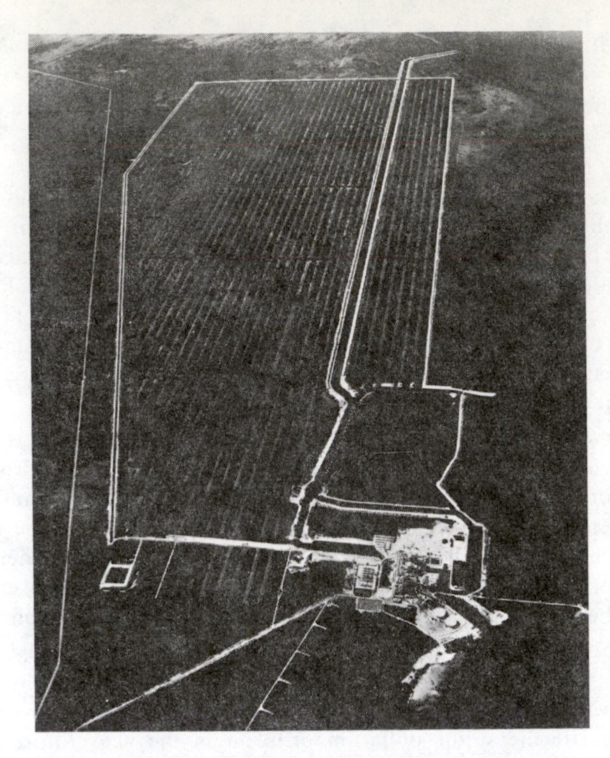

One way of cooling water used in power plants. The Turkey Point Power Plant uses an enclosed system of 168 miles of canals, representing a waterway more than four times the length of the Panama Canal. Cost was $35 million. Canals act as a giant radiator to cool water recirculated to the electrical generating units. An alternate way of cooling uses enormous cooling towers. Both methods depend on the concept of evaporation mentioned in chapter 7. (Thermal pollution is covered in chapter 15.) (Courtesy Florida Power & Light Company.)

Miscellaneous Industrial Air Pollutants

Industrial emissions of hydrocarbons, carbon monoxide, and nitrogen oxides are not as critical as those of particles and SO_x. Especially difficult problems with HC and CO can be handled by available methods like catalytic converters and thermal reactors.

Work on abatement of NO_x produced by combustion processes in industry and electric power plants has been limited. Research advances in the auto field may find use here, when required.

There is still much we do not know about pollution and even about chemistry of the air. We need continuing research in this area as in many others relating to the environment. And the word "environment" is all-inclusive. It certainly involves more than just the burning of gasoline, which was the opening subject of the first chapter and a good part of this chapter on pollution. You have seen how effectively people can use concepts to interpret the behavior of chemicals. But no matter how far we have gone, collectively or individually, there is always room for continued growth. Hopefully, the use of concepts will continue to make our lives safer, more meaningful, and more enjoyable.

PROJECTS AND EXERCISES

Experiments

1. This experiment will show how materials may be affected by acidic air pollutants. You will need a small piece of shell (clam, oyster, snail) and about half a glass of vinegar. Break off a small chip of shell if the shell you have is too large. Try to get a piece with some natural grooves or design feature. Or you can try scratching your initials in the shell with a sharp pointed object, like a nail or knife. Then place the piece of shell in half a glass of vinegar. (White vinegar is best because you can see through it easier, but any vinegar will work.) Observe the shell for a few minutes. What do you see? You may recall that shells are principally $CaCO_3$, as are also marble and limestone. Try to figure out what the bubbles are. (The vinegar supplies an acid, $HC_2H_3O_2$, which provides hydrogen ions.) After a while remove the shell and examine the design features, in comparison with a piece of shell you did not place in the vinegar. Repeat if necessary. Can you see how polluted air which contains SO_x might be involved in ruining statues and building structures? Explain. Vinegar is obviously not involved in air pollution. Yet we can use it as an indicator of an air pollution problem. How come?
2. This is a thought experiment on the automobile. Someone wants to eliminate the pollution of internal combustion engines and someone else wants to eliminate the slaughter on our highways (over 50,000 dead every year in the U.S. alone from auto accidents). Think of a new experimental approach to this dual problem. This is an open-ended, non-structured question. You may want to think in terms of entirely new inventions and life styles or use older ones.

Exercises

3. Describe one air pollution "disaster," giving suggestions on how a similar situation can be avoided in the future.
4. Briefly explain the nature of air, including the major components.
5. Do you consider carbon dioxide an air pollutant? If so, why is it not listed generally with the five primary pollutants? If not, why not?
6. Make a distinction between primary and secondary air pollutants. Give two examples of each.
7. Someone says that SO_x pollution is so dangerous that sulfur compounds should be outlawed. How would you best answer this?
8. A bright ten-year-old asks you where the smell of rotten eggs goes. Briefly explain, bringing in nature's cycling system.
9. Review the five major air pollutants. Pick out the one which is produced in greatest quantity by autos and the one produced in greatest quantity by industry. Are these the most dangerous? Explain.

10. Consider the problem of auto pollution. What do you suggest as short- and long-range solutions?
11. Discuss the effects of removing lead from gasoline, including reference to methods of accomplishing this while still providing a supply of gasoline adequate for the cars now on the road.

SUGGESTED READING

1. There are many topics in this chapter that are being discussed in current magazine and newspaper articles. In addition, you may want to refer to encyclopedia articles, as well as some of the general pollution references given in chapter 15 which are not repeated here.
2. Oort, A. H., "The Energy Cycle of the Earth," *Scientific American,* 223, No. 3, pp. 54–63 (September 1970).
3. Singer, S. F., "Human Energy Production as a Process in the Biosphere," *Scientific American,* 223, No. 3, pp. 174–190 (September 1970).
4. Squires, A. M., "Clean Power from Dirty Fuels," *Scientific American,* 227, No. 4, pp. 26–35 (October 1972).
5. Gregory, D. P., "The Hydrogen Economy," *Scientific American,* 228, No. 1, pp. 13–21 (January 1973). A very interesting proposal to use hydrogen as a general, multipurpose, non-polluting fuel.
6. Barnea, J., "Geothermal Power," *Scientific American,* 226, No. 1, pp. 70–77 (January 1972).
7. Summers, C. M., "The Conversion of Energy," *Scientific American,* 224, No. 3, pp. 148–160 (September 1971). This is one in a whole series of articles in the same issue on energy and power.
8. Esposito, J. C. and Silverman, L. J., *Vanishing Air,* Grossman Publishers, New York, 1970. One of the Ralph Nader Study Group reports.
9. Haagen-Smit, A. J., "The Control of Air Pollution," *Scientific American,* 210, No. 1, pp. 24–31 (January 1964).
10. Newell, R. E., "The Global Circulation of Atmospheric Pollutants," *Scientific American,* 224, No. 1, pp. 32–42 (January 1971).
11. McDermott, W., "Air Pollution and Public Health," *Scientific American,* 205, No. 4, pp. 49–57 (October 1961).
12. The Editors of Fortune, *The Environment—A National Mission for the Seventies,* Perennial Library Paperback, Harper and Row, New York, 1970.
13. Helfrich, H. W., Jr., *The Environmental Crisis, Man's Struggle to Live with Himself,* Yale University Press, New Haven, 1970.

Index

Bold entries indicate definitions.